T0222370

Lecture Notes in Civil Engineering

Volume 327

Lecture Notes in Civil Engineering (LNCE) publishes the latest developments in Civil Engineering—quickly, informally and in top quality. Though original research reported in proceedings and post-proceedings represents the core of LNCE, edited volumes of exceptionally high quality and interest may also be considered for publication. Volumes published in LNCE embrace all aspects and subfields of, as well as new challenges in, Civil Engineering. Topics in the series include:

- Construction and Structural Mechanics
- Building Materials
- Concrete, Steel and Timber Structures
- Geotechnical Engineering
- Earthquake Engineering
- Coastal Engineering
- Ocean and Offshore Engineering; Ships and Floating Structures
- Hydraulics, Hydrology and Water Resources Engineering
- Environmental Engineering and Sustainability
- Structural Health and Monitoring
- Surveying and Geographical Information Systems
- Indoor Environments
- Transportation and Traffic
- Risk Analysis
- Safety and Security

To submit a proposal or request further information, please contact the appropriate Springer Editor: - Pierpaolo Riva at pierpaolo.riva@springer.com (Europe and Americas);- Swati Meherishi at swati.meherishi@springer.com (Asia—except China, Australia, and New Zealand);- Wayne Hu at wayne.hu@springer.com (China).

All books in the series now indexed by Scopus and EI Compendex database!

Guangliang Feng

Editor

Proceedings of the 9th International Conference on Civil Engineering

 Springer

Editor
Guangliang Feng
Chinese Academy of Sciences
Institute of Rock and Soil Mechanics
Wuhan, Hubei, China

This work was supported by Hubei Zhongke Natural Science Research Institute Co., Ltd

ISSN 2366-2557 ISSN 2366-2565 (electronic)
Lecture Notes in Civil Engineering
ISBN 978-981-99-2534-6 ISBN 978-981-99-2532-2 (eBook)
https://doi.org/10.1007/978-981-99-2532-2

This Springer imprint is published by the registered company Springer Nature Singapore Pte Ltd.
The registered company address is: 152 Beach Road, #21-01/04 Gateway East, Singapore 189721, Singapore

Preface

Civil engineering is closely related to people's life, and the quality of civil engineering directly affects people's life and personal safety and affects the development of society to a great extent. Therefore, it is very necessary to analyze the development status of civil engineering, to improve the problems existing in the development of civil engineering.

This book contains the proceedings of the 9th International Conference on Civil Engineering (ICCE 2022) which was held on December 24, 2022, as a hybrid conference (both physically and online via Zoom) at Nanchang Institute of Technology in Nanchang, China. The conference is hosted by Nanchang Institute of Technology and Civil Engineering Academy of Jiangxi Province, co-organized by Journal of Rock and Soil Mechanics, Key Laboratory for Safety of Water Conservancy and Civil Engineering Infrastructure in Jiangxi Province, Key Laboratory of Sichuan Province for Road Engineering, Southwest Jiaotong University. More than 150 participants were able to exchange knowledge and discuss the latest developments at the conference. The book contains 53 peer-reviewed papers, selected from more than 200 submissions and ranging from the theoretical and conceptual to strongly pragmatic and addressing industrial best practice.

The book shares practical experiences and enlightening ideas from civil engineering and will be of interest to researchers in and practitioners of civil engineering everywhere.

Contents

Study on the Influence of Sand Mining in the Channel of Sima Bend in Yangzhong of the Yangtze River

Cao Hu, Kunpeng Fan, Minghua Li, Jie Wu, and Yiwen Xu

Abstract Taking the sand mining project in a local river section in the lower reaches of Yangtze River, as an example, this paper investigates the complex effects of sand mining project on river potential. Based on MIKE21 software to establish a two-dimensional water–sand mathematical model of the project river section, this study investigates the change of river potential in the local river section near the project after the implementation of different sand mining schemes through numerical simulation to explore the influence of this sand mining project on river potential. The results show that (1) except for slight changes in the project area, there are no significant changes in the rest of the section. Under each calculation condition, the mainstream line in the sand mining area is slightly rightward by about 5 m; (2) the implementation of sand mining project, the diversion ratio of the left branch of Taiping Island slightly increases, and the corresponding diversion ratio of the right branch of Taiping Island decreases, but the change of the diversion ratio is very small, such as the increase of the diversion ratio of the left branch of Taiping Island and the left branch of the inaugurated Island is generally within 0.01% (3) the implementation of the sand project, the water level The overall performance is slightly congested. The implementation of the project has a small impact on the change of the deep flood line in the project section, and the river pattern will not change significantly.

Keywords Sand Mining · Yangtze River · Influence

1 Introduction

When natural rivers are not disturbed by human activities, their evolution process follows certain rules under different incoming water and sediment conditions, which shows the adaptability to incoming sediment in this area [1]. The River type, width, depth, slope and section form remain relatively stable over a long period of time. When the river is disturbed by human activities, the original flow structure and

C. Hu (✉) · K. Fan · M. Li · J. Wu · Y. Xu
Nanjing Yangtze River Channel Management Office, Nanjin 210000, Jiangsu, China
e-mail: 706758101@qq.com

© The Author(s) 2023
G. Feng (ed.), *Proceedings of the 9th International Conference on Civil Engineering*,
Lecture Notes in Civil Engineering 327,
https://doi.org/10.1007/978-981-99-2532-2_1

1

sediment transport characteristics of the river may change. The type and degree of change may be long-term trend, short-time randomness, wide-range long reach or local reach. Sand mining is considered to be an intense process to intervene in the natural evolution of rivers. Proper and orderly sand mining is not only conducive to the flood and navigation of rivers, but also provides valuable sandstone resources for the construction industry [2]. The influence of sand excavation on river regime is determined by many factors, such as riverbed condition, evolution law, size of sand excavation, shape and position of sand excavation pits, exploitation time and so on. Large-scale sand excavation may change the original riverbed structure and flow trend, resulting in riverbed deformation, forced change of flow pattern, and ultimately affecting navigation safety. From the point of view of benefits and hazards removal, the effective utilization of sandstone resources in the river can not only enlarge the flow section of the river, improve the flow pattern of the flood, improve the standard of flood control, but also improve the conditions of ports and channels and improve the shipping capacity. The use of river sand in industrial, civil construction and other related industries will play a beneficial role in promoting the development of national economy [3–5].

In recent years, with the development of national economy and the change of thoughts on river management, the research on sand mining in China has begun to change from the early simple engineering study to the research on ecological environment impact. Taking the local engineering reach of the lower reaches of the Yangtze River as an example, this paper establishes a two-dimensional water–sediment mathematical model based on MIKE21 software to discuss and analyze the impact of sand mining project on river regime and the feasibility of sand mining by the project.

2 Overview of Research Area

2.1 River Profile

The Yangzhong River is a key section in the middle and lower reaches of the Yangtze River. The upper reaches of the Yangzhong River meet the Zhen Yang River at Wufeng Mountain and the lower reaches meet the Chengtong River at Jiangyin Goosebite [6, 7]. The Yangzhong River can be divided into two sections according to its morphology and flow characteristics: above the mouth of the river is the Taipingzhou branching section, which includes the Taipingzhou, Chengchengzhou, Lu'anzhou and Baozizhou river cores; below the mouth of the river to Goibizui is the Jiangyin waterway, which is a single straight and slightly curved section. The branching section of Taiping Chau is about 58 km long from Wufeng Mountain to Biaozi Chau, with an average river width of about 1700 m [8–10]. In recent decades, the sand ratio of the left and right branches of Taiping Zhou has changed relatively little, and the right branch diversion ratio has been around 10% for many years. The upper section of the left branch is the Hissing Horse Bend section, which runs from Wufeng Mountain to the mouth of Lao Yangwan River, with a length of about 15 km and an average river width of about 1700 m. The upper section of the left branch is the hissing horse bend section, and the main stream is close to the concave bank side of the hissing horse bend on the left bank, where the bank collapse used to be more violent. Near Hissing Horse, there are sandbars on the left edge of Taiping Island, including Chengcheng Island, Leigong Island and Xiaosha Island, and the diversion ratio of the right branch of Chengcheng Island has slightly increased recently [11]. The flow below the mouth of Lao Yang Wan River gradually transitions to the left edge of Taiping Island and enters the lower section of the left branch of Taiping Island. In recent decades, except for the bend section, the transition section and the local waters of the submerged island have been adjusted, and the plane of the deep water has oscillated, but the rest of the river has not changed much, and the overall river trend tends to be relatively stable.

2.2 Project Overview

The waterway mining area implements a sand mining project with blow-fill on the inner side of the main river bank. The sand mining and blow-filling project is divided into two areas, of which the area of the blow-filling project in the first area is about 542,000 m^2 and the area of the blow-filling project in the second area is about 443,000 m^2.

3 Numerical Simulation

3.1 Boundary Conditions

In this paper, the effect of changes in water temperature on the density of the water body is ignored and the density of the water body is a constant. As the implementation of sand mining project will cause changes in water level and flow velocity and flow pattern in the project river section, which may bring adverse effects on the flood control safety and navigation safety of the project river section as well as water ecology and environment. Therefore, a two-dimensional water flow mathematical model is proposed to calculate the water flow in the project river section and to analyse the impact of the sand mining project on the river flooding and river potential. Taking into account the river potential of the river section where the proposed project is located, the possible impact of the project and the hydrological data, a 68 km-long river section from the upstream of Wufeng Mountain (inlet) to the mouth of the Boundary River (outlet) was selected as the two-dimensional mathematical model verification and engineering impact calculation section. Due to the long time and high cost required to measure the topography of the entire river section, and the topographic changes in the river section in recent years are within the permissible range, the topography of the river section is calculated using the 1/10000 river topographic map measured in February 2006, and the topography of the sand mining local area is measured in September 2011. All the existing works in the vicinity of the proposed project are considered as inherent boundaries.

3.2 Numerical Basic Equations

$$\frac{\partial \eta}{\partial t} + \frac{\partial hu}{\partial x} + \frac{\partial hv}{\partial y} = 0 \tag{1}$$

$$\frac{\partial u}{\partial t} + u\frac{\partial u}{\partial x} + v\frac{\partial u}{\partial y} + g\frac{\partial \eta}{\partial x} - fv + \frac{gu|U|}{C^2 h} - \frac{F_x}{\rho_w h} - \nu\left(\frac{\partial^2 u}{\partial x^2} + \frac{\partial^2 u}{\partial y^2}\right) = 0 \tag{2}$$

$$\frac{\partial v}{\partial t} + u\frac{\partial v}{\partial x} + v\frac{\partial v}{\partial y} + g\frac{\partial \eta}{\partial y} + fu + \frac{gv|U|}{C^2 h} - \frac{F_y}{\rho_w h} - \nu\left(\frac{\partial^2 v}{\partial x^2} + \frac{\partial^2 v}{\partial y^2}\right) = 0 \tag{3}$$

where, u, v distribution for x, y direction of the water depth of the average flow velocity; g for the acceleration of gravity; h for the vertical direction of the water depth; for the flow velocity; C for the coefficient; ρ_w for the density of water; for the motion of viscous coefficient; where Fx, Fy are x, y direction of the wind stress (this paper does not consider the impact of wind, the item can be ignored).

3.3 Numerical Calculation Method

Using the finite volume method, the computational domain is divided into triangular grids, and the water volume and momentum balances are calculated separately for each triangular grid, yielding the flow and momentum fluxes along the normal input or output at each triangular grid boundary, and then calculating the average water depth and flow velocity for each triangular grid at the end of the time period.

3.4 Calculation Range and Grid

According to the model requirements and topographic features, the calculation area is gridded. The engineering river section is a tide-sensitive section, the tide is an informal semi-diurnal tide, the flood season mainly shows one-way flow and the dry season shows two-way flow. The calculation conditions are shown in Table 1.

The conditions for calculating the impact of the project are as follows: the flood control design flood and the flow conditions at the level of the Pingtan are calculated

Table 1 Calculated water flow conditions for engineering impacts

	Condition	Explanation
1	Design flood level	Flow rate 100400 m³/s, lower boundary water level 5.65 m
2	Low tide level	Flow rate 100400 m³/s, lower boundary water level 4.15 m
3	Flat beach water level flow	Flow rate 50,000 m³/s, lower boundary water level 2.85 m
4	Measured dry period tidal process conditions	The measured high tide process on March 10–11, 2020

Fig. 1 Mesh of computation zone

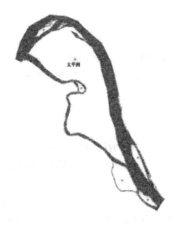

according to the given flow rate at the inlet of the stream (including the inlet of the upper Wufeng Mountain and the inlet of the Huai River into the river) and the given water level at the outlet; the measured high tide process during the dry season is calculated according to the given tide process at the inlet and outlet.

As shown in Fig. 1, the calculation grid consists of 61,254 grid cells, each with a scale of 2.0 × 1.0 m. In the initial stage of the calculation, the upstream high flow rate and the downstream normal water level were selected as the initial water level in Fig. 2. In the calculation process, the results of the above calculation after stabilisation were used as the initial conditions, the chezy coefficient was taken to be around 75 and the calculation time step was taken to be 3 s.

4 Mathematical Model Validation

During the hydrographic survey from 14 to 15 March 2021, a total of seven water scales were set up in the calculated river section, which were located at Wufeng Mountain, Sanjiangying, Hima River, Gaogang, Tongxing Gate, Yangsi Port and the

Fig. 2 Topographical
distribution

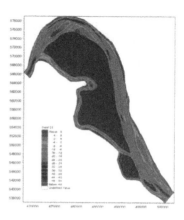

mouth of the boundary river. The flow rate is 20,300 m³/s during the flow measurement period. After the mainstream of the Yangtze River enters Yangzhong River section.

4.1 Verification of Flow Velocity Processes and Cross-Sectional Flow Distribution

The mainstream gradually transitions from close to the right bank to the left bank to the Zohan Hissingma Bend section, at the import of the Hissingma Bend, there is a completion continent to divide the river flow into two branches, the left branch is the main branch, near the downstream of the right branch of the completion continent outlet, there is a heart continent to divide the right branch flow into two branches, three streams of water converge between the Hissingma River and Yangwan Gate, the convergence of the mainstream between Yangwan Port and Erdun Port transition from close to the left bank to The main flow between Yangwan Port and Erdun Port transitions from close to the left bank to the right bank, and below Erdun Port is a straight section, with a submerged island in the middle dividing the main flow into two streams, and the two streams converge near Shengliang Port. From the calculated flow field, the calculated flow field is smooth, the branching and converging flows are well connected, and the position and direction of the main stream are consistent with the actual situation, indicating that the model can better simulate the complex water movement characteristics of the whole calculated river section.

From the validation results, it can be seen that the calculated flow velocity process of each flow measurement vertical line is in good accordance with the measured value, the phase change is consistent, the flow velocity distribution of the measured section is well verified, and the calculated and measured mainstream position is basically consistent. Statistically, the error between the calculated and measured values of each flow measurement plumb line is generally within 0.12 m/s, with a slightly larger error of 0.20 m/s at individual moments. From the above tidal process, flow velocity process, cross-sectional flow velocity distribution and branching channel divergence ratio verification results show that the river planar two-dimensional mathematical model used in this thesis can better simulate the flow movement of the whole calculated river section, and verify the calculation accuracy is high. Therefore, the mathematical model can be used for the calculation and analysis of the impact of sand mining on the water level and flow field of the river.

5 Analysis of the Impact of Sand Mining Projects on River Flows

The calculation results of the impact of the proposed project on the river flooding mainly include: water level, water depth and vertical average flow velocity at all two-dimensional calculation grid nodes in the river section before and after the construction of the project under the above three sets of flow conditions. By analysing the changes in water level and flow velocity at each monitoring point and section before and after the implementation of sand mining, and the changes in water level and flow velocity field in the river section near the sand mining area, the possible impact of sand mining activities on the water level and flow velocity of the river is studied.

5.1 Analysis of Plane Flow Velocity Field and Flow Pattern Changes

According to the comparison analysis of the mainstream line of the river section before and after sand mining: the mainstream line of the project river section tends to be straight with the increase of flow, and is constrained by the river channel and the two banks of the embankment protection project and the continental beach. There is no obvious change in the velocity field and flow pattern of the project river section. The left branch of Taiping Zhou, the inauguration of Zhou, Cannon Island and the left channel of Lu'an Zhou are the main navigation channels.

5.2 Analysis of Changes in the Velocity Field and Flow Regime

According to the comparison analysis of the mainstream line of the river section before and after sand mining: the mainstream line of the project river section tends to straighten with the increase of flow, and is constrained by the river channel and the two banks of the embankment protection project and the continental beach, the mainstream line position before and after sand mining, except for slight changes in the project area, the rest of the parts have no obvious changes, in each calculation condition, the maximum mainstream line in the sand mining area slightly to the right about 5 m; the project river section plane velocity field, flow pattern also basically did not There is no obvious change in the velocity field and flow pattern in the project section.

6 Conclusion

The results of the numerical model of water flow show that, after the implementation of the sand mining project, due to the increase of the over-water area in the project area, the flow velocity in the mining area is mostly reduced, and the water level is generally slightly higher. After the implementation of the sand mining project, the flow velocity in the mining area generally decreases by 2–10 cm/s under the calculated flow conditions. Under the calculated flow conditions, the flow velocity in the right side of the mining area generally decreases by 1–5 cm/s, with a small change in flow direction; the change in flow velocity in the main channel is around 0.1–0.7 cm/s; the flow velocity in the upstream of the mining area slightly increases, with a maximum increase of about 1.9 cm/s; the flow velocity in the downstream of the mining area decreases, with a maximum decrease of about 6.7 cm/s; the impact of the increased flow velocity is concentrated in the 500 m upstream of the mining area. The impact of the increase in flow velocity is concentrated within 500 m upstream of the quarrying area and 800 m downstream of the quarrying area. After the implementation of the project, there is basically no change in the water level at the mouth of the Hima River and the mouth of the Cuijiang River, and the change in flow velocity is within 0.1 cm/s.

References

1. Li J (2008) Numerical simulation of the impact of sand mining in river channels. Changjiang Academy of Sciences
2. Xu Y, Wang M, Zhu H, et al. (2021) Water Resour Hydropower Technol (in Chinese and English). 32(28):3
3. Pan C, Li W, Yao T (2012) Reflections on the management of sand mining in small and medium-sized rivers. Hubei Water Resour (2012)
4. Li J (2008) Numerical simulation study on the impact of sand mining in rivers. Changjiang Academy of Sciences
5. Feng Y, Lin, FB, Liu TH, et al. (2014) Study on the impact of the comprehensive improvement project of the Yangtze River shoreline in the Taohuagang section of Jiangyin waterway on the river potential. Zhejiang Water Conserv Sci Technol 42(5):5
6. Wei G, Liu J Scientific demonstration of local river sand mining in the lower reaches of the Yangtze River

7. Fan Y, Lu JY, Xu HTA (2009) Study of a planar two-dimensional sediment model of the river channel in the Wuhan section. J Shihezi Univ Nat Sci Edn, 27(2):4
8. Duan Dr (2016) Calculation of congestion analysis based on a two-dimensional mathematical model of water flow. Hunan Water Conserv Hydrop (2)
9. Lin M (2018) Research on safe river discharge under complex boundary conditions. Water Resour Sci Technol 2:5
10. Liu SH, Wei BQ, Huang L et al (2020) Research on the impact of reservoir congestion on the water environment of water sources. J Water Resour Water Eng 31(2):8
11. Hu Y (2009) Research and application of a planar two-dimensional water flow mathematical model visualization system. Chongqing Jiaotong University

Study on Fatigue Performance of Typical Fatigue Detail in Orthotropic Steel Deck

Bing Yan, Yaoyu Zhu, Cheng Meng, and Zhiyuan Yuanzhou

Abstract Orthotropic steel decks (OSDs) are easily subjected to fatigue cracking under the cyclic vehicle loading, and arc notch between diaphragm and U-rib is one of the typical fatigue details. The strain values of three measuring points at arc notch in a steel bridge were monitoring to obtain the stress time histories. Then, fatigue stress amplitudes and fatigue damage degrees were analysed. It was found that this typical fatigue detail was under tension–compression cyclic stress. The maximum stress amplitude appeared at U-rib weld toe, which indicated that fatigue cracks were more prone to propagate on this area, comparison of fatigue damage degrees also corroborated it.

Keywords Orthotropic Steel Deck · Fatigue Crack · Stress Amplitude · Field Monitoring · Fatigue Damage Degree

1 Introduction

Orthotropic steel decks (OSDs) are broadly used in steel bridges for its superior advantages, like light weight, reliable force and convenient construction [1]. However, various categories of fatigue cracks are found in orthotropic steel decks because of the repeated traffic loads, together with welding residual tensile stress and welding defects [2]. There are several typical fatigue details, such as the butt welds, arc notch, rib-deck welds [3,4]. Thereinto, the arc notch between U-rib and

B. Yan
Jiangsu Sutong Bridge Co., Ltd., No.1 East Jianghai Road, Nantong, China

Y. Zhu
China Communications Construction Company Highway Bridges National Engineering Research Center Co., Ltd., Huangsi Street A23#, Beijing, China

C. Meng (✉) · Z. Yuanzhou
College of Civil and Transportation Engineering, Hohai University, No.1 Xikang Road, Nanjing, China
e-mail: ycmc1995@163.com

© The Author(s) 2023   11
G. Feng (ed.), *Proceedings of the 9th International Conference on Civil Engineering*,
Lecture Notes in Civil Engineering 327,
https://doi.org/10.1007/978-981-99-2532-2_2

diaphragm has attracted wide attention due to its complex and special structure over recent years [5].

There are already extensive researches focus on evaluating the fatigue performance of the arc notch between diaphragm and U-rib. Cyclic vehicle loading causes the out-of-plane deformation of U-ribs and diaphragms, and large great out-of-plane bending stresses appear at the arc notch due to the relative constraint between U-rib and diaphragm [6, 7]. Fatigue tests show that out-of-plane stress could reach 20% to 35% of in-plane stress [8]. Based on large quantities of measured data, many fatigue cracks at arc notch mainly appear on U-rib weld toe, diaphragm weld toe and the fillet [9].

In this study, field monitoring in orthotropic steel deck of a real bridge was carried out and the strain data at arc notch between diaphragm and U-rib under vehicle loads were recorded. The fatigue performance of this typical fatigue detail was evaluated.

2   Field Monitoring

2.1   Measuring Points

Inspection records of a cable-stayed steel bridge show that there are a large number of fatigue cracks grow at arc notch between diaphragm and U-rib in OSD, and more than half of cracks occur below overtaking lane. As the traffic volumes at the upstream side are much more, arc notch of 17# U-rib at the Diaphragm NJ22-3 below truck lane at the downstream side was chosen as the monitoring target, seen in Fig. 1.

Tri-axial strain gauges were used and set at 10 mm away from the arc notch and the rib weld toe. And uniaxial strain gauges were arranged at the arc notch with a distance of 10 mm away from edge. The measuring points were named as G1-G3, G1 was arranged at diaphragm weld toe, G2 was arranged at U-rib weld toe, G3 was arranged at the fillet of arc notch, as shown in Fig. 2.

The resistance of these gauges was 120Ω, and the sensitivity ratio is from 1.0% to 3.0%. The strain data were captured by a dynamic strain indicator with a frequency of 512 Hz to obtain the strain cycles caused by vehicle loads. The strain data were recorded for 24 h.

2.2   Stress Time History

The strain data were converted into stress data by Hooke's Law $\sigma = E\varepsilon$. For the material of OSD in the studied bridge, the yield strength is about 345 MPa, Young's modulus E is about 2.06×10^5 MPa, and the Poisson's ratio is about 0.3.

Figure 3 plots part of stress time-history of G1. Under vehicle loads, the stress at 0°direction (i.e., parallel to the weld) is negative which means the structural

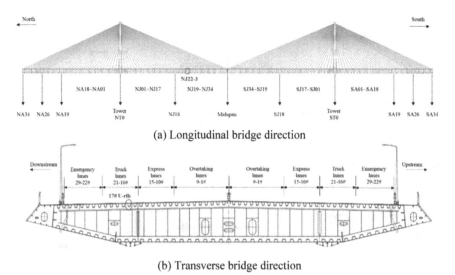

(a) Longitudinal bridge direction

(b) Transverse bridge direction

Fig. 1 Measuring points layout

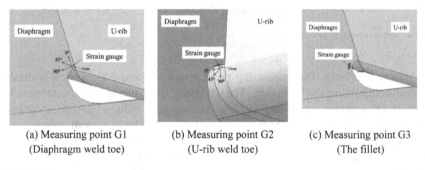

| (a) Measuring point G1 (Diaphragm weld toe) | (b) Measuring point G2 (U-rib weld toe) | (c) Measuring point G3 (The fillet) |

Fig. 2 Strain gauge arrangement

detail is under compressive stress at 0°direction. The mean and maximum value of compressive stress at 0°direction are 6.5 and 20 MPa respectively. The mean/peak value of tensile stress and compressive stress at 45°direction are about 7/12 and 4/12 MPa respectively, which mean the stress at 45°direction of this detail is dominated by tensile stress. The mean/peak value of tensile stress and compressive stress at 90°direction (*i.e.*, perpendicular to the weld) are about 5/13 and 10/20 MPa respectively, which mean the stress at 45°direction of this detail is dominated by compressive stress. The measuring point area is under tension–compression cyclic loading at 45°and 90°direction, which is the main reason for fatigue cracking. The maximum stresses occur at 90°direction,thus the fatigue cracks are more likely to propagate at 0° direction, which agrees well with the reality.

Figure 4 plots part of stress time-history of G2. Under vehicle loads, the stress at 0°direction is compressive stress with the mean and maximum value are 20 and

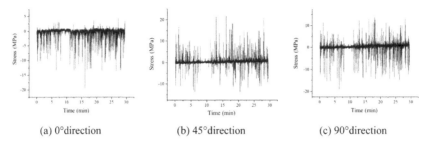

(a) 0°direction   (b) 45°direction   (c) 90°direction

Fig. 3 Part of stress time-history of G1

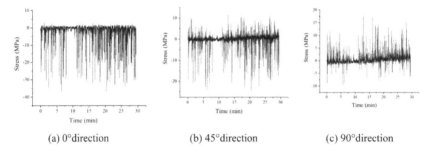

(a) 0°direction   (b) 45°direction   (c) 90°direction

Fig. 4 Stress time-history of G2

37 MPa respectively. This measuring point area at 45°and 90°direction is under tension–compression cyclic loading. The peak value of tensile stress and compressive stress at 45°direction are 10 and 25 MPa. And the peak value of tensile stress and compressive stress at 90°direction are 16.5 and 10 MPa. The maximum compressive stress occurs at 0°direction among three directions, which indicates that cracks are more prone to grow at 90°direction, consistent with reality.

Figure 5 plots part of stress time-history of G3. The fillet at arc notch is mainly under compressive stress with the peak value of 45 MPa, and tensile stress occurs less frequently with the peak value of 15 MPa. The measuring point area is prone to crack due to the combined action of cyclic compressive stress and welding residual tensile stress.

3 Analysis of Measurement Results

3.1 Fatigue Stress Amplitude

Based on the stress data in Sect. 2.2, the rain-flow counting method [10] was applied to get the stress range spectrum. The stress amplitude below 4 MPa was removed because the low-stress random amplitudes below 4 MPa had little contribution to the

Fig. 5 Stress time-history of
G3

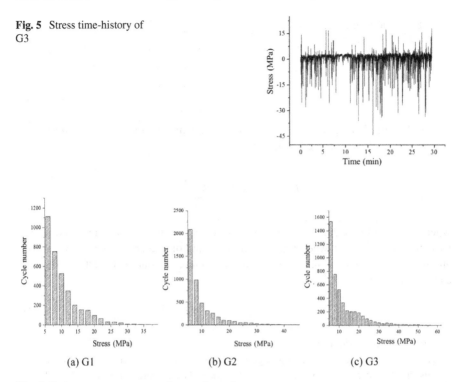

Fig. 6 Fatigue stress spectrums of measuring points

fatigue damage. The segment length of stress range is 2.0 MPa. The fatigue stress spectrums of measuring points are presented in Fig. 6.

As shown in Fig. 6, variation trends of stress spectrums are similar, appearing as the cycle numbers decrease gradually with the increase of stress amplitudes. More than 90% of stress amplitudes are below 10 MPa which means that the measuring points are under low cyclic stress for most of the time. Large stress amplitude could enlarge the initial defects of the components, thereby reduce the fatigue limit. Then, as time grows, fatigue damage increases under low stress amplitude.

3.2 Fatigue Damage

Fatigue damage degree refers to the damage accumulation of bridges under cyclic vehicle loading. Recommended in Chinese standard (JTG D64-2005), the value of design fatigue strength of arc notch between U-rib and diaphragm is 70 MPa, and of rib-to-diaphragm weld is 71 MPa. Miner's linear cumulative rule is often used in practice projects [11]. In the light of Miner's linear cumulative rule, fatigue damage caused by each stress amplitude can be expressed by $\Delta\sigma_i/N$, and superimposed linearly.

Table 1 Fatigue damage degrees of measuring points

Point number	Position	D
G1	Diaphragm weld toe	1.053E−05
G2	U-rib weld toe	4.048E−05
G3	Fillet at arc notch	3.668E−05

Hence, fatigue damage degree of any component under random stress amplitude ($\Delta\sigma_i$, $i = 1, 2, 3......$) can be defined by Eq. (1).

$$D = \sum_{i=1}^{\infty} \Delta D_i = \sum_{i=1}^{\infty} \frac{n_i}{N_i} \tag{1}$$

where D is the fatigue damage degree, n_i is the cycle number of stress amplitude $\Delta\sigma_i$. N_i is the cycle number when fatigue failure occurs on the component under stress amplitude $\Delta\sigma_i$ in fatigue tests. According to Chinese standard (JTG D64-2005), N_i can be calculated by Eq. (2).

$$N_i = 2 \times 10^6 \times (\sigma_0 / \Delta\sigma_{x,i})^3 \tag{2}$$

Fatigue damage degrees of three measuring points are given in Table 1.

As shown in Table 1, fatigue damage degree of U-rib weld toe is the largest, which means that U-rib weld toe is more prone to crack. The damage degree of the fillet at arc notch is also at a high level, suggesting that this part is more vulnerable to fatigue loads.

3.3 Conclusions

(1) Diaphragm weld toe and the fillet at arc notch subject mainly to cyclic compressive stress, while U-rib weld toe endures tension–compression cyclic stress, and residual welding stress also made contribution to fatigue cracking.

(2) The stress peak amplitude of U-rib weld toe is higher than diaphragm weld toe and the fillet at arc notch, so is the fatigue damage degree. Fatigue cracking is more easily appear at U-rib weld toe, matches the actual inspection of the bridge well.

Acknowledgements The research reported herein has been conducted as part of the research projects granted by the National Key Research and Development Project (2017YFE0128700), the Natural Science Youth Foundation of Jiangsu Province (BK20200511), Postdoctoral Science Foundation of Jiangsu Province (2021K564C) and Academician Project Foundation of CCCC (YSZX-03-2021-01-B). The assistances are gratefully acknowledged.

References

1. Ji B, Liu R, Chen C, Maeno H, Chen X (2013) Evaluation on root-deck fatigue of orthotropic steel bridge deck. J Constr Steel Res 90:174–183
2. Ya S, Yamada K, Shikawa T (2013) Fatigue evaluation of rib-to-deck welded joints of orthotropic steel bridge deck. J Bridg Eng 18:492–499
3. Wang Q, Ji B, Fu Z, Yao Y (2020) Effective notch stress approach-based fatigue evaluation of rib-to-deck welds including pavement surfacing effects. Int J Steel Struct 20:272–296
4. Liu J, Guo T, Feng D, Liu Z (2018) Fatigue performance of rib-to-deck joints strengthened with FRP angle. J Bridg Eng 23:04018060.1–04018060.14
5. Fisher J, Barsom J (2016) Evaluation of cracking in the rib-to-deck welds of the Bronx-Whitestone bridge. J Bridg Eng 21:04015065
6. Sim H, Uang C, Sikorsky C (2009) Effects of fabrication procedures on fatigue resistance of welded joints in steel orthotropic decks. J Bridg Eng 14:366–373
7. Tsakopoulos P, Fisher J (2003) Full-scale fatigue tests of steel orthotropic decks for the Williamsburg Bridge. J Bridg Eng 8:323–333
8. Fangjiang G, Ye Q, Fernandez O, Taylor L (1892) Fatigue analysis and design of steel orthotropic deck for Bronx-Whitestone bridge. Transport. Res. Rec. 1892:69–77
9. Cheng B, Ye X, Cao X, Mbako D, Cao Y (2017) Experimental study on fatigue failure of rib-to-deck welded connections in orthotropic steel bridge decks. Int J Fatigue 103:157–167
10. Matsuishi M, Endo T (1968) Fatigue of metals subjected to varying stress. Jpn. Soc. Mech. Eng. 68:37–40
11. Miner M (1945) Cumulative damage in fatigue. J Appl Mech 12:159–216

Comfort Behavior of High Performance Floor Based on Single-Jump Excitation Mode Considering Time–Space Effect

Jiang Yu, Haifeng Xu, and Weiyun Zhang

Abstract Because the traditional analysis methods cannot directly describe the comfort behavior of high performance floor through the dimensions of time and space, this paper presents a new method to derive Dirac delta functions and governing differential equations to assess its comfort behavior. On the basis of this, an improved single-jump excitation model is proposed for comfort behavior analysis. In addition, the paper analyzes distribution characteristic of sensitive parameters for the high performance floor. The results show that the peak acceleration is concentrated in the middle of the high performance floor, and then the acceleration response gradually weakened to the two sides, and its degree of weakening is greater along the two corners of the floor model from its spatial and temporal distribution characteristics. In future, it is recommended to change the strength grade of concrete materials to improve its comfort behaviors. The study also demonstrates the strong applicability of improved single-jump excitation model as an effective approach to analyzing the comfort behavior of high performance floor.

Keywords Comfort Behavior · High Performance Floor · Time–Space Effect · Experimental Verification · Parametric Sensitivity Analysis

1 Introduction

The degree to which the human body perceives dynamic vibrations of this kind of floor structure is called comfort behavior. Rapid developments in the design of modern floors have resulted in their span becoming bigger and bigger, which reduced the comfort behavior. The comfort behavior of floor structures has become an important topic for scholars and researchers in recent years [1–4]. This comfort behavior has an

J. Yu (✉) · H. Xu
Nanjing Hydraulic Research Institute, Nanjing 210024, China
e-mail: yuj@nhri.cn

W. Zhang
Nanjing Water Conservancy Planning and Design Institute Co., Nanjing 210022, China

G. Feng (ed.), *Proceedings of the 9th International Conference on Civil Engineering*,
Lecture Notes in Civil Engineering 327,
https://doi.org/10.1007/978-981-99-2532-2_3

important component on the mechanism characteristics of this type of floor structures and can directly lead to safety problem and human physiological discomfort reaction. Therefore, the rational analysis of the comfort behavior is critical for the floor structure. In relation to this, Ding [5] investigated vibration test and comfort analysis of environmental and impact excitation for wooden floor structure, and it has value in engineering applications, as it advances understanding concerning the vibration characteristics and comfort optimization of light-wood frame construction. Vibration testing and comfort analysis were investigated for floor structures by a large number of researchers [6–9]. To ensure the safety of floor structures, a number of practical engineering projects have also adopted a sensitive parameter and performance parameter to describe the comfort behavior during the design process. Zhang et. al. [10] researched the large span truss-corrugated steel deck RC composite floor by method of dynamic measurement and comfort parameter analysis. And then, the dynamic performance under crowd excitation for floors with different structural parameters was studied by He and Fu [11–13]. Vibration analysis of steel–concrete composite floors was finished when subjected to rhythmic human activities by Campista and Jiang [14–16].

The primary objective of this paper is to propose an improved single-jump excitation model for the analysis of comfort behavior of high performance floor that is based on the traditional simulation method. The process involved in deriving comfort behavior using the Dirac Delta function method is first of all presented in detail. The results of experimental research of an example of this type of high performance floor is then compared with the results generated using the new method to verify its accuracy. After this, the influence of spatial and temporal distribution characteristics on the comfort behavior is examined. That further suggests that it provides a more reference value for engineering design and structure optimization in some extent for the floor structures.

2 Establishing the Differential Equations

2.1 Improved Single-Jump Excitation Model

This type of the single-jump load excitation has been researched through the establishment of dynamic characteristic model and dynamic test by Allen, Rainer, and Kasperski[17–19] but the dynamic factors and other summarized parameters were quite different in the above research results, so the improved single-jump excitation model is proposed based on the modified half-sine square model by Chen[20], it is as follows:

$$F(t) = \begin{cases} K_\mathrm{p}G\sin(\pi t_\mathrm{p}^{-1}t), & (0 \le t < t_\mathrm{p}, f_\mathrm{p} \le 1.5\mathrm{Hz}); \\ K_\mathrm{p}G\sin^2(\pi t_\mathrm{p}^{-1}t), & (0 \le t < t_\mathrm{p}, 1.5\mathrm{Hz} < f_\mathrm{p} \le 3.5\mathrm{Hz}); \\ 0, & (t_\mathrm{p} \le t < T_\mathrm{p}). \end{cases} \quad (1)$$

where, K_p represents the correction for impulsiveness, G represents the weight of a single person, t_p represents the time spent touching the floor, T_p represents the time taken for a single bounce, f_p represents the hopping frequency.

The following equation could be obtained based on conservation of energy theorem during a single-jump for load excitation.

$$\int_0^{t_p} F(t)dt = G \cdot T_p \tag{2}$$

And Eq. (3) is obtained for the correction for impulsiveness through Eq. (1) and Eq. (2).

$$K_p = \begin{cases} \dfrac{\pi}{2.4\alpha}, & (f_p \leq 1.5\text{Hz}); \\ \dfrac{2}{\alpha}, & (1.5\text{Hz} < f_p \leq 2.0\text{Hz}); \\ \dfrac{\eta}{\alpha}, & (2.0\text{Hz} < f_p \leq 3.5\text{Hz}). \end{cases} \tag{3}$$

where, $\alpha = \frac{t_p}{T_p}$, $\eta = -0.332 f_p^2 + 1.908 f_p - 0.792$.

2.2 Governing Differential Equation

An analytical mode is as shown in Fig. 1 for the proposed model. To analyze the single-jump load excitation behaviors on this type of high performance floor, the following theoretical foundations are mentioned: (a) its properties of both all isotropic and linear elastic properties are satisfied; (b) the deflection is significantly less than its plate thickness during single-jump load excitation; (c) the absolute equal thickness principle is satisfied.

And then, its expressions of internal force are as followings for this type of high performance floor:

$$M_x(x, z, t) = -D[\frac{\partial^2 w(x, z, t)}{\partial x^2} + v\frac{\partial^2 w(x, z, t)}{\partial z^2}]; \tag{4}$$

$$M_z(x, z, t) = -D[\frac{\partial^2 w(x, z, t)}{\partial z^2} + v\frac{\partial^2 w(x, z, t)}{\partial x^2}]; \tag{5}$$

$$M_{xz}(x, z, t) = -D(1 - v)\frac{\partial^2 w(x, z, t)}{\partial x \partial z}; \tag{6}$$

$$M_{zx}(x, z, t) = -D(1 - v)\frac{\partial^2 w(x, z, t)}{\partial z \partial x}; \tag{7}$$

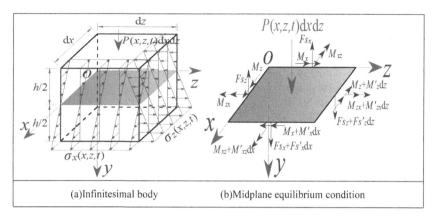

(a)Infinitesimal body	(b)Midplane equilibrium condition

Fig. 1 Section size and coordinate system of analytical mode

$$Fs_x(x, z, t) = D\frac{\partial \nabla^2 w(x, z, t)}{\partial x} \tag{8}$$

$$Fs_z(x, z, t) = D\frac{\partial \nabla^2 w(x, z, t)}{\partial z} \tag{9}$$

The following governing differential equations can be obtained by establishing the internal force balance equation from Eq. (4)–Eq. (9).

$$\frac{\partial M_x(x, z, t)}{\partial x} + \frac{\partial M_{xz}(x, z, t)}{\partial z} - Fs_z(x, z, t) = 0; \tag{10}$$

$$\frac{\partial M_z(x, z, t)}{\partial z} + \frac{\partial M_{zx}(x, z, t)}{\partial x} - Fs_x(x, z, t) = 0; \tag{11}$$

$$\frac{\partial Fs_x(x, z, t)}{\partial x} + \frac{\partial Fs_z(x, z, t)}{\partial z} + \overline{m}\frac{\partial^2 w(x, z, t)}{\partial t^2} = P(x, z, t). \tag{12}$$

2.3 Dirac Delta Functions and Its Boundary Conditions

In order to represent the position of the operating point of the single-jump load excitation behavior, Dirac delta functions are brought in for analysis of load excitation.

$$P(x, z, t) = \begin{cases} K_p G\delta(x - x_0)\delta(z - z_0)\sin(\omega_p t), & (0 \leq t < t_p, f_p \leq 1.5\text{Hz}); \\ K_p G\delta(x - x_0)\delta(z - z_0)\sin^2(\omega_p t), & (0 \leq t < t_p, 1.5\text{Hz} < f_p \leq 3.5\text{Hz}); \\ 0, & (t_p \leq t < T_p). \end{cases} \tag{13}$$

where, $\omega_p = \pi t_p^{-1}$.

For Eq. (13), its functions of $\delta(x)$ and $\delta(z)$ are expressed as followings in the process of analysis.

$$\delta(x - x_0) = \frac{2}{B} \sum_{j=1}^{\infty} \sin(\frac{j\pi}{B} x_0) \sin(\frac{j\pi}{B} x) \tag{14}$$

$$\delta(z - z_0) = \frac{2}{L} \sum_{k=1}^{\infty} \sin(\frac{k\pi}{L} z_0) \sin(\frac{k\pi}{L} z) \tag{15}$$

Based on the minimum potential energy principle, Differential equation of vibration could be obtained by means of variational approach from Eq. (10) –Eq. (12)[21, 22].

$$D\nabla^2\nabla^2 w(x, z, t) + \overline{m}\frac{\partial^2 w(x, z, t)}{\partial t^2} = P(x, z, t) \tag{16}$$

where, $D = \frac{Eh^3}{12(1-\nu^2)}$, D represents bending stiffness of high performance floor, ν represents its Poisson's Ratio, E represents Young's Modulus, $\overline{m} = \rho h$, it represents mass distribution per unit area of floor.

The mode shape function of this type is given to the boundary conditions for this high performance floor by Eq. (16).

$$\psi_{mn}(x, z) = X_{mn}(x) \sin(\omega_n z) \tag{17}$$

where, $\omega_n = n\pi L^{-1}$.

And then, the following equation could be obtained by going to plug in the formula of Equation.

$$\text{into } \nabla^2\nabla^2\psi(x, z) - \beta^4\psi(x, z) = 0 \tag{18}$$

$$\frac{d^4 X_{mn}(x)}{dx^4} - 2\omega_n^2\frac{d^2 X_{mn}(x)}{dx^2} + (\omega_n^4 - \beta^4)X_{mn}(x) = 0 \tag{19}$$

where, $\beta^4 = \omega_{mn}^2\frac{\overline{m}}{D}$.

And it $(X_{mn}(x))$ is as following by the formula of Eq. (19).

$$X_{mn}(x) = c_m \sin(\vartheta_{1mn}x) + d_m \cos(\vartheta_{1mn}x) + e_m \text{sh}(\vartheta_{2mn}x) + f_m \text{ch}(\vartheta_{2mn}x) \tag{20}$$

where, $\vartheta_{1mn} = (\beta^2 - \omega_n^2)^{0.5}$, $\vartheta_{2mn} = (\beta^2 + \omega_n^2)^{0.5}$.

So the mode shape function of this type is also solved by Eq. (20) into Eq. (17).

$$\psi_{mn}(x, z) = [c_m \sin(\vartheta_{1mn}x) + d_m \cos(\vartheta_{1mn}x) + e_m \text{sh}(\vartheta_{2mn}x) + f_m \text{ch}(\vartheta_{2mn}x)] \sin(\omega_n z) \tag{21}$$

And then, its boundary conditions are expressed as for this type of high performance floor:

$$[\frac{\partial^2\psi_{mn}(x,z)}{\partial x^2} + v\frac{\partial^2\psi_{mn}(x,z)}{\partial z^2}]|_{x=0} = 0; \tag{22}$$

$$[\frac{\partial^2\psi_{mn}(x,z)}{\partial x^2} + v\frac{\partial^2\psi_{mn}(x,z)}{\partial z^2}]|_{x=B} = 0; \tag{23}$$

$$[\frac{\partial^3\psi_{mn}(x,z)}{\partial x^3} + (2-v)\frac{\partial^3\psi_{mn}(x,z)}{\partial x\partial z^2}]|_{x=0} = 0; \tag{24}$$

$$[\frac{\partial^3\psi_{mn}(x,z)}{\partial x^3} + (2-v)\frac{\partial^3\psi_{mn}(x,z)}{\partial x\partial z^2}]|_{x=B} = 0. \tag{25}$$

2.4 Closed Form Solutions

The following matrix equation could be derived by Eq. (21) into Eq. (22) ~ Eq. (25).

$$[\Omega]_{4\times4}\begin{bmatrix} c_m \\ d_m \\ e_m \\ f_m \end{bmatrix} = 0 \tag{26}$$

where,

$$[\Omega]_{4\times4}\begin{bmatrix} 0 & -(\vartheta_{1mn}^2 + v\omega_n^2) & 0 & (\vartheta_{2mn}^2 - v\omega_n^2) \\ -(\vartheta_{1mn}^2 + v\omega_n^2)\sin(\vartheta_{1mn}B) & -(\vartheta_{1mn}^2 + v\omega_n^2)\cos(\vartheta_{1mn}B) & (\vartheta_{2mn}^2 - v\omega_n^2)sh(\vartheta_{2mn}B) & (\vartheta_{2mn}^2 - v\omega_n^2)ch(\vartheta_{2mn}B) \\ -(\vartheta_{1mn}^2 + v\omega_n^2)\sin(\vartheta_{1mn}B) & 0 & \vartheta_{2mn}[\vartheta_{2mn}^2 - (2-v)\omega_n^2] & 0 \\ \vartheta_{1mn}[\vartheta_{1mn}^2 - (2-v)\omega_n^2]\cos(\vartheta_{1mn}B) & -\vartheta_{1mn}[\vartheta_{1mn}^2 - (2-v)\omega_n^2]\sin(\vartheta_{1mn}B) & \vartheta_{2mn}[\vartheta_{2mn}^2 - (2-v)\omega_n^2]ch(\vartheta_{2mn}B) & \vartheta_{2mn}[\vartheta_{2mn}^2 - (2-v)\omega_n^2]sh(\vartheta_{2mn}B) \end{bmatrix}$$

A series of ϑ_{1mn} and ϑ_{2mn} could be solved depending on the values of m and n by Eq. (26).

So the mode shape function of $\psi_{mn}(x,z)$ is also to be expressed as followings.

$$\psi_{mn}(x,z) = \{\vartheta_{2mn}\eta_{1mn}\sin(\vartheta_{1mn}x) + \vartheta_{1mn}\eta_{2mn}sh(\vartheta_{2mn}x) - \varphi_{mn}[\eta_{2mn}\cos(\vartheta_{1mn}x) + \eta_{1mn}ch(\vartheta_{2mn}x)]\}\sin(\omega_n z) \tag{27}$$

where $\varphi_{mn} = \eta_{1mn}^{-1}\eta_{2mn}^{-1}[\cos(\vartheta_{1mn}B) - ch(\vartheta_{2mn}B)]^{-1}[\vartheta_{2mn}\eta_{1mn}^2\sin(\vartheta_{1mn}B) - \vartheta_{1mn}\eta_{2mn}^2 sh(\vartheta_{2mn}B)]\eta_{1mn} = \beta_{mn}^2\omega_n^{-2} + v - 1, \eta_{2mn} = \beta_{mn}^2\omega_n^{-2} - v + 1.$

It is difficult to solve Eq. (10) to Eq. (12), the mode shape function of this type is therefore used. The deflection of the composite floor could be expressed as the followings by using the function expansion method.

$$w(x, z, t) = \sum_{m=1}^{\infty} \sum_{n=1}^{\infty} \psi_{mn}(x, z) w_{mn}(t) \qquad (28)$$

So the following expression of $w_{mn}(t)$ is also obtained by Eq. (28) in Eq. (16).

$$w_{mn}(t) = a_{mn} \sin(\omega_{mn} t) + b_{mn} \cos(\omega_{mn} t) + w_{Pmn}(t) \qquad (29)$$

where, $w_{Pmn}(t) = \dfrac{\int_{0}^{t} P_{mn}(\tau) \sin(\omega_{mn}(t-\tau)) d\tau}{\omega_{mn} \iint_{s} \overline{m} \psi_{mn}^{2}(x, z) ds}$.

Further, the position of the operating point of the single-jump load excitation behavior can also be expressed as:

$$P_{mn}(t) = \begin{cases} K_p G \iint_{s} \sin(\omega_p t)\delta(x - x_0)\delta(z - z_0)\psi_{mn}(x, z)ds, & (0 \leq t < t_p, f_p \leq 1.5 \text{ Hz}); \\ K_p G \iint_{s} \sin^2(\omega_p t)\delta(x - x_0)\delta(z - z_0)\psi_{mn}(x, z)ds, & (0 \leq t < t_p, 1.5 \text{ Hz} < f_p \leq 3.5 \text{ Hz}); \\ 0, & (t_p \leq t < T_p). \end{cases} \qquad (30)$$

The deflection of the composite floor could be solved as following based on the initial conditions of both $w_{mn}(0){=}0$ and $\frac{\partial w_{mn}(0)}{\partial t}{=}0$.

$$w(x, z, t) = \sum_{m=1}^{\infty} \sum_{n=1}^{\infty} \psi_{mn}(x, z)[\omega_{mn}^{-1} w'_{Pmn}(0) \sin(\omega_{mn} t) + w_{Pmn}(t)] \qquad (31)$$

The dynamic steady state solution of the deflection can be get for this type of high performance floor under action of the single-jump excitation based on properties of Dirac Delta functions of $\int_{-\infty}^{\infty} F(x)\delta(x - x_0)dx = F(x_0)$ and $\int_{-\infty}^{\infty} F(x)\delta'(x - x_0)dx = -F'(x_0)$.

$$w(x, z, t) = \sum_{m=1}^{\infty} \sum_{n=1}^{\infty} \frac{G * \psi_{mn}(x_0, z_0)}{J_{mnp} \iint_{s} \psi_{mn}^{2}(x, z)ds} \psi_{mn}(x, z) \sin(\omega_p t) \qquad (32)$$

where, $G* = K_p \delta(x - x_0)\delta(z - z_0)G$, $J_{mnp} = \overline{m}(\omega_{mn}^2 - \omega_p^2)$.

Its analytical expression of acceleration can be solved for the high performance floor considering both time effect and space effect by this method of second order differential calculation.

$$a(x, y, t) = w(x, z, t) = \sum_{m=1}^{\infty} \sum_{n=1}^{\infty} \frac{-\omega_p^2 G * \psi_{mn}(x_0, z_0)}{J_{mnp} \iint_{s} \psi_{mn}^{2}(x, z)ds} \psi_{mn}(x, z) \sin(\omega_p t) \qquad (33)$$

3 Verification of the Experimental Method

3.1 Model Design

In order to explore comfort behaviors of this type of high performance floor considering time–space effect, a dedicated specimen of the floor model were designed. The length is 2.5 m, the width is 2.0 m and the thickness is 0.1 m. The longitudinal reinforcement is made of threaded steel bars with a spacing of 450 mm, and the transverse reinforcement is made of light circular steel bars with a spacing of 200 mm The mixing ratio of high performance concrete is: cement: sand: stone: water: 1.00:1.20:1.92:0.29, JM-8 content 1.8%, mineral admixture content 30.0%, steel fiber volume content 1.0%. A detailed mechanical property of HPC material is listed in Table 1, and physical single-jump excitation of experimental mode is shown in Fig. 2.

Table 1 Mechanical property of HPC material

Species	Poisson's ratio	Modulus of elasticity against compression (MPa)	Cube compressive strength (MPa)	Axial tensile strength (MPa)
HPC	0.21	38.25	68.22	2.62
		36.28	67.33	2.68
		38.16	65.96	2.62
	Average	37.56	67.17	2.64

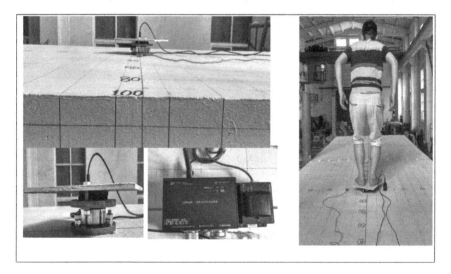

Fig. 2 Physical single-jump excitation of experimental mode

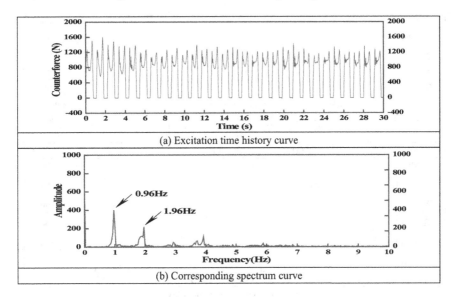

(a) Excitation time history curve

(b) Corresponding spectrum curve

Fig. 3 Mode 1 of test set-up

3.2 Test Set-Up and Items

To get the required parameters of t_p, T_p, and α, two different jumping frequencies were tested for the high performance floor based on the improved single-jump excitation model. The average value and standard deviation were obtained through 30 consecutive jumps. Two types of excitation time history curves and corresponding spectrum curves were shown in Fig. 3 and Fig. 4.

3.3 Calculation and Verification

In this section, a typical closed solution of the model is used to calculate the acceleration under two different ways of jumping frequencies. Based on "Code for design of steel structures" and "Code for design of concrete structures", its physical properties of system required in the comfort behaviors analysis are referenced and calculated, listed in Table 2 and Table 3.

In the process of analyzing features on comfort behaviors of this high performance floor model, comparison analyses were also launched between experimental research and theoretical derivation. Based on Eq. (33), its numerical solutions of acceleration are detailed compared withF those data measured in experimental investigation, as shown in Fig. 5 and Fig. 6. It was further shown that this kind of method, which were obtained based on the previous analytical expression of acceleration, had a

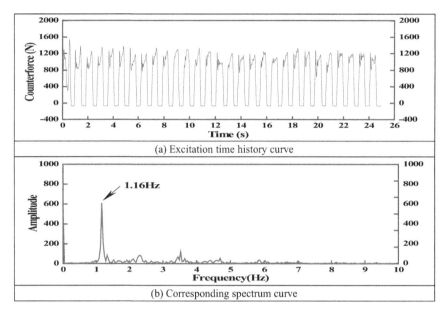

Fig. 4 Mode 2 of test set-up

Table 2 Physical properties of high performance floor, where deduced parameter is denoted by *

Net width B(m)	Net length L (m)	Net height H (m)	Density P (kg/m^3)	Elasticity modulus E_{cs} (GPa)	Poisson's ratio v_{cs}	Bending stiffness $D*$ (N·m)	Mass distribution per unit area $\overline{m}*$ (kg/m^2)
2.00	2.50	0.09	2450	36.0	0.25	2.2781E + 6	2.205E + 2

Table 3 Deduced parameters based on experimental test

Mode 1 0.96 Hz	x_0	z_0	G	α	f_p	t_p	T_p	ω_p	J_{11p}
	1.0	1.25	630	0.72	0.96	0.747	1.038	4.206	5.68E + 6
Mode 2 1.16 Hz	x_0	z_0	G	α	f_p	t_p	T_p	ω_p	J_{11p}
	1.0	1.25	630	0.68	1.16	0.588	0.859	5.343	5.67E + 6
Vibration parameters	ω_1	ω_{11}	ϑ_{111}	ϑ_{211}	η_{111}	η_{211}	φ_{11}		
	1.257	160.511	0.00147	1.77715	0.19942	1.79942	0.01387		

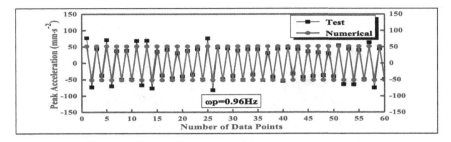

Fig. 5 Comparison between theoretical and test for mode 1

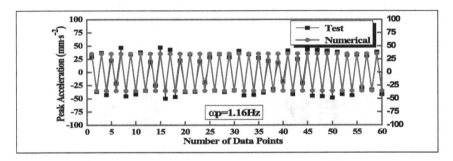

Fig. 6 Comparison between theoretical and test for mode 2

very good precision for solving the problem of comfort behaviors for two types of different jumping frequencies.

3.4 Spatial and Temporal Distribution Characteristics

To assess the accuracy of the proposed high performance floor model, we will launch analysis of spatial distribution of acceleration response at typical time for two types of different jumping frequencies. And then, two typical characteristics of the spatial distributions are shown in Fig. 7 and Fig. 8. It is show that the peak acceleration is concentrated in the middle of the high performance floor, and then the acceleration response gradually weakened to the two sides, but the degree of weakening was different, and its degree of weakening was greater along the two corners of the floor model.

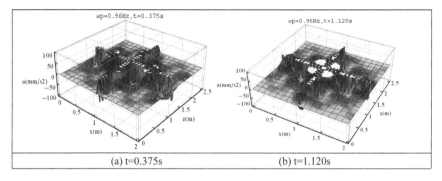

Fig. 7 Spatial distribution of acceleration response at typical time for mode 1

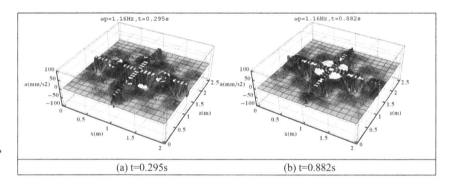

Fig. 8 Spatial distribution of acceleration response at typical time for mode 2

3.5 Parameter Analysis

Understanding the comfort behaviors of the high performance floor based on single-jump excitation mode is important for comfort performance designers. There is a particular need to study the influence of the material parameters on the comfort behaviors, especially strength grade. To be able to analyze the comfort behaviors more effectively, we will set up three different grades of concrete, including ordinary performance, high performance and ultra high performance. The values of strength of concrete can be as follows in Table 4 and Table 5.

Comparative analysis of acceleration response could be launched based on Eq. (33) for two different ways of jumping frequencies. Moreover, the peak acceleration of the high performance floor under different concrete grades is selected as the sensitive parameters, and its peak acceleration sensitive parameters were further compared under six grades of concrete materials. These are shown in Fig. 9 and Fig. 10.

Figure 9 shows its acceleration response process in one period ($w_p = 0.96$ Hz). It is shown that the acceleration response amplitude at the center of the high performance

Table 4 Sensitive parameters for the high performance floor

Performance level	Strength grade	Density ρ(kg/m^3)	Elasticity Modulus E_{cs}(GPa)	Poisson's ratio ν_{cs}	Bending stiffness D^*(N·m)	Mass distribution per unit area $\overline{m}^*$(kg/m^2)
Ordinary performance	C15	2400	22.0	0.20	1.3922E + 6	2.160E + 2
	C40	2430	32.5	0.22	2.0566E + 6	2.187E + 2
High performance	C60	2450	36.0	0.20	2.2781E + 6	2.205E + 2
	C80	2490	39.0	0.20	2.4679E + 6	2.241E + 2
Ultra high performance	C100	2500	43.08	0.19	2.7151E + 6	2.250E + 2
	C150	2500	49.53	0.14	3.0691E + 6	2.250E + 2

Table 5 Analysis parameters for the high performance floor

Performance level		C15	C40	C60	C80	C100	C150
Vibration parameters	ω_1	1.2566	1.2566	1.2566	1.2566	1.2566	1.2566
	ω_{11}	126.777	153.809	160.511	165.718	173.470	184.431
	ϑ_{111}	0.64821	0.00208	0.00147	0.00196	0.00182	0.00086
	ϑ_{211}	1.89173	1.77716	1.77714	1.77715	1.77715	1.77711
	η_{111}	0.46608	0.21990	0.19942	0.20002	0.19002	0.14000
	η_{211}	2.06608	2.06690	1.78002	1.79942	1.81002	1.86000
	φ_{11}	2.88737	0.01779	0.01387	0.01862	0.01832	0.01204
J_{11p}	0.96 Hz	3.468E + 06	5.169E + 06	5.677E + 06	6.150E + 06	6.767E + 06	7.649E + 06
	1.16 Hz	3.465E + 06	5.168E + 06	5.674E + 06	6.148E + 06	6.764E + 06	7.647E + 06

floor gradually changes from 79.446 to 38.571 mm•s^{-2} with the optimization of the high performance floor. And the acceleration response at the center of C40, C60, C80, C100, C150 grade for the high performance floor is 0.70496 times, 0.64753 times, 0.59547 times, 0.54261 times, 0.48550 times of the acceleration response at the center of C15 grade for the high performance floor.

Figure 10 shows its acceleration response process in one period ($w_p = 1.16$ Hz). There are two main regulars for the small discrepancies in the numerical values: (1) With the improvement of concrete performance grade, the acceleration response amplitude at the center decreases gradually; (2) The acceleration response at the center of C40, C60, C80, C100, C150 grade for the high performance floor is 0.70481

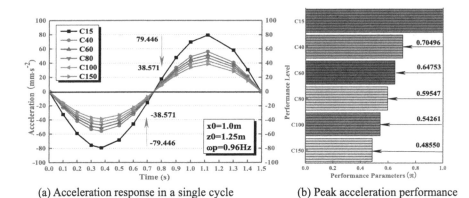

(a) Acceleration response in a single cycle   (b) Peak acceleration performance

Fig. 9 Comparative analysis of acceleration response for mode 1

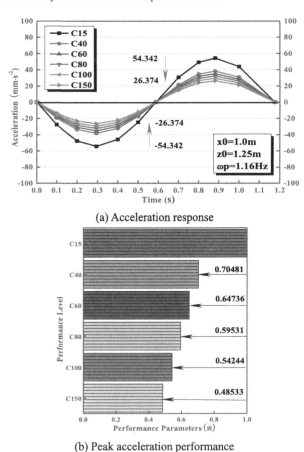

(a) Acceleration response

(b) Peak acceleration performance

Fig. 10 Comparative analysis of acceleration response for mode 2

times, 0.64736 times, 0.59531 times, 0.54244 times, 0.48533 times of the acceleration response at the center of C15 grade for the high performance floor. Through 1.16 Hz excitation mode, the obvious advantages of high performance concrete in high performance floor structure are illustrated again.

4 Conclusion

This paper has proposed comfort behaviors of high performance floor based on single-jump excitation mode considering the time–space effect. It is intended to help bridge designers understand the impact on comfort performance of high performance floor. The influence of the material parameters, spatial and temporal distribution characteristics on the comfort behaviors of this proposed high performance floor were analyzed. Based on the results from this study, the following conclusion can be drawn:

(1) Based on the improved single-jump excitation model, theoretical derivation is applied to study the comfort behaviors of this model considering both time effect and space effect by this method of second order differential calculation.
(2) Two different jumping frequencies (0.96 and 1.16 Hz) were tested for the high performance floor based on the improved single-jump excitation model, comparison analyses were completed between experimental research and theoretical derivation, and it had a very good precision for solving the problem of comfort behaviors for two types of different jumping frequencies.
(3) It shown that the peak acceleration was concentrated in the middle of the high performance floor, and then the acceleration response gradually weakened to the two sides, and its degree of weakening was greater along the two corners of the floor model from its spatial and temporal distribution characteristics.

Acknowledgements This study is funded by the National Key Research and Development Plan (Grant No. 2021YFB2600700), and the National Natural Science Youth Foundation (Grant No. 52109160).

References

1. Park B, Ryu SR, Cheong CH (2020) Thermal comfort analysis of combined radiation-convection floor heating system. Energies 13(6):1420
2. Lu Z, Lou Y (2912) Comfort retrofit design of the floor vibration in a gymnasium. Build Struct 42(3):45–48
3. Chen QJ, Zhao ZP, Zhang RF (2019) Comfort based floor design employing tuned inerter mass system. J Sound Vib 458:143–157
4. Lin W, Ye FF, Xiao ZB (2018) Vibration comfort control of long-span floor under excitation of pedestrian load. Spatial Struct 24(4):56–61

5. Ding YW, Zhang YF, Wang Z (2020) Vibration test and comfort analysis of environmental and impact excitation for wooden floor structure. BioResources 15(4):8212–8234
6. Yang WY (2018) Effects of indoor water sounds on floor impact noise perception and overall environmental comfort. Korean Soc Living Environ Syst 25(5):611–619
7. Zhou X, Liu YL, Luo MH (2022) Radiant asymmetric thermal comfort evaluation for floor cooling system-A field study in office building. Energy Build 260:1–12
8. Jin Q, Wu XY, Dai YX (2021) Analysis on floor vibration comfort of a reconstructed and expanded high school dining hall. Build Struct 51(16):102–109
9. Cen C, Jia YH, Geng RX (2018) Experimental comparison of thermal comfort during cooling with a fan coil system and radiant floor system at varying space heights. Build Environ 141:71–79
10. Zhang ZQ, Ma F, Chen X (2016) Dynamic measurement and comfort parameter analysis of large span truss-corrugated steel deck RC composite floor. J Build Struct 37(6):19–27
11. He HX, Yan WM (2013) Floor vibrations induced by human jumping and landing. J Vibrat Eng 26(2):220–225
12. He HX, Yan WM (2008) Human-structure dynamic interaction and comfort evaluation in vertical ambient vibration. J Vibrat Eng 21(5):446–451
13. Fu D., Wu HG (2014) Vibration analysis and human comfort evaluation of underground hydropower house. In: 4th International conference on civil engineering, architecture and building materials vol 4, pp 2055–2062
14. Campista FF, da Silva JGS (2018) Vibration analysis of steel-concrete composite floors when subjected to rhythmic human activities. J Civ Struct Heal Monit 8(5):737–754
15. Yang QZ, Ma KJ, He LX (2020) Vibration comfort analysis and measurement of steel-concrete composite open-web sandwich plate. Spatial Struct 26(3):66–74
16. Jiang L, Ma KJ, Huang JH (2016) Dynamic time-history analysis of large span open-web sandwich plate for vibration comfort based on walking route method. Spatial Struct 22(2):28–36,43
17. Allen DE, Rainer JH, Pernica G (1985) Vibration criteria for assembly occupancies. Can J Civ Eng 12(3):617–623
18. Rainer JH, Pernica G, Allen DE (1988) Dynamic loading and response of footbridges. Can J Civ Eng 15(1):66–71
19. Kasperski M (2002) Men-induced dynamic excitation of stand structures. In: 15th ASCE engineering mechanics conference, June 2–5
20. Cheng J, Wang L, Chen B, Yan SX (2014) Dynamic properties of human jumping load and its modeling: experimental study. J Vibrat Eng 27(1):16–24
21. Warren AG (1930) Free and forced oscillations of thin bars, flexible discs and annuli. PHIL Mag 9:881–901
22. Anderson BW (1954) Vibration of triangular cantilever plates by the rize method. JAM 21(4):365–366

Study on High Temperature Performance of Asphalt Mixture and Correlation of Its Evaluation Indexes

Dongbin Lv, Honggang Zhang, Lihao Zeng, and Jiechao Lei

Abstract In order to better evaluate the high-temperature performance of asphalt mixture, based on the conventional rutting test and Hamburg rutting test, this paper further studies the influence of temperature, grading and asphalt type on the high-temperature shear resistance of asphalt mixture by using uniaxial penetration test and uniaxial compression test. Based on Mohr Coulomb theory, the cohesive force and internal friction angle of mixture were calculated. Based on the analysis of variance, the significance of the influence of temperature and gradation on the high-temperature shear characteristics was studied. Linear regression analysis was used to reveal the correlation between the fractal dimension treated gradation and the shear resistance characteristics. Finally, the correlation between the high-temperature shear resistance performance index and the high-temperature rutting resistance evaluation index was analyzed. The results show that temperature has a significant effect on cohesion and internal friction angle of gradation pairs, and the fractal dimension Dc of coarse aggregate is positively correlated with cohesion and internal friction angle, with the strongest correlation. It indicates that the high-temperature performance of the mixture can be improved by increasing the passing rate of the sieve hole above 4.75 mm in the grading design. The correlation R2 between cohesive force index and high-temperature rutting resistance index of mixture at 35 °C is higher than 0.9, which belongs to extremely strong correlation. The shear resistance index of cohesion at 35 °C can be used to characterize the high-temperature rutting resistance of asphalt mixture.

D. Lv
Guangxi Expressway Investment Co., Ltd., Guangxi, Nanning 530021, China

H. Zhang (✉) · L. Zeng · J. Lei
Guangxi Transportation Science and Technology Group Co., Ltd., Guangxi, Nanning 530007, China
e-mail: 286601676@qq.com

Guangxi Key Lab of Road Structure and Materials, Guangxi, Nanning 530007, China

Research and Development Center On Technologies, Materials and Equipment of High Grade Highway Construction and Maintenance Ministry of Transport, Guangxi, Nanning 530007, China

© The Author(s) 2023   35
G. Feng (ed.), *Proceedings of the 9th International Conference on Civil Engineering*,
Lecture Notes in Civil Engineering 327,
https://doi.org/10.1007/978-981-99-2532-2_4

Keywords Road Works · Asphalt Mixture · Shear Resistance · Anti Rutting
Performance · Correlation Analysis

1 Introduction

High temperature is one of the important technical properties of asphalt pavement,
which directly affects the durability of pavement. The dynamic stability index in
conventional rutting test is mainly used for the evaluation of high temperature stability
in the current asphalt pavement design specifications in China. However, foreign
scholars mostly use Hamburg rutting test to evaluate the rutting resistance of asphalt
pavement under high temperature and humid environment during actual driving.
Both methods belong to the empirical test method. It belongs to mechanical testing
method to conduct stress–strain test on mixture to analyze the change of shear force.
Based on Mohr Coulomb theorem, the high temperature shear strength of materials is
mainly characterized by cohesion and internal friction angle [1]. At present, scholars
at home and abroad have made a lot of research achievements on the influencing
factors of high temperature shear resistance of asphalt pavement materials. TF Fwa
[2] establishes molar circles under different stresses through indoor triaxial test results
to calculate cohesive force and internal friction angle indexes of asphalt mixture.
Zhu [3] proposed that the grading type, asphalt content, asphalt property and test
temperature have significant effects on the shear strength of the mixture. Bi [4]
proposed to calculate the cohesion and internal friction angle based on the results of
uniaxial penetration test and uniaxial compression test, and verified the calculation
formula and found that its correlation is significant. Lu [5] established a relationship
model between the penetration strength of mixture and internal friction angle, and
the results show that the correlation is good.

The research results of domestic and foreign scholars found that the main method
for studying the high-temperature shear resistance of asphalt mixture is to use triaxial
shear test. However, triaxial shear test instruments and equipment are expensive and
rare in China. Therefore, it is feasible to replace triaxial shear test with uniaxial penetration test and uniaxial compression test proposed by Bi. At present, there is little
research on the correlation between rutting performance and high temperature shear
resistance of AC-13 graded asphalt mixture. Therefore, this paper selects different
asphalt types to prepare asphalt mixture from fine to coarse AC-13 for conventional
rutting test and Hamburg rutting test to study its high-temperature rutting resistance.
Then, the high-temperature shear resistance was studied by uniaxial penetration test
and uniaxial compression test. It can provide reference experience for future domestic
road network construction.

Table 1 Test results of basic performance indexes of asphalt

Performance index		Value		Methods method
		Base asphalt	Rubber asphalt	
Penetration(25°C,100 g,5 s)/0.1 mm		69	35.7	T0604
Softening point/°C		48.0	72.8	T0606
Ductility (15°C,5 mm/min)/cm		> 100	17.5	T0605
Rolling thin film oven (RTFO) aged	Quality change/%	−0.4	–	T 0610
	Residual penetration ratio/%	61.2	–	T 0604
	Residual ductility/cm	6.4	–	T 0605

Table 2 Test results of limestone aggregate density and water absorption

Performance index	Particle size			
	10–15	5–10	3–5	0–3
Apparent relative density	2.773	2.758	2.761	2.755
Gross volume relative density	2.738	2.723	2.707	–
Water absorption (%)	0.46	0.47	0.72	–

2 Materials and Methods

2.1 Materials

Limestone is used as aggregate for this test, which is taken from Xiangzhou-Longbang overhaul project site. Mineral powder is used as filler. Grade 70# A road petroleum asphalt and rubber asphalt are used as asphalt. Material testing indexes are shown in Table 1 and Table 2.

2.2 Grading Design

At present, the design system of asphalt mixture usually takes 4.75 mm sieve as the boundary size of coarse and fine aggregates. Academician Qinglin Sha divided the asphalt mixture structure into four types according to the 4.75 mm sieve hole passing rate, that is, the 4.75 mm sieve hole passing rate is about 30%, 35%, 40%, and more than 40%, which respectively correspond to the tight framework dense structure, the general framework dense structure, the loose framework dense structure, and the suspended dense structure. In order to study the high temperature rutting resistance of asphalt mixtures with different gradations, the key control screen hole of 4.75 mm with AC-13 grading is adjusted. Three mineral aggregate gradations with different

Table 3 Grading passing rate

Sieve hole(mm)	16	13.2	9.5	4.75	2.36	1.18	0.6	0.3	0.15	0.075
Grading A (Base asphalt)	100	95.2	75.2	43.0	32.0	23.7	17.0	13.1	10.3	6.3
Grading B (Base asphalt)	100	91.5	66.0	35.0	25.0	18.5	13.0	10.0	8.5	5.3
Grading C (Base asphalt)	100	90.5	65	31	21	13.5	10.5	8	6.5	4.5
Grading D (Rubber asphalt)	100	90.5	65	31	21	13.5	10.5	8	6.5	4.5

thickness are set for test. Among them, the passing rate of each sieve mesh of grading C and D is the same, but the matrix asphalt and rubber asphalt are respectively used as the binder. The passing rate of each sieve hole of grading is shown in Table 3.

In order to conduct quantitative analysis on the influencing factors of grading in the following paper, the grading is treated based on the fractal dimension theory. Fractal dimension theory is a new subject to quantitatively describe the complexity and space filling ability of geometric objects. It can better adapt to the characteristics of asphalt mixture, such as non-uniformity, nonlinearity, irregularity and fuzziness. Especially for aggregate gradation with outstanding self similarity, it can be described by fractal theory [6].

The double logarithmic scatter plot of three grading curves based on fractal theory is shown in Fig. 1. The overall fractal dimension D of the gradation is calculated by linear regression with the least square method. Then, only the part above 4.75 mm or below is taken from the grading curve to calculate the grading fractal dimensions D_c and D_f of coarse and fine aggregates respectively. The fractal index results of three grading types are shown in Table 4.

Fig. 1 Double logarithmic scatter plot of grading curve

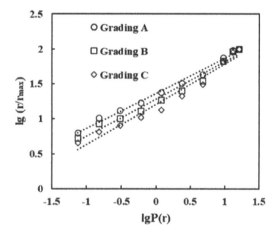

Table 4 Calculation results of Grading Fractal Dimension

Gradation types	D		Dc		Df	
	value	R^2	value	R^2	value	R^2
Grading A	2.4992	0.9896	2.28	0.9866	2.5539	0.9932
Grading B	2.4686	0.9756	2.1109	0.9948	2.5635	0.9916
Grading C	2.4245	0.9644	2.0099	0.9932	2.5519	0.9878

2.3 Test Method

(1) Rutting test

The rutting test is mainly used to determine the high-temperature rutting performance of asphalt mixture. The test sample size is 300 mm × 300 mm × 100 mm rut plate specimen. The test temperature is 60 °C. Conduct rutting test on the sample under the condition of 0.7 MPa wheel pressure. S1top the test when walking back and forth for 1h or the maximum deformation of the sample reaches 25 mm, and calculate the dynamic stability and other test results based on the standard formula.

(2) Hamburg rutting test

Hamburg rutting test is mainly used to test the high-temperature rutting resistance of asphalt mixture [7]. The size of the test sample is 260 mm × 320 mm × 40 mm track plate test piece. Load the test piece to the steel wheel under the immersion environment of 50 °C for 20000 times of reciprocating motion or until the rutting deformation of 20 mm is produced. In order to facilitate the analysis, the Hamburg rutting test results are evaluated by two indicators: the rolling times at 15 mm and the creep slope CS. The creep slope CS represents the rut development speed, which is the reciprocal of the slope at a certain stage on the rut development curve, and its physical meaning is the number of times of loading to produce 1mm deformation.

(3) Uniaxial penetration test

Based on MTS-810 material test system Uniaxial penetration test on Φ 100× 100 mm asphalt specimen. The test uses an indenter with a diameter of 28.5 mm. The loading rate is 1mm/min. Before the penetration test, the asphalt specimen is placed in a 35 °C and 50 °C water bath at constant temperature for 5 hours to explore the influence of different temperatures on the shear resistance of the mixture.

(4) Uniaxial compression test

In order to analyze the shear resistance of the mixture based on the results of the synergistic penetration test, based on the MTS-810 material test system Uniaxial compression test of Φ100 × 100 mm asphalt specimen. The test adopts a special indenter for uniaxial compression test. Head size diameter is 100 mm. Other test conditions are the same as that of uniaxial penetration test.

3 Results and Discussion

3.1 High Temperature Rutting Test

In order to study the high-temperature rutting resistance of different materials, the asphalt mixture prepared with the above grading and materials is used for rutting test and Hamburg rutting test. The test results are shown in Figs. 2 and Fig. 3.

As shown in Fig. 2 and Fig. 3, the data laws of conventional rutting test and Hamburg rutting test are basically the same:

(1) Within the same AC-13 grading range, fine grading is better than coarse grading in high-temperature rutting resistance. For example, the dynamic stability value of grading A is 31.2% and 78.8% higher than that of grading B and C. Because the fine grading is easier to compact than the coarse grading under the same compaction work, its porosity is smaller, and the density of the mixture is better, so its anti rutting performance is better.

(2) The rutting resistance of rubber asphalt is superior to that of base asphalt at high temperature.For example, the dynamic stability value of grading D is 2.5 times higher than that of grading C. It shows that asphalt type has more obvious influence on high-temperature rutting resistance. Because the addition of rubber modifier can effectively enhance the cohesion between asphalt molecules, increase the asphalt viscosity and improve the high-temperature performance of the mixture.

Fig. 2 Dynamic stability of rutting test at 60 °C

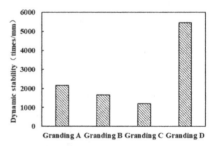

Fig. 3 Results of Hamburg rutting test at 50 °C

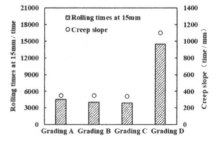

3.2 High Temperature Shear Resistance Test

Carry out uniaxial penetration test and uniaxial compression test respectively according to the above test methods to obtain the maximum load and maximum displacement of asphalt mixtures with different gradations, as shown in Fig. 4 and Fig. 5.

As shown in Fig. 4 and Fig. 5, the uniaxial penetration test and uniaxial compression test show a similar trend:

(1) The high temperature shear resistance of fine grading is better than that of coarse grading under the same asphalt type. The maximum load and displacement decrease with the coarsening of the grading. Because there are fewer fine aggregates in Grades B and C, with large voids, and less asphalt mortar between aggregates, which leads to a decrease in the cohesion of the mixture [8].

(2) Rubber asphalt mixture has obvious advantages over base asphalt mixture in high temperature shear resistance. It shows that the asphalt type has more significant influence on the high temperature shear resistance of the mixture than the grading type. Because the addition of rubber powder increases the viscosity

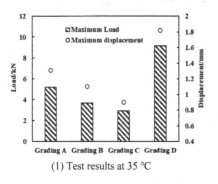
(1) Test results at 35 °C

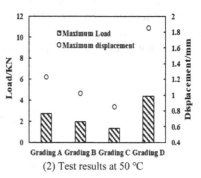
(2) Test results at 50 °C

Fig. 4 Results of uniaxial penetration test

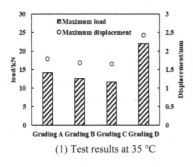
(1) Test results at 35 °C

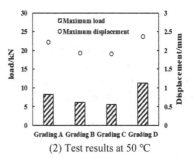
(2) Test results at 50 °C

Fig. 5 Results of uniaxial compression test

of asphalt, the force between asphalt molecules and the cohesion between aggregate and asphalt mortar.

(3) Increasing the test temperature will significantly reduce the shear strength of the sample. Because the increase of temperature will lead to the decrease of asphalt viscosity and then the decrease of mixture cohesion [9].

In fact, the strength of asphalt mixture is essentially determined by the cohesion (C) of binder and the internal friction angle with aggregate (φ) Common representation. Therefore, this paper refers to Bi's proposal to calculate the C of the sample and the φ. The conversion formula is as follows:

$$R_\tau = f_\tau \sigma_p \tag{1}$$

$$\sigma_p = \frac{P}{A} \tag{2}$$

$$\sigma_\mu = \frac{P}{A} \tag{3}$$

where: R_τ——shear strength (MPa); f_τ—— Penetration index 0.34; σ_p—— Uniaxial penetration pressure (MPa); σ_μ—— Is the uniaxial compressive strength (MPa); P——maximum load (N); A——Cross sectional area of indenter (mm^2).

Calculation of C and C based on shear test results φ. The formula is as follows:

$$\sigma_{P1} = 0.765\sigma_P \tag{4}$$

$$\sigma_{P3} = 0.0872\sigma_P \tag{5}$$

$$\varphi = \arcsin\left(\frac{\sigma_{p1} - \sigma_{p3} - \sigma_\mu}{\sigma_{p1} - \sigma_{p3} - \sigma_\mu}\right) \tag{6}$$

$$c = \frac{\sigma_\mu}{2} - \left(\frac{1 - \sin\varphi}{\cos\varphi}\right) \tag{7}$$

where: σ_{P1}——the first principal stress (MPa), σ_{P3}—— the third principal stress (MPa); C——Cohesion (MPa); φ—— Is internal friction angle (°)

According to the uniaxial penetration test and uniaxial compression test conducted in this paper, the shear strength, compressive strength, C and φ of different graded mixtures at different temperatures are calculated as shown in Table 5.

Table 5 shows that:

(1) The temperature rise will significantly reduce the C of the sample, but it has little impact on φ of the sample. The decrease of sample C at 50 °C is 66.7% higher than that at 35 °C, while φ only 1.9%. Because the cohesive is mainly represented by the adhesion of the sample binder, and the internal friction angle

Table 5 C, φ and shear resistance indexes of different specimens

Types	Temperature/°C	Uniaxial penetration strength/MPa	Uniaxial compression strength/MPa	Shear strength/MPa	C/MPa	φ/°
Grading A	35°C	8.108	1.798	2.749	0.360	46.33
	50°C	4.305	1.055	1.459	0.216	45.46
Grading B	35°C	5.734	1.594	1.944	0.337	44.13
	50°C	3.038	0.783	1.030	0.162	44.96
Grading C	35°C	4.603	1.478	1.565	0.328	42.18
	50°C	2.120	0.709	0.720	0.159	41.54
Grading D	35°C	13.350	2.805	4.665	0.549	47.26
	50°C	6.882	1.431	2.333	0.283	46.83

reflects the sliding friction caused by the rough texture of the aggregate surface and the ability of the aggregates to squeeze each other [10].

(2) The cohesion and internal friction angle of rubber asphalt are significantly higher than that of base asphalt mixture.The cohesion and internal friction angle of intermediate Grading D in the table are increased by 67.2% and 12.0% respectively compared with that of Grading C at 35 °C.This is consistent with the previous analysis, which shows that rubber asphalt not only improves the cohesion between the cements, but also increases the friction between the aggregates when the rubber powder is fully distributed between the aggregates, which has the effect of increasing the internal friction angle [11].

(3) Both C and φ of coarse and fine gradation show a downward trend. C and φ of Grading B relative to Grading A at 35°C reduced by 6.4% and 4.7% respectively. C and φ of Grading C relative to Grading A at 35°C reduced by 8.79% and 8.95% respectively. Gradation has a more significant impact on φ than other factors. To analyze the influence of temperature and grading on C and φ, the multi factor variance analysis in SPSS software is used to calculate the influence degree of grading A, B and C. The analysis results are shown in Table 6.

As shown in Table 6, the sig value of variance analysis of temperature on cohesion is 0.003, which is a significant impact. But the F value of variance analysis of

Table 6 Analysis of variance

Category of factors	Type of dependent variable	ANOVA Results		Significance
		F	Sig	
Grading	C	8.628	0.104	–
	φ	19.794	0.048	*
Temperature	C	293.642	0.003	*
	φ	0.186	0.709	–

Note: sig value < 0.05 is significant

Table 7 Correlation analysis of grading fractal dimension with C and φ

Fractal dimension index	Temperature (°C)	Fitting equation(C)	R^2	Fitting equation(φ)	R^2
D	35	y = 0.408x-0.6635	0.8762	y = 54.717x - 90.613	0.9808
	50	y = 0.7032x - 1.5534	0.6879	y = 54.268x - 89.736	0.9114
Dc	35	y = 0.1192x + 0.0875	0.9877	y = 15.112x + 11.971	0.9881
	50	y = 0.2196x - 0.2893	0.8861	y = 13.329x + 15.548	0.7261
Df	35	y = -0.2466x + 0.9721	0.0087	y = 42.303x - 63.931	0.016
	50	y = -1.648x + 4.3924	0.103	y = 185.24x - 429.57	0.2896

temperature on internal friction angle is 0.186, indicating that its impact is slight. In order to quantitatively analyze the correlation between the gradation fineness and the cohesion and internal friction angle of the mixture, the gradation D, Dc and Df after fractal treatment in Table 4 are linearly correlated with the data in Table 5, and the results are shown in Table 7.

Dc of coarse aggregate and C, φ have the strongest correlation. D and Dc are positively correlated with C and φ of samples at different temperatures. Their R^2 is greater than 0.6 and is strongly correlated. The correlation between Df and C, φ of fine aggregates is weak. The larger the D and Dc, the lower the proportion of coarse aggregates. To sum up, the shear resistance of the mixture can be improved by increasing the passing rate of the sieve openings above 4.75 mm in the grading design.

3.3 Correlation Analysis

In order to further reveal the high-temperature rutting resistance of the mixture and its C, φ, This paper compares the rutting test data with C and φ linear fitting. The results are shown in Fig. 6.

Figure 6 shows that the correlation between shear characteristic indicators and dynamic stability, rolling times at 15 mm and creep slope is as follows: C(35°C) > φ (35°C) > C(50°C) > φ (50°C). The R^2 values of C (35°C) and dynamic stability, number of rolls at 15 mm and creep slope are 0.9932, 0.9944 and 0.9857 respectively. It belongs to extremely strong correlation. Therefore, this index can be used as the relevant index of asphalt mixture high temperature rutting resistance in the evaluation of mixture high temperature performance.

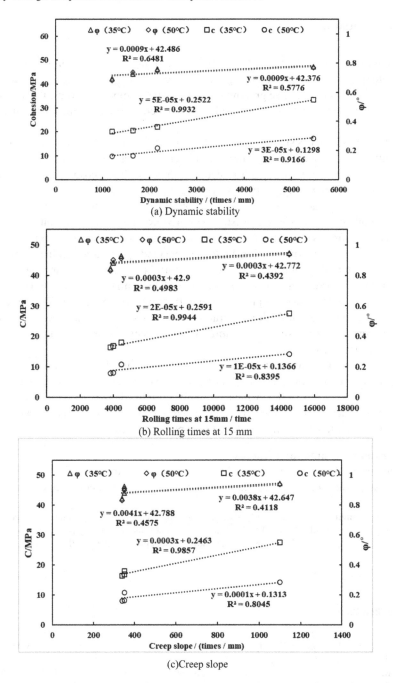

(a) Dynamic stability

(b) Rolling times at 15 mm

(c)Creep slope

Fig. 6 Correlation analysis of indicators

4 Conclusion

(1) Within the same AC-13 grading range, fine grading has better high-temperature rutting resistance than coarse grading. Because the fine grading is easier to compact than the coarse grading under the same compaction work, its porosity is smaller, and the density of the mixture is better, so its anti rutting performance is better.

(2) Rubber asphalt has significant advantages over base asphalt in high-temperature rutting resistance. The addition of rubber modifier can effectively enhance the cohesion between asphalt molecules, increase the viscosity of asphalt and improve the high-temperature performance of mixture.

(3) Higher temperature and lower proportion of graded fine aggregate will reduce the high temperature shear resistance of mixture. Based on the variance results, temperature has a significant impact on the cohesion of the mixture, but has little impact on the internal friction angle, and the grade pair has a significant impact on the internal friction angle. Based on the correlation analysis, it can be seen that the Dc of coarse aggregate is positively correlated with the cohesion and internal friction angle of the sample, and the correlation is very strong. Therefore, it is suggested that the high temperature shear resistance of the mixture can be improved by increasing the passing rate of the sieve openings above 4.75 mm in the grading design.

(4) Based on linear regression analysis, it is found that the R^2 of C (35 °C) of mixture and dynamic stability, rolling times at 15 mm, creep slope and other indicators are up to 0.9, which is highly correlated. It is recommended that the high-temperature shear resistance index can be used to characterize the high-temperature rutting resistance of asphalt mixture.

Declaration of Competing Interest
The authors declare that they have no known competing financial interests or personal relationships that could have appeared to influence the work reported in this paper.

Acknowledgements The research of this article is supported by Guangxi's major science and technology project (Guike AA18242032), Guangxi's science and technology plan project (Guike ZY21195043), and Guangxi's key research and development plan (Guike AB20297033)

References

1. Pu L (2018) Experimental Study on High Temperature Shear Resistance of GAC-16 Asphalt Mixture, Chongqing Jiaotong University
2. Fwa TF, Low BH, Tan SA (1994) A.S.f. Testing, Materials, STP, Behavior Analysis of Asphalt Mixtures Using Triaxial Test-Determined Properties, Engineering Properties of Asphalt Mixtures and the Relationship to their Performance, Phoenix, AZ, pp 97–110
3. Hao-ran ZHU, Jun Y, Zhi-wei C (2009) Research on test method of asphalt mixture's shearing properties. J Traffic Transp Eng 9(3):19–23

4. Yu-feng BI, Li-jun SUN (2005) Research on test method of asphalt mixture's shearing properties. J Tongji Univ (Nat Sci) 33(8):1036–1040
5. Zhi-wei C, Ping AN, Wei D, Xiangpeng YAN, Xiangpeng YAN (2021) Study on penetration strength of high modulus asphalt mixture. Subgrade Eng 3:65–69
6. Qun Y, Zhi-wei C, Li-jun SUN, Ju-liang MAO (2006) fractal analysis of gradation aggregate and its applications in pavement engineering. J Building Mater 9(4):418–422
7. Dong Z, Zhi-wei C, Ju-liang MAO (2022) Comparative study on rutting test and evaluation index of asphalt concrete in laboratory. J Railway Sci Eng 19(08):2287–2294
8. Wei J (2020) Study on the relationship between asphalt performance index and asphalt mixture cohesion, Shandong Jianzhu University
9. Zhou Y (2021) Experimental study on mechanical properties of asphalt mixture under triaxial torsional shear, Dalian University of Technology
10. Yuan, XU, Junqi GAO, Hongqiang LU, Zhaoqiang W (2016) Research on influence of long term immersion to cohesiveness of asphalt Mixture). J Wuhan Univ Technol (Transp Sci Eng) 40(6):963–967
11. Peng G (2008) Calculation and application of granule composite modified asphalt mixture interlocking granding. Highway Eng 33(5):121–124

Cases Study on Foundation Pit Design in Complex Environment of Urban Core Area

Lili Ma, Yun Chen, Liang Zhang, Gaigai Zhao, Fei Xia, Qinfeng Yang, and Zhiyuan Zhao

Abstract With the degree of urban underground space developing in China, the scale and excavation depth of foundation pit is much larger, and the surrounding environment is also more and more complex. Therefore, five cases of foundation pit design in complex environment have been studied in this paper. Then combined with local engineering experience, a design scheme of foundation pit adjacent to important facilities which is suitable for Qianjiang New Town has been summarized. Finally, based on the five cases, a simple model to assessment the complexity of the surrounding environment is proposed.

Keywords Foundation Pit Design · Surrounding Environment · Engineering Geology · Complexity Level

1 Introduction

With the degree of urban underground space developing in China, the scale and excavation depth of foundation pit is much larger, and the surrounding environment is also more and more complex, as shown in Fig. 1. The blue area is the foundation pit of a 258-m building, which is surrounded by dotted with dense buildings, subway tunnels and various underground pipelines. All these above make the contradiction between engineering construction and surrounding environment increasingly evident. As a result, the foundation pit design needs to meet not only the strength and stability requirements, but also the deformation control requirements of the

L. Ma (✉)
College of Civil Engineering and Architecture, Zhejiang University, Hangzhou 310000, China
e-mail: marry12@163.com

L. Ma · Y. Chen · Q. Yang · Z. Zhao
Zhejiang University Architectural Design and Research Institute Co., Ltd, Hangzhou 310000, China

L. Zhang · G. Zhao · F. Xia
Hangzhou CBD Construction Development Co., Ltd, Hangzhou 310000, China

© The Author(s) 2023
G. Feng (ed.), *Proceedings of the 9th International Conference on Civil Engineering*,
Lecture Notes in Civil Engineering 327,
https://doi.org/10.1007/978-981-99-2532-2_5

surrounding environment. Then the foundation pit design gradually changes from the traditional strength control to the deformation control. Finally, under the requirements of balance, safety, economy and convenient construction, the difficulty of foundation pit design gradually increases.

Amounts of research on foundation pit in simple environment have been published [2–7], but as for complex environments it should be further studied. Moreover, the effect of the characteristics of one single building on the foundation pit during excavation process has been discussed, and then a scheme of safety evaluation on foundation pit has been proposed [8]. However, as for various surrounding environments, it should be further extended and adjusted. Therefore, in this paper the authors have studied five typical engineering cases in Qianjiang New Town, the urban core area of Hangzhou City, Zhejiang Province, then summarized the characteristics of foundation pit design with complicated surrounding environment, and finally put forward a framework to distinguish the complex degree of surrounding environment.

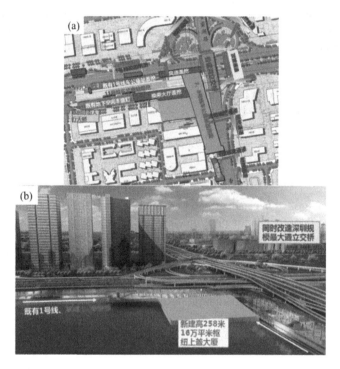

Fig. 1 Schematic diagram of surrounding environment of one foundation pit [1]: **a** plan sketch, **b** 3D diagram

2 Engineering Situation

2.1 Engineering Background

The cases in this paper are located in Qianjiang New Town, the urban core area of Hangzhou City, Zhejiang Province, China, as shown in Fig. 2. All these projects are in the north of Qiantangjiang, and the relationships between the foundation pits and there surrounding environments are also illustrated in Fig. 2.

Case a: Xinchen Business Center Project. This project consisting of two 18-story office buildings with a four-story basement, is located in the southwest of the intersection of Wangchao Road and Leiting Road. It is about 700 m away from Qiantangjiang.

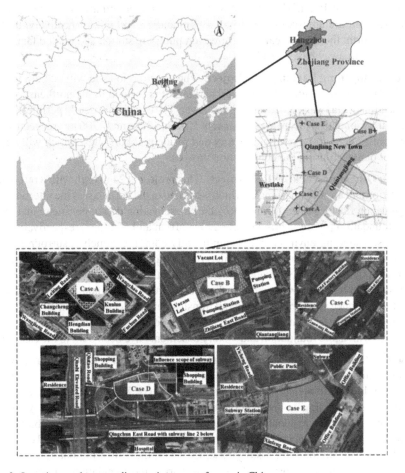

Fig. 2 Locations and surrounding environment of cases in China

The excavation depth, perimeter and area of its foundation pit are approximately 24 m, 392 m and 9690m^2, respectively.

In the east, there is Zhejiang Chouzhou Bank with a pile foundation and 3-story basement across Wangchao Road. In the north, there is Wangjiang International Building with a 2-story basement across Leiting Road. Changcheng Building with a pile foundation and a 2-story basement, Kunlun Center Building with a pile foundation and a 2-story basement, Hengdian Building with a pile foundation have been built in the west, south, and east of the construction site, respectively. Moreover, the three buildings are about 10.2 m, 18 m and 11 m away from the construction site, respectively.

Case B: Jiangchen Business Center Project. This project consisting of two tower buildings which are about 80 m high, is located in the northwest of the intersection of Hemu Port and Zhijiang East Road. Moreover, Qiantangjiang is acrossing the Zhijiang East Road. The excavation depth, perimeter and area of this project's foundation pit are approximately 15 m, 493 m and 13,100 m2, respectively. In the west and north, there are vacant lots, however, pumping station and related facilities which are 4 m away have been built.

Case C: Moyetang Parking Lot Project. The construction site will be built into a public park with a 2-story underground parking. It is located in the northeast of the intersection of Ganwang Road and Xinkai River. The excavation depth, perimeter and area of the foundation pit are approximately 10.4 m, 432 m and 7706 m^2, respectively. In the west, there are several old factory and residential buildings, which will be in a risk of damage during the construction process.

Case D: Qingchun Parking Lot Project. This project is 24 m high with a raft plate foundation and a 4-story basement, which is located in the northeast of the intersection of Qingchun East Road and Qiutao Road. The excavation depth, perimeter and area of its foundation pit are approximately 22 m, 310 m and 5547 m^2, respectively.

The surrounding environment is significantly complex, which is surrounded by roads, buildings, subway line and underground pipelines. In the west, it's Qiutao Road with Qiushi Elevated Road upside. In the south, it's Qingchun East Road with subway line 2 below it. What's more, the south part of the foundation pit is located with the influence scope of the urban rail transit safety protection zone [9]. A 5-story building with natural foundation has been built to the north side, while three 40-story buildings with a pile foundation and 4-story basement have been built to the east side.

Case E: Qiantang Business Center Project. This project consists of three 50 m high buildings with a 4-story basement and one 20 m high podium with a 2-story basement, located in the northwest of the intersection of Xinfeng Road and Yicheng Road. The surrounding environment in the north and west is less complex, where a public park and some residential buildings are found. However, two office buildings are building in the east and south, and the podium is just above the airport rail express, and the subway station with 3-story basement is on the left of the podium. Thus it is seriously important to protect the subway during the foundation pit design and construction progress.

2.2 General Situation of Engineering Geology and Hydrogeologic Situation

As known from the geological map of Hangzhou [10], it is divided into low hilly area, foothill valley area and plain area. Qianjiang New Town is located in the plain area, which is composed of three major layers from the ground down, namely soft and hard interbed, sand and gravel, and underlying bedrock. Combined with the properties of the soil and strata distributions in the field based on the above five cases, the strata of each case are simply reclassified, as shown in Fig. 3. Thus some conclusions can be drawn easily from Fig. 3.

Strata Characteristics. The strata of each case are similar, mainly composed of soft and hard soil layer. From the ground down, they are filled soil, silt soil, clay, sand and gravel, followed by bedrock layer. Silt soil layer is dominated by silt sand and sandy silt, clay layer is dominated by silt clay, while sand and gravel layer is dominated by silt sand and round gravel. It should be noted that in Case C bedrock layer has not exposed due to its shallow borehole depth. It can be found that for the five cases except for Case E, silt layer and sand and gravel layer share similar bury depths, respectively. The reason could be due to the relationship between the construction site distribution and Qiantangjiang.

Bearing Capacity Characteristics. The upper stratum is soft soil layer, which has weak foundation bearing capacity. Thus each building of the five cases adopts pile foundation, with sand and gravel or bedrock acting as the supporting layer.

Groundwater Characteristics. Each construction site has a high-pressure water level, and there are permeable layers such as filled soil, silt, sandy silt, and sand and gravel layer. The measured pressure head was buried about 8 m deep, and the static

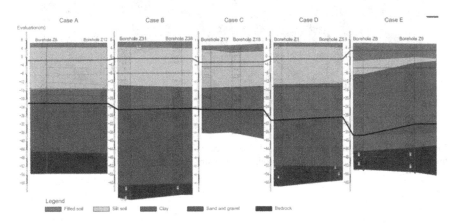

Fig. 3 Schematic diagram of engineering geological section (The blue line, the red line and the dark line represent the pressure water head, the bottom of foundation pit and the top surface of confined aquifer, respectively)

water level elevation was approximately 2 m underground. Based on calculation results, the foundation pit will suffer a risk by the confined water.

3   General Situation of Foundation Pit Design

3.1   Description

The surrounding environment of each case is diverse, the supporting scheme is different, and even the control requirements of deformation are different from each other. Figure 4. shows the comparison of each supporting system, where some design situations can be seen.

Case A, Case D and Case E adopt underground diaphragm walls and internal struts as the support structure. As for Case D, a row of TRD with shape steel inserted is added outside the foundation pit near the subway, which aims to control deformations. As for Case E, the north part is located just above the subway line, so high-pressure jet grouting piles are applied to reinforce the soil above the subway. Case B and Case C are supported by cast-in-place piles and internal struts, while double-row piles are constructed in the area near the old buildings and the pumping station to control the deformation, respectively.

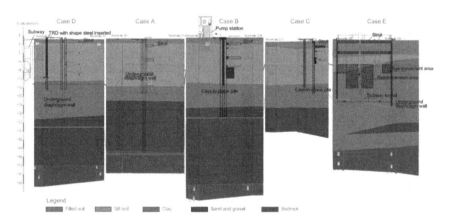

Fig. 4 Schematic diagram of support profile in each case (The red line and the yellow line represent the bottoms of foundation pit and the vertical support structure, respectively)

3.2 Summaries

Taking the excavation depth, surrounding environment, project characteristics and local engineering experience into account, the design and construction process of foundation pit should strictly control the deformation of the surrounding environment, which should refer to relevant specifications. Generally speaking, the vertical support structure for foundation pit can adopt cast-in-place pile, underground diaphragm wall, steel pipe pile, or sheet pile and so on. While the horizontal support structure usually utilize internal struts, which could be classified to reinforced concrete, steel pile and shape steel.

With the help of the unit (Zhejiang University Architectural Design and Research Institute Co., LTD.), the authors have summarized the design scheme of foundation pit adjacent to important facilities in recent projects of Hangzhou City, as shown in Table 1. Combined with the five cases in this paper and various local engineering experiences, the design scheme of foundation pit adjacent to important facilities which is suitable for Qianjiang New Town has also been presented in Table 2. What should be noted is that the scope of protection and relevant requirements of the important facilities are detailed in relevant policies and requirements, which are not described here.

4 Assessment Model of Surrounding Environment

Generally, the safety level of foundation pit is comprehensively determined according to the project characteristics, stratum situation, and surrounding environment, and

Table 1 Design scheme of foundation pit adjacent to important facilities in recent projects in Hangzhou

Adjacent facility	Excavation depth	Major strata characteristics	Supporting system	Control requirements of deformation
Subway	5m ~ 10m	Sandy silt	Cast-in-place piles and one internal strut of reinforced concrete	According to the related requirements (Zhejiang Provincial Department of Housing and Urban-Rural Development, 2017)
		Silt clay	Cast-in-place piles and two internal struts of reinforced concrete	
		Silt clay	Cast-in-place piles and one internal strut of reinforced concrete	
		Muddy clay	Cast-in-place piles and one internal strut of reinforced concrete	
		Silt clay	Slope setting design	
		Round gravel and strong-weathered rock	Slope setting design	
	10m ~ 15m	Silt clay	Cast-in-place piles and two internal struts of reinforced concrete	
		Round gravel	Cast-in-place piles and two internal struts of shape steel	
	15m ~ 20m	Muddy soil	(a) Underground diaphragm wall, one internal strut of reinforced concrete and two internal struts of shape steel. (b) Cast-in-place piles and three internal struts of reinforced concrete	
Electric power pipe	11m ~ 12m	Sandy silt	Cast-in-place piles and two internal struts of reinforced concrete	For underground high-voltage power pipelines, the safety distance must be greater than 2.0m and the deformation must be less than or equal to 20mm
Sensitive building	6m ~ 12m	Silt clay	Cast-in-place piles and one internal strut of reinforced concrete, low-pressure grouting reinforcement outside the foundation pit	For important buildings, the deformation must be less than or equal to 10mm
	14m ~ 15m	Muddy clay	Cast-in-place piles and three internal struts of reinforced concrete	
Railway	5m ~ 10m	Round gravel and strong-weathered rock	Cast-in-place piles and one steel incline strut, retain soils inside pit	Deformation is required to be controlled at 5mm

Table 2 Design scheme of foundation pit adjacent to important facilities in Qianjiang New Town

Excavation depth	Supporting system
<5 m, 5 m ~10 m	(a) If the geological condition is good and the site is permitted, slope, slope plus cement soil mixing wall, or composite soil nailing wall could be adopted (b) If the geological condition is general or the site is tight, piles plus one or two internal struts could be adopted (c) The deformation shall meet the relevant requirements
10 m~15 m, 15 m~20 m, 20 m~25 m	(a) Generally, piles and multi-internal struts should be adopted (b) If the geological situation is general or the site is tight, it is advisable to adopt the supporting structure of underground diaphragm wall with multi-internal struts or combined with the main structure (c) The deformation shall meet the relevant requirements

Note: (a) When there are important pipelines around or inside the foundation pit, the pipeline can be protected by soil reinforcement, pipeline relocation, additional support, pipe and soil isolation, isolation correction, information construction or other methods. (b) The partition method can be adopted to reduce the impact of foundation pit construction on the surrounding environment, and the partition wall can be reinforced by steel sheet pile, ground wall, root pile, stirring pile, grouting, etc.

then the supporting design and construction of foundation pit engineering are carried out. Among above, the complexity degree of surrounding environment greatly affects the safety level division of foundation pit. But existing specifications only classify it qualitatively [11, 12], or according to the facilities properties in the range of excavation effect [13]. However, this could be not economy if safety level is classified too high or too low. Therefore, based on the five cases above and the numerical model proposed by Feng Chunlei [8], this section proposes a simple model to quantify the complexity of the surrounding environment.

4.1 Parameters Description

The surrounding environment of each project in this paper is significantly complex, covering subway, pumping station, old building, underground pipeline, road and river. Due to the lack of a large number of data, statistical analysis of the data is not possible. Therefore, based on a simple two-dimensional finite element model, this section intends to analyse the excavation process of foundation pit with or without surrounding structures.

In the model, subway, pumping station, residence, road or pipeline is simplified into building. Building size A, stiffness E, and horizontal distance s from foundation pit are referred to as calculated parameters. The differential settlement, relative disturbance, tilt angle and torsion angle of the building during excavation process are referred to as the building deformation indexes, and the sum of their normalized values is the comprehensive deformation index, which could be used to measure the overall deformation situation of the surrounding buildings after excavation.

4.2 Model Framework Description

As shown in Table 3, in order to make the influence degree of each parameter on foundation pit excavation more intuitive, the influence degree of each parameter can be simply divided into three levels and assigned values. Considering different factors, the calculated values of three grades of each deformation parameter were averaged, and the absolute value of the difference between the maximum value and the minimum value of the average value was taken as the judgment standard of the influence of this factor on a certain deformation parameter.

Taking the differential settlement of building deformation index as an example, as shown in Table 4, Eq. (1) and Eq. (2), $\overline{a}_i(\delta^v)$, $\overline{b}_i(\delta^v)$ and $\overline{c}_i(\delta^v)$ are the average values of all calculated differential settlements of three influencing factors, namely building size A, stiffness E and horizontal distance s from foundation pit, at each level (I = 1, 2, 3) respectively. $r_A(\delta^v)$, $r_E(\delta^v)$ and $r_s(\delta^v)$ are the influence degrees of the four influencing factors on differential settlement. The influence degree of other factors on deformation parameters can be calculated accordingly.

$$\overline{a}_i(\delta^v) = \frac{\sum\limits_{i=1}^{3} s_i(\delta^v)}{3} \tag{1}$$

$$r_s(\delta^v) = Max\{\overline{a}_i(\delta^v)\} - Min\{\overline{a}_i(\delta^v)\} \tag{2}$$

After obtaining the influence degree of a factor on different deformation parameters, the average value of the influence degree of this factor on all deformation parameters is calculated, that is, the comprehensive influence index R of a factor on building deformation. For example, the comprehensive influence index R of the horizontal distance s between building and foundation pit on building deformation

Table 3 Influence degree classification of parameters on foundation pit excavation

Influent degree	A	E	s	Assignments
I	A_1	E_1	s_1	1
II	A_2	E_2	s_2	2
III	A_3	E_3	s_3	3

Table 4 Calculation value of the influence degree on differential settlement

i	A	E	s
1	$\overline{a}_1(\delta^v)$	$\overline{b}_1(\delta^v)$	$\overline{c}_1(\delta^v)$
2	$\overline{a}_2(\delta^v)$	$\overline{b}_2(\delta^v)$	$\overline{c}_2(\delta^v)$
3	$\overline{a}_3(\delta^v)$	$\overline{b}_3(\delta^v)$	$\overline{c}_3(\delta^v)$
$r(\delta^v)$	$r_A(\delta^v)$	$r_E(\delta^v)$	$r_s(\delta^v)$

Table 5 the relationship between Comprehensive deformation index and Safety evaluation index considering different parameters

Safety level L_j	A	E	s	Comprehensive deformation index F_j	Safety evaluation index S_j
L_1	3	3	2	F_1	S_1
L_2	3	2	3	F_2	S_2
L_3	2	3	3	F_3	S_3
L_4	2	2	1	F_4	S_4
L_5	2	1	2	F_5	S_5
L_6	1	1	3	F_6	S_6
L_7	3	1	1	F_7	S_7
L_8	1	2	2	F_8	S_8
L_9	1	3	1	F_9	S_9

is

$$R_s = \frac{r_s(\delta^\upsilon) + r_s(\Delta) + r_s(\alpha) + r_s(\beta)}{4} \tag{3}$$

where, $r_s(\Delta)$, $r_s(\alpha)$ and $r_s(\beta)$ are the influence degree of horizontal distance s on the relative disturbance, tilt angle and torsion angle of the building respectively.

By multiplying the assignment of the influence degree level in Table 4 with the comprehensive deformation index F_j of each influencing factor, as shown in Table 5 below, the safety evaluation index S_j of each scheme can be obtained.

4.3 Results

According to the safety level and the safety evaluation index of the surrounding environment under different schemes, as shown in Fig. 5., four regions can be divided into 3 types, namely, safe, relative safe and certain security risks. The specific descriptions can be seen in Table 6.

The above is only for the case that there is a single building around the foundation pit. When there are multiple construction facilities, it should be re-evaluated according to the above process, or the above safety evaluation indexes should be taken a weighted average according to the importance degree of subway, pumping station, road or other facilities, and then the safety level of the surrounding environment could be determined according to Table 6.

Fig. 5 Schematic diagram of the relationship between safety level and safety evaluation index

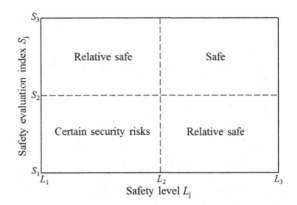

Table 6 Descriptions of Fig. 5

Classification of safety evaluation	Index		Description
	S_j	L_j	
Safe	$S_2 \leq S_j \leq S_3$	$L_2 \leq S_j \leq L_3$	The deformation of surrounding environment is **nearly not affected** by the excavation construction of foundation pit
Relative safe	$S_2 \leq S_j \leq S_3$	$L_1 \leq S_j \leq L_2$	The deformation of surrounding environment is **less affected** by the excavation construction of foundation pit
	$S_2 \leq S_j \leq S_3$	$L_2 \leq S_j \leq L_3$	
Certain security risks	$S_1 \leq S_j \leq S_2$	$L_1 \leq S_j \leq L_2$	The deformation of surrounding environment is **greatly affected** by the excavation construction of foundation pit

5 Conclusions

Based on several cases, this paper summarized the characteristics of engineering geology and foundation pit design in Qianjiang New Town. And then a simple model to assessment the complexity of the surrounding environment is proposed, which should be further discussed.

Acknowledgements The authors greatfully acknowledge the supports of "Study on bearing characteristics of pile foundation considering scour effect" (postdoctoral program, Zhejiang University Architectural Design and Research Institute Co., LTD.), and "Research on planning and construction of underground space in urban core area and whole-process control in complex environment" (horizontal project, Hangzhou CBD Construction Development Co., LTD.) The authors also thank the reviewers for their effort to improve the quality of the manuscript.

References

1. Xiangsheng C (2018) Build urban rail transit with maximum efficiency (Report). China Urban Rail Transit Key Technology Forum, Beijing
2. Xiaobing X, Qi H, Tianming H et al (2022) Seepage failure of a foundation pit with confined aquifer layers and its reconstruction. Eng Fail Anal 138:106366. https://doi.org/10.1016/j.eng failanal.2022.106366
3. Yang Y, Changhong Y, Baotian X, Shi L, et al (2018) Optimization of dewatering schemes for a deep foundation pit near the Yangtze River, China. J Rock Mech Geotech Eng 10:555–566. https://doi.org/10.1016/j.jrmge.2018.02.002
4. Shuaihua Y, Zhuangfu Z, Denqu W (2021) Deformation analysis and safety assessment of existing metro tunnels affected by excavation of a foundation pit. Underground Space 6:421–431. https://doi.org/10.1016/j.undsp.2020.06.002
5. Chavdar K (2016) Some geotechnical characteristics of the subway construction in Sofia city, Bulgaria. Procedia Eng 165:290–299. 10.1016/j.proeng.2016.11.703
6. Andrey B, Alexander K, Vladimir K et al (2016) Evaluation of deformations of foundation pit structures and surrounding buildings during the construction of the second scene of the state academic Mariinsky theatre in Saint-Petersburg considering stage-by-stage nature of construction process. Procedia Eng 165:1483–1489. https://doi.org/10.1016/j.proeng.2016.11.883
7. Xiaolin Y, Buyu J (2012) Analysis of excavating foundation pit to nearby bridge foundation. Procedia Earth Planetary Sci 5:102–106. https://doi.org/10.1016/j.undsp.2020.06002
8. Chunlei F (2019) Study on the safety and control of deep foundation pit engineering in subway station based on complex stratigraphic conditions (Doctoral Thesis). Beijing Jiaotong University
9. Zhejiang Provincial Department of Housing and Urban-Rural Development. Technical code for protection of urban rail transit structures (DB33/T 1139-2017). China Building Materials Press, Zhejiang (2017)
10. Zhejiang university architectural design and research institute Co., LTD., Urban geological survey of Hangzhou (2008)
11. Ministry of housing and urban-rural development of the people's republic of China. Technical specification for retaining and protection of building foundation excavations (JGJ 120–2012). China Architecture Publishing & Media Co., Ltd., Beijing (2012)
12. Zhejiang provincial department of housing and urban-rural development. Technical Specification for Building Foundation Excavation Engineering (DB33/T 1096–2014). Zhejiang Gongshang University Press (2014).
13. Guangdong Provincial Department of Housing and Urban-Rural Development. Technical Specification for Building Foundation Excavation (DBJ/T 15-20-2016). Guangdong. China City Publishing House (2017)

Investigation on Contact Force for Asphalt Mixture During Compaction Using DEM

Jian Feng, Weidong Liu, and Jiechao Lei

Abstract Contact force for asphalt mixture has a significant effect on mechanical properties, and discrete element method (DEM) is a feasible and efficient numerical simulation approach to investigate on skeleton structure of asphalt mixture from microscopic perspective. A three-dimensional (3D) DEM compaction model which can well describe morphology characteristics of mineral aggregate, compaction temperature, was established to investigate the contact force. It indicates that aggregate–aggregate (A-A) merely exists in contact pressure force, aggregate–mortar (A-M) and mortar–mortar (M-M) exists in contact pressure force and tension force. While contact force of two-dimensional (2D) sections is employed to characterize the contact force of 3D specimens, it needs to investigate rationality of the selected section. The 3D DEM compaction model can be utilized to investigate contact force for asphalt mixture.

Keywords Asphalt Mixture · Contact Force · Discrete Element Method (DEM)

1 Introduction

Asphalt mixture has been widely regarded as a heterogeneous multiphase granular material consisting of mineral aggregates, asphalt binder and air voids. According to Mohr–Coulomb strength theory, high-temperature stability of asphalt mixture is directly related to aggregate–aggregate (A-A) skeleton structure and adhesion property of asphalt binder [1]. Contact between aggregates, which affects the magnitude and transfer path of force between aggregate particles, and has significant influence

J. Feng · W. Liu (✉) · J. Lei
Guangxi Transportation Science and Technology Group Co., Ltd., Guangxi, Nanning 530007, China
e-mail: lwd200526066@163.com

Guangxi Key Lab of Road Structure and Materials, Guangxi, Nanning 530007, China

Research and Development Center on Technologies, Materials and Equipment of High Grade Highway Construction and Maintenance Ministry of Transport, Guangxi, Nanning 530007, China

© The Author(s) 2023

G. Feng (ed.), *Proceedings of the 9th International Conference on Civil Engineering*, Lecture Notes in Civil Engineering 327, https://doi.org/10.1007/978-981-99-2532-2_6

on micromechanical properties of asphalt mixture, plays an important role in asphalt mixture gradation. The evaluation index of contact characteristics of A-A is available to measure the internal structure of asphalt mixture at microscopic level, and contact behavior of aggregate to aggregate, asphalt mastic to aggregate and asphalt mastic to asphalt mastic is directly to contact force for each contact types [2]. According to current advances in research of contact property of aggregates, three kinds of prevalent approaches, namely, pure experimental tests, digital image processing (DIP) depending on laboratory tests, and numerical analysis method, have been summarized [3]. However, the first two approaches in rather costly and inefficient, a virtual simulation method, like discrete element method (DEM), has been confirmed that it is feasible to investigated on contact force for asphalt mixture. Liu et al. [4, 5] introduced a user-defined algorithm of coarse aggregate to investigate aggregate movement regulations during compaction with series of evaluation indices, which involved contact force, air voids, coordination number as well as motion angle. Ma. et al. [6, 7] and Xue et al. [8] developed a 3D DEM rutting model of asphalt mixture to describe aggregates movement properties with aggregate translation and rotation, contact number, and contact force.

Most previous studies are confined to contact behavior between aggregates but few published literatures focus on the contact behavior of M-M, A-M, and contact force during compaction, which is critical to disclosing compaction mechanism and controlling compaction quality. In this paper, a 3D DEM simulates experimental compaction, and investigates on contact force for asphalt mixture component including A-A, A-M as well as M-M. The research results can obtain an indeep understanding compaction mechanism, and provide reference for investigating contact behavior of loose asphalt mixture densification from micro-level.

2 Materials and DEM Compaction Model

2.1 Materials

This paper selects the stone matrix asphalt as research object, namely SMA-13, where SBS asphalt binder content is set to 5%, and the gradation is shown in Table 1. The loose asphalt mixture, whose initial porosity has an important influence on evolution of aggregates as well as contact behavior between aggregate and asphalt mastic, will be tightly packed in the compaction process. Therefore, the number of initial porosity equal to field compaction is 20.5%.

Table 1 SMA-13 and asphalt mastic gradation

Sieving size/mm	16	13.2	9.5	4.75	2.36	1.18	0.6	0.3	0.15	0.075
SMA-13	100	95	65	30	21	19	16	13	12	10
Asphalt mastic	–	–	–	–	100	90.5	76.2	61.9	57.1	47.6

2.2 Experimental Tests

According to the standard specification JTG E20 2011, the laboratory tests involving complex modulus and static compaction were performed to determine 3D DEM model parameters and demonstrate feasibility of the established compaction model, respectively. Moreover, the UTM was used to obtain dynamic modulus at different condition, including temperature: 5, 20 and 30°C, frequencies: 25, 20, 10, 5, 2, 1, 0.5, 0.2, 0.1 and 0.01 Hz.

2.3 DEM Compaction Model

The production of virtual mineral aggregates and asphalt mixture specimen can be introduced in details in the previous study [4, 5], as shown in Fig. 1. The linear contact model is utilized to characterize the particular contact of A-A, and the Burger's model is to act as the contact model of M-M as well as A-M. The parameters of compaction model can be obtained by performing dynamic modulus tests using UTM. The Burger's model parameters as following: $E_1 = 14.256$ MPa, $E_2 = 10.246$ MPa, $\eta_1 = 629.387$ MPa·s, $\eta_2 = 1.817$ MPa·s.

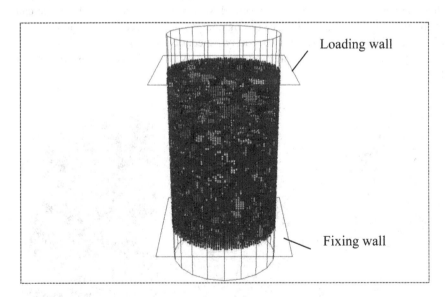

Fig. 1 DEM modelling of compaction

3 Results and Discussion

3.1 *Contact Force Distribution Law*

The internal contact of asphalt mixture mainly includes the following contact types: A-A, A-M, and M-M. The DEM results can well show the distribution characteristics of contact force of loose asphalt mixture during compaction. The irregular coarse aggregates herein are represented by clump, in which there is no contact force.

The contact force distribution characteristics of asphalt mixture are shown in Fig. 2, where contact pressure and contact tension are indicated by black and red, respectively. The force orientation is consistent with line segment direction, and the magnitude of force is direct proportional to width of line segment. The greater the contact force, the wider the line segment, and vice versa. It should be noted that the ratios of the line segments of Fig. 2 (a), (b) and (c) are not identical. Owing to the chaotic state of contact force in 3D DEM simulation, contact force can be qualitatively compared in the same figure. As shown in Fig. 2 (a), contact pressure is mainly between mineral aggregates, since the aggregates will have a movement, and the extrusion of A-A can form contact pressure force, instead of contact tension force. As plotted in Fig. 2 (b) and (c), both contact pressure force and contact tension force exist in A-M as well as M-M. Meanwhile, the contact tension force is greater than the contact pressure force between M-M. It indicates that asphalt mastic can share most of contact tension force during compaction. Moreover, the contact force distribution of asphalt mixture is disorderly, especially the contact of M-M. The most probable cause may be the contact number of A-A is smaller than those of contact types.

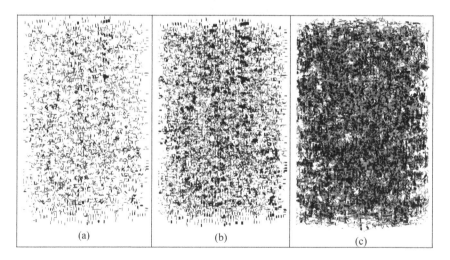

Fig. 2 Contact force distribution: (a) A-A, (b) A-M, (c) M-M

The qualitative analysis of the contact force distribution can't deeply disclose the mechanism of asphalt mixture during compaction from microscopic perspective, as dis-cussed above. In order to quantitatively analyse the distribution law of internal contact force of asphalt mixture, the internal contact force was extracted by user-defined FISH language programming. The distribution characteristics of contact force of asphalt mixture were obtained with statistical analysis method. It is necessary to introduce a sign convention for contact force. The contact pressure force is positive and the contact tension force is negative.

Figure 3 and Fig. 4 show the distribution of internal contact pressure force and contact tension force at the compaction displacement 7.5 mm. As plotted in Fig. 3, the contact pressure force of A-A, A-M and M-M share the same trend. The ratio of contact pressure force shows a similar with negative exponential tendency, and most of contact pressure force is from zero to a certain value. However, the contact pressure force value of all sorts of contact types are of different sizes. In terms of the A-A, those contact pressure force is about 66% in the interval of zero to 5N, and those ratio of the contact pressure force in the range of 5N to 10N is about 15%. Therefore, the contact pressure force is mainly distributed in the range of zero to 10N. The main range of A-M contact pressure force is from zero to 2N. Furthermore, the M-M contact pressure force is small, and the distribution is mainly concentrated in the section of zero to 0.2N. In summary, it holds that the contact pressure force between A-A is dramatically greater than these of A-M and M-M. The main reason is that modulus value of materials directly affects the contact pressure force value of different contact types, and the mineral aggregate modulus is significantly stronger than that of asphalt mastic.

As plotted in Fig. 4, the contact tension force of A-M and M-M is rather small compared to the contact pressure force of A-A, and the distribution range of these contact tension force has a similar discipline from −0.2N to zero. As stated above, the contact pressure force is directly related to the corresponding modulus, and the materials modulus the greater, the contact pressure forces the greater. However, these rules are inapplicable to the contact tension force distribution. In short, it indicates that there are migration and evolution of mineral aggregates in loose and hot asphalt mixture during compaction. The contact pressure force is more that the contact

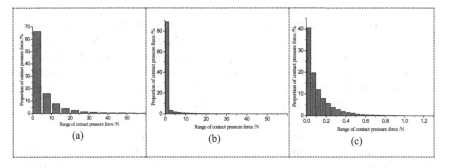

Fig. 3 Contact pressure force: (a) A-A, (b) A-M, (c) M-M

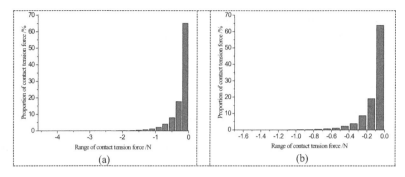

Fig. 4 Contact tension force: (a) A-M, (b) M-M

tension force. They exist in A-M as well as M-M, but there is only contact pressure force in A-A.

3.2   Contact Force Distribution Variability of Two-dimensional Section

It is commonly accepted that the 3D DEM is used to comprehensively and objectively investigate the internal microstructure of asphalt mixture, and a 3D virtual specimen can be easily cut into series of 2D planes to realize the visualization of internal contact force of all kinds of contact types within asphalt mixture. However, the multiphase asphalt mixture is heterogeneous granular material with a large variability in mechanical properties. So, this section will discuss the variability and complexity of contact force at different compaction displacement.

The 3D virtual specimen can be theoretically cut into numerous 2D sections, but too many or too few 2D sections will inevitably have a profound effect on studying the variability of contact force. Therefore, the virtual specimen was divided into 12 2D sections at intervals of 30 degrees in vertical direction, and Fig. 5 shows the maximum contact force distribution under different compaction displacements (7.5 mm, 15 mm and 22.5 mm).

It can be seen that the contact force with large compaction displacement contains the contact force curve with small compaction displacement, which shows that the maximum contact force is positively proportional to the magnitude of compaction displacement, namely the larger compaction displacement, the greater maximum contact force. Meanwhile, the maximum contact force fluctuates greatly in different cutting planes, particularly at the 30° cutting plane, and it shows that the 3D virtual specimen is cut into 2D sections which can be employed to evaluate the asphalt mixture variability by selecting ap-propriate indicators. However, it should be noted that the local anomaly of contact force in virtual specimen has a negative and great influence on the operating efficiency and simulation reliability with the 3D DEM.

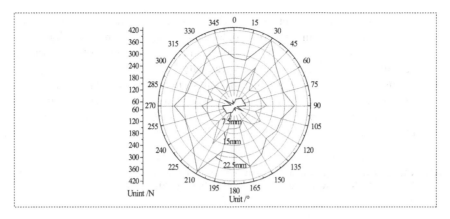

Fig. 5 Maximum contact force of specimen 2D sections

Table 2 Statistical analysis of maximum contact force of specimen 2D sections

Displacement /mm	Average /N	Standard deviation/N	Variation coefficient	Minimum /N	Median /N	Maximum /N
7.5	74.93	19.00	0.25	48.82	74.50	99.65
15	174.58	34.13	0.20	125.10	161.10	262.40
22.5	317.07	41.70	0.13	273.80	303.50	420.70

When the abnormal contact force exceeds the maximum allowable contact force of system, the particles will escape at a large speed and cause an interlocking effect on the surrounding particles, even the DEM model is out of equilibrium. Therefore, the escaped particles can be deleted without affecting the simulation results.

The contact force of asphalt mixture from 12 2D planes is utilized to perform statistics analysis to deeply investigate the variability of contact force, as shown in Table 2. Under the same compaction displacement, the maximum contact force of different sections shows a certain volatility, and the coefficient of variation changes from 10 to 30%. Therefore, when the contact force of a 2D section is used to represent the contact force of a 3D virtual specimen, it is necessary to evaluate the rationality of the selected 2D section. In addition, it indicates that the virtual cutting method can be judged to the homogeneity of asphalt mixture to some extent by selecting an appropriate evaluation index, such as contact force, contact number.

4 Conclusions

(1) The contact pressure force of aggregate to aggregate, aggregate to mortar, and mortar to mortar are mainly distributed in ranges of 0-10N, 0-2N and 0–0.2N, respectively.

(2) The absolute value of contact tension force of both aggregate to mortar and mortar to mortar are in range of 0–0.2N, and contact pressure or tension force can meet the regular that with the increase of contact pressure or tension force, the distribution frequency decreases gradually.

(3) The maximum contact force of different sections shows certain fluctuation. When contact force of a 2D section is used to represent the contact force of an asphalt mixture specimen, it needs to investigate rationality of the selected section.

Acknowledgements This research was funded by the Guangxi Science and Technology Project (ZY21195043), the Nanning Science and Technology Project (20223039).

References

1. Shi LW, Wang DY, Masley J, Zhang SW (2013) Comparison analysis of the aggregate contact characteristics between skeleton-dense and suspended-dense structure asphalt mixture. App. Mech. Mater. 470:889–892

2. Zhang Y, Ma T, Ling M, Huang X (2019) Mechanistic sieve-size classification of aggregate gradation by characterizing load-carrying capacity of inner structures. Eng Mech 145:04019069

3. Shi L, Wang D, Jin C, Li B, Liang H (2020) Measurement of coarse aggregates movement characteristics within asphalt mixture using digital image processing methods. Measurement 163:107948

4. Liu W, Gao Y (2016) Discrete element modeling of migration and evolution rules of coarse aggregates in the static compaction. Southeast Univ 32:85–92

5. Liu W, Gao Y, Huang X, Li L (2020) Investigation of motion of coarse aggregates in asphalt mixture based on virtual simulation of compaction test. Int J Pavement Eng 21:144–156

6. Ma T, Zhang D, Zhang Y, Hong J (2016) Micromechanical response of aggregate skeleton within asphalt mixture based on virtual simulation of wheel tracking test. Constr. Build. Mater 111:153–163

7. Ma T, Zhang D, Zhang Y, Wang S, Huang X (2018) Simulation of wheel tracking test for asphalt mixture using discrete element modelling. Road Mater Pavement 19:367–384

8. Xue B, Xu J, Pei J, Zhang J, Li R (2020) Investigation on the micromechanical response of asphalt mixture during permanent deformation based on 3d virtual wheel tracking test. Constr Build Mater 267:121031

Research on the Application of Small Caliber Pilot Jacking Method Under Silt Geology

Qixing Wu, Jieming Huang, Qijun Li, and Haoran Liu

Abstract With the rapid development of urban scale in China, the number of drainage pipes and new construction has also increased rapidly, and pipeline maintenance and new not suitable for excavation construction and silt formation construction due to limited space, and introduce the working principle, construction design, construct construction are facing problems such as environmental protection and narrow construction sites. The use of small-diameter pilot top pulling method can effectively overcome the difficulties of drainage pipeline projects in bustling urban areas that areion process and process of the construction method. Combined with the construction example of a small-caliber pilot top pull method of a drainage pipeline project in Zhongshan City, it shows that the construction technology has the advantages of high precision, space saving, small environmental impact and can be constructed on narrow roads, which can provide reference for similar projects and has greater promotion and application significance.

Keywords Pilot Jacking Method · Small Diameter Pipe · Silt Formations · Drainage Works · Trenchless Technology

1 Introduction

In China, local public infrastructure building has grown substantially since the dawn of the twentieth century [1]. According to the "2020 Urban and Rural Construction

Q. Wu (✉) · J. Huang · H. Liu
School of Mechanics and Construction Engineering, Jinan University, Guangzhou 510632, Guangdong, China
e-mail: wqx510632@126.com

Q. Wu
MOE Key Laboratory of Disaster Forecast and Control in Engineering, Jinan University, Guangzhou 510632, Guangdong, China

Q. Li
Foshan Sanxin Municipal Engineering Co., Ltd., Foshan 528000, Guangdong, China

© The Author(s) 2023   73
G. Feng (ed.), *Proceedings of the 9th International Conference on Civil Engineering*,
Lecture Notes in Civil Engineering 327,
https://doi.org/10.1007/978-981-99-2532-2_7

Statistical Yearbook" [2], there will be 687 cities in the nation in 2020, and investment in fixed assets of urban municipal public amenities will be $2 billion. Medium
drainage contributes 9%, gas accounts for 1%, and water supply accounts for 3%; by
2020, annual urban sewage discharge will total 57,136,330,000 cubic meters, with
drainage pipes totaling 802,721 km in length. As a result of the fast growth of the
national economy and the extension of the city size, the demand and quantity of
urban pipes are expanding, and their laying and maintenance are being hampered
by environmental protection pressure and limited building sites. Pipes with small
diameters. Traditional excavation construction affects traffic and the environment
and is not conducive to people's desire for a better life. At the same time, the most
wealthy cities are found around the coasts and rivers. Pipeline construction frequently
confronts weak stratum such as silt. Pipeline construction is prone to engineering
issues such as difficult height control, trajectory variation, and ground collapse [3].

As a function, for the laying, renewal, and maintenance of the pipeline network
when the construction site of the small-diameter pipeline is restricted and unsuitable
for excavation and the ground conditions are adequate, the trenchless construction
technology of the pilot top pulling technique is proposed. This study will conduct
important analysis and research on the construction technology and application of
the small-diameter pilot top-pull method.

2   Principle of Micro-slit Pilot Roof

The pilot jacking technique is a small-diameter mixed pipe jacking construction
method that was developed by refining the combination of directional drilling and
mud-water balancing pipe jacking construction technology for the common construction circumstances of sewage pipe networks. The pipe diameter is typically DN300-
600, which not only has the geological flexibility of directional drilling, strong
obstacle management ability, and minor secondary disasters, but also the benefits
of standard pipe jacking elevation control. The procedure works in tandem with the
self-sealing socket and the socket interface short tube [4]. The conventional pipeline
back-drag technique can be replaced with a top-pull process in which the pipe section
is inserted downhole at the end, the drill pipe is utilized to pass through the center
of the pipeline, and the pipe section is top-pulled at the pipeline's end. The mud
is allowed to enter the inside of the pipeline during the top-pulling step, which
reduces weight and resistance. Since this end hole's necessary diameter is near to
the pipeline's outer diameter, the annular gap is tiny or non-existent.

Following the completion of the well, the small-diameter pilot jacking method
employs the procedure of directional drilling and jacking. The wellbore height is
readily regulated, and the steel drill pipe is straightened and penetrated, as well as
the pipeline elevation. Figure 1. depicts the small-diameter pilot top pull method.

Since 2017, the small-diameter pilot top-drawing method and accompanying
pipes have been increasingly popular in towns and regions throughout Guangdong Province, including Foshan, Zhongshan, Dongguan, Shantou, Shenzhen, and

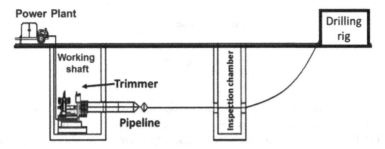

Fig. 1 Small-caliber pilot top pulling is indicated

Guangzhou. The total length of the single design and construction has topped 100 km, and it has also been expanded to Fujian, Hunan, Jiangxi, Anhui, Zhejiang, Shanghai, Jiangsu, and other locations. It is also being used to replace aged pipes as part of pipe network maintenance [5].

3 Construction Design of Small Diameter Pilot Top Pulling Method

3.1 Selection of Key Technical Parameters

The maximum mud pressure and nose pulling force are important technical elements of the small-diameter pilot top pulling technique.

Maximum Mud Pressure Analysis and Calculation. The mud circulation system must be utilized to discharge the created soil slag during the pipeline jacking operation under silt geology. To lessen the resistance between the pipeline, the machine head, the soil layer and clear water plus hydrolyzed polyacrylamide is used in the mud solution for reducing grouting drag in silt geology (HPAM).

The building hole wall formed by silt has weak stability, low shear strength, and high compressibility when compared to other soils. The pressure of the mud pump in the mud circulation system is too high, causing the mud to flow too quickly, resulting in formation damage and ground escape, which is easy to cause ground collapse and impairs the stability of nearby building foundations. As a result, determining the maximum mud pressure before construction is crucial to assure the formation's stability throughout construction. The maximum mud pressure is the sum of the mud pressure at the bottom of the hole and the head loss, and the maximum mud pressure at the bottom of the hole may be calculated using the Delft formula [6, 7].

$$P_{\max} = (p'_f + c\cot\varphi)[(\frac{R_0}{R_{\text{pmax}}})^2 + Q]^{\frac{-\sin\varphi}{1+\sin\varphi}} - c\cot\varphi \tag{1}$$

$$p'_f = \sigma'_0(1 + \sin\varphi) + c\cot\varphi \tag{2}$$

$$Q = \frac{\sigma'_0 \sin\varphi + c\cot\varphi}{G} \tag{3}$$

where: σ'_0 is the initial effective stress (kPa); φ is the internal friction angle (°); c is the cohesive force (kPa); R_0 is the initial aperture (m); $R_{p,max}$ is the maximum allowable The radius of the plastic zone is generally 2/3 of the height of the hole axis from the surface (m); G is the shear modulus of the formation (kPa).

Analysis and Calculation of the Pulling Force of the Nose. To carry out the pilot top pulling technique under silt stratum conditions, to construct securely and prevent ground damage, the maximum value of the back drag force (top pull force) of the pipeline crossing must first be determined.

The maximum pulling force of the nose during pipe jacking may be determined using the unloading arch algorithm [7, 8] as shown in Eq. (4). For silty strata, Eq. (5) can also be used to estimate the maximum top tensile force in engineering practice [7, 8].

$$T_{max} = [2P(1 + K_a) + P_0]f_e L \tag{4}$$

where: T_{max} is the maximum value of the return drag force of the pipeline crossing (kN); P is the earth pressure of the crossing pipeline per unit length (kN/m); K_a is the active earth pressure coefficient of the crossing soil layer, generally taken as 0.3; P_0 is the dead weight of the crossing pipe per unit length (kN/m); f_e is the friction coefficient between the pipe and soil, generally taken as 0.2 ~ 0.3; L is the length of the crossing pipe (m).

$$T_{max} = f_e L(4D_0 D_e + P_0) \tag{5}$$

where D_0 is the outer diameter of the crossing pipe (m); D_e is the maximum reaming diameter (m); the meaning of other parameters is shown above in Eq. (4).

3.2   Construction Process of Pilot Top Pulling Method

The underground pipeline survey, engineering design, well site construction, pilot hole construction, pipeline jacking and laying, landform restoration, and engineering acceptance are the primary construction operations of the small-diameter pilot jacking technique, as illustrated in Fig. 2.

Fig. 2 Construction process
of small-diameter pilot
jacking pipe

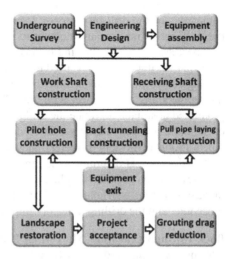

3.3   The Primary Process Design

The key processes of the pilot jack-up construction include well site construction, pilot hole construction and jack-up construction.

Well Construction. Working well construction and receiving well construction are both aspects of well construction. Wells are classified as circular, square, or round pits. The inner diameter of a circular working well should not be less than 2.0 m in general, but it should not be less than 1.8 m if the layout site is constrained. The inner diameter of the pit-type circular working well is not less than 2.8 m and the inner diameter of the pit-type square working well is not less than 2.8 m (length) 2.2 m (width). The inner diameter of the receiving well must be at least 2.0 m and, in severe instances, no less than 1.8 m. The well wall is often composed of the prefabricated wellbore, with caisson construction. Before constructing working wells and receiving wells, the site must be enclosed and the road and water elevation must be remeasured, and the road must be reduced to the minimal size necessary for construction. After cutting and fracturing the road surface, dig the soil using the telescopic boom excavator, then cycle to excavate and sink to the lower side in turn. Frequent inspections and measurements are performed during the excavation process to avoid over-excavation and disturbing the undisturbed soil, and the construction is strictly per the design and construction, and the wellbore and wellbore overlap are well waterproofed to ensure safety and quality.

Pilot Hole Construction. After the pipe jacking working well and receiving well are completed, the installation of the pipe jacking machine and horizontal directional drilling may begin, and the directional drilling equipment can be placed following the pilot hole trajectory design.

At the back end of the directional drilling bit, an electronic direction finder is installed in the drill pipe. During the drilling process, it continually measures the parameters of the drill bit's precise location, apex angle, and depth, and modifies the drill rig settings at any moment to regulate the drill bit to drill according to the specified trajectory. Drilling begins once all of the equipment is in place. The walking tracking technique is used to regulate the drilling according to the intended drilling curve, monitor the drilling direction, depth, drill bit plate status, and signal rod temperature, and connect with the driller. When drawing back, it is explicitly prohibited unsuitable to rotate the drill pipe counterclockwise. The incidence angle of the build-up section is less than 13 degrees, the build-up curvature fits the drill pipe's 1500D (drill stem diameter), and the curved drill pipe is not employed.

The design axis determines the direction and slope of steerable jacking. When jacking deviation occurs, it may be corrected using the center coordinate and elevation of the pipe jacking well hole. Also, The side station must oversee the pilot hole drilling procedure, and the drilling trajectory must be approved.

Back Towing Excavation, Mud-water Pressure Balance Construction. After the pilot hole is completed, the drill bit is withdrawn, and the back towed road-head, connecting piece, mud discharger head, connecting back drill pipe, pipe, roof plate, and hydraulic jack are placed in sequence. When the pipe section passes through the reinforcing region of the hole, the operator begins to pull the top into the hole whilst performing the anti-twist measures with the speed controlled at $0 \sim 10$ mm/min. The viscosity of the mud should be monitored often during the jacking operation, and the value should be regulated between 40-60 s, and the soil should be drained in time according to the pulling power of the drill pipe. Every time a portion is jacked, the pipe jacking machine's oil cylinder is pushed backwards, and another segment of the pilot pipe is inserted to continue jacking up until the receiving well is reached. When both the machine head and the pipe section enter the receiving well, the space between the pipe and the hole wall should be filled with water-stop sealing treatment and the water-stop sealing steel plate placed, as illustrated in Fig. 3.

3.4 Quality, Safety and Environmental Measures

In an attempt to adapt to the sustainable development of the city, reduce the degree of pollution to the environment, and reduce the damage to surrounding or adjacent buildings, the construction of the small-diameter drainage pipe top-drawing method should include: "Safety Inspection Standards for Municipal Engineering Construction" (CJJ/T275), "Technical Specifications for Temporary Electricity Safety at Construction Sites" (JGJ46), "Safety Technical Regulations for the Use of Construction Machinery" (JGJ33), "Technical Regulations for Pipe Jacking of Water Supply and Drainage Engineering" (CECS246: 2008), "Technical Specifications for Horizontal Directional Drilling Pipeline Crossing Engineering" (CECS382: 2014), "Flexible Sealed Self-locking Interface Polyethylene Wound Solid Wall Drainage Pipe and

Fig. 3 Schematic diagram of water-blocking steel plate

Fittings" (T/GDSTT 1–2021), "Expansion Lock" Sealing and connecting polyethylene solid wall drainage pipes and accessories" (Q/STXM 1–2021), "ductile iron castings" (GB/T 1348) and other national and industry standards to formulate construction quality, safety and environmental protection measures and other plans, and in accordance with "The project acceptance shall be carried out according to the standards for the construction and acceptance of water supply and drainage pipeline engineering (GB/T 50,268–2008).

3.5 Process Characteristics

When compared to typical pipe network excavation methods, small-diameter pilot pipe jacking construction technology offers the following important benefits:

(1) It offers a wide range of applications and uses a socket-type connecting mechanism that can be used with water and has no risk of leaking. Multiple winding and braiding extrusion form the HDPE solid wall winding pipe, which has high performance and is suited for most applications.

(2) High controllability, the pilot jacking method uses an electronic direction finder placed in the drill pipe at the rear end of the drill bit to continuously measure the parameters of the drill bit's position, apex angle, and depth, and adjust the drilling rig parameters at any time through the drilling rig to control the drill bit to drill according to the design trajectory, which can ensure drainage pipe elevation with high precision.

(3) The economic benefits are good. Only small equipment can be used for the laying of pipelines. The amount of soil and road restoration works produced is small, and a lot of operational costs are reduced.

(4) The environmental impact is minimal, the influence on ground traffic and nearby buildings is also minimal, the operating noise is relatively low, and inhabitants' daily traffic and lives are unaffected [9].

(5) Construction may be carried out on restricted site areas, narrow roadways, and densely packed urban settlements.

(6) The building efficiency is good, and no extensive excavation is required. A good positioning and deviation correction system may ensure that the pipeline trajectory is exact, saving time.

4   Engineering Case Study

4.1   Project Overview

A drainage pipeline project in Zhongshan City, Guangdong Province is located in an affluent region of the city [10]. According to the geological survey data, the strata inside the pipeline's buried depth range include plain fill, silty clay, silt, and medium sand from top to bottom. Following the technological demonstration, this area adopts the construction method of small-diameter pilot top-pulling pipe, the pipe buried depth is around 5 m, the pipe diameter is DN400500, and the pipe material is socket type HDPE solid wall winding pipe. The distance between functioning wells is between 22 and 31 m. Figure 4 depicts the pipeline's configuration.

Fig. 4 Schematic diagram of the project

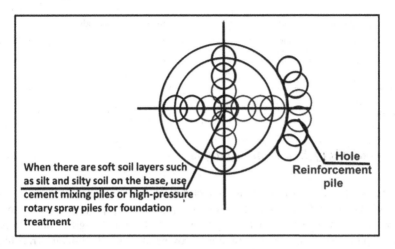

Fig. 5 Work well plan

Since the pipeline is placed in the silt stratum and the foundation's bearing capacity is low, the foundation of the working well is strengthened with cement mixing piles or high-pressure rotary jetting piles, and reinforcement piles are installed at the hole's entrance, as shown in Fig. 5.

4.2 Calculation of Maximum Mud Pressure and Nose Pulling Force

Using the DN400-29 section of the pipeline as an example, where the pipe length is $L = 29.0$ m and $D 0 = 0.4$ m, the maximum mud pressure during the pilot jacking method construction process is calculated using Eq. (1), (4), or (5) and the pulling force of the machine head, and the calculation results are shown in Table 1.

The viscosity of the mud was monitored and regulated on a regular basis by a Marsh funnel during the construction phase. The maximum mud pressure at the bottom of the hole was set at around 50 kPa, while the machine head's pulling force was set to around 8.5 kN. A compact directional drilling rig may be chosen for the top pulling equipment, and the machine's rated feed force and maximum pullback force can match the pipeline laying load requirements.

Table 1 The result of the calculation results in maximum mud pressure and nose pulling force

Maximum mud pressure at the bottom /kPa	Pulling force /kN
50.17	8.53

Table 2 Comparison of economic benefits of different construction methods

Construction method	Pipe laying material cost/(RMB/m)	Equipment fee for pipe laying personnel/(RMB/m)	Material cost of working well/(RMB/m)	Staff and equipment costs for working wells(RMB/m)	Work efficiency/(m/day)	Comprehensive unit price/(RMB/m)
Pilot top pull method	2847	748	1250	1627	28	6472
Pipe jacking	985	2214	3058	3702	16	9961

4.3 Analysis of Economic Benefits

The economic benefits of the small-diameter pilot jacking method and the regular pipe jacking method are contrasted using the pipe jacking project as an example. The costs are based on the "Guangdong Provincial Municipal Engineering Comprehensive Quota 2018" and the "Guangdong Provincial Drainage Pipeline Trenchless Repair and Renewal Project Budget Quota 2019." Table 2 shows the cost estimate comparison findings for the two building approaches.

The above table shows that the comprehensive cost of the pilot top pulling method is the lowest, and when considering factors such as construction period, quality and safety, and environmental protection, it is clear that this method has significant advantages in the construction of small-diameter pipelines.

5   Conclusions

The principle of the small-diameter pilot jacking method is analyzed, and the calculation methods of the maximum mud pressure and the maximum pulling force of the machine head in the pipe jacking construction process under silt formation are given, as well as the construction scheme design of the small-diameter pilot jacking method. The advantages of small-diameter pilot top-pull construction include a compact footprint, broad application, high efficiency, construction safety, environmental protection, and resource conservation. This approach has a wide range of applications and is particularly well suited for trenchless installation and maintenance of small-diameter pipes in densely populated metropolitan areas.

Acknowledgements The research work was supported by the Guangdong Province Key Field R&D Program Project (Item No. 2019B111105001); Guangdong Province Basic and Applied Basic Research Fund Project (Item No. 2020A1515010560).

References

1. Chao W (2018) Study on collapse mechanism of municipal buried pipeline. Jinan University
2. Ministry of Housing and Urban-Rural Development of the People's Statistical Yearbook of Urban and Rural Construction in 2020 (2020)
3. Kun G, Qijun L (2018) Discussion on the improvement of gravity flow drainage pipe directional drilling technology. Water Supply Drainage 54(06):102–105
4. T/GDSTT 1—2021 Flexible Sealing Self-Locking Interface Polyethylene Wound Solid Wall Drain Pipe and Fittings. Guangdong Trenchless Technology Association
5. Qijun L (2022) Technical reference for the design and construction of small diameter drainage pipe pulling
6. Shu-li L, Kai L, Shuning W (2018) Selection and control of key technical parameters of directional drilling process pipeline through Qin construction. People's Yellow River 40(07):124–128
7. Huijin C (2020) Research on the key technology of pilot top pipe drawing process. Jinan University
8. Jinlong A (2008) Calculation method and analysis of cross-return drag force of horizontal directional drilling. Petroleum Eng Constr 01: 21–26
9. Xinmin Z, Kun G, Zilin F (2014) Application of secondary pipe jacking method in laying small diameter drainage pipe. Water Supply Drainage 50(05):92–95
10. Tianjin Urban Construction Design Institute. Zhongshan City Center Group Black odorous (substandard) water remediation and upgrading project (project 3) Tugua Chung, Xiao Chung remediation and upgrading project construction drawing design (2020)

Detailed Analysis of Shrinkage and Creep Effect of Concrete in Prestressed Box Girder Bridge

Donglian Tan, Wenyuan Ding, Yue Zhao, and Chuqin Yan

Abstract For a 60 m prestressed concrete box girder bridge using structural analysis software as ANSYS and MIDAS/civil, combined with several creep prediction models, a reasonable creep prediction model has been obtained. Considering the influence of prestressed tendon relaxation on the shrinkage and creep of concrete, a refined finite element model is established for numerical analysis and compared with the experimental data. The comparison results show that the calculated value of the refined finite element model with reasonable prediction models and considering the effect of prestress relaxation is closer to the measured value. Finally, the finite element model of solid elements considering the effect of prestress relaxation with the GL2000 creep prediction model is used to analyze the shrinkage and creep effect of a prestressed concrete continuous rigid frame bridge.

Keywords Shrinkage and Creep · Prediction Models · Prestressed Concrete Box Girder · Continuous Rigid Frame Bridge · Accurate Analysis

1 Introduction

Prestressed concrete box girder has been widely used in bridges for its advantages of large structural stiffness, small deformation and beautiful shape. However, the inherent performance of concrete as shrinkage and creep has obvious impact on bridge deflection. It will lead to the bridge alignment cannot reach the expected design alignment. The shrinkage and creep of concrete has time-varying characteristics. Its laws are complex and changeable. It is one of the most uncertain characteristics of concrete. At present, there are many creep prediction models. The calculation results of various models are quite different. There are many factors influencing the shrinkage and creep of concrete. In this paper, combined with the measured values of concrete shrinkage and creep deformation of a 60 m prestressed concrete box

D. Tan (✉) · W. Ding · Y. Zhao · C. Yan
School of Railway Transportation, Shanghai Institute of Technology, 100 Haiquan Road, Fengxian District, Shanghai, China
e-mail: tdl021@126.com

© The Author(s) 2023   85
G. Feng (ed.), *Proceedings of the 9th International Conference on Civil Engineering*, Lecture Notes in Civil Engineering 327, https://doi.org/10.1007/978-981-99-2532-2_8

girder bridge, the refined analysis method of shrinkage and creep deformation of a prestressed concrete box girder bridge is studied. Then, the concrete shrinkage and creep effect of a prestressed concrete continuous rigid frame bridge is analyzed using the method proposed in this paper.

2   Prediction Model of Concrete Shrinkage and Creep

Based on the analysis of the establishment mechanism of the concrete creep prediction model, the creep prediction model can be divided into two categories.

One is to construct prediction formulas such as hyperbolic function, power function and exponential function by observing and studying the distribution phenomenon of many experimental data. Many prediction models according to this method have been established in the early stage. The most influential one is the hyperbolic power function prediction model proposed by Professor Ross in 1937, which was introduced by ACI in 1970. It was adopted by the 209 Committee after revision. Up to now, the ACI 209 committee still maintains the framework of this prediction model, only modifying the parameters.

The other is to establish the framework of a prediction model based on theoretical analysis and determine parameters by regression based on experimental data. This kind of model generally has a clear physical meaning and develops and improves with the development of theory. CEB-FIP (1978), BP-KX and B3 models are all such models. CEB-FIP (1990) creep prediction model has made a great adjustment to CEB-FIP (1978) model, from the form of continuous addition to the form of continuous multiplication. From the form of the formula, it is closer to the former one.

For the establishment mechanism of concrete shrinkage prediction models, almost all prediction models are constructed according to hyperbolic function or hyperbolic power function, and the fitting parameters of the models are adjusted through experiments and theoretical analysis. From the current research reports, no matter which form of prediction model is used, if the parameters are accurate and appropriate, the prediction results are in good agreement with the actual situation. The following four main forecasting models are introduced, which are CEB-FIP series model, ACI 209 model, B3 model and GL2000 model.

2.1   CEB-FIP Series Forecasting Model

The CEB-FIP (1978) model was jointly launched by CEB and FIP in 1978. In this model, the formula of creep coefficient is expressed in the form of continuous addition. Creep is composed of irrecoverable creep and recoverable creep. Irrecoverable creep is divided into initial creep and lag creep under loading. The shrinkage function of CEB-FIP (1978) model is expressed as the product of basic shrinkage coefficient and shrinkage time function.

CEB-FIP (1990) model was put forward by CEB-FIP organization in 1990. The model does not specifically distinguish various types of creep. The formula is expressed in the form of continuous multiplication. The variation law of creep coefficient with time is fitted as a hyperbolic power function. The formula is obtained by multiplying three correction coefficients.

$$\varphi(t, \tau) = \varphi_0 \beta_c(t, \tau) = \varphi_{RH} \beta(f_{cm}) \beta(\tau) \beta_c(t - \tau) \tag{1}$$

where, φ_0 is the nominal creep coefficient; φ_{RH} is the correction factor of environmental relative humidity; $\beta(f_{cm})$ is the concrete strength correction factor; $\beta(\tau)$ is the loading age correction factor; $\beta_c(t - \tau)$ is the creep process time function.

The contraction function of CEB-FIP (1990) is as follows:

$$\varepsilon_{cs}(t, t_s) = \varepsilon_{cs0} \beta_s(t - t_s) = \varepsilon_s(f_{cm}) \beta_{RH} \beta_s(t - t_s) \tag{2}$$

where, ε_{cs0} is the nominal shrinkage factor, $\varepsilon_s(f_{cm})$ is the concrete strength correction factor, β_{RH} is the correction factor of environmental relative humidity, $\beta_s(t - t_s)$ is the shrink process time function.

In the CEB-FIP (1990) model, the effects of loading age, environmental relative humidity, member size and concrete strength are considered in creep, which greatly improves CEB-FIP (1978) model.

2.2 ACI209 Series Forecasting Model

ACI209 is a prediction model recommended by American Concrete Institute (ACI), which is adopted by American state specifications, and is also used as a reference comparison standard for Canada, New Zealand, Australia and some Latin American countries.

ACI209 Committee has launched ACI 209–1978 model, ACI 209–1982 model, ACI 209r-1992 model [9]. The creep coefficient of ACI 209 series model is expressed as follows:

$$\varphi(t, \tau) = \varphi(\infty) \frac{(t - \tau)^{0.6}}{10 + (t - \tau)^{0.6}} \tag{3}$$

where, $\varphi(\infty)$ is the ultimate value of creep coefficient, $\varphi(\infty) = 2.35\gamma_c$, 2.35 is the ultimate value of creep coefficient under standard state, γ_c is the correction coefficient of various factors affecting creep in deviating from the standard state.

The contraction function of ACI 209 mode is as follows:

$$\varepsilon_{\text{sh}}(t, t_0) = \frac{t - t_0}{H + t - t_0} \varepsilon_{\text{sh}\infty} \tag{4}$$

where, $\varepsilon_{\text{sh}}(t, t_0)$ is Shrinkage strain at t; $\varepsilon_{\text{sh}\infty}$ is the ultimate shrinkage strain, $\varepsilon_{\text{sh}\infty} = 780 \times 10^{-6} \gamma_{sh}$, 780×10^{-6} is the ultimate value of shrinkage strain under standard conditions, γ_{sh} is the correction coefficient of each influencing factor when it deviates from the standard state; t_0 is the age of completion of maintenance; H is the coefficient related to maintenance.

ACI 209 series model considers that the size of the component has no effect on the ultimate value of shrinkage and creep, which is inconsistent with the test results. Moreover, the parameters of the shrinkage and creep time process function are taken as constants in this series of models, which cannot reflect the development law of shrinkage and creep of concrete with different strength grades over time. Therefore, ACI 209 series model needs to be further revised.

2.3 B3 Prediction Model

B3 prediction model, which is simpler and more theoretical, was proposed in 1997 by Professor Z.P.Bažant on the basis of BP model (proposed in 1978) and BP-KX model (proposed in 1991). B3 model is a shrinkage creep prediction model recommended by RILEM. The B3 model of concrete creep is established based on the solidification theory of concrete, which combines the elastic theory, viscoelastic theory and rheological theory to simulate the new theory that the macroscopic physical and mechanical properties of concrete change with time due to the hydration of cement and the increase of solid phase. It holds that the dependence of macroscopic material parameters of concrete on time is the result of the increasing volume of viscous phase and viscoelastic phase of concrete material, while the mechanical properties, elastic phase volume and non-bearing phase volume (such as pores, colloids, water, etc.) are constantly changing.

The creep coefficient expression of B3 prediction model:

$$\varphi(t, \tau) = \frac{J(t, \tau) - 1/E(\tau)}{1/E(\tau)} = E(\tau)J(t, \tau) - 1 \tag{5}$$

where, $J(t, \tau)$ is a creep function; $E(\tau)$ is the elastic modulus of concrete at τ.

The contraction function expression of B3 prediction mode is as follows:

$$\varepsilon_{\text{sh}}(t, \tau) = \varepsilon_{\text{sh}\infty} k_{\text{h}} S(t) \tag{6}$$

where, $\varepsilon_{\text{sh}\infty}$ is the ultimate shrinkage strain, k_{h} is the humidity correction factor, when $h \leq 0.98$, $k_h = 1 - h^3$; when $h = 1.00$, $k_h = -0.2$; when $0.98 \leq h \leq 1.00$, it is obtained by linear interpolation; $S(t)$ is the time process function.

B3 prediction model is a semiempirical and semi theoretical formula. Bezant pointed out that in order to improve the theoretical and prediction accuracy of the model, the calculation formula of various material parameters should also be established on a certain theoretical basis, and the research in this field needs to be further carried out.

The accuracy of B3 prediction model is limited by many conditions. Compared with CEB-FIP model and ACI 209 mode, B3 mode needs more parameters and more calculation.

2.4 GL2000 Prediction Model

Based on some criteria that should be satisfied by the shrinkage creep prediction model adopted by ACI 209 Committee in 1999, Gardner and Lockman revised the atlanta97 model (proposed by Gardner and Zhao in 1993) and proposed GL2000 model. The model can be applied to high strength concrete, and the influence of concrete drying before loading on creep deformation after loading is considered by a single item. The GL2000 model is described as follows:

Creep Coefficient:

$$\varphi_{28} = \Phi(t_c) \left\{ \begin{array}{c} 2\left[\dfrac{(t-t_0)^{0.3}}{(t-t_0)^{0.3}+14}\right] + \left(\dfrac{7}{t_0}\right)^{\frac{1}{2}}\left[\dfrac{t-t_0}{t-t_0+7}\right]^{\frac{1}{2}} \\ +2.5\left(1-1.086h^2\right)\left[\dfrac{t-t_0}{t-t_0+0.15\left(\frac{V}{S}\right)^2}\right]^{\frac{1}{2}} \end{array} \right\} \tag{7}$$

$$\Phi(t_c) = \left\{ \begin{array}{ll} \left[1-\left(\dfrac{t-t_0}{t-t_0+0.15\left(\frac{V}{S}\right)^2}\right)^{\frac{1}{2}}\right]^{\frac{1}{2}}, & t_0 > t_c \\ 1, & t_0 = t_c \end{array} \right. \tag{8}$$

where, φ_{28} is the creep coefficient of concrete at 28 days age. t is the calculation concrete age. t_0 is the loading age of concrete. t_c is the age of concrete at the beginning of drying. V/S is the ratio of volume to surface area of concrete components.

Shrinkage strain:

$$\varepsilon_{sh} = \varepsilon_{shu}\beta(h)\beta(t) \tag{9}$$

$$\beta(h) = 1 - 1.18h^4 \tag{10}$$

$$\varepsilon_{shu} = 900k\left(\frac{30}{f_{cm28}}\right)^{\frac{1}{2}} \times 10^{-6} \tag{11}$$

$$\beta(t) = \left(\frac{t - t_c}{t - t_c + 0.15\left(\frac{V}{S}\right)^2}\right)^{\frac{1}{2}} \tag{12}$$

where, h is the environmental relative humidity. t is the calculation concrete age. t_0 is the loading age of concrete. t_c is the age of concrete at the beginning of drying. k shrinkage coefficient which depends on cement type. V / S is the ratio of volume to surface area of concrete components. f_{cm28} is the average cube compressive strength of concrete at 28 days of age.

3 Comparison of Prediction Models

3.1 Calculation Method

Midas /civil has two methods for creep analysis, one is to directly define the creep coefficient of each stage of the element, and the other is to use the creep function to calculate by integral. The influence of early shrinkage and creep of concrete in construction stage can be considered in the software. Moreover, the software provides the calculation standards of many countries, which can fit the creep coefficient and shrinkage strain conveniently [4–10].

3.2 Finite Element Model

In this section, in order to compare CEB-FIP, ACI209 and GL2000 prediction models, the finite element model of the test beam is established by the bridge structure analysis program Midas for numerical simulation analysis. The calculation model is shown in Fig. 1.

The box girder is made of C55 concrete. According to the material property test results of concrete specimens, the elastic modulus is 4.65×10^4 MPa, and the 28-day cube compressive strength is 65.6 MPa; the Poisson's ratio of concrete material is 0.2, and the bulk density is 26.5kn/m^3.

The prestressed steel specific parameters are shown in Table 1.

According to the measured results, the influence coefficient K of local deviation of prestressed pipe is 0.001454, and the friction coefficient μ of pipeline is 0.1470.

Fig. 1 Finite element model of test beam

Table 1 Mechanical property index of prestressed steel strand

Elastic modulus (MPa)	1.95×10^5
f_{ptk} (MPa)	1860
Nominal diameter(mm)	15.2
Nominal section area(mm^2)	139
Nominal mass(kg/m)	1.101
Relaxation rate	2.5%
Tension control stress (MPa)	1395

3.3 Comparison of Calculated Results and Measured Results

The creep deflections of aci209, CEB-FIP and gl2000 are calculated by using the above finite element models and compared with the measured results. The results are shown in Fig. 2.

From the comparison chart, it can be seen that the calculation results of GL2000 creep prediction model are close to the measured values, except that the difference of some measuring points reach 18%, the rest are less than 10%. Therefore, the following analysis uses GL2000 prediction model.

4 Fine Analysis of Shrinkage Creep Deformation of Concrete Box Girder

According to the comparison results in Sect. 2, the calculation results of GL2000 prediction model are the closest to the measured values. The GL2000 concrete shrinkage and creep prediction model is used for the refined analysis of the shrinkage and creep deformation of the concrete box girder. MIDAS/civil finite element model using three-dimensional beam element, ANSYS solid model without considering the influence of prestressed tendon relaxation, and ANSYS solid model considering the influence of prestressed tendon relaxation are used respectively. Three kinds of refined finite element models are analyzed and compared.

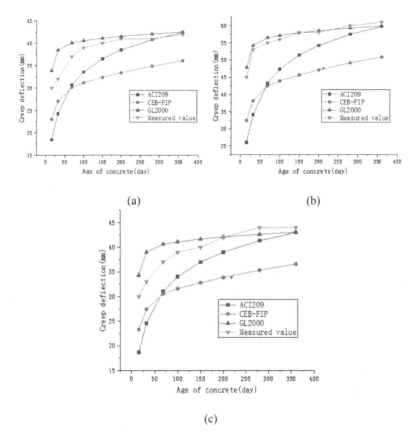

(a)   (b)

(c)

Fig. 2 Comparison between the calculated results and the measured ones **a** 1/4 span; **b** Middle span; **c** 3/4 span

4.1 MIDAS/civil Beam Element Model

The finite element model of 60 m concrete beam was set up according to the construction stage using MIDAS/civil. The finite element model is shown in Fig. 3.

Fig. 3 Three-dimensional beam model of Midas

4.2 ANSYS Solid Model without Considering the Influence of Prestressed Tendon Relaxation

Without considering the influence of the relaxation of prestressed tendons on the shrinkage and creep, the prestress is applied with the initial strain in the deduction of prestress loss. The concrete is built with SOLID65 element, which can consider the creep of concrete. There are 2610 concrete units in the whole bridge. The prestressed reinforcement is built with link8 unit, with a total of 386 prestressed reinforcement units. In the ANSYS analysis. It is considered that the concrete does not crack and there is no sliding between the prestressed reinforcement and the concrete.

4.3 ANSYS Solid Model Considering the Influence of Prestressed Tendon Relaxation

In order to consider the influence of prestress relaxation on concrete shrinkage and creep, the time development process of prestress loss caused by prestress relaxation is considered in the analysis. Finite element models are the same as those in the previous paragraph.

4.4 Comparison of Calculated Values and Experimental Values of Three Models

The comparison between the theoretical calculation results of the three finite element models and the test results is shown in Fig. 4. Method 1 in the table refers to MIDAS/civil space beam finite element model, method 2 means ANSYS solid model without considering the effect of prestressed tendon relaxation, method 3 is ANSYS solid model considering the influence of prestressed tendon relaxation.

Fig. 4 Comparison of calculated and experimental values

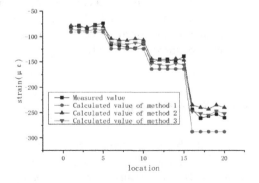

In method 2, the restraint effect of prestressed tendons on concrete is considered, but the influence of relaxation of prestressed tendons on concrete shrinkage and creep is not considered. The prestress loss caused by relaxation of prestressed tendons is deducted at the initial time, which makes the theoretical value of initial strain of concrete less than the experimental value, so the theoretical value of strain generated by shrinkage and creep is also less than the experimental value. In method 3, not only the restraint effect of concrete is considered, but also the influence of prestress relaxation on concrete shrinkage and creep is considered. The theoretical calculation value is closest to the experimental value.

5 Detailed Analysis of a Prestressed Concrete Continuous Rigid Frame Bridge

5.1 Bridge Overview

The span layout of a bridge is 90 m + 2 × 160 m + 90 m. The beam section is single box single cell box girder. The height of the box girder is changed from 9.0 m at the root to 3.5 m in the middle of the span. The width of the top plate of the box girder is 12.0 m, the width of the bottom plate is 6.50 m. The thickness of the top plate is 0.30 m. The thickness of the bottom plate is changed from 0.32 m in the middle of the span to 1.10 m in the root according to the quadratic parabola. The thickness of webs is 0.45 m and 0.60 m respectively. The prestress of the girder includes longitudinal prestress, transverse prestress and vertical prestress.

The piers are all double thin-walled hollow piers, which are 6.5 m across the bridge and 2.5 m along the bridge. The wall thickness is 0.5 m along the bridge direction and 0.8 m in the transverse direction.

The beam is made of C50 concrete, whose elastic modulus is 3.45×10^4 MPa. The pier is made of C40 concrete, whose elastic modulus is 3.25×10^4 MPa. The high-strength relaxation steel strand is used for the longitudinal prestressing of the bridge. Its elastic modulus and the relaxation rate are 1.95×10^5 MPa and 0.3.

The bridge is in the inland cool subtropical mountain climate zone. The annual average temperature is 15.1°C. The extreme maximum temperature is 41.7°C, while the extreme minimum temperature is - 10.2°C. The annual rainfall is 800–1000 mm. The relative humidity is 0.50–0.60.

5.2 Finite Element Model

Because the bridge is a symmetrical structure, in order to improve the calculation efficiency and reduce the calculation cost, only the finite element model of half bridge is established. The treatment on the boundary is to impose longitudinal symmetric constraint on the beam section at pier. It can ensure the consistency between the results and the whole bridge model. Because the main purpose is to analyze the shrinkage and creep effect of the main girder, the pier is established by beam element. The treatment method is to set up a rigid arm to restrain the deformation of pier top and girder bottom. It ensures the deformation coordination of pier top and girder bottom.

In order to better simulate the actual internal force state of the bridge when the bridge is completed, the key construction stage of the bridge is considered by using the unit life and death function provided by ANSYS software in the solid finite element model. The key construction stages are the maximum cantilever - side span closure - bridge completion. Because the relaxation of prestress mainly occurs in the period from tension to 40 days after tension, and concrete creep develops rapidly in the early days. In order to reduce the error caused by the division of time steps, the method of first small and then long is adopted in the time step division. A total of 23 time steps are divided into a total of 4210 days.

The pre-stressed steel bars are simulated by link8 element. The area and initial strain of reinforcement are considered in the real constant of the element.

5.3   Deflection Caused by Shrinkage and Creep

Considering the influence of prestressed tendon relaxation on concrete shrinkage and creep, the deflections of the centerline of half bridge deck caused by shrinkage and creep in one year, three years and ten years are analyzed. Their results are shown in Fig. 5.

It can be seen from Fig. 5 that the deflection of girder develops rapidly within one year after the completion of the bridge, while it gradually slows down a year later. The deflection of the middle span is greater than that of the side span. In order to analyze the structural deformation caused by the relaxation of prestressed tendons

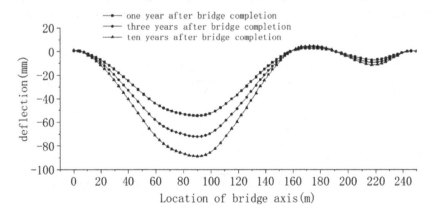

Fig. 5 Deflection by shrinkage creep (half bridge)

Table 2 Creep deformation over time (mm)

Section	Finished State	One year	Three years	Ten years
mid-span deformation of middle span	−33.4	−88.9	−102.0	−118.9
axial displacement of 1# pier top	−20.2	−76.5	−88.6	−103.8

on the shrinkage and creep of concrete, the vertical displacement in the middle span of the girder and the horizontal displacement of the pier are analyzed, as shown in Table 2.

It can be seen from the figures and tables that when considering the influence of prestressed relaxation on shrinkage and creep. Shrinkage and creep deflection of concrete in middle span is -55.5 mm in one year after completion and −68.6 mm in the third year after completion. It can be seen that the vertical displacement of the box girder due to shrinkage and creep is nearly 3 times of that at the completion time. It has reached 2 times of the completion time in one year.

5.4 Shrinkage Creep Stress

The shrinkage and creep stress of box roof and box floor at the time of completion, 1 year, 3 years and 10 years are shown in Fig. 6. and Fig. 7.

It can be seen from the figure that the roof stress decreases from the pier position to the middle of the span. The stress of the bottom plate increases from the support point to the middle of the span, while it decreases near the middle of the span. It is indicated that the stress change of the bottom plate is more complex. Moreover, the creep in the first year accounts for 80% of the total. Therefore, the stress state of some key sections should be accurately analyzed, such as the section at pier and mid-span section of middle span.

Because the shear stress and normal stress of the web are relatively large, the main stress state of concrete directly affects the failure mode of the structure. Therefore,

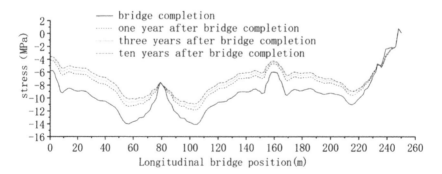

Fig. 6 Roof stress (half bridge)

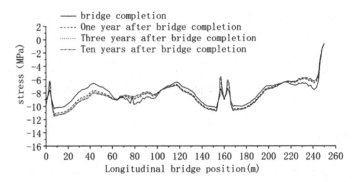

Fig. 7 Floor stress (half bridge)

it is of great significance to study the influence of concrete shrinkage and creep on the main stress of box girder web.

5.5 Prestress Loss

In long-span bridges, the prestress loss of different positions are quite different. The top plate steel tendons T20, web tendons W6, bottom plate steel tendons B7 and D17 of middle span closure section are selected for analysis.

In order to study the law of prestress loss, the absolute and relative values of prestress loss calculated by different calculation methods are compared, as shown in Table 3, Table 4, Table 5 and Table 6. In the table, the method I refers to the solid element model considering the effect of prestressed tendon relaxation on concrete shrinkage and creep, and the method II refers to the beam element model.

Table 3 Prestress loss of tendon T20

Method	One year		Three years		Ten years	
	value (MPa)	rate (%)	value (MPa)	rate (%)	value (MPa)	rate (%)
method I	−83.41	6.97	−92.33	7.72	−107.75	9.08
method II	−84.59	7.13	−94.93	8.00	−109.89	9.15

Table 4 Prestress loss of tendon W6

Method	One year		Three years		Ten years	
	value (MPa)	rate (%)	value (MPa)	rate (%)	value (MPa)	rate (%)
method I	−75.08	6.33	−84.24	7.1	−95.61	8.06
method II	−79	6.65	−92	7.74	−113	9.51

Table 5 Prestress loss of tendon B7

Method	One year		Three years		Ten years	
	value (MPa)	rate (%)	value (MPa)	rate (%)	value (MPa)	rate (%)
method I	−93.5	8.36	−105.84	9.45	−121.1	10.81
method II	−117.06	10.07	−136.12	11.71	−163.17	14.03

Table 6 Prestress loss of tendon D17

Method	One year		Three years		Ten years	
	value (MPa)	rate (%)	value (MPa)	rate (%)	value (MPa)	rate (%)
method I	−114.17	10.32	−130.80	11.82	−151.42	13.69
method II	−154.54	14.23	−179	16.54	−210.41	19.41

It can be seen from Table 3, Table 4, Table 5 and Table 6 that the loss value of top plate is greater than that of web at pier. This is because the stress of concrete at the top plate is higher than that at the web. The shrinkage and creep effect at the top plate will be greater than that at the web. The loss of prestress calculated by beam element is greater than that by solid element considering the effect of relaxation in prestressed reinforcement on the shrinkage and creep of concrete.

6 Summary

The comparison results of the three prediction models show that the variability of the prediction models is more than 20%. In the prediction of structural shrinkage and creep. It is better to select the appropriate prediction model according to the short-term experimental data, so as to reduce the prediction error. In the absence of data, GL2000 prediction model is recommended.

The finite element analysis of a 60 m prestressed concrete box girder is carried out. The theoretical value is compared with the measured value. The results show that the calculated value by beam element is larger than the measured value. The theoretical value is closest to the measured value by solid element considering the effect of prestressed tendon relaxation on shrinkage and creep.

The following conclusions have been drawn from the detailed analysis of concrete shrinkage and creep effect of a prestressed concrete continuous rigid frame bridge.

The mid-span deflection of the middle span is twice as much as that of the completed bridge in one year. It is three times of that in the ten years after completion of the bridge. The axial displacement of the pier top is also significant. The displacement of the pier top is 104 mm deflecting to mid-span direction when the bridge is completed for 10 years.

The results show that the stress of shrinkage and creep roof gradually decreases from the top of the pier to the middle of the span, while the stress of the bottom plate increases gradually from the top of the pier to the middle of the span. Because of Shrinkage and creep of concrete, the principal compressive stress decreases and the tensile stress increases.

The loss of prestress caused by shrinkage and creep is developing rapidly at the beginning. It gradually slows down after 3 years of completion. The loss value at pier top is less than that at mid span. The value of prestress loss calculated by solid element is less than that calculated by beam element.

References

1. Bazant ZP (2001) Prediction of concrete creep and shrinkage: past, present, and future. Nuclear Eng Des 203(1):27–38
2. Prediction of Creep, Shrinkage and Temperature Effect in Concrete Structure. ACI Committee 209 (1982)
3. Di H (2005) Creep effect theory of concrete structure. The Science Publishing Company, New York City
4. Jian-Ping L (2002) Evaluation of Concrete Shrinkage and Creep Prediction Models. A Thesis for the Degree Master of Science, San Jose State University
5. Gardner NJ (2005) Shrinkage and Creep of Concrete. Hermes Science, London
6. Zhi-hua C (2006) Comparative analysis of concrete creep prediction model. Bridge Constr 5:76–78
7. Wen-sheng, D, Zhi-tao, L, Shao-ping M, Zhao L (2004) Analysis and comparison of prediction models for concrete shrinkage and creep. Bridge Constr 6:13–16 (2004)
8. Shou-hui, C.: Analysis of shrinkage and creep effect of long-span prestressed concrete continuous box girder bridge. Railway Eng 8:15–17 (2009)
9. Zhi-min L (2002) The creep control of pc continuous beam bridge of rail transit. Bridge Constr 3:25–28
10. Fangzhi W (2008) Analysis and field measurement of concrete box girder bridges for shrinkage and creep effects. China Civil Eng J41(1):70–81

Research on Carbon Footprint Calculation and Evaluation in Assembled Building Phase

Miaomiao Zhao, Jiwei Zhu, Pengwei Han, and Chengling Liu

Abstract This study takes the carbon footprint of prefabricated buildings as the research object, and divides its physical and chemical process into four stages: building materials mining (production), prefabricated component production, material transportation and on-site construction. According to the carbon footprint sources of each stage, a carbon footprint model is established, and the carbon footprint factors and consumption factors required in the model are analyzed. Build a three-dimensional model based on BIM technology, and convert the consumption in combination with the project consumption quota to provide a data basis for calculation. The carbon footprint concentration produced in each stage of prefabricated buildings and cast-in-place buildings is analyzed by cases, and corresponding countermeasures and suggestions are put forward.

Keywords Fabricated Buildings · BIM · Materialization Stage · Carbon Footprint

1 Introduction

After China announced the goal of "carbon peaking and carbon neutralization" in September 2020, it rekindled the global attitude towards climate change. Now, China has become an innovator, practitioner and leader of global green, low-carbon and environmental protection. The carbon emission of China's construction industry increased from 1.354 billion tec in 2009 to 2.126 billion tec in 2018, with a growth rate of 57.02%. The energy consumption situation of the construction industry is still severe [1]. Assembled buildings characterized by energy conservation and emission reduction, low carbon environmental protection and high efficiency have become

M. Zhao · J. Zhu (✉)
School of Civil Engineering and Architecture, Xi'an University of Technology, Xi'an 710048, Shaanxi, China
e-mail: xautzhu@163.com

P. Han · C. Liu
China Construction Fourth Engineering Division Co., Ltd., Guangzhou 510665, Guangdong, China

© The Author(s) 2023
G. Feng (ed.), *Proceedings of the 9th International Conference on Civil Engineering*,
Lecture Notes in Civil Engineering 327,
https://doi.org/10.1007/978-981-99-2532-2_9

the focus of the world today, playing a pivotal role in the green development of the construction industry.

The prefabricated building is a green building with sustainable development. Its main characteristics are standardized design, factory production, prefabricated construction, information management and intelligent application. Quantitative analysis on energy conservation and emission reduction of prefabricated buildings is of great scientific significance [2]. Wang Guangming et al. [3] conducted a comparative study on the data of energy consumption, noise emissions and carbon emissions of traditional cast-in-place buildings and prefabricated buildings, and analyzed the social and economic benefits. Liu Meixia et al. [4] discussed and studied the energy-saving benefits and carbon emissions of prefabricated residential buildings, taking Building 5, the demonstration base of Zhengfangli Civilian Industrialized Construction Group, as an example.

It can be seen that at present, the domestic research on energy conservation and emission reduction of prefabricated buildings mainly focuses on the establishment of their carbon emission models, while the comparative analysis between them and traditional cast-in-place buildings is rare. This study is aimed at a specific prefabricated residential project, studying the building stage, through the carbon emission factor method to quantify the carbon footprint evaluation model, comparative analysis of the main differences of different structural construction of the same building. The BIM model is used to derive the project quantity, and the carbon footprint of the two building structures is calculated and analyzed based on the consumption quota and carbon footprint factor combined with the actual case, so as to provide a practical basis for energy conservation and emission reduction of prefabricated buildings.

2 Construction of Carbon Footprint Model for Assembled Building Process

2.1 System Boundary

Carbon Footprint was originally proposed by British experts, but it evolved from the "ecological footprint" proposed by scholars from Columbia University and is a quantitative description of carbon emissions in the whole life cycle of buildings [5]. According to the whole life cycle process, the project is mainly divided into three stages: materialization, use and demolition. In this study, the materialization stage is subdivided into building material mining (production) stage, prefabricated component production stage, material transportation stage and on-site construction stage, as shown in Fig. 1.

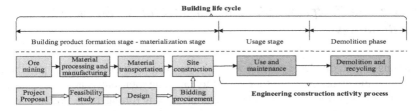

Fig. 1 Phase division of the whole life cycle of the building

2.2 Analysis on Consumption of Prefabricated Buildings

In this study, the combination of quota and work quantity has realized the conversion of material consumption and machine shift consumption. The specific statistical framework of project consumption is shown in Fig. 2.

According to the quantities of subdivisional works, the appropriate consumption quota is selected for resource consumption statistics. In this study, the straight stairs of cast-in-place projects are taken as an example for material consumption statistics. The BIM export volume is 43.5484 m^2. Since the project is located in Shaanxi, the 2019 Consumption Quota of Housing Construction and Decoration Works is selected as the quota standard for quantity conversion. Quota 5–56 is selected. See Table 1 for material consumption statistics.

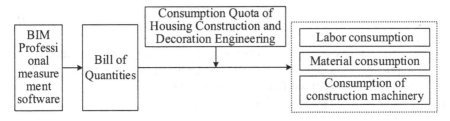

Fig. 2 Statistical Framework of Engineering Consumption

Table 1 Calculation Process of Construction Quantities of Straight Stairs

Quota No	Export Quantities	Company	Quota Details	Name	Quota consumption	Company	Actual consumption
5–56	43.5484	10m² horizontal projected area	Man-day	man-day	2.138	Man days	9.311
			Material	Concrete C35	2.586	m³	11.262
				Plastic film	11.529	m²	50.207
				Water	0.722	m³	3.144
				Power	1.560	kWh	6.794

2.3 Carbon Footprint Calculation Model for Prefabricated Buildings

According to the divided system boundary, the construction process of prefabricated buildings is divided into four stages to calculate their physical and chemical carbon footprint, respectively: building materials mining (production), prefabricated component production, materials (building materials and prefabricated components) transportation and on-site construction [6]. The calculation formula is expressed as:

$$E = E_r + E_p + E_t + E_s \tag{1}$$

$$UCE = {}^E\!/_A \tag{2}$$

where: E—total carbon footprint concentration in the prefabricated building stage; E_r—carbon footprint in the exploitation (production) stage of building materials; E_p—carbon footprint in the production stage of prefabricated components; E_{tp}—carbon footprint in component transportation stage; E_{tu}—carbon footprint in the transportation stage of building materials; E_s—carbon footprint of building construction; UCE—carbon footprint in the materialization stage of unit building area; A—Building area.

(1) Mining (production) stage of building materials

$$E_r = \sum_{j=1}^{j} M_j \times f_j \times (1 + \theta_j) \tag{3}$$

where: M_j—consumption of the jth building material; f_j—The jth carbon footprint factor of building materials considering recovery coefficient; θ_j—The loss rate of mining production of the jth material.

(2) Prefabricated component production stage

$$E_p = \sum C_p \times AU_i \times f_{ni}(i = 1, 2) \tag{4}$$

where: C_p—volume of prefabricated components; AU_1—Power consumption of prefabricated components per unit volume; AU_2—Fuel consumption of prefabricated components per unit volume; f_{n1}—electric power carbon footprint factor; f_{n2}—diesel carbon footprint factor.

(3) Material transportation stage

$$E_t = Q_p \times D_l \times H_p \times f_{n2} \tag{5}$$

where: Q_p—the number of means of transport; D_l—The distance from the material factory to the construction site; H_p—Fuel consumption per kilometer of material transport vehicles.

Table 2 Energy carbon footprint fact

Name	Carbon footprint factor	Company
Gasoline	3.5	$kgCO_2/kg$
Diesel oil	3.67	$kgCO_2/kg$
Natural gas	1.56	$kgCO_2/kg$
Power (Northwest)	0.94	$kgCO_2/kWh$

(4) Site construction stage

$$E_s = \sum_{v=1}^{n} MU_i \times M_i \times f_{ni} (i = 1, 2) \tag{6}$$

where: MU_i—consumption of machinery required in the construction process of item i; M_i—Energy consumption per unit shift of machinery used in the construction process of item i.

3 Carbon Footprint Factor Analysis

3.1 Energy Carbon Footprint Factor

As the basis of all research data, the accuracy of energy carbon footprint factor selection is crucial. In the selection, some selected the relevant data provided by IPCC and other institutions, and the other part was calculated by referring to the domestic energy calorific value and energy default emission factor. The obtained energy carbon footprint factor is shown in Table 2.

3.2 Material Carbon Footprint Factor

The carbon footprint factors of finished or semi-finished materials are calculated in combination with the carbon emission data of some raw materials in IPCC. The summary of calculated building carbon footprint factors is shown in Table 3. This study mainly analyzes six different carbon footprints of concrete, cement mortar, steel, water, welding rod and block. The use of other materials is small, which has little impact on the carbon footprint calculation results, so they are not included in the calculation.

Table 3 Summary of Carbon Footprint Factors of Common Materials

Material name		Carbon footprint factor
Concrete	Concrete C25	$262.50 kgCO_2/m^3$
	Concrete C30	$278.80 kgCO_2/m^3$
	Concrete C35	$290.50 kgCO_2/m^3$
	Concrete C40	$301.40 kgCO_2/m^3$
	Concrete C45	$315.40 kgCO_2/m^3$
Steel products	Medium and small section steel	$1420 kgCO_2/t$
	Hot rolled ribbed bar	$1777 kgCO_2/t$
	Cold rolled ribbed bar	$2133 kgCO_2/t$
Welding rod		$20.5 kgCO_2/kg$
Autoclaved aerated concrete blocks		$365.05 kgCO_2/m^3$
Cement mortar		$393.65 kgCO_2/m^3$
Water		$0.26 kgCO_2/m^3$

3.3 Carbon Footprint Factor of Transport Machinery

The carbon footprint of the material transportation phase in this study refers to the carbon footprint generated during the transportation of building materials, including prefabricated components, from the production plant to the construction site. The carbon footprint factor of transport means the fuel consumption standard per kilometer of the transport means under the rated load, which is determined in combination with the energy carbon footprint factor.

3.4 Carbon Footprint Factor of Construction Machinery

The carbon footprint of mechanical equipment is generated by energy consumption in the construction process, rather than energy consumption due to the operation of machinery itself [7]. In this study, the carbon footprint factor of mechanical equipment is determined by the shift energy consumption and energy carbon footprint factor in the construction process of mechanical equipment. Since there are too many types of machinery, they will not be listed here one by one.

4 Case Analysis

4.1 Project Overview

This study takes a residential project in Xixian New Area of Xi'an as a case, and selects a single project. The building has 25 floors above the ground (3.15 m high) and 2 floors underground, with a building height of 79.15 m. The foundation structure is raft foundation, the pile foundation is CFG composite pile, and the building area is about 14,237.3 m^2. SPCS fabricated system technology is adopted for construction, and the main types of prefabricated components include prefabricated walls, composite slabs and prefabricated stairs. The overall assembly rate reaches 50%.

4.2 Carbon Footprint Calculation

In order to analyze and compare the carbon footprint difference between the prefabricated construction and the traditional cast-in-place construction in the physical and chemical stage, this study designs two different structures for the same building. The first scheme is the traditional cast-in-place structure; The second scheme is prefabricated structure.

According to the carbon footprint model of each phase of the building, the carbon footprint of the exploitation (production) phase of building materials in Scheme I is calculated as shown in Table 4.

It should be noted that in the material transportation stage, the transportation machinery and transportation distance of the same material in the two schemes are the same, while prefabricated buildings need to be transported. The carbon footprint calculation in other stages is the same, and the carbon footprint concentrations in each stage of the two schemes are shown in Table 5.

4.3 Comparative Analysis of Carbon Footprint Concentration

The carbon footprint concentrations generated by two different structures of the same building are different. Compared with previous studies, the carbon concentrations of the two cases in this study are at a moderate level. The carbon footprint concentrations per unit area in the materialization stage of the two building structures are respectively: traditional cast-in-place building structures: $352.33kgCO_2/m^2$, and prefabricated building structures: $321.81kgCO_2/m^2$. The carbon footprint per unit area of prefabricated structures is $30.52kgCO_2/m^2$ less than that of traditional cast-in-place structures.

Table 4 Calculation of carbon footprint in the exploitation stage of building materials in scheme I

Building material	Company	Actual consumption	Carbon footprint(kgCO$_2$)
Autoclaved concrete block	m^3	1291.27	471,378.11
Concrete C20	m^3	165.77	43,514.23
Concrete C30	m^3	3355.60	935,541.75
Concrete C35	m^3	899.87	261,413.22
Concrete C40	m^3	116.63	35,151.11
Concrete C45	m^3	966.27	304,762.69
Deformed bar	t	796.10	1,414,669.70
Medium and small section steel	t	25.67	36,451.40
Cement mortar	m^3	1507.58	593,460.44
Dinas	m^3	3708.85	20,732.45
Gravel	m^3	931.60	2906.61
Welding rod	kg	5575.93	114,306.53
Iron piece	t	29.08	66,874.06
Coating	t	3.69	13,284.00
Waterproof roll	m^2	2842.11	36,805.32
Water	m^3	10,491.26	2727.73
Gasoline	kg	2660.56	9311.95
Diesel oil	kg	40.33	148.01
Quick lime	t	216.94	438.85
Aluminum alloy square tube	m	4054.36	14,352.41379
Summary			4,378,230.60

Table 5 Carbon Footprint Summary of Two Schemes kgCO$_2$

Stage	Carbon footprint of traditional cast-in-place buildings	Carbon footprint of prefabricated buildings
Mining (production) of building materials	4,378,230.60	4,015,708.00
Prefabricated component production	—	5507.83
Material transportation	55,634.01	45,805.38
Site construction	582,304.24	514,701.62

By comparing the total carbon footprint of different building structures in different stages of the same building, it is found that the largest contribution to the carbon footprint is in the building materials mining (production) stage, and the carbon footprint concentration in each stage is as shown in Fig. 3.

When analyzing the carbon footprint in the production stage of building materials, the main materials are classified into six categories: masonry, concrete, steel,

Fig. 3 Proportion of carbon footprint in each stage

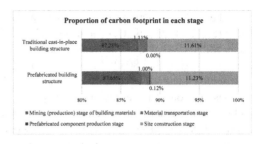

Fig. 4 Carbon footprint concentration of six main materials

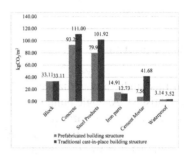

cement mortar, iron parts and waterproof materials. The carbon footprint of some materials accounts for a small proportion of the total carbon footprint. This study will not conduct a detailed analysis temporarily. The carbon footprints of six types of materials are calculated respectively, as shown in Fig. 4. Through analysis, it can be concluded that:

(1) From the analysis of the figure and the material consumption table, it can be seen that the amount of steel is not the largest, but its carbon footprint concentration is at a higher level. Therefore, the carbon footprint concentration of materials with large amount of steel is not necessarily high. On the contrary, the carbon footprint concentration of materials with small amount of steel is not necessarily low. Therefore, the carbon footprint of buildings cannot be reduced by reducing the amount of steel.

(2) As a building material with a large carbon footprint, reducing the carbon footprint of steel can effectively control the carbon footprint of the project. According to the recyclability of steel, improve its production process and recovery rate, reduce its carbon footprint, and achieve emission reduction.

In the material transportation stage, due to prefabricated components in the prefabricated building structure, the amount of steel bars, concrete and other materials that need to be transported to the site is greatly reduced. In this case, the prefabricated component factory is closer to the construction site, so the carbon footprint concentration generated during the transportation process is reduced.

At the site construction stage, the total carbon footprint of prefabricated building structure is $514701.62kgCO_2$, and that of traditional cast-in-place building structure is $582304.24kgCO_2$. The total carbon footprint of prefabricated building structure is reduced by $104,803.85kgCO_2$, because the prefabricated building adopts the construction methods of factory production and on-site assembly, which reduces the consumption of materials and energy. The concrete pumping, vibrating, rebar binding, welding and other operations are significantly reduced compared with the traditional cast-in-place structure, thus effectively reducing the carbon footprint concentration in the on-site construction stage.

5 Conclusion

In this study, a quantitative model of carbon footprint in the building stage was established. A single building of a prefabricated residential project was selected as a case, and a comparative analysis of carbon footprint of different structures was conducted based on the same building, namely prefabricated building structure and traditional cast-in-place building structure. The main differences between the two and some influencing factors were analyzed, and the following conclusions were obtained:

(1) Through case analysis, it is found that the carbon footprint is the highest in the building materials mining (production) stage, followed by the on-site construction stage, and the proportion of prefabricated components in the production stage is the lowest, which can be ignored compared with the first three items. Therefore, there is a great potential for carbon emission reduction during the production and on-site construction of building materials. Through carbon footprint calculation, the carbon footprint of prefabricated buildings is calculated in advance, the design scheme is deepened, the materials are selected reasonably, and the carbon footprint of prefabricated buildings is reduced.

(2) In this case, the three building materials with the largest carbon footprint in the building materials mining (production) stage are concrete, steel and block. For materials, the carbon footprint concentration cannot be reduced by reducing the consumption. Therefore, improving the recycling rate of building materials is an effective way to reduce carbon emissions in the stage of prefabricated buildings.

(3) Compared with the carbon footprint of two different structures in the same building, the carbon footprint per unit building area of prefabricated structure is reduced by $30.52kgCO_2/m^2$ compared with that of traditional cast-in-place structure. It shows that prefabricated buildings have obvious carbon emission reduction advantages.

In conclusion, empirical research has proved that the application of prefabricated construction technology can reduce building carbon emissions to a certain extent. Therefore, in order to further reduce emissions, it is very important to expand the scale of prefabricated buildings. In addition, since the largest source of carbon footprint is the construction mining (production) stage, further promoting low-carbon materials and improving the recovery rate of building materials in the process of building industrialization is also an effective way to reduce carbon emissions.

Acknowledgements This work was supported by the Science and Technology Project of China Construction Fourth Engineering Bureau Co., Ltd. (107-441222063).

References

1. Zhao WX (2022) Research on carbon emission accounting model and low-carbon path in prefabricated building stage (Anhui Jianzhu University) chapter 1, 1–2:(2022)
2. Cao X, Miao CQ, Pan HT (2021) Comparative analysis and research on carbon emission of prefabricated concrete and cast-in-place buildings based on carbon emission model. Building Struct 51:1233–1237
3. Wang GM, Liu MX (2017) Empirical analysis and research on comprehensive benefits of prefabricated concrete buildings. Building Struct 47:32–38
4. Liu MX, Wu Z, Wang JN, Liu HE, Wang GM, Peng X (2015) Evaluation on energy saving benefits and carbon emission of residential industrialized assembled construction. Architecture 45:71–75
5. Weidema BP, Thrane M, Christensen P, Schmidt J, Løkke S (2008) Carbon footprint: a catalyst for life cycle assessment? J Ind Ecol 12:3–6
6. Gao X, Zhu JJ, Chen M, Shen S (2019) Research on carbon footprint calculation model of fabricated concrete building phase. Building Energy Conser 47:97–101
7. Li J, Bao YP (2016) Research on the fast calculation model of carbon footprint in building physicochemical. Phase Building Econ 37:87–91 (2016)

Numerical Simulation on Smoke Control for Extra-Long Tunnel Fires

Wenbo Liu, Junmei Li, and Yanfeng Li

Abstract Based on an actual project in Beijing, this article investigates the effect of smoke control strategies on smoke extraction efficiency under different fire source locations of the point smoke extraction system in extra-long tunnels using Airpak software. The results show that when a fire occurs in a tunnel, the smoke extraction efficiency of the tunnel smoke extraction system varies greatly depending on the location of the fire source and the adoption of different smoke extraction strategies. Due to the suction of the smoke exhaust shaft fan, the relative distance between the electric smoke exhaust valve and the entrance of the tunnel is close, which will cause the smoke exhaust valve within a certain range to be plug-holing, seriously affecting the smoke exhaust effect of the smoke exhaust system. Smoke exhaust valve beyond this range, although not occurring plug-holing the smoke exhaust efficiency is also relatively low, by changing the opening strategy of the smoke exhaust valve can effectively improve the smoke exhaust valve plug-holing, so as to improve the smoke exhaust efficiency.

Keywords Smoke Control · Extra-long · Fire · Simulation

1 Introduction

With the continuous improvement of China's transportation capacity and tunnel construction level, a large number of highway tunnels have been built nationwide, a significant portion of which are extra-long tunnels. According to the relevant specifications, the length of the tunnel is greater than 3,000 m that belongs to the extra-long tunnel. The structure of extra-long highway tunnels is relatively complex, long in depth, confined and narrow in space, once a fire occurs, the high temperature and smoke generated by the fire is difficult to discharge in time, which will not only affect the structural safety of the tunnel itself, but also pose a great threat to the escape and

W. Liu · J. Li (✉) · Y. Li
Beijing Key Laboratory of Green Built Environment and Energy Efficient Technology, Beijing University of Technology, Beijing, China
e-mail: lijunmei@bjut.edu.cn

© The Author(s) 2023
G. Feng (ed.), *Proceedings of the 9th International Conference on Civil Engineering*,
Lecture Notes in Civil Engineering 327,
https://doi.org/10.1007/978-981-99-2532-2_10

rescue of tunnel personnel. Therefore, it is crucial to vent the hot smoke from the tunnel in time when a fire occurs [1].

Scholars at home and abroad attach great importance to the safety of tunnel fires and have conducted a lot of research, including theoretical analysis, numerical simulation and experimental studies, which provide theoretical support for tunnel fire smoke control as well as safety protection.

Hu et al. [2] conducted an experimental study on the smoke temperature and stratification height distribution characteristics of highway tunnel fires and found that the smoke temperature decays exponentially with power below the tunnel vault and the smoke layer settles as the distance of smoke spread increases, posing a threat to personnel evacuation.

Du et al. [3] found in a study of the temperature distribution of the tunnel strong plume driven roof jet that the maximum temperature in the fire region without longitudinal ventilation in the tunnel is not much related to the heat release rate of the fire source, which is around 820 °C. However, the range of the high temperature region increases with the increase of the heat release rate and the decrease of the relative distance between the fire source and the roof.

Xu [4] analyzed and studied the smoke decay rate of fire in the exhaust vent of long tunnels and found that the smoke temperature below the top plate of the near-wall fire source is the highest and the value is higher than the same fire source located in the center position. J. Ji and R. Huo et al. [5] found that the smoke extraction efficiency of mechanical smoke extraction is influenced by the relative position of the fire source and the smoke vent, and the smoke flow condition in long tunnels is different from that of ordinary buildings, and the smoke vent should not be set in the thought spreading stage, and the distance from the fire source should not be greater than 1.33 times the width of the tunnel. In one end of the open channel for mechanical smoke evacuation, not the more smoke venting is started, the more evenly distributed the better the smoke evacuation effect, it is appropriate to start the smoke vent on the opposite side of the filler vent alone or simultaneously start the smoke vent on both sides of the fire source, while in the two ends of the open channel for mechanical smoke evacuation, it is appropriate to start the smoke vent on both sides of the fire source.

In the study of centralized smoke exhaust in extra-long highway tunnels, Zhang. [6] conducted a numerical simulation of the effect of the size of the smoke vent and the spacing of the smoke vent on the smoke exhaust effect of a centralized smoke exhaust system in a highway tunnel in Zhejiang, and found that the smoke exhaust effect of point smoke extraction is better, and the smoke can be controlled within an effective distance, the larger the area of the smoke vent the larger the effective smoking area, and the better the smoke exhaust effect; the spacing of the smoke vent has a relatively small effect on the smoke exhaust effect, and both 25 m and 50 m can meet the smoke exhaust requirements.

Although a lot of research has been conducted on focused smoke extraction, there are still many issues that require continuous in-depth research to ensure that smoke extraction systems can operate safely and effectively in actual projects, and that smoke is controlled in a safe range during fires, while supplementing and theoretically supporting existing codes.

2   Method

2.1   Physical Model

The physical model of the tunnel is shown in Fig. 1. and Fig. 2. The tunnel section is 9096 m long, 13 m wide, and 8.5 m high, with a smoke venting mezzanine height of 1.5 m. There are three smoke shafts at the top of the tunnel, with an effective smoke exhaust size of 5×6 m. The distance between shaft 1 and shaft 2 is 2567.159 m, and the distance between shafts 2 and 3 is 4769.134 m, which are asymmetrically distributed. The exhaust volume of the fan used in shaft 1 is 200 m^3/s and the wind pressure is 2700 pa, the exhaust volume of the fans in shaft 2 and shaft 3 is 180 m^3/s and the wind pressure is 2417 pa. The electric smoke exhaust valve on the smoke exhaust roof is 3 m long and 2 m wide, and the spacing between smoke exhaust valves is 60 m.

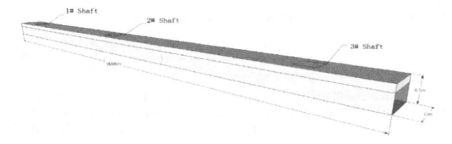

Fig. 1 Geometry of the tunnel model

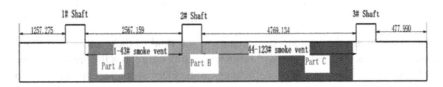

Fig. 2 Overview map of the actual tunnel project

2.2 Fire Scenarios

According to the survey of the type of vehicles passing through the tunnel, and with reference to domestic and international norms on the fire size of different types of vehicles inside the tunnel, and considering that the tunnel is closed to some large tankers, the fire power of the study is set at 50 MW.

2.3 Governing Equations

The flow of fluids is controlled by the laws of physical conservation, the basic control equations include, Continuity equation, Momentum equation, Energy equation and k-ε equation.[7] According to the conservation of mass in the micro-element per unit time, The continuity equation:

$$\frac{\partial u}{\partial x} + \frac{\partial v}{\partial y} = 0 \tag{1}$$

The rate of change of the momentum of a fluid in a micro-element with respect to time is equal to the sum of the various external forces acting on the micro-element. According to this law, the conservation of momentum equation is

$$\frac{\partial(\rho u)}{\partial t} + \frac{\partial\left(\rho u^2\right)}{\partial x} + \frac{\partial(\rho uv)}{\partial y} = -\frac{\partial p}{\partial x} + \mu\left(\frac{\partial^2 u}{\partial x^2} + \frac{\partial^2 u}{\partial y^2}\right) \tag{2}$$

The rate of increase of energy in the microelement is equal to the net heat flux into the microelement plus the work done on the microelement by the body force and the surface force. This law is actually the first law of thermodynamics. According to this law, the energy equation is derived as follows:

$$\frac{\partial(\rho T)}{\partial t} + \frac{\partial(\rho uT)}{\partial x} + \frac{\partial(\rho vT)}{\partial y} = \left(\frac{\lambda}{C_P}\right) + \left(\frac{\partial^2 T}{\partial x^2} + \frac{\partial^2 T}{\partial y^2}\right) \tag{3}$$

where u and v are the velocity components of the fluid in the x and y directions, respectively, ρ is the fluid density, T is the fluid temperature, p is the pressure of the fluid, μ is the dynamic viscosity coefficient, λ is the thermal conductivity of the fluid and C_p is the constant pressure specific heat capacity.

k Turbulent kinetic energy transport equation:

$$\frac{\partial(\rho k u_x)}{\partial x} = \frac{\partial}{\partial x}\left[\left(\mu + \frac{u_t}{\sigma_k}\right)\frac{\partial k}{\partial x}\right] + G_k + G_b + \rho\varepsilon, \ G_k = -\rho u_x' u_y' \frac{\partial u_y}{\partial u_x} \tag{4}$$

ε Turbulent kinetic energy dissipation rate transport equation:

$$\frac{\partial(\rho\varepsilon u_x)}{\partial x} = \frac{\partial}{\partial x}\left[\left(\mu + \frac{u_t}{\sigma_\varepsilon}\right)\frac{\partial\varepsilon}{\partial x}\right] + C_{1\varepsilon}\frac{\varepsilon}{k}(G_k + G_{3\varepsilon}G_b) - C_{2\varepsilon}\rho\frac{\varepsilon^2}{k}, G_b = \beta_g\frac{\mu_t}{\Pr_t}\frac{\partial T}{\partial x}$$

$$(5)$$

where, k is the turbulent kinetic energy, μ_t is the turbulent viscosity, σ_k, σ_ε are the turbulent Prandtl numbers of k and ε, respectively. G_k is the turbulent kinetic energy generated by the mean velocity gradient. G_b is the turbulent kinetic energy generated by buoyancy. ε is the turbulent kinetic energy dissipation rate. C_μ, $C_{1\varepsilon}$, $C_{2\varepsilon}$, $C_{3\varepsilon}$ are constants. β is the expansion coefficient.

2.4 Boundary Condition Setting

The wall is adiabatic condition, the entrance at both ends of the tunnel is set as a free boundary, and the smoke exhaust shaft is set with both exhaust air volume and wind pressure conditions, and the outdoor temperature is 20°C.

2.5 Grid Independence Test

When the software performs simulation calculations, the density of the mesh and the quality of the mesh will directly affect the calculation results. After several simulations of models with different grid sizes, the maximum temperature of the top plate at different locations from the fire source was counted as a reference value, and the results are shown in Fig. 3. The grid size is $1 \times 1 \times 1$ m and $0.7 \times 0.7 \times 0.7$ m calculation results basically match. However, because the tunnel model is too large, the grid size is set too small, which will lead to too many total grids increasing the unnecessary calculation time, so the model grid size is set to $1 \times 1 \times 1$ m and the number of grids is 2800000.

2.6 Working Conditions Setting

In the event of a fire, this study investigates the effect of different smoke evacuation strategies on the smoke evacuation efficiency of the tunnel by adjusting the location of the fire source in the tunnel, changing the opening status of fans in different shafts or changing the opening strategy of electric smoke exhaust valves. region A and region C have similarity, and only region A will be studied in the article, 8 working conditions are set in the simulation, as shown in Table 1. Conditions 1–4 simulate the fire source is located in region A, using shaft 1 with different smoke exhaust valve opening strategy; Conditions 5–8 simulate the fire source is located in region

Fig. 3 Temperature distribution of top plate with different grid sizes

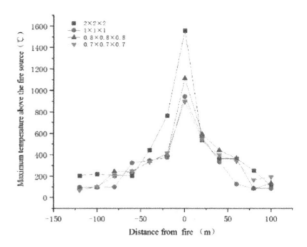

Table 1 Simulated working conditions setting

NO	Fire Location	Code of open smoke exhaust valve	Distance of fire source from left tunnel entrance	Smoke vent opening strategy
1	Region A	1–6	1451 m	3–3
2		2–7	1451 m	2–4
3		7–12	1811 m	3–3
4		8–13	1811 m	2–4
5	Region B	38–43	3647 m	3–3
6		37–42	3647 m	4–2
7		44–49	4021 m	3–3
8		45–50	4021 m	2–4

B using shaft 2 smoke exhaust. The 3-3, 2–4, and 4-2 in the working condition are the number of electric smoke exhaust valves opened on both sides of the fire source.

3   Results and Discussion

3.1   Fire Source Located in Region A

Smoke Exhaust Valve Opening Strategy for 3-3. In the event of a fire in this area, three smoke exhaust valves are opened upstream and downstream of the fire source for smoke exhaust, using smoke exhaust valves No. 1–6 for smoke exhaust, No. 1 smoke exhaust valve will plug-holing obviously, as the fire source position right, the use of 3–8, 7–12, 8–13 groups of smoke exhaust valve for smoke exhaust,

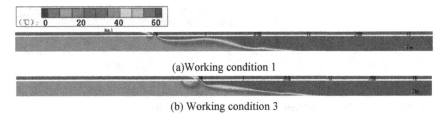

(a)Working condition 1

(b) Working condition 3

Fig. 4 Smoke spread under different fire source locations in region A

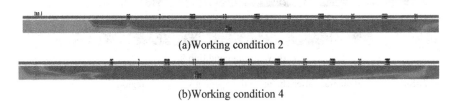

(a)Working condition 2

(b)Working condition 4

Fig. 5 Smoke spread diagram after changing the opening strategy in region A

the leftmost end of the smoke exhaust valve closer to the No. 1 shaft will also occur suction through the phenomenon, until the beginning of the 9th smoke exhaust valve, smoke exhaust valve suction through the phenomenon began to disappear. However, their efficiency will be relatively low, mainly due to the fact that this section is closer to the exit on the left side of the tunnel, and the axial fan in Shaft 1 has a relatively large smoke discharge, the simulation results are shown in Fig. 4.

Smoke Exhaust Valve Opening Strategy for 2–4. Compare the fire source upstream and downstream each open three smoke exhaust valves, the third smoke exhaust valve on the left side of the fire source will be closed, while in the right side of the fire source and then open a smoke exhaust valve, at this time the spread of smoke in the tunnel as shown in Fig. 5.

According to the simulation results, with the smoke exhaust strategy of 2–4, the overall distance of smoke spread is significantly reduced compared with Case 1 and Case 3 under the opening strategy of 3–3, while the smoke exhaust valve's suction penetration phenomenon is also improved and the smoke exhaust efficiency is increased.

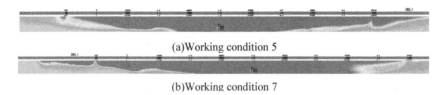

(a)Working condition 5

(b)Working condition 7

Fig. 6 Smoke spread under different fire source locations in region B

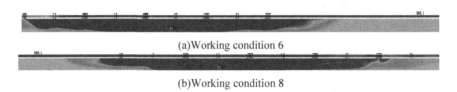

(a)Working condition 6

(b)Working condition 8

Fig. 7 Smoke spread diagram after changing the opening strategy in region B

3.2 Fire Source Located in Region B

The fire source is in area B, using the fire source on each side of the opening of three smoke exhaust valves, shaft 2 on both sides of the nearest smoke exhaust valve will not occur suction through the phenomenon, but in a critical suction through the state, at this time the smoke exhaust efficiency of the smoke exhaust valve will be relatively low. The simulation results are shown in Fig. 6.

Smoke Exhaust Valve Opening Strategy for 3-3.

Smoke Exhaust Valve Opening Strategy for 2–4 (4-2). The smoke exhaust valve close to the shaft, although no suction through the phenomenon, but close to the critical state, the smoke exhaust efficiency is very low, the smoke spreads a long distance, is not conducive to the evacuation of people in the tunnel, when the same smoke exhaust strategy as above, that is, close the smoke exhaust valve close to the shaft, while opening one on the other side of the fire source, so that the smoke exhaust efficiency of the smoke exhaust valve will improve, while the smoke spreads a distance will be controlled in a limited This way, the smoke exhaust valve will be more efficient and the smoke spread will be controlled within a limited distance. The results of the smoke spread simulation are shown in Fig. 7.

4 Conclusion

In this study, the efficiency of smoke exhaust valves under different smoke exhaust strategies during focused smoke exhaust in tunnels was investigated by means of Airpak numerical simulations, with the following conclusions.

(1) When the fire occurs in area A or C, within a distance of the fire source close to the shaft, the smoke exhaust valve is relatively close to the smoke exhaust shaft, which will cause some of the smoke exhaust valves to be absorbed through the phenomenon or in a critical state of absorption through, Severe reduction in smoke extraction efficiency and excessive distance of smoke spread in the tunnel, it is not conducive to the evacuation of people in the tunnel.

(2) Due to the long depth of the extra-long tunnel, when in area B, the tunnel longitudinal wind speed is reduced, at this moment the smoke exhaust valve closer to the shaft suction through the phenomenon is not obvious, but the smoke exhaust efficiency will still be affected, the same change in smoke exhaust strategy to improve the smoke exhaust efficiency.

(3) In the actual project, the design of the key smoke exhaust system should fully consider the impact of the location of different fire sources and the relative distance between the smoke exhaust valve and the shaft on the smoke exhaust efficiency, and can take different smoke exhaust strategies according to different flame zones, or take intelligent control of the smoke exhaust system to flexibly adjust the smoke exhaust strategy.

References

1. Xu FQ, Du ZG, Chen C (2022) Distribution and development characteristics of urban road tunnels in China. Mod Tunn Technol 59(06):35–41
2. Hu LH, Huo R, Wang HB Yang RX Experiment of fire smoke temperature and layer stratifi cation height distributi on characteristic along highway tunnel. China J Highway Transp 2006(06):79–82
3. Du C (2018) Experimental study on temperature distribution and flame extension length driven by strong plume in longitudinal ventilated tunnel. MA thesis
4. Xu L, Chang J Wang Z (2015) Analysis of decrease rate of fire smoke near extraction vents of long tunnel. J Shandong Jianzhu Univ 30(01):19–24
5. Ji J, Huo R, Hu LH Wang HB (2009) Influence of relative location of smoke exhaust opening to fire source on mechanical smoke exhaust efficiency in a long channel. Eng Mech 26(05):234–238
6. Zhang YC, He C, Zeng YY Wu DX (2009) Characteristics and counter measures of traffic accidents in expressway tunnel. J Southwest Jiaotong Univ. 44(05):776–781
7. Ai S (2014) Comparative study on heating system of high space industrial building. MA thesis

Surface Vibration of Throw-Type Blast in an Open-Pit Mine

Yong Wang, Chen Xu, Changchun Li, Xiaofei Yao, Xingbo Xiang, and Haoxuan Huang

Abstract In the process of Open-pit mining in China, throwing blasting is an important method, which is very likely to cause serious damage to the slope of the discharge field under the action of vibration load of throwing blasting. With the background of throwing blasting process in Heidaigou open-pit mine in Ordos, vibration velocity data were collected from the discharge field near the throwing blast, and the vibration signal of throwing blast was analyzed by means of Fourier transform to obtain the characteristics of throwing blast vibration velocity wave and the attenuation law and prediction formula in the process of propagation. The results show that: 100 ~ 300 m away from the blasting area, the radial direction (X direction) of the blasting area produces the largest vibration velocity of 26.8 cm/s, but at the same time, the decay rate of the peak vibration velocity of the survey line 2 in each direction is small, and the decay percentages of 56, 75 and 70% are smaller than that of the survey line 1 and survey line 3 in the lateral direction of the blasting area, and the decay rate of the velocity is smaller as the propagation As the distance increases, the decay rate of the velocity decreases. The curve gradually tends to flatten and the vibration velocity of the three directions gradually close. The frequency band of the blast vibration is distributed within 200 Hz and the frequency and energy are mainly distributed in the low frequency stage (0–20 Hz), accounting for more than 50% of the total energy.

Keywords Open-Pit Mine · Throwing Blasting · Vibration Signal · Fourier Transformation

Y. Wang (✉) · H. Huang
Shenyang Design and Research Institute of Sino-Coal International Engineering Group, Shenyang, China
e-mail: 25986814@qq.com

C. Xu
Heidaigou Open Pit Mine of Zhunneng Company, Inner Mongolia, China

C. Li · X. Yao · X. Xiang
College of Civil and Transportation Engineering, Hohai University, Nanjing, China

© The Author(s) 2023
G. Feng (ed.), *Proceedings of the 9th International Conference on Civil Engineering*,
Lecture Notes in Civil Engineering 327,
https://doi.org/10.1007/978-981-99-2532-2_11

1 Introduction

As one of the most primary and important energy sources in China, coal plays a vital role in China's economic development and energy security. Although the proportion of coal resources in the energy consumption structure has been declining due to the country's promotion of green energy and the requirements of reducing carbon emissions, coal resources still account for 57.7% of the energy consumption structure according to the 2019 China Mineral Resources Report [1]. In the future, coal resources will still be one of the most important and largest energy sources in China.

China's mines have developed rapidly and their scale and efficiency have been continuously improved due to the importance of mineral resources and the high demand for social and economic development. According to statistics, about two-thirds of the world's solid mineral resources are mined by open-pit mining, which plays an important role in the mining industry. There are about 1200 open-pit mines in China, and the throwing blasting is widely used as an efficient mining method in open-pit mines. The vibration generated by throwing blasting has become a problem that must be concerned and studied. It should be noted that the throwing blasting in the mine will have a negative impact on the stability of the slope in the mine, so it is of great significance to study the characteristics and propagation laws of the vibration caused by the blasting load.

The paper is based on the mining project of Heidai George Open-pit Mine in Ordos. The on-site monitoring test of blasting vibration and numerical simulation are carried out in order to analysis the dynamic response of rock and soil mass in Heidai George Open-pit Mine under blasting vibration. The characteristics and the attenuation law of throwing blasting seismic wave are obtained in the process of propagation.

2 Engineering Condition

Heidai George Open-pit Coal Mine of Zhungeer Energy limited ability company is located in the Zhungeer Banner of Nei Monggol Autonomous Region, where is located on the west bank of the Yellow River (111°11'–111°25' E, 39°25'–39°59'N).

Heidai George Open-pit Coal Mine is located in the northeast of Ordos Plateau, facing the Yellow River in the east, and the surface is covered with thick Quaternary clay. As the climate in this area is dry and rainless, vegetation is scarce, rainfall is concentrated, and the loss of water and soil is serious. Dendritic gullies and river valleys dominated by the Yellow River are developed on the surface, so that platform gullies are cut into vertical and horizontal gullies and the terrain is fragmented and extremely complex, forming a source hill landform with gentle ridge valley and Gaoliang valley terrain. Heidai George Open-pit Coal Mine in Ordos is a large-scale Open-pit coal mine designed and constructed by China, which belongs to SHENHUA GROUP ZHUNGEER ENERGY CO., LTD. The mine was commenced in 1990 and put into trial production in 1998, and then officially handed over for production in 1999 and reached production capacity the next year. In June 2006, the annual output of raw coal of Heidai George Open-pit Coal Mine after capacity expansion and transformation reached 25 million tons, becoming the largest Open-pit coal mine in China. In 2011, its annual output exceeded 31 million tons which rank first in terms of capacity and scale in Asia.

3 Materials and Methods

3.1 Experiment Instrument

TC-4850 blasting vibration meter is used for on-site blasting vibration monitoring test. The instrument can record the blasting vibration velocity, acceleration and other data, and is suitable for data acquisition of the scene of throwing blasting vibration.

The technical indicators and advantages of TC-4850 blasting vibration meter are listed as follows: the acquisition channels are X, Y and Z channels for parallel acquisition, and the sensors are three-dimensional, which is convenient for burial. The sampling frequency is 1–50 kHz, which can meet the requirements of this vibration measurement. The error is less than 0.5% and the reading accuracy can reach 1, which can meet the accurate measurement and recording of the vibration signal.

3.2 The Blasting Site

The blasting area is located at the elevation of 1095–1130 m in the south of Heidai George Open-pit Mine with a length of 400 m, a width of 85 m and a bench height of 43– 49 m whose average height is 46.6 m. The on-site explosive holes with a depth of 45–54 m, an average hole depth of 50 m and a blasting amount of 1,511,580 m^3 and blasting conditions are shown in Fig. 1 and Fig. 2, respectively. The hole spacing of throwing blasting is 12 m and the row spacing is 6–7 m. The total rows of holes and the total number of throwing blasting holes are 12 and 388, respectively. The average unit consumption is 0.753 kg/m^3. The presplitting holes were blasted first with a hole spacing of 3.5 m. There are 114 and 23 presplitting holes in the back row and the south end, respectively. The holes are 137 in total with a unit consumption of 1.3 kg/m^2. The total number of holes is 525 with a charge of 1138 tons, including 551 tons of heavy ammonium oil explosives and 587 tons of ammonium nitrate oil explosives.

Fig. 1 The throwing
blasting holes

Fig. 2 The blasting site

3.3 The Arrangement of Measured Points

The whole history of blasting vibration is recorded by the site survey in order to study the attenuation law of blasting seismic wave and achieve the detection purpose. In addition, the test scheme of blasting seismic is determined according to the layout principle of vibration measured points, the characteristics of throwing blasting and the actual geological and topographic conditions of the blasting area. The relative position of measured points is shown in Fig. 3.

(1) The survey Line 1 is arranged in order to monitor the blasting seismic intensity and developed law on the left side of the blasting area. The measured points are arranged on the same steps of the blasting area, which are located in the south of the blasting area. The measured points of Line 1 are numbered 1 #, 2 #, 3 # and 4 # respectively.

(2) The survey Line 2 is arranged in order to monitor the blasting seismic intensity and its developed law behind the blasting area. The measured points are arranged on the same steps in the blasting area. The measured points of survey Line 2 are numbered 5 #, 6 #, 7 # and 8 # respectively.

(3) The survey Line 3 is arranged in order to monitor the propagation law of blasting seismic wave on the right side of the blasting area. The number of measured

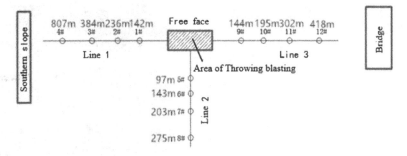

Fig. 3 The schematic diagram of relative position of measuring points

points is 9 #, 10 #, 11 # and 12 # of survey Line 3 respectively. The measured points are arranged on the same steps.

Each measured point can simultaneously monitor the vibration velocity in X, Y and Z directions. The horizontal radial vibration velocity of vibration propagation is recorded in X direction, the tangential vibration velocity in Y direction is perpendicular to the vibration propagation direction, and the vertical vibration velocity in Z direction is perpendicular to the platform ground.

The embedding method of the sensor is crucial in order to accurately record the whole process vibration of the throwing blasting and the main characteristics of the vibration waveform. The vibration speed sensor needs to form an integral part with the platform soil. Therefore, it is necessary to dig a trench on the test platform and bury the sensor according to the layout plan of the measured points.

On the site, a trench with a width of about 40 cm and a depth of about 30 cm is opened on the designated platform by the slotting machine in the direction of the preset measuring line. The sensors are placed at the corresponding positions according to the design distance, as shown in Fig. 4. The surrounding soil is used to fill the trench and compact it in order to reduce the impact of other interference vibration and the sensor's own movement on the data collection of blasting vibration.

4 Result and Discussion

Vibration amplitude, main vibration frequency and duration are three parameters to describe blasting vibration. Peak vibration velocity refers to the maximum vibration amplitude of medium particle, and main vibration frequency refers to the frequency of wave The amplitude of the medium particle reaches the maximum. In China's Code for Blasting Safety GB 6722–2014 [3], vibration velocity and main vibration frequency of slope particle are taken as the basis for safety discrimination. Li et al. [4] thinks that the influence of frequency should also be fully considered in the safety evaluation system of blasting vibration except for vibration amplitude, Wang et al. [5] introduced the necessity and feasibility of using particle vibration

Fig. 4 The arrangement of
on-site measuring points

velocity and frequency as safety criteria for blasting vibration, and the calculation formula of blasting vibration frequency. Yang et al. [6] Zhuang et al. [8] proposed to incorporate the amplitude, frequency spectrum and duration of blasting earthquake into the blasting earthquake safety criterion on the basis of on-site blasting monitoring, so as to establish a multi parameter safety criterion. Previous studies have shown that particle vibration velocity, that is, the magnitude of vibration amplitude and main vibration frequency, will have an impact on structures. The two should be combined to analyze the slope in blasting vibration. At the same time, obtaining the vibration amplitude and main vibration frequency can help us simulate the vibration wave generated in the blasting process. Therefore, the paper will analyze the peak vibration velocity in the blasting test.

4.1 Monitoring Results of Throwing Blasting Vibration

Figure 5 is the measured vertical and horizontal vibration velocity waveform diagram of each measurement point on each survey line of 5# under the blasting vibration in the same time period. (Note: v is ordinate vibration velocity in cm/s).

Table 1 and Table 2 are summary tables of blasting vibration data of each measured line. The comparison of peak vibration velocities of measured points on each measured line in X, Y and Z directions is shown in Fig. 6. It can be seen from the figure that, in general, the vibration velocities of survey line 2 are greater than those of the other two measuring lines in X, Y and Z directions. The closer to the blasting area, the faster the maximum vibration velocity of throwing blasting decays with the increase of distance. In addition, the variation of vibration speed tends to be gentle with the increase of the propagation distance and the vibration speed of different measuring lines tends to be consistent.

The safe range of blasting vibration can be controlled by exploring the attenuation law of blasting vibration wave in the process of propagation in geotechnical media.

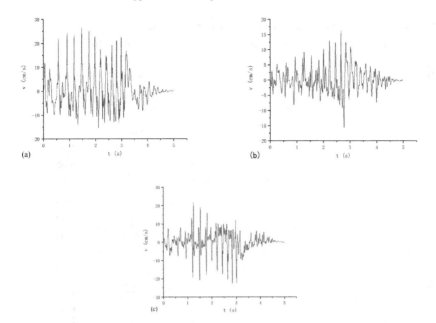

Fig. 5 Waveform Chart of Vibration Velocity of Measuring 5#. **a** Channel X; **b** Channel Y; **c** Channel Y

Table 1 Summary of blasting data of survey line 2

Measuring line	Measuring point	Distance R/m	Direction	Propagation distance
2#	5	97	X	4.68
			Y	
			Z	
	6	143	X	6.89
			Y	
			Z	
	7	203	X	9.78
			Y	
			Z	
	8	275	X	13.27
			Y	
			Z	

Many scholars have proposed some empirical formulas with good correlation by using mathematical methods based on field blasting tests [7, 8]. Sadovsky formula is the most widely used and traditional formula for predicting blasting vibration.

Table 2 Summary of blasting data of survey Line 1 and survey Line 3

Measuring Line	Measuring point	Distance R/m	direction	Propagation distance	V_{max}[a]/cm·s^{-1}
1#	1	142	X	6.84	18.13
			Y		13.44
			Z		14.29
	2	236	X	11.38	8.01
			Y		5.84
			Z		5.90
	3	384	X	18.51	2.65
			Y		4.55
			Z		2.92
	4	807	X	38.90	0.99
			Y		1.55
			Z		1.18
3#	9	144	X	6.94	10.17
			Y		6.96
			Z		9.39
	10	195	X	9.40	6.22
			Y		4.60
			Z		4.63
	11	302	X	14.56	2.47
			Y		3.17
			Z		2.21
	12	418	X	20.15	2.24
			Y		2.87
			Z		1.34

[a]Note: Vmax is the maximum vibration speed

$$V = K \cdot (\frac{\sqrt[3]{Q}}{R})^{\alpha} \qquad (1)$$

where V is the vibration speed of particle in cm/s; Q is the maximum initiation quantity of a single section in turn in kg, R is the horizontal distance from the test point to the explosion source in m; K is the coefficient related to factors such as geology and blasting methods and α is the attenuation coefficient of vibration wave related to geological conditions.

Figure 7 shows the relationship between the vibration velocity of each survey line in the X direction and the proportional distance ($\frac{\sqrt[3]{Q}}{R}$). Points of Figures are fitted by Sadowski's formula and then K and α are obtained. The attenuation prediction formula of the peak vibration velocity in each direction is obtained by taking values

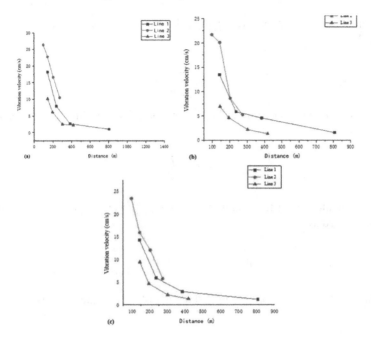

Fig. 6 Comparison of vibration velocity of each survey line. **a** X direction; **b** Y direction; **c** Z direction

of K and α so that the peak vibration velocity in any distance around the blasting perimeter can be predicted. The two survey lines are analyzed together because survey lines 1 and 3 have symmetrical similarity. It can be seen from the figure that the formula fits well with the point. Therefore, the formula can be used to predict the attenuation of vibration velocity at different positions.

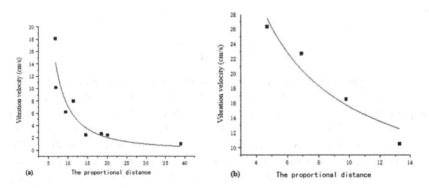

Fig. 7 Fitting regression of vibration velocity in X direction. **a** survey line 1 and line 3. **b** survey line 2

The vibration attenuation formula in X direction of survey line 1 and 3 can be explained as:

$$v = 467.9 \cdot (\frac{\sqrt[3]{Q}}{R})^{1.816} \tag{2}$$

The vibration attenuation formula in Y direction of survey line 1 and 3 can be explained as:

$$v = 99.6 \cdot (\frac{\sqrt[3]{Q}}{R})^{1.204} \tag{3}$$

The vibration attenuation formula in Z direction of survey line 1 and survey line 3 can be explained as:

$$v = 447.6 \cdot (\frac{\sqrt[3]{Q}}{R})^{1.891} \tag{4}$$

The vibration attenuation formula in X direction of survey line 2 can be explained as:

$$v = 87.8 \cdot (\frac{\sqrt[3]{Q}}{R})^{0.752} \tag{5}$$

The vibration attenuation formula in Y direction of survey line 2 can be explained as:

$$v = 156.2 \cdot (\frac{\sqrt[3]{Q}}{R})^{1.222} \tag{6}$$

The vibration attenuation formula in Z direction of survey line 2 can be explained as:

$$v = 130.4 \cdot (\frac{\sqrt[3]{Q}}{R})^{1.103} \tag{7}$$

It can be analyzed that the propagation attenuation law of the vibration velocity in all directions of the throwing blasting based on the prediction formula and the speed comparison curve. The maximum vibration velocity in the X direction on the survey line 2 is about 26.8 cm/s within the range of 100–300 m from the throwing blasting area The vibration velocity decreases to about 11.7 cm/s within 200 m and the attenuation reaches about 56%. However, the maximum vibration velocity of survey line 1 and line 3 is about 26.8 cm/s and attenuates to about 3.7 cm/s respectively at the same distance, which attenuates about 86%. The maximum vibration velocity of survey line 2 in Y direction is about 22.8 cm/s and attenuates to 6 cm/s within 200 m, which attenuates to about 75%; while the maximum vibration velocity of survey line 1 and line 3 is about 15 cm/s and attenuates to about 3 cm/s respectively at the same distance, which attenuates about 80%. The maximum vibration speed on survey line 2 in the Z direction is about 23 cm/s and attenuates to 6.8 cm/s within 200 m, which attenuates about 70%, while the maximum vibration speed of line 1 and survey line 3 is about 23 cm/s and attenuates to 2.9 cm/s within the same distance, which attenuates about 87%. In conclusion, the vibration velocity in X direction is greater than that of the other two directions during the propagation of throwing blasting vibration, which indicates that the vibration generated by throwing blasting has a greater impact on X direction. At the same time, Vibration generated at the front of blasting area is close to the vibration generated laterally to the throwing blasting area when the distance to the throwing blasting area is relatively close. However, with the increase of the distance, the vibration attenuation rate generated laterally of the throwing blasting area is significantly higher than that generated directly to the blasting area. This is because there exist free faces between the area of survey line 2 and throwing blasting area. Survey line 1 and line 3 are first affected by splitting blasting vibration, and then affected by throwing vibration of block stone. Survey line 2 is affected by splitting blasting and throwing vibration at the same time, which lasts for a long time. Therefore, it can be judged that the vibration generated by blasting in the direction opposite to the blasting area is maximum and the attenuation rate is small.

4.2   Fourier Transformation

Fourier transformation is one of the widely used methods in the analysis of blasting vibration signals. It can transform the time domain of blasting vibration signals into the frequency domain for analysis. Song et al. [9] analyzed the blasting experiment of large caverns with Fourier spectrum. The result indicates that there is a significant difference between the Fourier spectrum of the waveform in the source area and that in the far area of the explosion in the frequency domain.

The collected velocity signal of blasting vibration is subject to fast Fourier transformation in combination with the blasting monitoring data in the mining area. Figure 8 is the Fourier transformations of 1#.

It can be seen intuitively from the FFT spectrum of the vibration velocity that the frequency band of the vibration generated by the throwing blasting is wide, the frequency is mainly distributed within 200 Hz, and the energy is mainly concentrated in the low frequency band. Compared with the general natural seismic wave, the vibration frequency of the natural seismic wave is lower, generally within 10 Hz, while the vibration frequency of the throwing blasting is significantly higher than the natural earthquake. Furthermore, the energy distribution in different frequencies in three directions of each measuring point can be obtained by integrating the spectrum diagram curve. Taking measured point 5 (5#) as an example, Table 3 shows the energy proportion distribution in each frequency range in all directions of measuring point 5. It can be seen that the vibration generated by throwing blasting is mainly

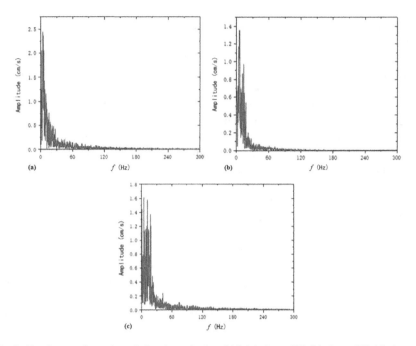

Fig. 8 Fourier transformation of vibration velocity of 1#. (a) channel X; (b) channel Y; (c) channel Z

Table 3 Distribution table of energy proportion of each frequency

Measured point	Channel	0–10 Hz	10–20 Hz	20–30 Hz	30–40 Hz	40–50 Hz	50–60 Hz	≥60 Hz
5#	X	38.9%	12.3%	8.0%	4.4%	3.5%	3.0%	29.9%
	Y	45.6%	25.2%	2.7%	2.9%	2.4%	1.6%	19.6%
	Z	26.9%	26.4%	8.2%	4.0%	2.2%	1.8%	30.5%

concentrated in 0-20 Hz, and its energy accounts for more than 50% of the total energy, and from the total integration in all directions, the energy proportion of channel X is the highest, it conforms to the above analysis of vibration velocity in three directions.

5 Conclusion

(1) Among the three measuring lines, the vibration speed of survey line 2 facing the blasting area in each direction at approximately the same distance is higher than that of the other two measuring lines. At the initial stage, the vibration speed of the measuring point attenuates rapidly. The attenuation rate of the speed decreases with the increase of the propagation distance. The vibration

speed distance curve gradually tends to be flat and the vibration speed in the three directions gradually approaches.

(2) In the process of equal charge blasting, within the range of 100 –300 m from the throwing blasting area, the radial (X direction) vibration velocity produced by the survey line 2 directly opposite the blasting area is the largest, which is 26.8 cm/s. The attenuation rate of peak vibration velocity of the measuring line 2 in each direction is small. The attenuation percentages within 200 m are 56, 75 and 70%, respectively, which is less than those of the measured lines 1 and 3 on the lateral side of the blasting area.

(3) In the frequency spectrum after Fourier transformation, the frequency band of throwing blasting vibration is relatively wide and distributed within 200 Hz, while the frequency and energy are mainly distributed in the low frequency stage (0–20 Hz), accounting for more than 50% of the total energy. It can be judged that throwing blasting is a low frequency vibration. In the later protection work, the protective measures to reduce low frequency vibration shall be taken as the main measure.

Acknowledgements Thanks for CCTEG Science and Technology Innovation and Entrepreneurship Fund Special Key Project "Research on Key Technologies of Comprehensive Management and Integrated Utilization of Fushun West Open-pit Mine" (Project No. 2019-ZD004).

References

1. Ministry of Natural Resources of the People's Republic of China 2020 China Mineral Resources Report. Geological Publishing House, Beijing, 5 January 2020
2. Xitai S, Jianmin R Explosion wave analysis and its spectrum. Explosion Shock 1982(4):33–42
3. National Bureau of Standards Blasting Safety Regulations (GB6722-2014) (2014). China Standards Publishing House, Beijing
4. Xiaolin L, Taisheng M, Xin D, Dongliang Z, Sidu X The role of frequency in blasting damage and its influencing factors analysis. Eng Blasting 2001(3):15–18
5. Xuguang W, Aaron Y Several issues on safety criteria for blasting vibration. Eng Blasting 2001(2):88–92
6. Shengquan Y, Xiankui L, Baochen L Defects and improvement of safety criteria for blasting earthquake. Explosion Shock 2001(3):223–228
7. Shengquan Y, Xiankui L, Baochen L Discussion on several issues of safety criteria for blasting earthquake. Min Metall Eng 2001(3):24–27
8. Jinzhao Z, Renshu Y, Xiaolin L, Tongshe X, Jingfeng H Discussion on safety standards for blasting vibration of structures. Min Metall Eng 2003(2):11–13+17
9. Traditional Z, Honggen L, Jinyu M Selection of formulas for the propagation law of seismic wave parameters along slope surfaces. Blasting 1988(02):30–31

Study on Failure Process of Freeze–Thaw Fractured Rock Under Multistage Cyclic Loads

Liu Peng

Abstract In order to further study the failure characteristics of freeze-thawed rocks in the alpine region under multistage cyclic loads, a numerical simulation analysis was carried out with RFPA2D software, taking the natural fractured granite from the Beizhan Iron Mine in Hejing County, Xinjiang Province as an example. The results show that the degree of natural fracture determines the fracture form of rock, and when the degree of natural fracture is large, the rock will eventually undergo shear slip failure along the natural fracture. When natural fissure rock is subjected to load, its initial structural deterioration occurs at the fissure, and tensile failure occurs. When the natural fracture expands to a certain extent, the rock begins to undergo large-scale compressive shear failure, which eventually leads to shear-slip failure of the fractured rock. The failure mode of fractured rock is affected by the degree of fracture development, the degree of penetration and the inclination Angle.

Keywords Cyclic Load · Fissured Rock Mass · Frozen and Thawed Rock · RFPA2D · Destructive Process

1 Introduction

With the deep implementation of the western development strategy, China has made greater efforts to develop mineral resources in the western high-altitude cold regions. In the process of mine construction, the problem of freezing and thawing disasters of the slopes in the cold regions has become increasingly prominent and has been paid attention to [1]. Among them, freeze–thaw cycle, mining, blasting, etc. are the main factors causing slope disasters in cold regions. It is of great significance to carry out research on the damage and failure process of freeze–thaw fissured rocks under multi-stage cyclic load for promoting the development of slope treatment in high cold regions in China, it is of great significance to prevent slope geological disaster and ensure mine safety production.

L. Peng (✉)
University of Science and Technology Beijing, Beijing 100083, China
e-mail: liupeng@ustb.edu.cn

© The Author(s) 2023   137
G. Feng (ed.), *Proceedings of the 9th International Conference on Civil Engineering*,
Lecture Notes in Civil Engineering 327,
https://doi.org/10.1007/978-981-99-2532-2_12

   Many scholars have carried out number of studies on the physical and mechanical characteristics of rock and the law of crack growth under the action of freeze–thaw cycle, and obtained the results of reference value. Wen Lei et al. [2] studied the influence of freeze–thaw cycles at different temperature intervals on the mass loss rate, saturated water absorption rate, uniaxial compressive strength and other parameters of granite in water-saturated state. Zhang Huimei et al. [3] studied the influence of freezing and thawing cycles on red sandstone under different confining pressures. The results showed that with the freezing and thawing cycles, the mass and density of the sample increased first and then decreased, while the longitudinal wave velocity continued to decrease. Song Yongjun et al. [4] carried out uniaxial cyclic loading and unloading tests of red sandstone with different freeze–thaw cycles. The results showed that, with the increase of freeze–thaw cycles, the peak strength and elastic modulus of rock gradually decreased, while the peak strain and Poisson's ratio gradually increased, and the rock failure showed a trend of transition from brittleness to ductility. Zhou Shengtao et al. [5] studied the fracture morphology of sandstone under uniaxial compression with different times of freeze–thaw cycles. Liu Quansheng et al. [6] pointed out that the dynamic frost heaving force in fractured rock mass was related to the physical parameters, strength parameters, freezing temperature and fracture morphology of the rock mass through the laboratory freeze–thaw test and the establishment of a theoretical model, and the frost heaving force value was considered under the condition of water migration and non-migration. By preforming fractures in red sandstone, Renliang et al. [7] analyzed the evolution law of frost heaving force in penetrating fractures, and verified the damage and deterioration effect of frost heaving force on fractured rock mass. Li Ping et al. [8] prepared sandstone-like samples with similar materials and prefabricated double cracks with different rock bridge angles. Through a series of laboratory tests, they obtained the propagation characteristics of rock cracks caused by frost heaving. Zhao Jianjun et al. [9] studied the effects of fractures of different lengths on freeze–thaw damage deterioration of rocks using rock-like materials, and the results showed that freeze–thaw cycling had significant effects on pore fracture compaction stage and main fracture expansion stage in rock uniaxial compression test.

   At present, scholars focus on the failure of freeze–thaw cyclic rock under monotonic loading, and rarely study the failure mechanism of freeze–thaw fractured rock under multistage cyclic loading. Therefore, based on the engineering background of Beizhan Iron Mine in Hejing County, Xinjiang, using the rock dynamic fracture analysis system RFPA2D, this paper aims to study the characteristics of the activation process and fracture morphology of the structural planes of natural fracture samples after freezing and thawing.

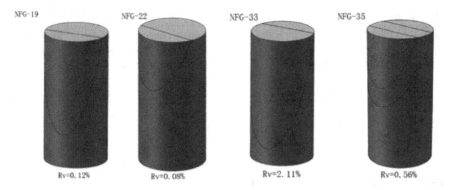

Fig. 1 Model of freeze–thaw granite sample with natural cracks

2 Model and Test Scheme

2.1 Model Establishment

The model test of this paper, natural fissure granites from the Beizhan Iron mine in Hejing County, Xinjiang Province were selected, which are mainly distributed in the west slope of the open-pit mine, and the representative rock mass structural plane inclination is 290° and inclination is 70°. Further select typical crack samples NFG-19, NFG-22, NFG-33 and NFG-35 for numerical analysis, and the sample model size is 50 × 100 mm, the natural fissure volume ratio Rv of each sample is 0.12, 0.08, 2.11 and 0.56% respectively, as shown in Fig. 1.

The corresponding analysis model is built through the rock dynamic fracture analysis system RFPA2D. The model is divided into two parts: sample and cushion block. The parameters of the two materials are shown in Table 1. Mohr Coulomb model is selected as the constitutive model, The expression is:

$$f_s = \sigma_1 - \sigma_3 \frac{1 + \sin \varphi}{1 - \sin \varphi} - 2c\sqrt{\frac{1 + \sin \varphi}{1 - \sin \varphi}} \tag{1}$$

In the formula, σ_1 and σ_3 stand for maximum and minimum principal stresses, c and φ stand for cohesion and internal friction angle.

2.2 Simulation Scheme

Combined with the engineering practice, the test was carried out by multistage cyclic loading, loading mode and loading path of the model, as shown in Fig. 2.

Table 1 Sample Model Parameters

Name	Elastic modulus E/GPa	Uniaxial compressive strength/MPa	Poisson's ratio μ	Density ρ/kg·m^{-3}	Internal friction angle φ/°	Pressure coefficient	Evenness
Granite sample	6	100	0.25	2600	60	200	3
Cushion block	900	800	0.3	7800	-	-	100

Fig. 2 Numerical simulation loading path

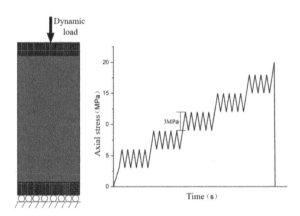

During the simulation, the mechanical simulation test of cyclic loading and unloading was carried out on the model specimen with a stress increase of 3 MPa. The stress loading time step was $1 \times e-5$ s. The plane stress model was used in the calculation and simulation, and the total loading step was set as 160 steps. In the first loading stage, the sample model is loaded and unloaded for 5 cycles. In each subsequent stage, the average stress increases by 3 MPa, and the sample model is loaded again for 5 cycles. In this way, the multi-stage cyclic loading is continued until the instability calculation of all models is stopped, and the stress map and acoustic emission map in the simulation process are compared and analyzed.

3   Analysis of Numerical Calculation Results

As the loading time goes on, the gradual fracture process of the sample model can be clearly observed through the rock dynamic fracture analysis system RFPA2D software during the loading process. The representative stress field diagram and acoustic emission dynamic distribution diagram of each loading stage in the sample model are selected for analysis. Limited to space, only the test results of NFG-19 and NFG-33 are displayed.

3.1   Analysis of NFG-1 Simulation Results

The NFG-19 specimen was subjected to graded loading with a 3 MPa stress increase cycle until the specimen was damaged, and the fracture damage process is shown in Fig. 3.

As shown Fig. 3(a), it can be seen that at the beginning of loading, it is obvious that stress concentration occurs at the fissures of the specimen model, and with the increase of loading steps, the specimen model firstly changes at the ends of the fissures, and the fissures may expand or new pores or micro fissures may sprout by the load excitation, and finally, with the continuous expansion of the fissures in the model leading to the destabilization of the specimen model, the upper part of the model shows signs of slippage along the fissures, and a slight rupture of the rock matrix at the lower part of the model can be found.

As shown Fig. 3(b), it can be seen that the model was firstly damaged at the pre-existing fracture, and the damage point was close to the model surface, with the loading step, the specimen model underwent deeper rupture, but the damage point was always at the pre-existing fracture. There are red and white circles with different diameters in the dynamic distribution of acoustic emission, and the diameter

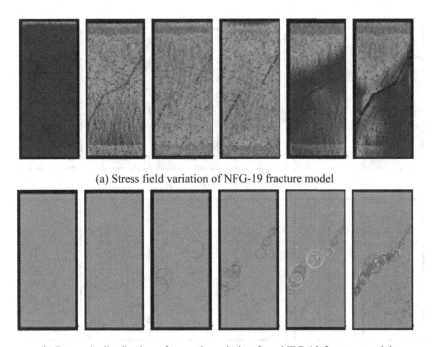

(a) Stress field variation of NFG-19 fracture model

(b) Dynamic distribution of acoustic emission from NFG-19 fracture model

Fig. 3 Cloud diagram of the damage process of NFG-19 specimen

of the circles in the Figure represents the relative size of acoustic emission intensity. With the expansion scale of the original natural fracture increasing, the model undergoes large-scale shear slip along the natural fracture, and the acoustic emission phenomenon of compression-shear damage gradually increases, so the final damage of the model is in the form of shear slip along the original fracture.

3.2 Analysis of NFG-33 Simulation Results

The NFG-33 specimen was subjected to graded loading with a 3 MPa stress increase cycle until the specimen was damaged, and the fracture damage process is shown in Fig. 4.

Comparing Fig. 3(a) and Fig. 4(a), the NFG-33 model has a strong similarity with NFG-19 in terms of overall morphology during the loading process, both of which are the gradual expansion of the pre-existing fracture by the load, which eventually leads to the shear slip damage of the specimen model, but the NFG-33 model loading step occurs near the ring-breaking stage when the expansion occurs at the fracture tip and the new fracture sprouts inside the specimen model.

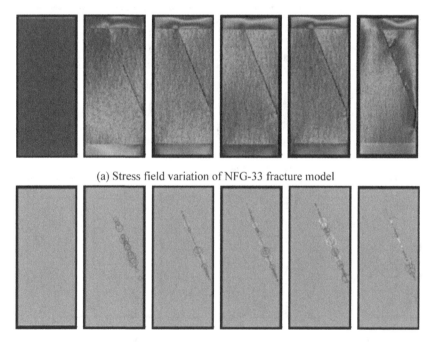

(a) Stress field variation of NFG-33 fracture model

(b) Dynamic distribution of acoustic emission from NFG-33 fracture model

Fig. 4 Cloud diagram of the damage process of NFG-33 specimen

Fig. 5 Freeze–thaw granite like damage pattern containing natural fissures (red line is natural fissures, green line is new fissures)

(a) NFG-19 damage pattern   (b) NFG-33 damage pattern

As shown Fig. 4(b), the NFG-33 specimen model also produces tensile damage at the beginning of loading, which leads to the expansion of pre-existing fractures, and as the loading step proceeds, the degree of fracture expansion gradually increases and compression-shear damage of different sizes occurs, similar the NFG-19 specimen model, and the final damage of the model is in the form of shear slip along the original fractures.

Due to the large initial extension and better penetration of the fracture in specimen NFG-33, the final acoustic emission distribution of the loading step shows that specimen NFG-19 also ruptures at the non-fracture, while specimen NFG-33 only extends at the tip of the fracture. When the structural face of the rock is developed to a certain extent, it gradually controls the strength of the whole rock, which makes the rock more prone to shear damage along the structural face during rupture, and in Fig. 4(b), the number of white circles is significantly more, i.e., more compression-shear damage occurred.

Figure 5 shows the actual damage pattern of the freeze–thaw granite sample containing natural fissures under multi-stage cyclic loading, the test results show that the specimen produced new fissures during the cyclic loading process, and the final destabilization damage occurred in the form of shear slip damage along the original natural fissures, which verifies the reasonableness of the numerical simulation.

4 Conclusion

In this study, using the rock dynamic rupture analysis software RFPA2D, the structural surface characteristics of representative granite rocks from Beizhan Iron Mine were selected for modeling and numerical simulations were performed using a multi-cycle loading path written by ourselves, and the structural surface activation process characteristics and rupture morphology of natural fracture specimens after freeze–thaw were studied in a targeted manner, and the main conclusions reached were as follows:

(1) The development degree of natural fissures determines the rupture form of the rock, when the development degree of natural fissures is large, the rock will eventually shear slip damage along the natural fissures, the less developed fissured rocks, the damage form is close to the intact rock.

(2) When the natural fractured rocks are subjected to loading, the initial structural deterioration occurs at the fractures, and tensile damage often occurs, making the natural fractures continue to expand, and when the natural fractures expand to a certain degree, the rocks begin to undergo large-scale compression-shear damage, which eventually leads to shear-slip damage of the fractured rocks.

(3) The damage form of fractured rocks is not only related to the development degree of fractures inside the rocks, but also related to the penetration degree and tilt angle of fractures.

Acknowledgements The authors would like to thank the editors and the anonymous reviewers for their help fand constructive comments.

References

1. Li C, Xiao Y, Wang Y (2019) Research status and trend of deformation and failure mechanism of rock slope in cold area of High altitude. J Eng Sci 41(11):1374–1386
2. Wen L, Li X (2017) Study on physical and mechanical properties of rock subjected to freeze-thaw in variable temperature range and its application in engineering. Chin J Eng Mech 34(5):247–256
3. Zhang H, Xia H, Yang G (2018) Experimental study on effects of freeze-thaw cycle and confining pressure on physical and mechanical properties of rock. J China Coal Soc 43(2):441–448
4. Song Y, Zhang L, Ren J (2019) Experimental study on Mechanical properties of red sandstone under cyclic loading after freeze-thawing. Coal Eng 51(2):112–117
5. Zhou S, Fang W, Jiang N (2020) Fractal study on fracture characteristics of sandstone under uniaxial compression under freeze-thaw cycle. Bull Geol Sci Technol 39(5):61–68
6. Liu Q, Huang S, Kang Y (2015) Research progress and consideration on freeze-thaw damage of fractured rock mass. Chin J Rock Mech Eng 34(3):452–471
7. Shan R, Bai Y, Sun P (2019) Experimental study on frost heaving force of fractured red sandstone. J China Coal Soc 44(6):1742–1752
8. Li P, Tang X, Liu Q (2020) Study on frost heaving fracture characteristics and strength loss of sandstone with double fractures. Chin J Rock Mech Eng 39(1):115–125
9. Zhao J, Xie M, Yu J (2019) Experimental study on mechanical characteristics and damage evolution law of fractured rock under freeze-thaw. J Eng Geol 27(6):1199–1207

Optimization Design of Sand and Loess High Slope Based on Combination of Wide and Narrow Platfom——A Case Study of a High Slope in Yulin City

Xingya Lu, Peng Li, Shengrui Su, Haibo Jiang, and Fu Dong

Abstract Due to the limited land resources available for engineering construction in Northern Shaanxi, engineering activities such as mineral resources development, basic engineering construction, slope reduction and building houses, and farming are often accompanied by the formation of a large number of Manually Excavated high slopes. Effective high slope design can reduce the waste of land resources, mitigate the damage to the natural environment, and reduce the project cost to a certain extent. Therefore, this paper takes a high sand loess slope in Yulin City, Shaanxi Province as an example, through field investigation, system theoretical analysis and numerical simulation calculation, comprehensively considering the anti scouring property, overall stability and excavation volume of the slope, the design scheme that meets the requirements and consumes the least capital is compared and selected, so as to obtain the optimal slope shape. The conclusion shows that the optimal single slope height of sand loess high slope in Northern Shaanxi is 7–8 m, and the slope ratio is 1:0.75. The setting of wide platform can change the stress distribution of high slope and weaken the continuity of effective plastic strain. The optimal slope shape of the high side slope includes the single slope of 8 m, the slope ratio of 1:0.75, the narrow platform of 3 m wide, 3 wide platforms are arranged, the location of ③④⑤, the width of 10 m, and the excavation volume of 696 m^3 (per linear meter).

Keywords Scour Resistance · Overall Stability · High Slope · Wide and Narrow Platforms · Optimized Design

X. Lu (✉) · P. Li · S. Su
School of Geology Engineering and Geomatics, Chang'an University, Xi'an, China
e-mail: 1150548500@qq.com

S. Su
Geotechnical Institute of China Coal Xi'an Design and Engineering Co., Ltd., Xi'an, China

H. Jiang
China Coal Xi'an Design Engineering Co., Ltd., Xi'an, China

F. Dong
Zhejiang East China Construction Engineering Co., Ltd, Zhejiang, China

G. Feng (ed.), *Proceedings of the 9th International Conference on Civil Engineering*,
Lecture Notes in Civil Engineering 327,
https://doi.org/10.1007/978-981-99-2532-2_13

1 Introduction

With the development of the national economy and the continuous promotion of the western development strategy, engineering construction projects in the western loess region have been increasing, and the limited available land resources have been unable to meet the current demand. Therefore, people try to solve this problem by changing the topography and landscape, but with this comes the formation of more and more artificial excavation of high slopes during the construction of engineering projects [1, 2]. According to research, the number of slope damages in China has accounted for more than half of the number of geological disasters since the twentieth century, and the problem of high slopes has become increasingly prominent [3]. It is generally considered that rocky slopes with a height greater than 30 m and soil slopes with a height greater than 20 m are high slopes, and their stability is mainly determined by the basic characteristics of the geotechnical body itself and the degree of human modification [4, 5]. In recent years, railroad, water conservancy and hydropower and highway construction industries have conducted more systematic research on loess slopes below 30 m, and achieved good results [6–8]. And at present, the slope height is no longer satisfied with 30 m, and the figure keeps breaking new records. Although in the process of engineering construction, emphasis has been placed on avoiding engineering construction in adverse geological locations, but high filling and deep excavation of roadbed, artificial slope reduction to build houses and other acts are inevitable, especially in some areas where available land is scarce is difficult to avoid. Northern Shaanxi is a typical loess area, and due to its special physical and mechanical properties, the loess structure is easily damaged when encountering precipitation or earthquake effects, resulting in serious geological disasters such as collapse, landslide, and slope scour [9, 10]. During September–November last year alone, 11 geological disasters occurred in northern Shaanxi under the action of heavy precipitation, resulting in 8 deaths, hundreds of people affected, and direct economic losses of more than 6 million yuan. It can be seen that it is necessary to conduct an in-depth study of certain areas, especially in such areas with special geological and environmental conditions, and the study of slope stability is particularly important.

The current analysis methods for slope stability research can be broadly divided into two types of analysis methods: deterministic and uncertainty [11]. Among them, the deterministic analysis methods mainly include limit theory equilibrium method, plastic limit analysis method, finite element method, Monte Carlo method, etc.; the uncertainty analysis methods mainly include fuzzy comprehensive evaluation method, gray analysis method, information quantity simulation method; besides, there are some other uncertainty analysis methods, such as quantitative theory method, quantitative table method, etc.

With the continuous improvement and development of slope prevention technology, nowadays there are more and more kinds of slope management methods, among which, the common form of high slope protection is slope reduction + interception and drainage + greening, among which the slope shape of slope reduction mainly includes "equidistant step slope" and "wide and narrow platform slope The shape of slope reduction mainly includes "equidistant step slope" and "wide and narrow platform slope". According to the survey, the protection effect of "wide and narrow platform slope" is better in terms of overall stability of slope. However, whether it is "equidistant stepped slope" or "wide and narrow platform slope", the problems of slope scour and overall stability are still significant and there are large safety hazards, and the unreasonable setting of slope reduction platform will lead to the increase of excavation volume, thus destroying the original terrain and wasting cost. Therefore, on the basis of slope stability research, it is a new issue worth thinking about how to optimize the design of slopes for different engineering fields.

Currently, many scholars have achieved good results in slope optimization design. Qian Gao, Wanjun Ye, Yanan Zheng et al. optimized the design of high slopes based on reliability theory analysis and proposed an effective and reliable design scheme [12–14]; Xinli Hu et al. designed the optimal slope angle so as to establish a slope optimization model [15]; Linhai Wan et al. proposed an open-air slope design scheme by establishing a radial basis function (RBF) neural network model, which solved the actual slope of the iron ore mine of water plant problem [16]; Yuming Xu used FLAC3D for slope stability evaluation so as to optimize the design of slopes [17]; Xuan Zhou proposed an optimal design scheme suitable for slopes based on sensitivity analysis [18]; Hongwei Fang et al. proposed a new slope optimization design method based on the limit curve method, giving objective quantitative indicators of slope instability [19]; Nengpan Ju et al. relied on a highway, put forward a set of operational highway slope optimization design research program [4].

In the existing slope research, most of the slopes with the same width platform are used as the research object to carry out the slope optimization design, but little research has been done on the "wide and narrow platform slope". The current research on this type of slope is mainly based on engineering experience and engineering analogy, and lacks practical basis and theory, in which the safety and stability of slopes still threaten people's life and property safety and social development to a certain extent. Therefore, based on a sandy loess high slope in Fugu County, Yulin City, Shaanxi Province, this paper analyzes the single-stage slope scour resistance and overall stability of the slope based on the idea of combining wide and narrow platforms, so as to propose a feasible slope optimization design scheme and provide some ideas for the optimization design of sandy loess high slope.

2 Profiles

2.1 Study Area Profile

The study area belongs to Fugu County, Yulin City, Shaanxi Province, and is located in the northernmost part of Shaanxi Province. It has a temperate continental monsoon climate, with cold and long winters and hot and short summers, low precipitation in spring and winter, high precipitation in summer and autumn, mostly heavy rainfall or continuous rains, low rainfall and high evaporation, severe spring droughts, frequent sand and wind, cold and dry sandy areas and mainly concentrated in the northwest, large temperature differences between day and night, and many early frosts and hail. The surface rivers in Fugu County are relatively dense, with dense gullies, of which there is only the Shagoucha Gully on the east side of the study area site, which belongs to the Yellow River water system, with heavy rise and sediment content in the rainy season and broken flow in the dry season. The study area is located in the border area between Inner Mongolia Plateau and the northeastern part of the Loess Plateau in northern Shaanxi Province, which is a loess gully landscape with micro-geomorphology divided into loess residual beams and river valleys. The overall topography shows a trend of high in the south west and low in the north east.Loess is widely distributed in the region, and its strong wet-sink nature, loose soil, porosity, and vertical fissure development make it prone to slope scouring and slope destabilization under heavy or persistent rainfall, and in severe cases, it will trigger geological disasters such as landslides [20–23].

Due to the constraints of the topography, a large number of artificially excavated high slopes have been formed in the study area due to engineering activities such as mineral resources development, infrastructure construction, slope reduction for housing and farming, as shown in Fig. 1. and Fig. 2.

All the high slopes seen in this survey have been graded, and the overall stability is better with "wide and narrow platform slope". However, when the slope surface is not protected, the slope surface is prone to serious scouring damage under rainfall conditions, especially heavy or continuous rainfall. In addition, the wide platform will lead to an increase in excavation volume, thus increasing the investment of project funds. Therefore, it is especially important to optimize the factors affecting the scouring and stability of high slopes.

Fig. 1 Photo of Shagoucha sandy loess high slope

Fig. 2 Photo of High slope of Shenghai coal mine

2.2 Side Slope Profile

The slope studied in this paper is the high slope of Shagoucha coal mine located in the west of Fugu County, Yulin City, Shaanxi Province. The maximum slope height of this slope is about 50 m, the longitudinal length is about 220 m, the slope gradient is 35–60°, the slope surface generally tends to the north-east, grass and other plants are planted on the slope surface, the slope is relatively intact, and no water fall hole is developed.

The slope is mainly composed of Quaternary Upper Pleistocene loess (Q^3_{eol}), and its physical and mechanical indexes are determined by indoor geotechnical tests. The basic physical and mechanical indexes are shown in Table 1, and the cumulative gradation curve of grain size is shown in Fig. 3. Among them, the mass fractions of clay particles (particle size < 0.005 mm), powder particles (particle size 0.005–0.075 mm) and sand particles (particle size > 0.075 mm) are 8.20, 71.80 and 20.00%, respectively.

Loess slope, due to the special nature of its material composition, is prone to deformation and damage under the action of rainfall, and the main form of damage is slope scouring, which will affect the stability of the slope and thus make the slope destabilized and damaged, therefore, it is necessary to carry out the research of slope scouring resistance of the slope.

Table 1 Basic physical properties of the soil bodies in the study area

Soil name	Water content $\omega/\%$	Specific gravity Gs	Volumetric weight $\gamma/(kN \cdot m{-}3)$	Plasticity Index Ip/%	Liquidity index IL/%	hydraulic conductivity k/(m · s−1)	Elastic Modulus E/MPa	Poisson's ratio υ	Cohesion c/kPa	internal friction angle $\varphi/(°)$
Sandy loess	8.2	2.69	16.6	8.3	-1.08	3.32×10^{-6}	30	0.3	13.43	32.2

Fig. 3 Accumulated cascade curve of powdered loess particle size

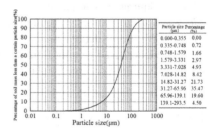

Particle size (µm)	Percentage (%)
0.000-0.355	0.00
0.335-0.748	0.72
0.748-1.579	1.66
1.579-3.331	2.97
3.331-7.028	4.93
7.028-14.82	8.42
14.82-31.27	21.73
31.27-65.96	35.47
65.96-139.1	19.60
139.1-293.5	4.50

3   Optimal Design of Scour Resistance Based on Single-stage Slope

The process of scour damage of loess slopes is more complex, and scour damage is one of the most harmful and common phenomena in slope design [20, 24]. Tests have shown that there are many factors affecting the scouring of loess slopes [25, 26], and in addition to natural factors such as precipitation and earthquakes, the morphological characteristics of the slope are also important influencing factors. In order to investigate the influence of slope morphology such as slope length, slope height and slope rate on scour resistance, the limit equilibrium method is used to obtain the optimal single-stage slope rate by considering changing the slope length of the single-stage slope under the most unfavorable precipitation working conditions with the help of field investigation and previous research results.

4   Slope Water Infiltration Analysis

The slope water infiltration problem has been developed and studied for a long time, and numerous empirical and theoretical formulations have been formed. In this paper, the Green-Ampt model is selected, which has low requirements for parameters, clear physical meaning, strong applicability, fast and simple calculation and meets the accuracy requirements [20, 27, 28]. The model formulation is:

$$i = k_{s1}[1 + (h_0 + h_f)\frac{1}{z_f}]$$ (1)

Among them:

i—— Infiltration rate cm/min;

k_{s1}——Saturated hydraulic conductivity, cm/min;

h_0——Soil surface water accumulation depth, cm;

h_f——Wet front suction, cm;

z_f——Generalized wetting front depth, cm;

Since the infiltration time is relatively short, Eq. 1 can be simplified as:

$$i = k_{s1}\left(\frac{h_f}{z_f}\right)$$ (2)

The rainfall lasted for 12 h, and the parameters were obtained by indoor infiltration test k_{s1}, taking 3.32×10^{-6} m/s, h_f and z_f the field experiment results with reference to similar soil properties, of which z_f 236.6 cm was taken and h_f 150 cm was taken.

Fig. 4 Schematic diagram
of soil elements on slope

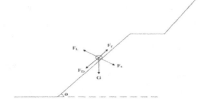

4.1 Slope Soil Force Analysis

According to the principle of dynamics, the stability of the soil unit on the slope is analyzed as the object of study, so as to ensure the stability of the slope as a whole. The force analysis of the soil unit is shown in Fig. 4:

According to the conditions of equilibrium of forces, parallel to the direction of the slope:

$$F_D + G \sin \theta = \mu(F_s + G \cos \theta - F_L) \tag{3}$$

Among them:

$$F_D = 1.9\pi \rho C_D D^2 \mu^{1/9}[(p-i)L]^{2/9} \sin^{-1/9} \theta \cos^{1/3} \theta d^{2/3} \tag{4}$$

$$F_L = 1.9\pi \rho C_L D^2 \mu^{1/9}[(p-i)L]^{2/9} \sin^{-1/9} \theta \cos^{1/3} \theta d^{2/3} \tag{5}$$

$$G = (\gamma_s - \gamma_w)\pi D^3/6 \tag{6}$$

$$F_S = \gamma_S J_S \pi D^3/6 \tag{7}$$

G——Gravity; d——Sharmov's formula particle diameter; Same value as D;
ρ——Density; C_D, C_L——Resistance coefficient, Lifting force coefficient;
p——Rainfall intensity; θ——Slope angle;
I——Infiltration rate; L——Slope length;
γ_s, γ_w——Saturated capacity, Watercapacity; μ——Roughness coefficient;
D——Diameter; J_s——Hydraulic gradient

4.2 Parameter Selection

The physical and mechanical parameters of the soil required for the model were obtained from indoor tests of soil samples in the study area; the single-width rainfall intensity was obtained from the statistics of annual rainfall in the study area; the

Table 2 Table of slope design parameters

Rainfall intensity $p/(m\cdot s{-}1)$	Infiltration rate $i/(m\cdot s{-}1)$	Roughness coefficient μ	Diameter D/mm	Saturated capacity $\gamma_s/(N\cdot cm{-}3)$	Density $\rho/(g\cdot cm{-}3)$	Water capacity $\gamma_w/(N\cdot cm{-}3)$	Resistance coefficient C_D	Lifting force coefficient C_L
5.032×10^{-6}	2.1×10^{-6}	0.05	0.08	26.9×10^3	1	10×10^3	0.2	0.8

infiltration rate of the slope surface was calculated from the infiltration analysis.CD and CL were calculated by referring to the research results of Jemenyev Idgiazarov [26], and the values of the calculated parameters of the final force analysis are shown in Table 2.

4.3 Single-stage Slope Scour Analysis

In measuring the limit equilibrium state of the slope soil against scouring, the slope soil is exactly in the limit state when the difference between the slip resistance and the decline force is 0, i.e.

$$F_D + G \sin\theta - \mu(F_s + G \cos\theta - F_L) = 0 \qquad (8)$$

According to the calculation combined with the field survey results, the single slope height of the slope in the study area is mainly 7–8 m, and 1:0.75 is selected as the optimal anti-scouring slope rate for the single slope height.

5 Optimized Slope Design Based on Overall Stability

When discussing the scour resistance of slopes, slope morphology is one of the decisive factors, and reasonable slope morphology plays an important role in reducing scour damage. However, in the actual slope design, the overall stability of the slope is another key indicator.

For artificial excavation of loess high slope, the most common method is slope rate method, which involves many factors, not only affects the overall stability of slope, but also affects construction period, engineering capital investment, environmental protection, etc. Therefore, it is crucial to design a safe, effective and economical slope reduction optimization plan. According to the field survey, the loess high slopes in northern Shaanxi can be mainly divided into two categories: one is slow slope and narrow platform, and the other is wide platform and narrow slope. Among them, the single slope height and slope rate of slow slope and narrow slope are small, with poor anti-scouring performance and large volume of works, and serious damage to the environment; while the volume of works of wide platform and narrow slope

Fig. 5 Schematic diagram of two wide platform high slope models

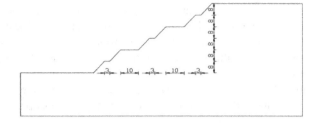

is relatively small, and it has great advantages in resistance to slope damage and overall slope damage. Since the design is only based on practical experience, the optimization design theory is lacking. Therefore, in this paper, considering the antiscouring performance of the slope and taking the overall stability of the slope as the premise, the width, location and quantity of the wide platform of the slope are studied with the help of MIDAS numerical simulation software.

According to GB 50,330–2013 Technical Specification for Construction Slope Engineering, the high slope of coal mine industry belongs to Grade III slope, and the safety factor is required to be greater than 1.25, therefore, 1.25 is used as the stability evaluation index in the numerical simulation analysis.

5.1 Geometric Models

The basic assumptions for establishing the model are: firstly, the geotechnical body is regarded as a uniform isotropic elastic body; secondly, the plane strain state is considered; thirdly, the geotechnical body obeys the Moore-Coulomb damage criterion.

The total slope height of the high side slope in this study is 48 m. According to the actual survey and previous engineering experience, the single slope rate of this model is designed to be 1:0.75; the slope height is 8 m; the width of the narrow platform is 3 m; the wide platform is taken to be 6–16 m, and the interval is 2 m in turn; the number of wide platforms is 1–4, and the position is freely combined. The following is an example of two 10-m wide platforms, modeled as shown in the Fig. 5.

5.2 Boundary Model

The simulation parameters are selected in Table 2.

The strength discounting method is used in the calculation, and the convergence condition in the analytical control is set to the displacement standard 0.001 m. The boundary conditions are horizontal and vertical constraints at the lower boundary and horizontal constraints at the left and right boundaries. Quadrilateral cells are used to divide the mesh. The specific model is generalized in the calculation, and each layer of the geotechnical body is considered as an isotropic material. The analysis considers the influence range of the model, the model extension dimension is 50 m from the top of the slope to the X axis, 40 m from the foot of the slope to the left, and 30 m from the foot of the slope to the Y axis.and the actual distance of the slope height above the foot of the slope to the top of the slope.

5.3 Analysis of Results

Stress-strain Analysis. Due to space limitation, the simulation results for one and two wide platforms are listed here. Simulation results for three and four wide platforms are similar to them and are not shown separately here.

(1) One wide platform

Take the 10 m wide platform as an example, the wide platform is set at ①, ③ and ⑤ respectively. When the wide platform is located at ①, the maximum effective plastic strain is located at the toe of the slope above the wide platform(Fig. 6(a)); when the wide platform is located at ③(Fig. 6(b)), the effective plastic strain is distributed in the whole slope, and the maximum value is located at the foot of the slope; when the wide platform is located at ⑤(Fig. 6(c)), the effective plastic strain is only distributed in the slope below the wide platform.

Take the location of wide platform ③ as an example, the width of wide platform is 8, 12 and 16 m respectively, with the increase of wide platform width, the maximum effective plastic strain decreases and the strain concentration area is all located at the foot of the slope, but the distribution of effective plastic strain area in the slope is gradually discontinuous, which indicates that the overall stability of the slope has improved (Fig. 6(d-e)).

It can be concluded from this that when there is only one wide platform, the width of the wide platform is large enough and the location is close to the foot or the top

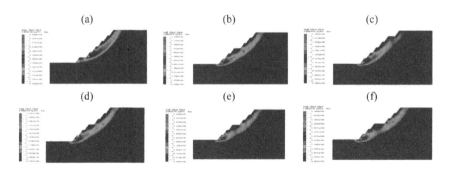

Fig. 6 Plastic strain zone of one wide platform

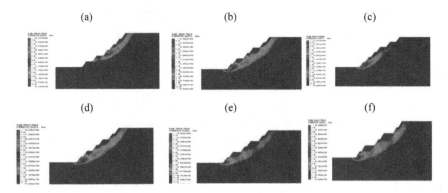

Fig. 7 Plastic strain zone of two wide platforms

of the slope, the whole slope can be analyzed as two independent slopes above and below with the wide platform as the boundary.

(1) Two wide platforms

Take the 10 m wide platform as an example, the wide platform is set at ①②, ②④ and ④⑤ respectively.When the wide platform is located in ①②, the maximum effective plastic strain is located at the foot of the slope above the wide platform (Fig. 7(a));when the wide platform is located at ②④, the effective plastic strain is distributed throughout the slope, and the maximum value is located at the foot of the slope (Fig. 7(b)); when the wide platform is located at ④⑤, the effective plastic strain is concentrated only in the slope below the wide platform (Fig. 7(c)).

Take the location of wide platform ②④ as an example, the width of wide platform is 8, 12 and 16 m respectively, with the increase of platform width, the maximum effective plastic strain decreases first and then increases (Fig. 7(d-e)).

It can be concluded from this that when there are two wide platforms, the changes of effective plastic strain zone, maximum stress strain and stability are similar to one wide platform, and the strain zone has been obviously divided by the wide platform. The maximum effective plastic strain at the lower side of the wide platform position is greater than that at the upper side of the wide platform position, and the strain concentration phenomenon gradually appears at the slope corner of the upper slope, and the overall stability of the slope improves and basically meets the requirements after the width is greater than 14 m.

Analysis of Stability Simulation Results. The simulation results of the wide platform model with different widths are shown in Fig. 8.

For the 48 m high slope in this study, when there is only one wide platform (Fig. 8(a)), as the wide platform gradually moves from the foot of the slope to the top, the stability coefficient shows an overall trend of first increasing and then decreasing, and the stability coefficient is the largest when the wide platform is at position ③. However, no matter changing the position of wide platform or the width of wide platform (6–16 m), it cannot meet the stability requirement. When there are

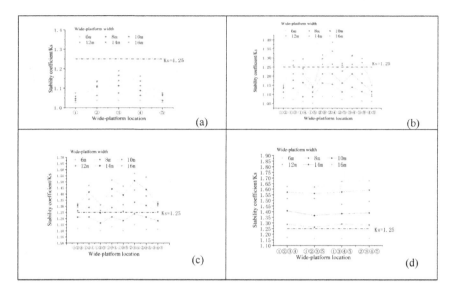

Fig. 8 Relationship between wide platform position and stability coefficient Optimized slope design combining excavation volume and stability

two wide platforms, the stability of high side slope meets the requirements when the width of wide platform is 16 m and the location of wide platform is ①③, ①④, ②③, ②④, ②⑤, ③④, ③⑤; the width of wide platform is 14 m and the location of wide platform is ①③, ②③, ②④, ②⑤, ③④, ③⑤; the width of wide platform is 12 m and the location of wide platform is ②③, ②④, ③④. Among them, the stability coefficient is the highest when the location of wide platform of the same width is ②④. When there are three wide platforms, except for the 6 and 8 m wide platforms, most of the other wide platforms can meet the stability requirements. Among them, the stability coefficient is the highest when the position of wide platform of the same width is ②③④. When there are four wide platforms, only the 6 m wide platform does not meet the stability requirements, and the rest all meet. Among them, the stability factor is the highest when the position of the wide platform of the same width is ②③④⑤.

As the excavated wide platform gradually moves from the foot of the slope to the top of the slope, the stability coefficient changes obviously. When there is only one wide platform, the stability of the middle position is often the highest; when there are multiple wide platforms, the stability coefficient increases when the wide platform positions are adjacent to each other, especially the stability of the wide platform position located in the middle adjacent to each other is the highest.

As can be seen from Fig. 9, the stability coefficient changes with the excavation volume, and the width, location and number of wide platforms change. Under the premise of satisfying the overall stability requirement of the slope, the smaller the excavation volume, the less the corresponding project economic consumption and the less damage to the original ecological environment. In this simulation, there are 67 slopes that meet the overall stability of slopes.

Fig. 9 Relation diagram between excavation amount and stability coefficient

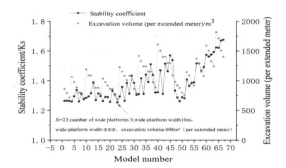

Through comprehensive consideration and analysis, 3 wide platforms with 10 m width and location ③④⑤ are the optimal slope design for the 48 m high slope in this study.

6 Conclusions

(1) The optimal single slope height of 7–8 m and slope rate of 1:0.75 for sandy loess high slopes in northern Shaanxi Province are studied on the artificial excavation slope of Shagoucha coal mine in western Fugue County, Yulin City, Shaanxi Province.

(2) The setting of wide platform can change the stress distribution of high slope and weaken the continuity of effective plastic strain: when it is located at the foot and top of the slope, the distribution of potential slip surface is bounded by the location of wide platform and distributed on the slope above and below the wide platform; when it is located in the middle of the slope, the distribution of potential slip surface is discontinuous at the wide platform, and when the wide platform is continuously distributed and the width is large, the high slope can be considered as two independent slopes for analysis Calculation.

(3) Considering only the overall stability of the slope, setting 1 wide platform cannot meet the slope stability requirements; setting 2 wide platforms, when the size of wide platform is larger than 12 m, only a few slope shapes meet the requirements; setting 3 wide platforms, when the size of wide platform is larger than 10 m, the number of slope shapes that meet the stability requirements is larger; setting 4 wide platforms, except when the size of wide platform is 6 m, all other slope shapes meet the stability requirements.

(4) Combining the scour resistance and excavation volume, considering the overall stability of the slope, the optimal slope shape of this artificial excavation slope is: single slope height of 8 m, slope rate of 1:0.75, narrow platform width of 3 m, arrangement of 3 wide platforms, position ③④⑤, width of 10 m.

The research results can compensate to a certain extent the defects of the traditional loess high slope stability analysis and evaluation in which the potential slip surface is assumed to run through the whole slope, and also solve to a certain extent the difficulties of high slope shape selection mainly by experience, and provide guidance basis for future slope management projects in the area, which has important practical significance.

Acknowledgements This research was supported by the National Natural Science Foundation of China (42041006, 41831286 and 41672285)

References

1. Zheng XH (2017) Discussion on deformation and failure characteristics of loess high slope. Technol Innov Appl 14:149–150
2. Zhao FC (2018) Stability calculation and optimization design of a loess high slope. Electr Power Surv Des 10:4–7
3. Yuan Z, Yan CG, Tao Y et al (2022) Experimental study on erosion model of loess slope protected by skeleton. J Eng Geol 2022:1–9
4. Ju NP, Zhao JJ, Deng H et al (2009) Stability evaluation and support optimization design of highway high slope. Chin J Rock Mech Eng 28(06):1152–1161
5. Wang GX (2003) Discussion on design and reinforcement of high slope. J Gansu Sci 2003(s1):002
6. Qiu JR, Yu RL, Zhou H (2018) Study on slope design of loess high excavation. Site Invest Sci Technol 2018(s1):44–46+50
7. Ye WJ, Wang DY, Zhe XS et al (2005) A new method for high loess slope design. J Eng Geol 3:415–418
8. An M, Yang HL, Su WB (2007) Slope design and selection in loess region of northwest China. J Wuhan Univ (Engineering Edition) 6:66–71
9. Pu XW, Wang LM, Wu ZJ et al (2016) Engineering geological problems and stability analysis of excavated loess high slope in hilly and gully region of Lanzhou. China Earthq Eng J 38(05):787–794
10. Su SR, Peng JB (2003) Study on major engineering geology problems in northwest China. J Eng Geol 1:105–110
11. Sheng LF, Liao JY, Zhang YL et al (2005) Review of slope stability analysis and evaluation methods. Min Res Dev 1:24–27
12. Gao Q, Wang SJ (1991) Reliability analysis of high slope of lock of Longtan Hydropower Station. Chin J Rock Mech Eng 1:83–95
13. Ye WJ, Zhe XS, Chen ZX et al (2005) Optimization design of loess high slope based on reliability theory. J Earth Sci Environ 2:82–85
14. Zheng YN, Hou XK, Li P et al (2014) Study on the reliability of loess high slope in Xi County and Lishi Area. J Eng Geol 22(03):372–378
15. Hu XL, Tang HM, Chen JP (2001) Optimization design method of highway bedding cutting slope. Sci China Earth Sci 4:373–376
16. Wan LH, Wang P, Cai MF (2004) Optimization design method of open-pit slope based on RBF neural network. China Min Mag 7:51–54
17. Xu YM (2015) Optimization design of high slope support scheme of mountain expressway based on FLAC(3D). Highway Eng 40(06):145–148+152
18. Zhou X (2014) Optimization design method of reinforced soil slope based on sensitivity analysis. Chongqing Jiaotong University

19. Fang HW, Chen YJ, Deng XW (2019) A new slope optimization design method based on limit curve method. J Cent South Univ 26(07):1856–1862
20. Dong F (2021) Study on optimum design of artificial excavation loess High slope in Northern Shaanxi. Chang'an University, 001443
21. Ma X (2021) Development characteristics and susceptibility evaluation of cave-slip geological hazards in Fugu County. Xi'an University of Science and Technology, 000284
22. Zhang TY (2016) Study on zoning method of geological hazard susceptibility in Fugu County. Xi'an University of Science and Technology
23. Zhang. B (2009) Research on the establishment of geological disaster database and evaluation of prone area in Fugu County based on ArcGIS. Chang'an University
24. Jiang C (2021) Numerical simulation of erosion of loess excavated slope. Chang'an University, 001169
25. Wang GQ, Li TJ, Xue H et al (2006) Mechanism analysis of sediment process in watershed. J Basic Sci Eng 04:455–462
26. Zhang DL, Wang BL (2006) Study on calculation method of slope plant protection. J East China Jiaotong Univ 1:52–55
27. Su YH, Li CC (2020) Slope stability analysis based on Green-AMPT model under heavy rainfall. Rock Soil Mech 41(02):389–398
28. Dou HQ, Han TC, Gong XN et al (2016) Slope reliability analysis considering variability of saturated permeability coefficient under rainfall conditions in loess region of Northwest China. Rock Soil Mech 37(04):1144–1152

Key Construction and Control Technology of Long Span Self-anchored Suspension Bridge with Cable Before Beam

Chengming Peng, Zhihui Peng, Jiaqi Li, and Junzheng Zhang

Abstract Shatian Bridge is a self-anchored suspension bridge with a main span of 320 m. It is constructed by the overall construction technology of "cable before the beam". The main cable is temporarily fixed through the temporary anchorage system, and the main beam construction is based on the main cable. After the main beam is hoisted and welded, the main cable is temporarily fixed, and the tensile force of the main cable is transferred to the main beam to complete the system conversion. The bridge adopts permanent-temporary combined with temporary anchorage, effectively saving the cost. The lifting of the stiffening beam adopts inverted lifting technology. For the area of the short sling in the middle of the span, a non-full-length joist is designed to solve the problem of main beam lifting in the area of the short sling. During the construction, the steel beam of the anchorage section and the auxiliary pier are temporarily consolidated. The temporary cable actively balances the tension of the main cable with clear stress, which is convenient for construction control. Temperature welds are set at both ends of the closure beam section, which not only makes room for the hoisting of the closure beam section but also avoids the structural safety problems caused by the temperature deformation of the steel beam. The slip control method of cable strands based on water bag weight ensures that the main cable does not slip during steel beam hoisting. The length of the sling is increased through the extension rod, and the horn-shaped guide device is added to avoid sling damage caused by the sling colliding with the conduit mouth. Generally speaking, the construction scheme of "cable before beam" adopted by the bridge is reasonable and feasible, which enriches the construction technology of self-anchored suspension bridges and can provide a reference for similar bridge construction in the future.

C. Peng · Z. Peng · J. Li (✉) · J. Zhang
CCCC Second Harbor Engineering Company Ltd., Building A, China Communications City, Wuhan, Hubei, China
e-mail: 196356311@qq.com

Key Laboratory of Large-Span Bridge Construction Technology, Wuhan, Hubei, China

Research and Development Center of Transport Industry of Intelligent Manufacturing Technologies of Transport Infrastructure, Wuhan, Hubei, China

G. Feng (ed.), *Proceedings of the 9th International Conference on Civil Engineering*,
Lecture Notes in Civil Engineering 327,
https://doi.org/10.1007/978-981-99-2532-2_14

Keywords Self-anchored Suspension Bridge · Cable before Beam · Permanent Temporary Combined Temporary Anchorage · Reverse Lifting · Strand Slip

1 Introduction

Self-anchored suspension bridges have strong competitiveness in bridge engineering in the range of 100 ~ 400 m span because they do not need large anchorage, occupy a small area, and have beautiful structure and shape [1–3]. Since the main cable of the self-anchored suspension bridge is directly anchored on the main beam, the construction sequence of "beam before cable" is usually adopted [4–6], which is precisely opposite to the construction sequence of "cable before beam" adopted by ground anchored suspension bridge. It is challenging to apply the traditional "beam before cable" technology when crossing the busy waterway, with high navigation requirements, high environmental protection requirements, considerable water depth, thin overburden, and rugged construction of steel pipe piles into rock. The technology of "cable before beam" will solve this problem perfectly. There are few pieces of research on this technology, and its practical engineering application is rare.

Wen Shudong et al. [7] put forward the idea of constructing self-anchored suspension bridges with "cable before beam" technology in 2005. Their erection scheme was temporary consolidation of pier beams, transferring the horizontal force in the construction process to the pier, and discussing the design of the pier section. The system conversion was realized by relaxing the temporary consolidation tie rod. Tian Hanzhou [8] and Jia Guang [9] completed the construction of the Beijing Road Bridge (main span 132.5 m) in Huai 'an for the first time by setting temporary anchorage and using a temporary cable to control the displacement of the beam's end.

Generally speaking, the research on constructing self-anchored suspension bridges with "cable first and beam second" is still limited. There is only one small span mixed beam self-anchored suspension bridge in practical engineering application, and the side span has been erected on the support in advance. Only the middle span beam section is lifted and erected, and the side span balances the weight of the middle span erection beam section by tensioning the sling. Generally speaking, it is not difficult to control. Currently, the construction technology of "cable before beam" has not been adopted on self-anchored suspension bridges with a span of more than 200 m at home and abroad.

The main span of south branch of Dongjiang River Harbor Bridge (now renamed Shatian Bridge) in Dongguan is 320 m, spanning the busy waterway, and the designed navigable clearance is 294 × 34 m. The original design is the construction technology of cable first and beams later. In order to reduce the influence of the construction process on the waterway and ensure the safety of the construction process, the construction technology of "cable first and beam later" is adopted after modification. In this paper, the critical construction and control technologies in the construction process of the bridge are summarized to provide a reference for the subsequent construction of similar projects.

2 Engineering Situation

Shatian Bridge is located in Shatian Town, Dongguan City. It is an important channel to cross the tributaries of East and South China and connect all the villages of Shatian Town with Nizhou Island. The main bridge is a double-tower, five-span steel box girder self-anchored suspension bridge, as shown in Fig. 1. The span of the bridge is $60 + 130 + 320 + 130 + 65 = 705$ m, and the semi-floating structure system is adopted with a two-way six-lane arrangement. The ratio of the main cable to the span is 1:5, the theoretical sag of the middle span is 64 m, and the distance between the center lines of the main cable is 28 m. A single main cable is composed of 37 cable strands, and each cable strand is composed of 91 galvanized high-strength steel wires with a diameter of 5.0 mm, and the ultimate tensile strength of the steel wire is 1770 MPa. The standard distance of the sling is 12 m, and the bridge tower adopts the portal frame tower with a height of 117.59 m.

The 320 m long main beam of the middle span and 130 m main beam of the side span are arranged as steel box girders with slings, and the 60/65 m anchor span is arranged as steel box girders without slings. The steel box girder of the whole bridge is divided into class A segments such as A1-A17, B, C1, C2, D, E, F, G, H, I, J, and K, among which class A segment is the standard segment, D segment is the tower section, and G segment is the main cable anchorage section. There are 55 segments in total, including four beam segments for a single anchor, nine beam segments for a single edge, 25 beam segments for a middle span, one anchor beam segment, and one tower beam segment at the auxiliary pier position.

The stiffening beam adopts a flat streamline steel box girder, orthotropic slab bridge surface structure, the entire width of the cross-section is 38.5 m (including air nozzle), the standard beam height at the center line of the bridge is 3.5 m, and the steel beam height at the anchor section is uniformly transitioned from 3.5 m to 6 m. Iron sand concrete with a weight of 1000t is pressed on the steel beam at the anchor position of the main cable. The standard segment length is 12 m, the beam height is 3.5 m, the roof thickness is 18 mm, the bottom plate thickness is 14 mm, the outer web thickness is 16 mm, the inner web thickness is 14 mm, and the weight is 238.3 t. The standard section of the 1/2 main beam is shown in Fig. 2.

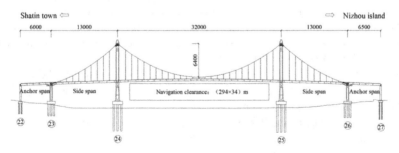

Fig. 1 South branch of Dongjiang River Shatian Bridge

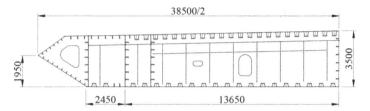

Fig. 2 1/2 Standard section layout of main beam

3 Overall Construction Technology of "Cable before Beam"

Shatian bridge construction adopted the "cable before beam" overall construction technology. Its main idea is to the temporary anchorage of the main cable temporary fixed, and then apply a similar mode of construction of suspension Bridges by anchor cable, main cable, and saddle, catwalks stiffening girder construction, stiffening girder hoisting, welding, after the completion of main cable release temporary fixed, transfer the main cable tension to the stiffening girder, complete system transformation. Temporary cables are set between the temporary anchorage and the main cable to form a temporary anchorage system to balance the tension of the main cable during construction. The overall process is divided into the following seven steps:

1. Construction of cable pylon, auxiliary pier, handover pier, and temporary anchorage, erection of tower area, side span, and anchor span range of beam support.
2. The steel beam of the tower area and anchor span is hoisted to the bracket of the retaining beam by a floating crane. It is slipped to the design position, pieced together into a whole, and the anchor is temporarily locked across the steel beam and auxiliary pier.
3. Install tower top gantry, hoist the main tassel saddle, install the hashing base, construction catwalks, install temporary cable, erection of the main cable
4. The main span is from the middle of the span to the cable tower, and the side span is from the anchor to the cable tower. The stiffening beam hoisting is carried out symmetrically, and the cable saddle pushing and temporary cable tension adjustment are carried out simultaneously during the hoisting process.
5. Hoist the closing beam section near the cable tower in the order of the first middle and second span, complete the construction of the stiffening beam closing, and weld each steel beam section.
6. Release the temporary lock between the anchor span steel beam and the auxiliary pier, relax the temporary cable, transfer the tension of the main cable to the steel beam, and complete the system conversion.
7. Carry out subsequent ancillary work such as asphalt paving, main cable winding wire, cat path removal, etc.

Typical overall construction process steps 3–5 of "cable before beam" are shown in Fig. 3(a)-(c).

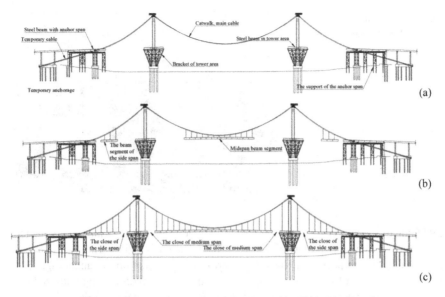

Fig. 3 Typical flowchart of overall construction technology of "cable before beam"

4 Key Construction Technology of "Cable Before Beam"

4.1 Permanent-Temporary Combined Temporary Anchoring System

The design of temporary anchorage system is the key to the construction technology of "cable before beam" for long-span self-anchored suspension bridge, which is directly related to the safety and economy of the whole project. The temporary anchorage system of this project is mainly composed of permanent-temporary combined with temporary anchorage and temporary stay cable, the main cable is anchored to the steel beam in the anchorage section, and the temporary stay cable connects the steel beam in the anchorage section with the temporary anchor as a whole, jointly bearing the huge main cable tension during the construction process. The site arrangement of temporary anchorage system is shown in Fig. 4.

Temporary Anchorage. For this project, Huang Jianfeng [10] compared and analyzed three temporary anchoring schemes: independent anchoring, combined anchoring without increasing pile diameter and pile number, and combined anchoring with increasing pile diameter and pile number, and finally adopted permanent, temporary anchoring with increasing pile diameter and pile number from the aspects of force safety, economy, and construction convenience.

Vertical elevation and plan of temporary anchorage are shown in Fig. 5(a) and (b), in the implementation process, a single temporary anchorage uses the pile foundation and cap of the original approach bridge 20 and 21# (side of Shatian Town) or 28 and

Fig. 4 Site layout of
temporary anchorage system

Fig. 4 Site layout of
temporary anchorage system

29# (side of Nizhou Island) and the new pile foundation and tie beam are connected between the two piers into a whole. The cap and tie beam together form a back-shaped structure, which improves the overall performance of the temporary anchorage. The plane dimensions of the whole temporary anchorage are 37.5 m long and 18.75 m wide, one for the upper and lower reaches of the bridge, a total of 8 piles for a single pier, and four root beam piles are added. The diameter of the approach bridge pile foundation is 1.8 m, the thickness of the cap is 5 m, the width is 7.5 m, and the length is 18.75 m. The newly constructed tie beam is 5 m thick, 5 m wide, and 22.5 m long. The diameter of the tie beam pile is 1.8 m, and the elevation of the pile bottom is the same as that of the bridge pier pile foundation.

The soil around the temporary anchorage is strengthened (as shown in the shaded part of Fig. 5.). High-pressure rotary jet grouting pile is used to handle the height below 7 m of cap mark, and filling is used to handle the height above the cap mark. The soil displacement rate in the reinforced range of high-pressure rotary jet grouting pile is 60%, the ordinary silicate water is 42.5 MPa, and the cement incorporation ratio is 25–30%. The unconfined compressive strength of the reinforced foundation soil reaches at least 1 MPa.

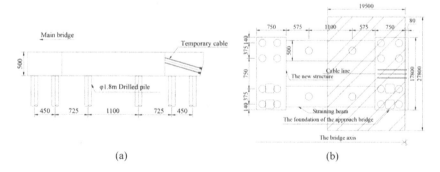

(a)   (b)

Fig. 5 Temporary mooring arrangement plan

Temporary Cable. The main cable is anchored to the steel beam in the anchoring section, and the temporary cable is connected with the steel beam in the anchoring section to jointly bear the considerable tension of the main cable during the construction process, as shown in Fig. 6. A single temporary anchor cable is arranged in double rows up and down, left and right, with a total of 4 parallel steel wire cables. A single cable contains 301 galvanized steel wires of low relaxation and high strength with a diameter of 7 mm, and the standard tensile strength is 1670 MPa. Both the upper and lower piers are cold cast anchors.

A fork-shaped lug plate is arranged at the upper end of the temporary cable, which is connected with the lug plate reserved on the steel beam of the anchorage section through a pin. The anchor cup is set at the tension end of the temporary stay cable. The temporary stay cable passes through the reserved conduit hole of the temporary anchorage. It is anchored on the anchor plate at the rear end of the temporary anchorage through the nut. Figure 7(a) and (b) show the arrangement of temporary cable beam segment and tensioning end.

After the closing of the stiffening beam, the single-side main cable temporarily pulls the maximum pulling force, and the maximum pulling force is 2380 t. During the construction, the maximum horizontal deformation of the temporary anchorage is 17 mm. After the temporary cable is relaxed and the system is converted, it will return to the zero-displacement state.

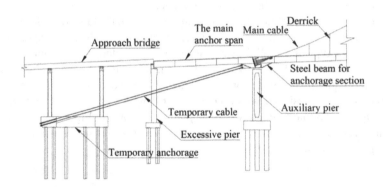

Fig. 6 General layout of temporary cable

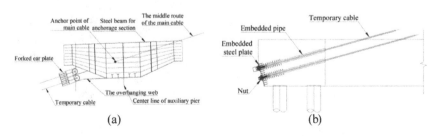

Fig. 7 Temporary cable beam end and tensioning end layout

4.2 Key Technology of Reverse Lifting Construction

Unlike the ground-anchored suspension bridge, the self-anchored suspension bridge has a sizeable carrier-span ratio, a slight stiffness of the main cable, and a large inclination Angle of the main cable. It is calculated that the inclination angle of the main cable is greater than 30° in the span of nearly 200 m during the hoisting of the main beam. According to the survey, the maximum working slope of existing cable-borne cranes in China is not more than 30°, and the conventional cable-borne cranes may not meet the climbing requirements. The input cost of a single set of cable-borne cranes is high. The erection of the main beam by cable crane has a large amount, an extended period, and the site construction is complicated, and the construction cost is high. The use of floating crane construction needs to cross the main cable, which requires high crane height, high requirements on the selection of floating crane equipment, and a long time to occupy the channel, great impact on navigation, and the economic benefits of floating crane is poor.

When the reverse lifting equipment carries out the main girder hoisting, the reverse lifting equipment is mainly composed of a temporary cable clip, anchor head seat, steel strand, continuous lifting jack, active coil of steel strand, pump station, and control operation table. The temporary cable clip is supported on the main cable to provide the anchor point for the reverse lifting equipment. The steel rope of the lifting jack is connected to the temporary cable clip through the anchor head seat. The typical arrangement of the reverse lifting condition is shown in Fig. 8.

Due to the short sling of the mid-span beam section and the large overall structure size of the reverse lifting equipment, the steel beam cannot be lifted to the designed height when the typical overall lifting arrangement is adopted. The bottom joists are designed to increase the distance between the main cable and the lifting point to meet the hoisting requirements of steel girders in the mid-span and short-span sling area. The bottom joist is designed with a non-through-length structure and anchored to the steel beam's bottom through two hinge points, as shown in Fig. 9. The design of non - the through - length bottom joist structure saves material, has a good economy, and is light weight easy to install.

Fig. 8 Layout of typical working conditions of reverse lifting

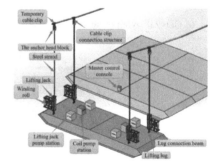

Fig. 9 Short cable area reverse lifting arrangement

5 Key Control Technology of "Cable Before Beam"

5.1 Consolidating the Steel Beam of the Anchorage Section with the Auxiliary Pier

The main cable is anchored to the anchor section steel beam, which is installed on the auxiliary pier. The temporary cable transfers the reverse tension of the main cable to the temporary anchor by connecting the anchor section steel beam with the temporary anchor. According to the consolidation of the steel beam and auxiliary pier in the anchorage section and the tension mode of temporary cable during construction, it can be divided into two working conditions: 1. The anchor section steel beam and auxiliary pier are temporarily consolidated, and the increasing main cable tension is balanced by tensioning the temporary cables in batches during construction. During the process, the unbalance horizontal force of the auxiliary pier is too large, which affects the structure's safety; 2. The whole anchor span girder can move horizontally freely during the construction period. The steel beam of the anchor span is pre-skewed first. In lifting the beam, the main cable will be moved to the designed position due to the increase in tension of the main cable, and the cable will not be tensioned temporarily. In the process of moving with the steel beam, the cable will be passively increased to balance the increasing tension of the main cable.

The two schemes are analyzed and compared by finite element calculation. The consolidation scheme of the auxiliary pier balances the main cable force by tensoring the temporary cable several times, ensuring the maximum horizontal displacement of the auxiliary pier within 9 mm and ensuring structural safety. The unconsolidated scheme of the auxiliary pier and main beam reduces the tension times of the temporary cable. However, the displacement of the steel beam in the anchorage section is significant in the construction process, and the steel beam is active, so the temporary cable is passively stressed, and the force in the process is complicated. By comprehensive comparison, although the auxiliary pier consolidation scheme has more tensioning times, the stress is more apparent, more conducive to construction control, and safer, so this scheme is adopted. The comparison of consolidation schemes of piers and beams is shown in Table 1.

Table 1 Comparison of consolidation schemes of piers and beams

Main Analysis Results	Temporary consolidation	No temporary consolidation
Amount of saddle advance deviation	565 mm	900 mm
Number of saddle thrusts	3	3
Number of temporary cable tensioning	6	For the first time
Maximum horizontal displacement of auxiliary pier/steel beam	9 mm	361 mm

Fig. 10 Temporary consolidation diagram of steel beam and auxiliary pier in anchorage section

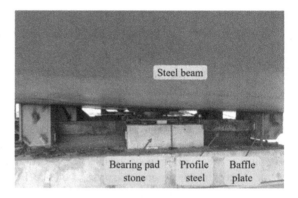

By welding the retaining block on the steel beam and using the shaped steel to resist the retaining block and support cushion stone, the steel beam and auxiliary pier in the anchorage section are temporarily consolidated, as shown in Fig. 10. During the construction process, the displacement of the auxiliary pier top is continuously measured to indirectly reflect the unbalance force of the pier top and observe the safety of the whole structural system.

5.2 The Temperature Weld and the System Conversion

Temperature Welds. After the steel beam hoisting is completed, the steel beam welding shall be carried out, and the temporary ground anchor system shall be converted to the self-anchor system after the welding is completed, which is the key to the whole construction control.

Steel beam welding is a continuous process, and many beam joints must be welded. Since the anchor section of the steel beam at the auxiliary pier has consolidated with the pier, the consolidation at the pier and beam should be relieved immediately after the welding is completed. Otherwise, the steel beam will cause excessive deformation of the auxiliary pier due to temperature deformation and cause structural damage.

In order to prevent structural safety problems caused by temperature deformation of steel beams during welding, welds between hoisted steel beams are divided into two types: conventional welds and temperature welds, and welded in a particular order, as shown in Fig. 11. After the completion of conventional welds, all temperature welds should be welded simultaneously, and the pier beam constraints should be lifted immediately after the completion of temperature welds.

The so-called temperature weld refers to the weld on both sides of the sealing section, and the width of the weld is 4 cm. The total length of the steel beam between the middle span closure section is about 280 m, and the total width of the temperature weld between the two closure sections is 16 cm. According to the temperature expansion coefficient of 1.2×10^{-5} line, the total width of the temperature weld can adapt to the temperature change of 47.6 °C of the 280 m-long steel beam in the middle of the span and meet the field requirements. The 4 cm wide weld can meet the requirements of the corresponding construction code after making the welding process evaluation test.

The setting of temperature weld can ensure that the steel beam welding process will not cause safety problems due to temperature deformation. At the same time, because the steel beam at both ends of the anchorage section is installed in the design position when the closing beam section is hoisted, the width of the closing gap can't be increased by pre-deflection. After setting the temperature weld, with the help of the surplus width of 4 cm at both ends and supplemented with pulling measures, the closing beam section can be successfully lifted.

The System Conversion. After the completion of conventional welding, temperature welding should be carried out at night when the temperature is stable. Due to the large seam width, large quantity, and heavy workload, it is challenging to complete at once. After the temperature weld web welding and the bottom laying and partial filling welding of the upper and lower bridge panels are completed, the auxiliary pier consolidation shall be lifted immediately, and the remaining welding shall be completed later. After all the welding is completed, the temporary cable is relaxed, and the system is converted. The temporary cable relaxation scene is shown in Fig. 12.

The specific construction steps are as follows:

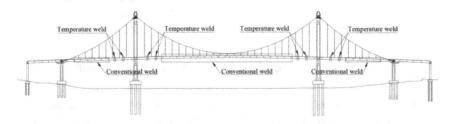

Fig. 11 General and temperature weld layout of Shatian Bridge

Fig. 12 Temporary cable
scene relaxation diagram

(1) Complete welding of other steel beam welds besides temperature welds. The middle span weld is carried out from the middle to both sides, and the edge span weld is carried out from the anchor to the tower;

(2) Select a cloudy day with a stable temperature for the temperature weld's welding, and polish the temperature weld in advance to save time. After the temperature stabilized in the afternoon, a steel box girder code plate was constructed for eight temperature welds.

(3) After the completion of the temperature weld code plate, all temperature welds should be welded synchronously at the time of the day when the temperature is stable (evening to early morning). First, the web welding, and then synchronous symmetric welding roof and bottom plate, complete the roof, bottom welding, and partial filling welding;

(4) After the temperature weld is welded, the consolidation of the auxiliary pier will be relieved immediately. The consolidation of the auxiliary pier will be relieved before the temperature rises at 9 a.m., and the consolidation force will be small before the release through the deformation size of the bridge pier;

(5) After the constraint of the pier beam is lifted and the construction of temperature weld top and bottom plate filling and the cover surface is completed, all welding construction of the steel box girder of the main bridge is completed.

(6) After all welds are welded, the temporary cable is temporarily unloaded in stages, the stress system conversion of the superstructure is completed, and the load is unloaded in 13 stages. During uninstallation, ensure that the inner side and the upstream and downstream are uninstalled simultaneously.

5.3 Main Cable Slip Control and Cable Conduit Collision

Main Cable Slip Control. Since the rise span ratio of the self-anchored suspension bridge is more significant than that of the ground-anchored suspension bridge, it is difficult to control the slip of the main cable strand at the saddle position and meet the code's requirements by adopting the conventional erection method of synchronous edge-middle lifting. At the same time, the first beam segment of the middle span adopts the overall lifting form of large sections (three beam segments) with a weight of 637.6 t. However, the tension of the main cable is slight in the early stage, and the compressive stress reserve of the cable pylon is insufficient, which brings great trouble to the stress control of the cable pylon.

The slip of the main cable and the stress of the main tower can be solved by using water bags. When the safety factor of the main cable strand slip due to the increase of unbalance force does not meet the specification requirements and affects the force safety of the cable tower during lifting of the middle span beam section, water bag pressure is arranged on the installed beam section of the side span to adjust the unbalanced force of the main cable at the side span and middle span side. With the erection of the side beam section, the water bag is unloaded synchronously to reduce the unbalance force of the main cable on both sides of the saddle. The water bag pressure is used to solve the slip problem of the main cable strand in the hoisting process. The steel beam hoisting process is mainly simulated by the finite element method, and the trial calculation is carried out for each hoisting condition to determine the weight of water bag pressure so as to ensure that the slip safety factor of the main cable strand and the structural stress of each hoisting condition meets the specification requirements. The implementation of water bag pressure and weight on site is shown in Fig. 13.

Cable Conduit Collision. The diameter of the anchorage conduit of the suspension cable of the self-anchored suspension bridge is usually smaller, which makes the anchor plate safer and more reliable. However, the small inner diameter may easily cause the sling to bend at its anchorage conduit port during construction [11]. It has been mentioned in the construction of the Foshan Pingsheng Bridge, Guangzhou Liede Bridge, and Peach Blossom Valley Yellow River Bridge that a small cable anchorage conduit port will lead to a collision conduit port in the process of cable tension, resulting in more complex system conversion process. In order to avoid a collision, this problem is often solved by split-tension, which is also the difficulty in the conversion control of the "beam first and cable second" self-anchored suspension bridge system.

This project adopts "cable before beam" construction, and the sling is in place by one tension (installed to the stress-free length of the completed bridge). A temporary

Fig. 13 Implementation of water bag pressure on site

hinge connection is adopted between girder segments after steel girder lifting. In the early stage of erection, the bottom mouth of the steel box girder is in the open state, and the girder segment is significantly inclined. Especially in the middle span beam section, the two ends of the main beam tilt angle are more significant, and the sling collision cable conduit problem is more prominent; if no measures, the maximum impact force will reach 205 KN.

However, this problem is more prominent only when there are four pairs of mid-span slings in the early stage of erection, and there is no such problem when the line shape is smooth in the late stage of erection. For several pairs of slings at risk of collision, lengthen the derrick, increase the "sling length," put the main beam horizontally, reduce the angle of the trabecular section, and avoid slings colliding with the conduit port. As shown in Fig. 14, increasing the sling length by lengthening the derrick can significantly improve the collision problem. At the same time, a guide device in the shape of a horn is added to the cable guide opening, as shown in Fig. 15. The contact surface between the sling and the cable guide wall is enlarged to avoid scratching the PE sheath. In this project, except for four pairs of mid-span slings due to collision, the lengthening derrick is tensioned several times (to increase the length of the slings), and the rest are tensioned in place at one time.

Fig. 14 Maximum impact force of sling during construction

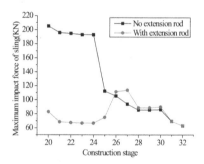

Fig. 15 Horn-shaped device for cable conduit

Conventional self-anchored suspension bridges usually adopt the construction technology of "beam before cable." For the need for suspension cable tension, the cable anchor box structure of the anchorage conduit and anchor plate is generally adopted at the lower end of the suspension cable, which will inevitably lead to a collision. For the self-anchored suspension bridge with the construction technology of "cable first and beam later," the design form of hinged suspension cable at the lower end of the ground-anchored suspension bridge can be used for reference to solve the above collision problems. The construction is not only efficient and convenient but also does not affect the appearance of the bridge. The optimization design in this aspect can be considered in the subsequent construction of similar projects.

6   Conclusion

The main span of Shatian Bridge is 320 m, and the river channel in the bridge location area is busy and close to the Pearl River Estuary's anchorage area. During construction, navigation closure (or traffic restriction) significantly interferes with local navigation and Taiwan prevention. This project abandoned the commonly used construction technology of "beam before cable" for conventional ground-anchored suspension bridges. It proposed the construction technology of "cable before beam" to realize the stent-free construction of the edge-middle span main beam.

In the construction process, permanent-temporary combined temporary anchorage is adopted. The stiffening beam hoisting adopts reverse lifting technology; For the short and medium-span sling area, the non-through-length bottom joists are designed. Temporary consolidation of steel beam and auxiliary pier in anchorage section; Temperature welds are arranged at both ends of the closing beam section; The water bag weight was used to control the main cable strand slip, and the sling length was increased by the extension rod to reduce the risk of collision between the sling and the catheter, and the horn-shaped guide device was added. These key construction and control technologies effectively ensure the smooth implementation of the bridge.

The bridge completed the system conversion in September 2021, and the construction process is progressing smoothly. The construction scheme of "cable before beam" is reasonable and feasible, which enriches the construction technology of self-anchored suspension bridges, further enhances the competitiveness of self-anchored suspension bridges, and can provide a reference for similar bridge construction in the future.

References

1. Cai YC, Wan C, Zheng YX (2013) Summary of the development of self anchored suspension bridges in China. Sino Foreign Highway 33(04):143–147

2. Wang ZJ, Zhou JB (2019) Design of main bridge of Weihe River Bridge on Baoji Lianhe Road. World Bridg 47(05):1–6
3. He YB, Shao XD, Zhang X (2021) Zhao H and Wang Y Mechanical performance and economic analysis of steel UHPC composite beam self anchored suspension bridge. Bridg Constr 51(01):51–57
4. Li CX, Ke HJ, Yang W, He J, Li HL (2014) Comparative study on system transformation schemes of Taohuayu self anchored suspension bridge on the Yellow River. J Civil Eng 47(09):120–127
5. Yuan Y, Yi LX (2019) Design and key technology of Wuhan Gutian bridge self anchored suspension bridge. Bridg Constr 49(02):80–85
6. Li YS, Song WJ, Chen NX, Gong GF (2018) Study on key technologies of cable-stayed construction of long-span self anchored suspension bridge. Bridg Constr 48(04):108–112
7. Wen SD, Zheng KF, Huang J (2005) Study on the construction of self anchored suspension bridge by "cable before beam" construction. J Southwest Jiaotong Univ 06:750–753
8. Tian H Z (2006) Research on design calculation and temporary anchorage analysis of Beijing Hangzhou canal super major bridge. Dalian Dalian University of technology
9. Jia G (2008) Construction control of Huai'an Beijing Road Bridge. Dalian University of technology, Dalian
10. Huang JF, Xu Z, Peng P (2021) Study on temporary anchorage scheme of "cable before beam" construction of self anchored suspension bridge. Bridg Constr 51(05):130–137
11. Ke HJ, Zhou Q, Li CX (2017) Study on optimization of system transformation scheme of small aperture sling anchor pipe double tower single span plane main cable self anchored suspension bridge. Sino Foreign Highway 37(05):88–93

Numerical Simulation for the Dynamic Response of Step Topography Subjected to Blasting Load

Lixiang Yang, Yaobin Han, Xu Zhang, and Zhiwen Li

Abstract The vibration velocity and frequency of rock mass are closely related to the instability and failure of the building. The step topography has great influence on blasting vibration wave propagation. Therefore, it is of great significance to study the dynamic response of step topography subjected to blasting load. In this paper, the FLAC2D program is used to study the dynamic response of step topography subjected to blasting load. The plane strain mode is adopted in the calculation, and the blasting load is assumed to be a triangular pulse wave which is applied to one side of the calculating mode. The amplification effect of the height of the step is first studied. The calculation result shows that the amplification factor increases first and then decreases with the increase of the step height. The maximum value of the amplification factor is consistent with the results of the field test, which indicates the feasibility of the numerical simulation.

Keywords Step Topography · Blasting Vibration · Amplification Effect · Numerical Simulation

1 Introduction

When a blasting vibration wave encounters a step topography during its propagation, the phenomenon of wave scattering occurs, which significantly affects the dynamic response of the near region of the step. The velocity and frequency of particle vibration of rock masses are closely related to the damage and structural instability of nearby buildings [1]. Therefore, the study of the dynamic response of the step topography under blasting load is very important.

Currently, numerous field tests and numerical simulations have been conducted to investigate the dynamic response of the step topography under blasting load. Zhu and Liu proposed a formula for blasting vibration and concluded that the dynamic

L. Yang · Y. Han · X. Zhang · Z. Li (✉)
College of Civil Engineering and Architecture, Nanchang Institute of Technology, No. 289, Tianxiang Avenue, High Tech Development Zone, Nanchang, China
e-mail: 2227638024@qq.com

© The Author(s) 2023
G. Feng (ed.), *Proceedings of the 9th International Conference on Civil Engineering*, Lecture Notes in Civil Engineering 327,
https://doi.org/10.1007/978-981-99-2532-2_15

response on a slope is related to the height of the slope [2]. By monitoring the blasting vibration, Wang and Lu found that the blasting vibration increase with the height of the slope, but not monotonically, and there is a maximum value [3]. Similarly, Xia declared that the top of a step topography exhibits dynamic amplification compared to a flat ground, with the amplification coefficient first increasing and then decreasing with increasing step height [4]. Hong studied the propagation of a blasting vibration wave in a step topography using a two-dimensional finite difference program, and found that the bottom of the step has a dynamic reduction effect, while the top of the step has a dynamic amplification effect [5]. Jiang and Zhou used the dynamic finite element method to simulate the blasting vibration of a slope. The results show that the height of the slope can significantly affect the vibration velocity on the slope, and the dynamic amplification effect mainly occurs in the vertical direction [6]. Tang used the UDEC software to study the blasting vibration of a step topography and found that the dynamic response of the step has an amplification effect, and that the amplification factor is related to factors such as the step height and the distance of blasting source [7].

In summary, there are many research results on the dynamic response for step topography and much valuable information has been obtained. However, most of these studies focus on specific parts of the step (e.g., the top and bottom corners) and rarely consider the entire dynamic response of the step. In this paper, the finite difference program FLAC2D is used to study the dynamic response of step topography under blasting load, with the aim of better understanding the propagation law of blasting vibration waves in step topography.

2   Numerical Model

2.1   Blasting Load

Because the FLAC2D program cannot simulate explosion process using equation of state, a triangular pulse load is applied to the blast hole wall to excite the blasting vibration wave. Since we focus mainly on the dynamic response of the step without considering the rock damage around the blast hole, the peak value of blasting load is taken as 50 MPa, which is smaller than the compressive strength of the rock of 50 to 200 MPa. Moreover, according to the theoretical calculation, the duration of the blasting pressure is generally hundreds of microseconds, so the rise time of the blasting load is taken as 100 μs, and the overpressure time is taken as 400 μs [8]. The time history curve of the blasting load is presented in Fig. 1.

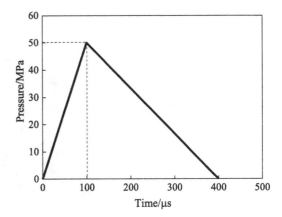

Fig. 1 The time history
curve of the blasting load

2.2 FLAC2D Calculation Model

The calculation model is mainly based on the field test conducted by Xia [4]. The
size of the calculation model is 100×50 m, the top of the model is set as a stress-free
boundary to simulate step surface and flat ground, and the bottom and both sides of
the model are set as absorbing boundaries to simulate infinite space. The blasting
load is applied perpendicularly on the left side of the model, and the range of the
loading depth is $2.5 \sim 3.0$ m. Since the duration of the blasting load is only hundreds
of microseconds, the frequency of the input wave can reach thousands of hertz.
To ensure the accuracy of numerical simulation, the calculation model is divided
into uniform square grids, and the grid size is set as 0.25 m. Besides, to facilitate the
analysis of the blasting vibration distribution law of the step topography, 13 vibration
monitoring points are placed on the step surface, with 8 monitoring points located at
the top and bottom of the step at a distance of 1 m, and 5 monitoring points located
at the side of the step at equal distances. The schematic diagram of the calculation
model and its grid division are shown in Figs. 2 and 3.

Since the numerical simulation here does not take into account the rock damage,
the rock is considered as an elastic medium, and the absorption attenuation of blasting
vibration wave by the medium is simulated by setting a Rayleigh damping. The

Fig. 2 The schematic
diagram of the calculation
model

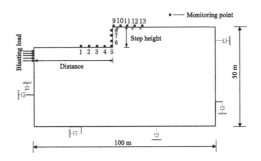

Fig. 3 The grid division of
the calculation model

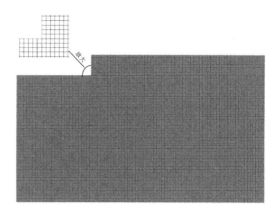

mechanical parameters of the rock refer to the field test in [4]: the rock density is
2610 kg/m³, the elastic modulus is 49GPa, and the Poisson's ratio is 0.19. Considering
that the rock mass in the field is relatively complete, the damping ratio of the medium
is set as 0.01.

3 Results and Discussion

3.1 Amplification Effect of Step Topography

To represent the amplification effect of the step topography, an amplification coefficient is defined as the amplitude ratio between the top corner of the step (i.e.,
monitoring point 9) and the flat ground at the same distance. Figure 4 compares
the variation of the amplification coefficient with the step height in the numerical
simulation and in the field test.

Fig. 4 The variation of the
amplification coefficient with
the step height

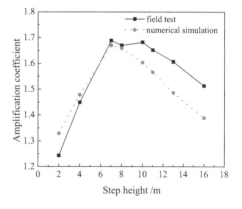

As illustrated in Fig. 4, the numerical results agree well with the test results. Both the numerical results and the test results show that the amplification coefficient first increases and then decreases as the step height increases, and the amplification effect is most significant at the step height is 7–8 m. However, there are still some differences between the numerical results and the field test results. When the step height is small, the numerical results are larger than the field test results, but when the step height is large, the numerical results are smaller than the field test results. This is because the numerical simulation uses a two-dimensional model, while the field test is a three-dimensional problem.

3.2   Vibration Distribution on Step Surface

As shown in Fig. 1, 13 vibration monitoring points are arranged at equal intervals on the step surface to study the overall dynamic response of the step topography. The amplitude at each monitoring point is represented by a dashed line perpendicular to the surface, while the amplitude at two corners is represented by a dashed line along the diagonal. Figure 5 shows the law of vibration distribution the step surface at step heights of 2, 4, 7 and 10 m, respectively.

As presented in Fig. 5, the step height has a significant effect on the dynamic response of the step topography. For example, when the step height is 2 m, the amplitudes at the top and bottom of the step do not differ significantly, indicating that the step topography has little effect on the propagation of the blast vibration wave in this case. However, as the step height increases, the amplitude difference between the top and bottom of the step gradually increases. For example, when the step height is 4, 7 or 10 m, the amplitude of the top corner of the step is the largest, showing a local dynamic amplification effect, while the amplitude of the rest of the top of the step is smaller than the bottom of the step, indicating a dynamic reduction effect. Moreover, the degree of dynamic reduction effect increases with increasing step height. This is mainly because as the height of the step increases, the propagation distance of the blasting vibration wave to the top of the step increases, so the degree of geometric attenuation and absorption attenuation of the blasting vibration wave increases.

Figure 5 also shows that regardless of the step height, the amplitude near the bottom corner of the step is significantly smaller than in the rest of the step, indicating a dynamic reduction effect. This is because the angle at the bottom corner of the step is 270°. When the blasting vibration wave propagates to the bottom corner, the wave front suddenly changes from an angle of 180° to 270°, resulting in a sudden increase in geometric attenuation and a significant decrease in amplitude.

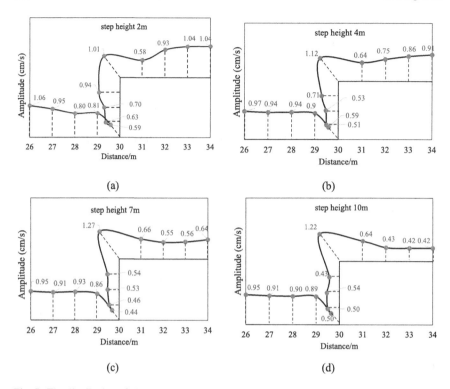

Fig. 5 The distribution of vibration amplitude on the steps of different heights

3.3 Frequency Variation of Blasting Vibration Wave

When the blasting vibration wave passes through the step topography, not only the amplitude but also the frequency changes. Obviously, the change in frequency is also related to the step height. To show the influence of the step height on the frequency of the blasting vibration wave, Fig. 6 compares the amplitude spectra at the bottom and top corners of the step at different step heights.

First, it can be seen in Fig. 6(a) that the amplitude spectrum of the bottom corner of the step contains many middle and low frequency components when the step height is small, but as the step height increases, the middle and low frequency components show a decreasing trend. The reason is that the larger the step height, the smaller the influence of the wave reflected from the top of the step on the bottom corner. Second, Fig. 6(b) indicates that as the step height increases, the low-frequency component of the top corner of the step gradually increases, while the high-frequency component gradually decreases, i.e., the energy ratio of the low-frequency component increases, which is consistent with the results of the field test in [9].

To sum up, when the blasting vibration wave passes through a step topography, the amplitude and low-frequency components of the top corner of the step are larger than

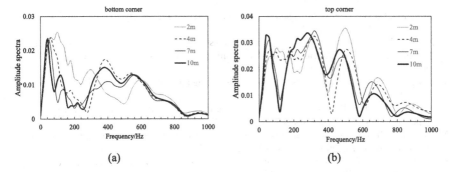

Fig. 6 The amplitude spectrum of the bottom corner and the top corner of the steps with different heights

those of the other parts, making buildings (structures) more susceptible to damage. Therefore, the buildings (structures) should be located as far as possible from the top corner of the step topography.

4 Conclusion

In this paper, the dynamic response of the step topography under blasting load is investigated using the finite difference program FLAC2D, and the feasibility of the numerical analysis is illustrated by comparison with field tests. Based on the numerical results, the following conclusions are drawn.

(1) Under the same blasting load, the amplification coefficient first increases and then decreases with increasing step height, reaching a maximum value at a step height of 7–8 m.

(2) The dynamic amplification effect of the step topography is mainly concentrated near the top corner, while there is a significant dynamic reduction effect near the bottom corner.

(3) When a blasting vibration wave passes through the step topography, the amplitude and low-frequency components of the top corner of the step are larger than those of the other parts, thus the buildings (structures) should be located as far as possible from the top corner of the step topography.

References

1. The National Standards Compilation Group of People's Republic of China (2015) GB6722—2014 Safety regulations for blasting. Chinese Standard Press, Beijing
2. Zhu CT, Liu HG, Mei JY (1988) Equation selection of thetrans mission rule of seismic wave parameters along slope surface. Blasting. 10(2):30–31

3. Wang ZQ, Lu WX (1994) Propagation of blasting vibration and quality control of high slope in excavation by blast. Blasting 11(3):1–4
4. Xia X, Li H, Niu J, Li J, Liu Y (2014) Experimental study on amplitude change of blast vibrations through steps and ditches. Int J Rock Mech Mining Sci 71:77–82. https://doi.org/10.1016/j.ijrmms.2014.03.016
5. Hong MX (1987) Numerical modelling of wave propagation and interaction. J Rock Soil Mech 3:23–31
6. Jiang N, Zhou CB, Ping W et al (2014) Altitude effect of blasting vibration velocity in rock slopes. J Central South Univ 45(1):237–243
7. Tang H, Li JR (2010) Numerical simulation of influence of protruding topography on blasting vibration wave propagation. Rock Soil Mech 29(6):1540–1544
8. Lu WB, Tao ZY (1994) Theoretical analysis of the pressure-variation in borehole for pre-splitting explosion. Explos Shock Waves 2:140–147
9. Li XH, Long Y, Ji C et al (2013) Influence of height difference based on wavelet packets transformation on blasting vibration signals. J Vib Shock 32(4):21–44

Study on the Calculation Mode of Rod Piece Reliability

Baohua Zhang, Peng Zheng, and Zhenkang Zhang

Abstract In this paper, the strength stability and stiffness reliability calculation model of the member bar under the basic deformation of the online elastic range was studied. JC method was used to analyze the reliability of the member considering the different distribution of random variables, and the correlation of failure modes was considered to calculate the failure probability-general range.

Keywords Random Variables · Member Bar · Reliability · JC Method

1 Introduction

The traditional safety factor method is still used in the calculation of the strength, stiffness and stability of the material mechanical rods. This method has obvious disadvantages. First, there is no quantitative consideration of the randomness of load effect, material strength and component size, often based on engineering experience or a measured value, so it is inevitable to have human factors, even subjective assumptions; Secondly, safety factor $k = m_R / m_s$ depends only on the relative positions of R and S, and depends on their dispersion degree (σ_R, σ_S) has nothing to do [1], This is not consistent with objective reality. In recent years, there have been some reliability analyses of member bar under basic deformation, but the correlation of failure modes of strength, stiffness and stability has not been considered comprehensively, which is somewhat biased with the actual situation. In this paper, the reliability calculation mode of the component's online elastic range strength, stiffness and stability

B. Zhang (✉)
School of Civil and Architectural Engineering, Nanchang Institute of Technology, Nanchang, China
e-mail: 2351459436@qq.com

P. Zheng
Institute of Technology, Nanchang Institute of Technology, Nanchang, China

Z. Zhang
Nanchang Road Bridge Engineering Corporation, Nanchang, China

© The Author(s) 2023
G. Feng (ed.), *Proceedings of the 9th International Conference on Civil Engineering*,
Lecture Notes in Civil Engineering 327,
https://doi.org/10.1007/978-981-99-2532-2_16

is considered simultaneously and the different distribution of random variables is considered. JC method is used to calculate its single reliability, and then the correlation of failure mode is considered, and the general limit range of member bar failure probability is obtained.

2 Equation of Limit State under Basic Deformation of Online Elastic Range of Member Bar

2.1 Selection of Functional Functions, Constants and Random Variables (Taking Plastic Materials as an Example) [2]

Axial Stretching. Generally, it is controlled by strength conditions. Take circular rod as an example, and its function function is

$$g = \frac{1}{4}\pi d^2 \sigma_s - F \tag{1}$$

Select $\frac{1}{4}\pi = A$, and diameter $d = x_1$, yield limit $\sigma_s = x_2$, pull $F = x_3$, then

$$g = A x_1^2 x_2 - x_3 \tag{2}$$

Axial Compression. Take the round rod for example:

1) Large flexibility bar $(\lambda \geq \lambda_p)$ is generally controlled by stable conditions, and its function is

$g = \pi^2 E I_{\min} - (\mu L)^2 F$ namely $g = \frac{\pi^3 E}{64} d^4 - \mu^2 l^2 F$.

Take $\frac{\pi^3 E}{64} = A$, $d = x_1$, length coefficient $\mu^2 = B$, the length $l = x_2$, $F = x_3$, then

$$g = A x_1^4 - B x_2^2 x_3 \tag{3}$$

In the large flexibility bar $\lambda = \frac{4\mu m_L}{m_d}$, corresponding to the proportional limit flexibility $\lambda_p = \sqrt{\frac{\pi^2 E}{\sigma_P}}$, where the length coefficient is μ, and the mean length is m_L, mean diameter m_d, elastic modulus E, proportional limit σ_p are obtained from statistics, so all are taken as constants.

2) Medium flexibility $(\lambda_p > \lambda > \lambda_S)$ is generally controlled by stability conditions, and its functional function is

$g = (a - b\lambda)\frac{\pi d^2}{4} - F$ namely $g = \frac{a\pi}{4} d^2 - \pi\mu bLd - F$.

Take $\frac{a\pi}{4} = A$, $d = x_1$, $-\pi\mu b = B$, $L = x_2$, $d = x_1$, $F = x_3$ then

$$g = Ax_1^2 + Bx_1x_2 - x_3 \tag{4}$$

In the middle compliance bar $\lambda_y = \frac{a-\sigma_S}{b}$, λ_S is the compliance corresponding to the yield limit, a and b are the coefficients related to materials, which are obtained from statistics, so they are taken as constants.

3) The small flexibility bar ($\lambda \leq \lambda_S$) is generally controlled by strength conditions and its functional function is

$$g = \frac{1}{4}\pi d^2\sigma_S - F \tag{5}$$

Take $\frac{\pi}{4} = A$, $d = x_1$, $\sigma_S = x_2$, $F = x_3$, then

$$g = Ax_1^2x_2 - x_3 \tag{6}$$

Use Rivets and Ordinary Bolts for Tension and Compression Connection. Generally controlled by shear and extrusion strength conditions, the tensile strength conditions of the motherboard are solved by structural measures, and its functional function is

$$g_1 = n_1n_2\frac{1}{4}\pi d^2\tau_S - F$$

Type: n_1 is the number of shear surfaces of a rivet or bolt; n_2 is the number of rivets or bolts.

Take $n_1n_2\frac{\pi}{4} = A$, $d = x_1$, shear yield strength $\tau_S = x_2$, $F = x_3$, then

$$g_1 = Ax_1^2x_2 - x_3 \tag{7}$$

$$g_2 = n_2d\sum t \cdot \sigma_{jy} - F$$

Take $n_2 = A$, $d = x_1$, $\sum t = x_2$ is the minimum plate thickness,and the extrusion yield limit $\sigma_{jy} = x_3$, $F = x_4$, then

$$g_2 = Ax_1x_2x_3 - x_4 \tag{8}$$

Pure Torsion. It is generally controlled by strength and stiffness conditions, and its functional function is

$$g_1 = \frac{\pi d^3}{16}\tau_S - M_T \tag{9}$$

Take $\frac{\pi}{16} = A, d = x_1, \tau_s = x_2$, torque $M_T = x_3$, then

$$g_1 = Ax_1^3 x_2 - x_3 \qquad (10)$$

$$g_2 = GI_p \cdot \frac{\pi}{180°}[\theta] - M_T l = \frac{\pi^2 G[\theta]}{5760} d^4 - M_T l$$

Take $\frac{\pi^2 G[\theta]}{5760} = A$, where the shear elastic modulus is obtained by statistics, and $[\theta]$ is given by design requirements, both of which are constants, $d = x_1, M_T = x_2$, $l = x_3$, then

$$g_2 = Ax_1^4 - x_2 x_3 \qquad (11)$$

Plane Bending. A simple beam with rectangular section under uniformly distributed load is taken as an example (the analysis method of non-rectangular beam is similar). It is mainly controlled by normal stress and shear stress intensity conditions, and the deformation and stability problems can be solved by structural measures. Its functional function is

$$g_1 = \frac{1}{6} bh^2 \sigma_S - \frac{1}{8} pl^2$$

Take $\frac{1}{6} = A$, width $b = x_1, h = x_2, \sigma_S = x_3, \frac{1}{8} = B$, load concentration $p = x_4$, $l = x_5$, then

$$g_1 = Ax_1 x_2^2 x_3 - Bx_4 x_5^2 \qquad (12)$$

$$g_2 = \frac{2}{3} bh \tau_S - \frac{1}{2} pl$$

Take $\frac{2}{3} = A$, width $b = x_1, h = x_2, \tau_S = x_3, \frac{1}{2} = B$, load concentration $p = x_4$, $l = x_5$ then

$$g_2 = Ax_1 x_2 x_3 - Bx_4 x_5 \qquad (13)$$

2.2 Equation of Limit State

In structural reliability analysis, the limit state of the structure is generally the case that the function function is equal to 0, namely

$$Z = g(x_1, x_2, ..., x_n) = 0$$

For the basic deformation of the rod, the limit state is (1) to (13) equal to 0.

3 Reliability Calculation Mode in the Case of Different Distribution of Random Variables and Correlation of Failure Modes

3.1 JC Method's Equivalent Normal Mean and Standard Deviation and Component Failure Probability p_f the Bounds Interval Estimation Method

JC method is suitable for solving structural reliability index under random distribution of random variables. The basic principle of this method is to replace the non-normal distribution random variables with normal distribution, but it requires to replace the normal distribution function in the design of checking point x_i^*, the cumulative probability distribution function (CDF) and probability density function (PDF) values are the same as the original distribution function (CDF) and (PDF) values According to the above two conditions, The mean and standard deviation of the equivalent normal distribution ($\overline{X}_i'$ and $\sigma_{X_i'}$) are calculated according to the above two conditions. Finally, the reliability index and failure probability of components are calculated by the second-order matrix method. If a component is designed with two or more functional functions and the functional functions have the same random variables, there is a correlation between failure modes. and the following formula can be used to find the limit range of component failure probability [3].

$$\max_i p_{f_i} \le p_f \le 1 - \prod_{i=1}^{n} \left(1 - p_{f_i}\right) \tag{14}$$

The probability density functions corresponding to random variables commonly used in engineering structural component design are normally normal distribution lognormal distribution, extreme value I type distribution, Γ distribution, etc. The mean and standard deviation of their equivalent normal distribution are listed below [4]

1) the variable x_i is normal distribution, directly take the standard deviation and mean of the variable as "equivalent normal" standard deviation and mean, namely

$$\sigma_{X_i}' = \sigma_{X_i} \tag{15}$$

$$\overline{X}_i' = m_{x_i} \tag{16}$$

2) the variable x_i for lognormal distribution, the standard deviation and mean of the equivalent normal distribution are

$$\sigma'_{X_i} = x_i^* \left[\ln\left(1 + v_{x_i}^2\right) \right]^{\frac{1}{2}} \tag{17}$$

$$\overline{X}'_i = \overline{X}_i^* - s_i^* \sigma'_{X_i} \tag{18}$$

Type: $s_i^* = \dfrac{\ln x_i^* - \left[\ln m_{x_i} - \frac{1}{2} \ln\left(1 + v_{x_i}\right)^2 \right]}{\left[\ln\left(1 + v_{x_i}^*\right) \right]^{\frac{1}{2}}}$

Type: m_{x_i} is mean; v_{x_i} is the coefficient.

3) the variable x_i is the extreme value I distribution, its equivalent normal distribution standard deviation and mean is h

$$\sigma'_{X_i} = \varphi\left[\Phi^{-1}\left(F_{x_i}\right)\right] / f_{X_i}\left(x_i^*\right) \tag{19}$$

$$\overline{X}' = x_i^* - \sigma'_{X_i} \Phi^{-1}\left[F_{X_i}\left(x_i^*\right)\right] \tag{20}$$

Type: x_i is check points for design.

$$f_{X_i}\left(x_i^*\right) = a \exp\left[-a\left(x_i^* - k\right) - e^{-a\left(x_i^* - k\right)}\right] \tag{21}$$

is variable x_i, the original probability density is the value of the design check point.

$$F_{Xi}\left(x_i^*\right) = \exp\left[-\exp\left(-a\left(x_i^* - k\right)\right)\right] \tag{22}$$

is variable x_i, the original cumulative probability distribution function is the value of the design check point.

$$a = 1.2825 / \sigma_{X_i} \quad k = m_{x_i} - 0.5772 / a$$

4) variable x_i is Γ distribution, the standard deviation and mean of the equivalent normal distribution are

$$\sigma'_{X_i} = \phi\left[\Phi^{-1}\left(F_{x_i}\right)\right] / f_{X_i}\left(x_i^*\right) \tag{23}$$

$$\overline{X}'_i = x_i^* - \sigma'_{X_i} \Phi^{-1}\left[F_{x_i}\left(x_i^*\right)\right] \tag{24}$$

Type: $f_{X_i}\left(x_i^*\right) = \dfrac{\lambda\left(\lambda x_i^*\right)^{k-1} e^{-\lambda x_i^*}}{\Gamma(k)}$, $x_i \geq 0$, $\Gamma(k) = \int_0^\infty e^{-u} u^{k-1} du$; λ and k are two parameters that can be derived from the random variable x_i the original mean μ standard deviation σ formula find out. Formula: $\mu = k / \lambda$,

$$F_X\left(x_i^*\right) = \frac{1}{\Gamma(k)} \int_0^{\lambda x_i^*} e^{-u} u^{k-1} du \tag{25}$$

$f_X\left(x_i^*\right)$, $F_X\left(x_i^*\right)$ are the value of the original probability density and accumulated probability distribution function at the design checking point x_i^*.

3.2 Jc Method Analysis Steps and Component Failure Probability p_f Bound Interval Estimation Method [5]

Functional functions (1) to (13) JC method analysis steps are the same, axial tension and axial compression rod design only one failure mode; And rivets, ordinary bolt connection, pure torsion, plane bending have two related failure modes. The latter analysis steps include the former, and the latter failure probability JC method analysis steps, failure probability p_f the estimation methods of the boundary interval are the same [6], so the rivets or ordinary bolts are used as examples to illustrate.

1) JC method is used to analyze the limit state equation of function function (7) and obtain its reliable index β:

① Assume a beta value β.

②For all i values, the initial value of the design checking point is selected $x_i^* = m_{x_i}$.

③Calculate σ'_{x_i} and $\overline{X}'_i$ (The formulas for σ'_{x_i} and $\overline{X}'_i$ for various basic deformation have been found)by $\sigma'_{x_i} = \varphi\left[\Phi^{-1}\left(F_{x_i}\right)\right] / f_{x_i}\left(x_i^*\right), \overline{X}'_i = x_i^* - \sigma'_{X_i} \Phi^{-1}\left[F_{x_i}\left(x_i^*\right)\right]$.

④Calculate $\partial_g / \partial_{x_i}|_{x^*}$, taking Eq. (7) as an example

$$\frac{\partial g_1}{\partial x_1} = 2Ax_1x_2 \qquad \frac{\partial g_1}{\partial x_2} = Ax_1^2 = -1$$

⑤Calculate the sensitivity coefficient a_i from $a_i = \dfrac{\sigma'_{x_i} \cdot \frac{\partial g}{\partial x_i}\big|_{x^*}}{\sqrt{\sum\limits_{i=1}^n \left(\sigma'_{x_i} \cdot \frac{\partial g_1}{\partial x_i}\big|_{x^*}\right)^2}}$ $(i = 1, 2, 3)$.

⑥Calculate the new design checking point $X_i^* = \overline{X}' - a_i \beta \sigma'_{x_i}$. Repeat steps ③ ~ ⑥ until x_i^* the two differences are within the allowable range.

⑦Take x_i^* into the limit state equation $g\left(x_i^*\right) = 0$, calculate the value of β, take the limit state equation of function (7) as an example:

$$g_1 = C\beta^2 - D\beta + E = 0$$

Type: $C = Aa_1a_2\sigma'_{x_1}\sigma'_{x_2}$; $D = Aa_1\sigma'_{x_1}\overline{X}'_2 + Aa_2\sigma'_{x_2}\overline{X}'_2 - a_3\sigma'_{x_3}$; $E = A\overline{X}'_1\overline{X}'_2 - \overline{X}'_3$, then $\beta = \frac{D\pm\sqrt{D^2-4CE}}{2C}$, take a reasonable value of β.

⑧Repeat steps ③ ~ ⑦ until the absolute value of β difference is very small ($\Delta\beta \geq 0.01$). Finding the failure probability p_f by $p_f = 1 - \Phi(\beta)$.

2) The general limit range of component failure probability when the limit state function is related.

$$\max_i p_{f_i} \leq p_f \leq 1 - \prod_{i=1}^{n} \left(1 - p_{f_i}\right)$$

The limit state equations corresponding to functional functions (7) and (8), (10) and (11), (12) and (13) are related by random variables respectively, and the reliability limit under basic deformation of components can be calculated from the above equation [7].

4 Conclusion

In this paper, the failure modes of strength stability and stiffness under basic deformation in the online elastic range of member bar are comprehensively analyzed. the equivalent normal mean and standard deviation of JC method are calculated when random variables are distributed differently, and the reliability limit analysis method of member bar is given when functional functions are related. The results provide a preliminary conclusion for further analysis of reliability calculation of member bar under basic deformation in the elastic–plastic range,and further analysis of reliability calculation of member bar under combined deformation in the online elastic range and elastic–plastic range. It can be used for reference in the analysis of structural reliability.

References

1. Gong JX (2003) Calculation method of engineering structure reliability. Dalian University of Technology Press, Dalian
2. Sun XF (2009) Mechanics of Materials, 5th edn. Higher Education Press, Beijing
3. Wu SW (1990) Structural Reliability analysis. People's Communications Press, Beijing
4. Zhao G (1984) Engineering Structure Reliability. Water Conservancy and Electric Power Press, Beijing
5. Chen LM, Jiang ZC (2020) Construction Error Analysis of cable rod pretension structure node based on reliability theory. Spatial Structure
6. Wu Y (2018) Study on the calculated length coefficient of the tree-like structure rod piece. J Build Architect
7. Yao SY (2021) Study on bearing capacity of single rod. Build Struct

Research on Numerical Simulation Analysis and Engineering Application of Prestressed Anchor Cable Construction

Yanfang Zhu and Bing Xiong

Abstract The prestressed anchor cable can improve the overall stability of the rock and soil mass of the slope, thereby improving the mechanical properties of the rock and soil mass, and can better control the structural displacement, and achieve the purpose of reducing landslides, dangerous rocks and dangerous rocks. At the same time, the pre-stressed anchor cable construction is convenient, fast in progress, and economical, and has a broad application space. Based on the actual engineering background, this paper analyzes the key technologies in the construction process, and uses finite element software to conduct an overall numerical analysis of the soil after the use of prestressed anchor cables. It analyzes the impact of different slope top loads and finds that for non-prestressed anchor rods, the depth of action is limited, excessively increasing their length will not only cause difficulties in construction, but also have no practical significance. When prestress is applied, the horizontal displacement of the baffle is not uniform under the action of different prestress and slope top load, showing a twisting phenomenon. Along with the increase of the slope top load, the lateral pressure also increases, which causes the displacement of the baffle to decrease with the increase of the slope top load. The research in this paper will have certain guiding significance for the engineering application of prestressed anchor cables.

Keywords Prestress · Anchor Cable · Finite Element Analysis · Baffle Displacement

Y. Zhu
College of Civil and Architectural Engineering, Nanchang Institute of Technology, Nanchang 330099, China

B. Xiong (✉)
Shen Zhen Capol Interational & Associatesco, Ltd. Nanchang Branch Office, Nanchang 330000, China
e-mail: 55374782@qq.com

G. Feng (ed.), *Proceedings of the 9th International Conference on Civil Engineering*, Lecture Notes in Civil Engineering 327, https://doi.org/10.1007/978-981-99-2532-2_17

1 Introduction

With the further development of urbanization in my country, various projects such as urban rail transit, super high-rise buildings, water conservancy projects, bridge projects, tunnel projects, and expressways are developing vigorously. During the construction of these projects, it is often necessary to solve the problem of high slopes. Therefore, how to ensure the stability of the rock and soil mass in the construction process has become an urgent problem to be solved.

The emergence of prestressed anchoring technology has successfully solved the stability problem of high slope rock and soil. Rock-soil anchoring is an important branch of the field of geotechnical engineering. The technology is fixed to the slope surface through the outer end, and the other end is anchored in the stable rock mass within the sliding surface. The pre-stressed steel strand passes through the sliding surface to produce anti-sliding resistance on the sliding surface of the rock and soil, which increases the anti-sliding friction and makes the sliding surface of the rock and soil in a compressed state, thereby improving the slope. The integrity of the rock and soil body, which fundamentally improves the mechanical properties of the rock and soil body of the slope, and also enables the displacement of the rock and soil body of the slope to be well controlled, thereby achieving the reduction of landslides, dangerous rocks, and dangerous rocks purpose. The pre-stressed anchor cable has the following characteristics: the pre-stressed anchor cable can reduce the amount of excavation work, enter the hole in advance, and greatly shorten the excavation period; In the soil reinforcement and heightening construction, it does not affect the normal operation of the original project; It is better to repair the cracks or defects of concrete, which can disperse the excessive concentrated load in a larger range; It can improve the seismic and impact resistance of the soil without increasing its own weight. In short, prestressed anchor cables have the advantages of reasonable design, simple construction, significant benefits, and economic viability. These unique advantages have made them widely used at home and abroad.

Chen G et al., investigate the factors that influence variations in anchored cable prestess in EUD strata, and proposed a variation prediction model based on the equal-strain assumption for the anchored cable and the rock mass. One direct application of the new method is to predict the long-term variations in anchored cable prestress and evaluate the anchoring conditions, which can provide theoretical guidance for the safe operation and management of prestressed anchor engineering projects [1]. Sun X et al., analyzes the deformation mechanism of surrounding rock before and after prestressed anchor cable support through numerical simulation. A new tunnel support method using high prestressed constant resistance and large deformation anchor cable was proposed. The field monitoring results show that the constant resistance and large deformation anchor cable support can well control the deformation of surrounding rock, at the same time, the risk of damage to the steel arch due to local compression is reduced [2]. Li F et al., through long-term systematic experiments and theoretical analysis, a series of time-varying models of instantaneous corrosion rate, corrosion weight loss rate and tensile capacity were established based on the Hill function, and the corrosion process of the internal bonding section of the anchor cable under chloride erosion was studied [3]. Kim S H et al., through research and investigation, the structural performance of a long-span partially anchored cable-stayed bridge based on a new key-segment closure method based on thermal prestressing technology was investigated, and the three-dimensional finite element (FE) model of the partially anchored cable-stayed bridge was used to compare with the free cantilever. Method (FCM) matching detailed construction sequence analysis [4]. Qu H L et al., did some tests, the test results reveal the distribution pattern of earth pressures, seismic response of displacement and cable prestress, dynamic characteristics and amplification effect of reinforced slopes under earthquakes, and they provide a reliable basis for understanding the seismic behaviors and mechanism of the prestressed anchor sheet pile wall [5]. Bi J et al., a slope stability evaluation method is proposed to evaluate the stability of reinforced complex rock slopes, and two main reinforcement measures are considered, prestressed anchor cables and shear tunnels [6]. Li Z et al., studied the influencing factors of slope stability during excavation. The results show that the slope stability will change with the dynamic changes of the design parameters of anchor rods and lattice beams, and a stability calculation model for slope reinforcement with prestressed anchor rods and lattice beams is proposed [7].

Jia X Y et al., studied the influence of train vibration load on anchor cable prestress loss, pile sinking and horizontal displacement through model tests. The research results show that with the increase of the vibration frequency, the prestress loss of the anchor cable and the sinking and horizontal displacement of the pile body increase greatly [8]. Jian WX et al., established an optimal deformation coordination equation for pile-anchor cable structures on the basis of conventional deformation coordination conditions and modified deformation coordination conditions, combined with the virtual work principle of structural mechanics and the principle of graph multiplication, compiled a calculation program, and calculated prestress through examples and the total tension of the anchor cable. The results of the deformation coordination conditions show that the optimized deformation coordination conditions make the anti-slide pile in a good state of stress and deformation in the prestress stage [9]. Luo Q et al., studied the bond strength of the interface between grouting and rock during the design of concentrated tension prestressed anchor cable structure. The failure mode of the prestressed anchor cable is obtained through the laboratory pull test. The test results show that the main failure mode of prestressed anchor cables under concentrated tension is debonding along the cable-grouting material interface [10].

Li J et al., established a calculation model for high and steep slopes and calculated the changes in slope toe displacement and cable force increments under different prestressing actions. The results show that the reinforcement effect of the prestressed anchor cable on the slope effectively limits the displacement of the sliding body in the slope [11]. Chen Y et al., based on a certain soft rock slope support project, obtained a quantitative prestress loss law through field measurement and analysis. Prestress loss can be divided into instantaneous loss and time loss. The instantaneous loss is caused by the retraction of the steel strand, which can reach about 8% of the initial tensile load. Time loss is divided into short-term and long-term [12]. Kropuch S et al., measured these new defect mirror methods and their results for the first time. The application of the new method in the inspection of steel wires at anchor points will be able to increase the safety measures for the wire rope anchorages in the construction of bridges, roofs, towers and aerial ropeways [13]. He Z Y et al., proposed a new soft soil prestress loss analysis model and calculation formula. Applying the calculation formula to the analysis of the actual prestress loss law, the model essentially explains the prestress loss mechanism and provides a solution for the analysis of the prestress loss characteristics of anchor cables in soft soil areas [14].

Briskin ES et al., use anchor cable thrusters to ensure the high stability and energy efficiency of the entire system. They not only consider the resistance of exchange rates to movement, but also consider the current impact. Depending on the direction, it can either promote the movement of the platform or can offset [15]. Yu haikuo1 et al., in order to study the different control effects of grouting anchor cables, high-strength anchor cables, constant resistance and large deformation anchor bolts on the surrounding rock of coal roadways, the stress monitoring and analysis of the surrounding rock of the roadway is carried out by using drilling instruments and bolt dynamometers. The cable support changes the rules of the anchor cable during the tunneling and shutdown under the conditions of deformation and stress [16]. Maione A et al., uses the possibility of combining injection anchors made of inno-vative materials (such as pultruded CFRP pipes) with traditional grouting. For this purpose, a macroblock model is established [17]. Ahmed Lenda T et al., performed a finite element analysis of the shear behavior of adhesive anchors. The shear resis-tance and the corresponding dynamic increase factor (DIF) under different strain rates ($\dot{\varepsilon}$) and different design parameters were studied, and the failure mode of the adhesive anchor was examined. The results show that as the strain rate increases, the shear capacity and DIF of the adhesive anchor increase [18]. Moore Ellie et al., drags the micro-anchor through the sand of different densities in the centrifuge to measure the maximum penetration of the anchor. It is determined that the biggest factor affecting the penetration depth of the anchor is the size of the anchor by completing the centrifuge test at different g levels. The linear relationship between anchor size and penetration depth is recorded and used to develop new cable laying guidelines [19]. Adrian Batugin et al., developed and analyzed the combined support technology of adjacent roadways with respect to the bolt and cable mechanism using the spatial structure of the external staggered staggered slab layout. Subsequently, the optimal supporting parameters of side rock bolts and cables were determined, and the supporting effect of the gob-side roadway in the mining scene was verified [20].

However, the current understanding of the mechanism of prestressed anchorage is not very deep, which has a certain impact on its engineering impact. The key to anchoring technology lies in the reinforcement effect of anchoring on the rock and soil and the stress of a single anchor rod itself; The current engineering application mainly relies on experience, and the lack of relevant theoretical research cannot provide support for engineering applications. In this paper, combined with the actual project, the three-dimensional finite element simulation analysis of the anchor cable retaining wall in the project is carried out. The horizontal displacement of the retaining plate and the axial stress of the anchor cable under different slope top loads are emphatically

Fig. 1 Pore-forming
construction at the scene

analyzed, and the problems in the construction process are summarized and analyzed, which has good engineering application significance.

2 Project Profile and Key Technologies

2.1 Project Profile

The engineering example is a section of an anchor-cable retaining wall (shown in Fig. 1. of a city's flood control project, and the section of the retaining wall is shown in Fig. 2. The physical and mechanic al parameters of the loose layer of the embankment foundation are shown in Table 1, the physical and mechanical indexes of the rock mass of the embankment foundation are shown in Table 2, and the size of the anchor cable of the retaining wall is shown in Table 3.

2.2 Key Technology

Pore-forming. The anchor rod construction must first be drilled in accordance with the requirements of the construction drawings. The bore hole of the anchor rod should meet the design requirements of the aperture, length and inclination angle. The appropriate drilling method should be selected to ensure the accuracy, so that the rod body is inserted and the grouting operation and other subsequent constructions can proceed smoothly. Generally, it is required that the horizontal error of the anchor hole entry point is less than 50 mm, and the vertical error is less than 100 mm. In order to avoid the deviation of the inclined hole, the drilling rig parameters should be adjusted at any time according to the actual formation changes. Pay attention to the role of core picking in the process of drilling, and pick up as many cores as possible to ensure that the strata can be accurately divided at any time, the thickness of unstable rock masses can be understood at any time, and the locations of fractured

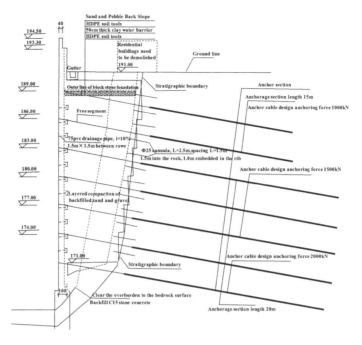

Fig. 2 Anchor retaining wall section

fracture zones, slip surfaces, and weak structural layers can be judged. Thickness to verify the accuracy of the geological survey data and modify the design if necessary.

Anchor Rod Production and Installation. The steel cable is cut according to the design requirements, and the nut is installed on the exposed end processed into a threaded shape. In order to ensure that the anchor rod is centered in the anchor hole, spacers must be installed on the anchor rod every 2 ~ 3 m, and finally anti-corrosion treatment is performed; The anti-corrosion and positioning treatment of the free section shall be carried out according to the design requirements. In order to ensure the unobstructed passage of the anchor cable, no residual debris in the passage and the integrity and defect-free work of the anchor cable body shall be completed before the anchor cable enters the hole, and the anchor cable construction shall be carried out according to the design requirements. During the installation of the cable body, to avoid torsion and bending of the anchor bundle, the grouting pipe and the cable body should be inserted into the hole at the same time. The bottom of the grouting pipe and the hole bottom should be maintained at 50 ~ 100 mm. When the cable body enters the hole, the central device should not move to prevent damage to the non-bonded sheath and anti-corrosion system of the free section. The length of the cable body entering the hole is greater than 95% of the length of the anchor bundle. The structure diagram and cross-section diagram of the prestressed anchor cable are shown in Fig. 3 and Fig. 4 respectively.

Table1 Physical and mechanical parameters embankment loose bed

Stratum lithologic		Specific gravity	Natural density	Void ratio	D_{85}	Internal friction angle	compression modulus	Permeability coefficient	Allowabe hydraulic Slope of landside	Allowabe bearing capacity	Friction coefficient
		-	g/cm^3	-	mm	-	MPa	m/d	-	kPa	-
Q_4^{3al} Medium-fine sand		2.60	1.8	0.8	-	20	7.0	5–15	0.35	80–100	0.25–0.30
Q_4^{2al} Pebble gravel	Baishi River	2.65	1.85	0.6	30–50	32–34	14.0	70–90	0.15–0.20	250–280	0.45–0.50
	Han jiang	2.67	1.86	0.5	80–110	33–35	15.0	80–100	0.10–0.15	270–300	0.50–0.55

Table 2 Dike rock physical and mechanical indexes

Strata	Saturation density	Saturated Compressive strength	Allowable bearing capacity	Softening factor	Elastic Modulus	Deformation modulus	Poisson's ratio	Shear strength f'	Coefficient of friction c'	Impact Resistance coefficient f	Strata
	g/cm^3	MPa	MPa	-	GPa	GPa	-	-	MPa	-	-
ε_{1+2}	2.68	25.0	1.0–1.4	0.6	7.0	3.5	0.29	0.65	0.50	0.55	1.20

Table 3 Retaining wall anchor cable size

Anchor number	1	2	3	4	5	6	7
Average length of free segment(m)	14.0	13.0	12.0	12.0	11.0	11.0	10.0
Length of anchoring section (m)	15.0	15.0	20.0	20.0	20.0	20.0	20.0
Total length (m)	29.0	28.0	32.0	32.0	31.0	31.0	30.0
Design anchoring force (kN)	1000	1000	1500	1500	1500	2000	2000

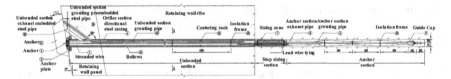

Fig. 3 Prestressed anchor cable structure

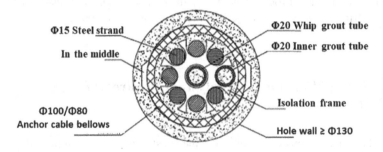

Fig. 4 Restressed anchor cable structure profile

Grouting Construction. Cement slurry or cement mortar is usually used to infuse the anchor hole. The bond strength and anti-corrosion effect of the anchor rod are largely affected by the composition of the slurry, the quality and the method of infusion. The compressive strength of the grouting slurry is required to reach the design strength at the age of 28 days. The slurry should have good stability after the preparation is completed, and can be stored for a long time under normal temperature and normal pressure, its basic performance remains unchanged, and no strong chemical reaction occurs. The grouting process cannot be interrupted to ensure that the operation is completed when the first injected slurry is still plastic. In the grouting process, the grouting pipe should be injected and lifted to ensure that the slurry liquid level is 50-80 cm higher than the grouting pipe head, and the situation that the grouting pipe head is higher than the slurry surface is prohibited.

Tension and Locking of Anchor Rod. The tensioning equipment is determined according to the material of the anchor rod and the size of the load lock, and the anchor clamps must be strictly inspected one by one before installation. Tension construction can be carried out only after the concrete strength of the anchorage section and the bearing platform reaches the design strength. In addition, it is necessary to ensure the smoothness of the bearing support member surface, install the platform and anchorage, and ensure that the vertical deviation from the axial direction of the anchor cable is not more than 5 degrees. At the same time, it must be tensioned in accordance with the prescribed procedures and the designed tensioning speed (usually 40 kN/min). The second pre-tensioning must be performed before the formal tensioning, and the tensioning force is about 10% to 20% of the design force.

Monitoring of Anchor Cables. After the construction of the anchor rod is completed, in order to ensure the working stability of the anchor rod, the pre-stress loss and displacement change of the anchor rod should be known at any time. Therefore, the anchor rod should be monitored for a long time, and the continuous observation time is usually longer than 24 h as long-term observation. During the monitoring process, if the working performance of the anchor rod is reduced and the anchoring force cannot be fully borne, it is necessary to adopt methods such as secondary tensioning of the anchor rod and increasing the number of anchor rods according to the monitoring results to ensure the stability of the anchoring project.

3 Numerical Calculation Analysis of Prestressed Anchor Cable

The Mohr–Coulomb model is selected as the constitutive relation of backfill gravel, and this plastic model is mainly applicable to materials characterized by particle structure under monotonic loading. The three-dimensional finite element model is used to analyze the anchor cable retaining wall. The components of the model are composed of concrete retaining plate, prestressed anchor cable, rock mass and backfill gravel, and each part adopts solid elements. The height (Z axis direction) of the anchor cable retaining wall is 28.0 m, the horizontal width (X axis direction) is 36.0 m, and the longitudinal width (Y axis direction) is 4.0 m between the prestressed anchor cables. The model is meshed by hexahedral elements and structured grid technology (STRUCTURED), and the calculation model is shown in Fig. 5. The boundary conditions include: the displacement of Y direction is constrained by two sides of the model, the displacement of Z direction is constrained by the bottom, and the displacement of X direction is constrained by the horizontal. The load includes: the weight of each unit component, the prestress of the anchor cable and the load on the top surface of the retaining wall. The top surface load is taken as 10 kPa, 30 kPa, 50 kPa, 100 kPa.

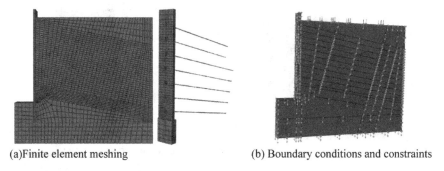

(a)Finite element meshing   (b) Boundary conditions and constraints

Fig. 5 Finite element model

Simulation of the construction process: 1) Excavation of the underground contin-uous wall and concrete pouring, concrete pouring to an elevation of 169.0 m; 2) After the concrete of the continuous wall and the wall panel reaches more than 85% of the design strength, the replacement part is excavated, and the C20 rubble concrete is replaced to the elevation of 169.0 m; 3)Carry out the drilling construction of No. 5, 6, and 7 anchor cables, and install No. 7 anchor cables (including corrugated pipes, position steel pipes, anchor pads, etc.); 4)C20 rubble concrete is poured to the elevation of 171.0 m, and then the retaining wall panel and the rib column are constructed to the elevation of 178.0 m (at the same time, No. 5 and No. 6 anchor cables and bellows in the rib column, positioning steel pipe, anchor pad plate, etc.). The panel and the rib column should be poured as a whole; 5) When the concrete of wall panel and rib column reaches more than 85% of the design strength, the back of the wall is backfilled. When backfilled to the elevation of 174.0 m, the No. 7 anchor cable is grouted and graded tensioned to 2000kN. When backfilled to the elevation of 177.0 m, the No. 6 anchor cable is grouted and graded tensioned to 2000 kN; 6)Carry out the drilling construction of No. 1, 2, 3, 4 anchor cables, and then pour the retaining wall panels and rib plate at the elevation of 178.0 ~ 187.0 m (at the same time, install the 2, 3, 4 anchor cables and the corrugated pipes in the ribs. Positioning steel pipes, anchor pads, etc.); 7) After the concrete strength of the retaining wall panel and rib column reaches more than 85% of the design strength, backfill the wall back. When the backfill reaches an elevation of 180.0 m, the No. 5 anchor cable is grouted and tensioned in stages, tensioned to 1500kN, and backfilled at an elevation of 183.0 m, grouting and graded tensioning of No. 4 anchor cable are carried out to 1500 kN, and when backfilling to an elevation of 186.0 m, grouting and graded tensioning of No. 3 anchor cable are carried out to 1500kN; 8)Pouring the retaining wall panels and ribs at an elevation of 187.0 m ~ 194.5 m (at the same time, the No. 1 anchor cable and the corrugated pipe in the rib column, positioning steel pipes, anchor pads, etc. should be installed); 9)After the retaining wall panel and rib column concrete reaches more than 85% of the design strength, backfill the wall back. When back-filled to an elevation of 189.0 m, grouting and graded tensioning of the No. 2 anchor

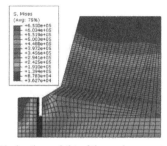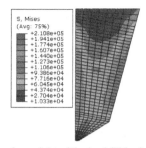

(a) Under the weight of the rock mass stress   (b) Under the weight of the backfill body stress

Fig. 6 Finite element calculation results

cable, tensioning to 1000kN, backfilling to the design at an elevation of 191.0 m, the No. 1 anchor cable is grouted and stretched to the designed anchoring force in stages; 10) Re-test the axial force of No. 2, 3, 4, 5, 6, and 7 anchor cables. If the design anchoring force is not reached, it shall be tensioned to the design anchoring force in stages; 11) Anchor sealing construction. The simulation results are shown in Fig. 6 (Figs. 7 and 8).

Fig. 7 Stress distribution of anchor cable

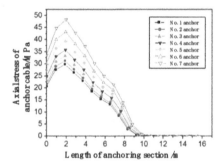

Fig. 8 Horizontal displacement of the baffle

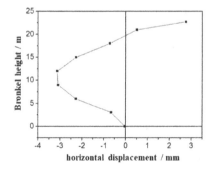

4   Numerical Simulation of Different Slope Top Loads

Numerical simulations are carried out for 10 kPa, 30 kPa, 50 kPa and 100 kPa slope top loads. The comparison results are shown in Fig. 9.

Because the designed prestress is applied to the anchor cable, the axial stress of the anchorage section changes greatly. The axial stress of the anchorage segment increases first and then gradually decreases along the length of the anchor rope, which is different from the rapid decrease to zero when there is no prestress. The axial stress of No. 1 anchor rope is small, and the axial stress of No. 7 anchor rope is the largest. Due to the self-weight of the rock mass and the backfill, the axial stress variation range of the anchoring section of the anchor rope gradually increases from top to bottom. The horizontal displacement of the baffle does not show a negative value, it changes unevenly along the height direction, and it also presents a distortion phenomenon. As the load on the top of the slope increases, the displacement of the baffle in the positive X direction decreases. In short, the axial stress of the anchorage section of the anchor rope presents a changing law that first increases and then decreases from the starting point of the anchorage section; Excessively increasing the length of the non-prestressed anchor rope is meaningless; The horizontal displacement of the baffle decreases with the increase of the slope top load (Fig. 10).

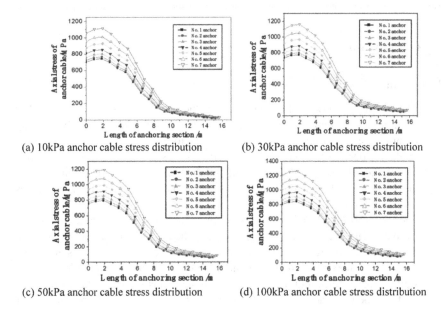

(a) 10kPa anchor cable stress distribution   (b) 30kPa anchor cable stress distribution

(c) 50kPa anchor cable stress distribution   (d) 100kPa anchor cable stress distribution

Fig. 9 Comparison of stress distribution of anchor cables under different slope top loads

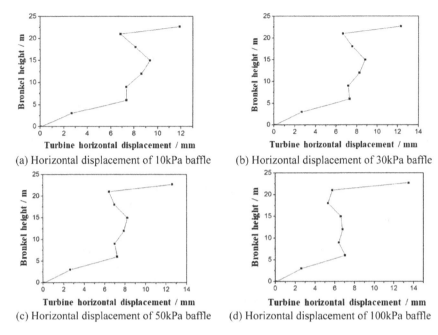

(a) Horizontal displacement of 10kPa baffle   (b) Horizontal displacement of 30kPa baffle

(c) Horizontal displacement of 50kPa baffle   (d) Horizontal displacement of 100kPa baffle

Fig. 10 Comparison of horizontal displacement of load baffles on different slope tops

5 Conclusion

Through the research of this paper, the following results are mainly obtained:

(1) The distribution law of the axial stress of the anchoring section of the anchor cable along the length of the anchor cable: a certain value from the starting point of the anchoring end rapidly increases to the maximum value, and then gradually decreases along the length direction, showing a nonlinear change. Due to the self-weight of the backfill and the rock mass, the variation range of the axial stress of the anchoring section of the anchor cable gradually increases from top to bottom.

(2) For non-prestressed anchor rods, the depth of action is limited, and excessively increasing its length will not only cause difficulties in construction, but also have no practical significance. Under the action of different prestress and slope top load, the horizontal displacement of the baffle of the anchor-cable retaining wall is not uniform and appears to be twisted. This is due to the co-restraint of the slope top load and the side pressure of the backfill body and the anchor cable.

(3) When the anchor cable is not prestressed, the horizontal displacement of the baffle is negative, that is, it deforms along the negative direction of the X axis under the action of lateral pressure; After prestress is applied, the effect of prestress is greater than the effect of lateral pressure, which the deformation

changes along the positive direction of the X-axis; With the increase of the slope top load, the lateral pressure also increases, which in turn causes the displacement of the baffle to decrease with the increase of the slope top load.

Acknowledgements Science and Technology Project of Jiangxi Provincial Education Department (Project Approval No. GJJ211908)

Humanities and Social Science Research Youth Project of Universities in Jiangxi Province in 2021 (Project Approval No. JC21223)

Key Research Base of Humanities and Social Sciences in Universities of Jiangxi Province (Project Approval No. JD21096)

References

1. Chen G, Chen T, Chen Y et al (2018) A new method of predicting the prestress variations in anchored cables with excavation unloading destruction. Eng Geol 241:109–120
2. Sun X, Zhang B, Tao Z et al (2020) Research on supporting measure at intersection of inclined shaft and major tunnel in highway. Adv Civil Eng 2020(5):1–15
3. Li F, Liu Z, Yu Z et al (2014) Experimental study on corrosion progress of interior bond section of anchor cables under chloride attack. Constr. Build. Mater. 71:344–353. https://doi.org/10.1016/j.conbuildmat.2014.08.063
4. Kim SH, Won JH (2016) Structural behavior of a long-span partially earth-anchored cable-stayed bridge during installation of a key segment by thermal prestressing. Appl Sci 6(8):231–231
5. Qu HL, Zhang JJ, Wang FJ (2013) Seismic response of prestressed anchor sheet pile wall from shaking table tests. Chin. J. Geotechn. Eng. 35(2):313–320
6. Bi J, Luo X, Zhang H et al (2019) Stability analysis of complex rock slopes reinforced with prestressed anchor cables and anti-shear cavities. Bull Eng Geol Env 78(3):2027–2039
7. Li Z, Wei J, Yang J (2014) Stability calculation method of slope reinforced by prestressed anchor in process of excavation. Sci World J 2014:1–7. https://doi.org/10.1155/2014/194793
8. Jia XY, Zhu YQ, Jia C (2018) Dynamic model test on anchoring effect of prestressed anchor cables. Zhongguo Gonglu Xuebao/China J Highway Transp 31(10):350–358
9. Jian WX, Deng XH (2014) Application of optimized deformation consistence condition to anchor cable tensile force calculation of pile-anchor cable structure. Rock Soil Mech 35(8):2171–2178
10. Luo Q, Liang LI (2011) Investigation into failure mechanism of prestressed anchor cables under concentrated tension in contrlling landslide. J Mt Sci 29(3):348–355
11. Li J, Chen S, Yu F, Jiang L (2019) Reinforcement mechanism and optimisation of reinforcement approach of a high and steep slope using prestressed anchor cables. Appl Sci 10(1):266. https://doi.org/10.3390/app10010266
12. Chen Y, Yin J, Hu Y (2013) Research on prestress quantitative loss law of soft rock slope anchor cable. Chin J Rock Mech Eng 32(8):1685–1691
13. Kropuch S, J Krešák, Peterka P (2012) Testing methods of steel wire ropes at the anchor. Acta Montanistica Slovaca. 17(3):174–178
14. He ZY, Xin A (2012) Prestress loss model and experimental study of anchor cables in soft soil area. Adv Mater Res 433–440:2769–2773
15. Briskin ES, Sharonov NG, Kalinin YV, et al (2021) About the geometrical form of the under-water robot platform with non-parallel system the anchor-rope propulsion. IOP Conf Ser Mater Sci Eng. 1129(1):012053. https://doi.org/10.1088/1757-899X/1129/1/012053

16. Yu H, Zhang X, Li Y, et al (2019) Comparative test and study on different types of anchor cable support in high stress deep rock mass. Am J Geogr Res Rev. 2(14). https://doi.org/10.28933/ajgrr-2019-09-2805
17. Maione A, Casapulla C, Di Ludovico M, et al (2021). Efficiency of injected anchors in connecting T-shaped masonry walls: a modelling approach. Constr Build Mater. 301(5):1–16. https://doi.org/10.1016/j.conbuildmat.124051
18. Ahmed Lenda T, Braimah (2021) A Shear behaviour of adhesive anchors under different strain rates. Eng Struct. 244(5):112763. https://doi.org/10.1016/J.ENGSTRUCT.112763
19. Moore E, Haigh SK, Eichhorn GN (2021) Anchor penetration depth in sandy soils and its implications for cable burial. Ocean Eng. 235(5):109411
20. Adrian B, Wang Z, Su Z, et al (2021) Combined support mechanism of rock bolts and anchor cables for adjacent roadways in the external staggered split-level panel layout. Int J Coal Sci Technol 4. https://doi.org/10.1007/S40789-020-00399-W

Stress State of Asphalt Mortar Based on Meso-Scopic Finite Element Simulation and Verification

Xiangye Tan, Yangpeng Zhang, Zhipeng Chen, and Qinglin Guo

Abstract The composition and structure of asphalt mixture have a significant impact on its mechanical performance, and the randomly distributed coarse aggregate impose a significant impact on the stress state of asphalt mortar. However, the traditional method can only evaluate the mechanical properties of asphalt mixture from macro scopic indexes, cannot show the stress state of asphalt mortar in details. Therefore, based on digital image processing technology, a meso-finite element model of asphalt mixture is established, and the stress state of asphalt mortar is analyzed by using finite element method. The correlation between the distance from mortar to aggregate and the stress state is established. Results show that the stress of asphalt mortar basically approaches to the external applied stress or zero when the distance from aggregate surface to mortar exceeds 4 mm. The closer it is to the aggregate surface, the greater the stress of the mortar is. Under compressive conditions, asphalt mortar will be affected by tensile stress in its vertical direction, and the maximum tensile stress is about 0.2–0.25 times of the maximum compressive stress. The results of this paper have a good reference for explaining the complex stress state of asphalt mortar.

Keywords Asphalt mortar · Stress state · Aggregate · Digital Image Processing · Meso-finite Element model

1 Introduction

The service life of asphalt pavement closely relates to the material properties of asphalt mixture, and the mixing degree and uniformity of asphalt mixture will affect

X. Tan
Expressway Development Center of Guangxi Zhuang Autonomous Region, Nanning, China

Y. Zhang (✉)
Guangxi Transportation Science and Technology Group Co., Ltd, Nanning, China
e-mail: zyp_engineering@outlook.com

Z. Chen · Q. Guo
School of Civil Engineering, Hebei University of Engineering, Handan, Hebei, China

© The Author(s) 2023   213
G. Feng (ed.), *Proceedings of the 9th International Conference on Civil Engineering*,
Lecture Notes in Civil Engineering 327,
https://doi.org/10.1007/978-981-99-2532-2_18

its performance. In the design of the material composition, the designed asphalt proportion is generally determined according to the macro mechanical parameters of the mixture [1–4], but the mechanical parameter is essentially the statistical average value which is determined through the test. These mechanical indexes cannot effectively characterize the stress transmission mechanism of different components in the mixture. Moreover, the stress state of asphalt mortar inside of the mixture is very complex due to the random distribution of coarse aggregates, and the stress state of the mortar will affect the travelling performance of asphalt pavement. In traditional research, the stress state of asphalt pavement is generally analyzed from the macro perspective, but it cannot reflect the internal stress state of asphalt mixture. This increases the risk of asphalt pavement damage [5–9]. On the other hand, the random internal structure of asphalt mixture will affect its macro mechanical properties. It takes a lot of money and people to test the performances of asphalt mixture through a lot of experiments, which is not a green and sustainable way. With the development of digital image processing technology, the internal structure of asphalt mixture can be obtained through digital image processing technology, then mesoscopic finite element simulation can be conducted on the mixture to determine the stress state of asphalt mortar. This is expected to become an efficient, green and sustainable research method in the future.

Therefore, digital image processing technology is used to obtain the internal structure of asphalt mixture in this paper, a 2D meso-finite element model of asphalt mixture is established, and the stress state of asphalt mortar is determined by simulation analysis. Moreover, the stress state of asphalt mortar on the surface of asphalt mixture is measured by digital image correlation (DIC) technology, which is compared with the theoretical results finally.

2 Theoretical and Experimental Methods

2.1 Digital Image Processing Method

For the convenience of image acquisition, this work cuts the asphalt mixture sample using the rock cutter, then one scans the section with a scanner and obtains a clear RGB image. After that, the image is denoised and segmented to obtain a binary image of coarse aggregate. The specific processing procedure is shown in Fig. 1.

2.2 Analysis Method on the Distance from Asphalt Mortar to Aggregate Surfac

In this paper, Euclidean distance is used to calculate the nearest distance from mortar to aggregate surface. The calculated result of Euclidean distance between a unit pixel

Fig. 1 Specific processing
procedure of digital image

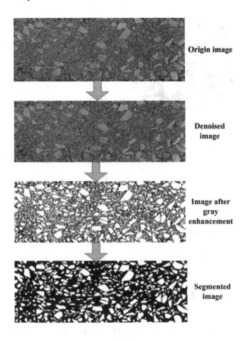

Origin image

Denoised image

Image after gray enhancement

Segmented image

Fig. 2 Euclidean pixel
distance

$$\begin{matrix} \sqrt{2} & 1 & \sqrt{2} \\ 1 & 0 & 1 \\ \sqrt{2} & 1 & \sqrt{2} \end{matrix}$$

O and its periphery is shown in Fig. 2. In other words, the coarse aggregate is used as the referenced pixel, so the Euclidean distance between the surrounding mortar and the aggregate surface can demonstrate the mortar away from the aggregate.

2.3 Finite Element Modeling Method

Import the binary image processed by MATLAB into software R2V for vectorizing processing. Geometric information of aggregate is saved as dxf file, and then dxf file is imported into ABAQUS to build the geometric sketch of aggregate. Finite element model is built based on the sketch. Finally, the finite element model is established by the default meshing of ABAQUS, the whole modeling process as shown in Fig. 3. Material parameters of aggregate and asphalt mortar in finite element simulation are listed in Table 1.

Binary image   Geometric sketch   FE model   Meshing

Fig. 3 Modeling process of FE model for asphalt mixture

Table 1 Material parameters

Type	Apparent density(g/cm^3)	Young's modulus(Mpa)	Poisson's ratio
Aggregate	2.79	20,000	0.2
Asphalt mortar	2.32	710	0.3

2.4 Uniaxial Compression Test

In order to verify the simulation results of this paper and determine the feasibility of meso-finite element analysis, Marshall specimen of dense asphalt mixture AC-16 was prepared according to standard for Asphalt and Asphalt Concrete of Highway Engineering (JTG E-20–2011). The cuboid specimen of 75 mm × 60 mm × 25 mm was made by cutting the marshall specimen. the apparent density of the specimen was 2.414 g/cm^3, air void percentage is 4.8%. Uniaxial compression test was carried out by electrical universal testing machine at 20°C, and the surface strain of asphalt mixture was tested by DIC technology. The loading rate is 5 mm/min. Test procedure is shown in Fig. 4.

Fig. 4 Uniaxial compression test

Fig. 5 Histogram of the nearest distance between the mortar and aggregate edge

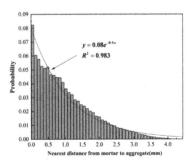

3 Stress State of Asphalt Mortar

3.1 Distribution of the Distance from Asphalt Mortar to Adjacent Aggregate Surface

Due to the random distribution of aggregate and irregular shape, the nearest distance between mortar and aggregate in the mixture also presents a random distributed form. In order to determine this distribution law, this paper makes statistical analysis on the nearest distance between mortar and aggregate of AC-16. Taking the nearest distance between each element of mortar and aggregate edge, and the area of each element as the statistical index. Result is shown in Fig. 5.

It can be seen from Fig. 6. that the statistical value of the nearest distance between mortar and aggregate edge in asphalt mixture shows a negative exponential distribution trend, and its distribution frequency decreases with the increase of the distance to aggregate edge, and the smaller the distance is, the higher the proportion is. According to the changing law of frequency curve, the maximum nearest distance from asphalt mortar to aggregate is about 4 mm.

3.2 Correlation Analysis between Mortar Stress and Distance

In this paper, the correlation between the stress of asphalt mortar and the nearest distance is analyzed. Because the stresses and the nearest distances are discrete, they cannot be fitted by commonly functions. But these data approximately obeys Gaussian distribution. Therefore, a Gaussian function is applied to contain these points as much as possible. The proportion of points falling below the Gaussian function in all points is taken as parameter T instead of coefficient R^2. If T belongs to the interval $(0.975, 1)$, it is considered that the Gaussian function can approximately represent the distribution. Gaussian function in this paper has the following form.

$$f(x) = a \exp(-\frac{(x-b)^2}{c}) \tag{1}$$

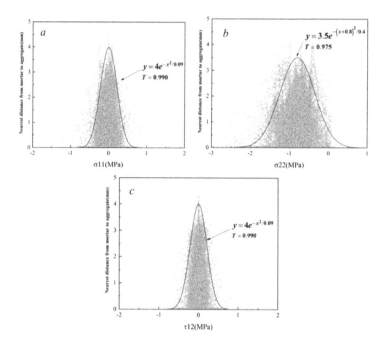

Fig. 6 Relationships of different stresses to nearest distance

where, a, b and c are regression analysis parameters.

Results of different stresses under the uniaxial compression condition are shown in Fig. 6. σ_{11} represents the stress perpendicular to the compressive direction, σ_{22} represents the compressive stress, and τ_{12} represents the shear stress.

As can be seen from Fig. 6, the distribution of the stress of asphalt mortar to the nearest distance is also approximately in accordance with the Gaussian trend. The maximum amplitude of stress appears at the aggregate edge, this shows that there is a high stress area near the interface between bitumen and aggregate, the maximum compressive stress can reach 1.45Mpa, while the maximum tensile stress and shear stress can reach 0.3Mpa. This shows that tensile stress still exists in its vertical direction even if the mixture is subjected to compression.

3.3 Influence of Loading Mode on Stress Characteristics of Asphalt Mortar

The loading conditions in compressive and shear mode are shown in Table 2. After simulating and analyzing different models, the stress of asphalt mortar is extracted and statistically analyzed. The regression results of σ11 after analysis are listed in Table 3.

Table 2 Load case for different loading mode

No	1#	2#	3#	4#	5#	6#
Shear stress (MPa)	0.72	0.576	0.432	0.288	0.144	0
Compressive strength (MPa)	0	0.144	0.288	0.432	0.576	0.72
Ratio of compression to shearing	0	0.25	0.666	1.5	4	∞

Table 3 Gaussian parameters for σ_{11} at different conditions

Parameters	a	b	c	T
1#	3.8	0	0.25	0.982
2#	3.8	0	0.23	0.982
3#	3.8	0	0.20	0.983
4#	3.8	0	0.13	0.982
5#	3.8	0	0.03	0.985
6#	3.8	0	0.02	0.987

As can be seen in Eq. (1) and Table 3, a represents the maximum distance that the stress in asphalt mortar becomes to 0. b represents the stress value corresponding to the maximum value of probability, which determines the position of Gaussian function on the x axis. c reflects the deviation degree of the stress for asphalt mortar away from the average stress. Based on the data in Table 3, it can be seen that the shortest distance for stress to zero is 3.8 mm, and there is a big change, while $b =$ 0 indicates that the horizontal stress is symmetrically distributed, and the value of c gradually decreases with the increase of the compression-shear ratio. It means that the horizontal stress tend to be concentrated, and the horizontal tensile effect caused by pressure is getting weaker and weaker.

3.4 Actual Stress Characteristic of Asphalt Mortar under Uniaxial Compression

Compared with the simulation results as shown in Fig. 7, it can be seen that the stress distance distribution characteristics calculated by finite element model are more concentrated, while those measured by uniaxial compression test are more discrete. This is because there are inevitably fine aggregates and voids in asphalt mixture mortar, which makes the stress transmission mechanism and deformed mechanism extremely complicated, so the stress is more random and discrete, which is different from the theoretical model to some extent. However, the stress and distance distribution characteristics of FE simulation and uniaxial compression test approximate to Gaussian distribution, and it is better to adopt Gaussian function for the next evaluation.

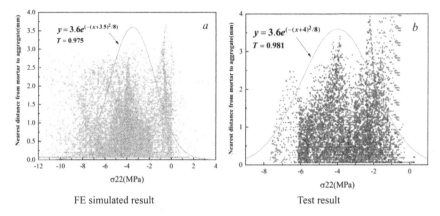

Fig. 7 Compressive stress characteristics of asphalt mortar for simulation and test

4 Conclusion

The following conclusions can be drawn based on the above analysis of this paper.

- The nearest distance between mortar and aggregate edge in asphalt mixture is decrease as a negative exponential trend. The stress decrease to zero when this distance reaches 4 mm for the dense asphalt mixture.
- The distribution of the stress of asphalt mortar to the nearest distance follows Gaussian function, there will be a certain tensile stress in its vertical direction while one direction is compressed. The maximum compressive stress is 4–5 times of the tensile stress.

References

1. Masad E, Saadeh S, AlRousan T et al (2005) Computations of particle surface characteristics using optical and X-ray CT images. Comput Mater Sci 34(4):406–424
2. Menapace I, Masad E, Bhasin A (2016) Effect of treatment temperature on the microstructure of asphalt binders: insights on the development of dispersed domains. J Microsc 262(1):12–27
3. Mahmoud E, Masad E (2010) A probabilistic model for predicting the resistance of aggregates in asphalt mixes to fracture. Road Mater Pave Des 11(11):335–360
4. Wang LB, Lai JS (2009) Quantification of specific surface area of aggregates using an imaging technique. Int. J Pave Res Technol 2(1):102–106
5. Wang L, Sun W, Tutumluer E et al (2013) Evaluation of aggregate imaging techniques for quantification of morphological characteristics. Transp Res Rec J Transp Res Board 233(51):39–49
6. Wang LB, Lane DS, Lu Y et al (2009) Portable image analysis system for characterizing aggregate morphology. Transp Res Rec J Transp Res Board 2104(1):3–11
7. Khattak MJ, Khattab A, Rizvi HR et al (2015) Imaged-based discrete element modeling of hotmix asphalt mixtures. Mater Struct 48(8):2417–2430

8. Moon KH, Falchetto AC (2015) Microstructural investigation of hot mix asphalt (HMA) mixtures using digital image processing. J Civ Eng 19(6):1727–1737
9. Arshadi A, Bahia H (2015) Development of an image based multi scale finite element approach to predict mechanical response of asphalt mixtures. Road Mater Pave Des 16(s2):214–229

Research on Key Technology of Rapid Integrated Construction of Fully Prefabricated Rigid Frame Bridge

Chao Yuan, Feng Li, Min Wang, and Fei Tian

Abstract Based on the reconstruction and expansion project of an expressway, a set of rapid construction technology for integrated erection of fully prefabricated rigid frame bridges is proposed, which solves the problems of unmatched work efficiency of each working face in the conventional integrated bridge erectors and long construction period of traditional suspension assembly technology. This method realizes the rapid construction of this type of bridges in highly urbanized areas. In this method, a new bridge girder integrated erection machine is used to synchronously install precast segmental beams, pier top blocks and precast pier column. At the same time, precast segmental beams near the middle pier are assembled by conventional cantilever assembly method, and the installation method of the precast segmental beams near the transition pier is optimized from the half span suspension assembly method to the cantilever assembly method by temporarily fixing the pier top block and arranging temporary prestressed tendons. This new integrated construction method improves the construction efficiency from 39.5d/unit to 32d/unit, and reduces the interference to traffic and environment.

Keywords Assembled Bridge · Rigid Frame System · Integrated Construction · Full Cantilever Assembly Technology

C. Yuan (✉) · F. Li · M. Wang · F. Tian
CCCC Second Harbor Engineering Company Ltd., Building A, China Communications City, No.668 Chunxiao Road, Wuhan, Hubei, China
e-mail: 415511610@qq.com

F. Li
Key Laboratory of Large-Span Bridge Construction Technology, Wuhan, Hubei, China

F. Tian
Research and Development Center of Transport Industry of Intelligent Manufacturing Technologies of Transport Infrastructure, Wuhan, Hubei, China

M. Wang
CCCC Highway Bridge- National Engineering- Research Centre Co., Ltd., Wuhan, Hubei, China

223

G. Feng (ed.), *Proceedings of the 9th International Conference on Civil Engineering*,
Lecture Notes in Civil Engineering 327,
https://doi.org/10.1007/978-981-99-2532-2_19

1 Introduction

At present, the construction of concrete bridges in China is still dominated by cast-in-situ technology, which has many disadvantages such as great environmental impact, long construction period, and not guaranteed construction quality. For this reason, the superiority of prefabricated bridges are more and more prominent in municipal projects with strict environmental regulations, high traffic safety demand and high construction efficiency demand [1, 2]. With the active promotion of the policy of developing prefabricated structures in China and the mature development of prefabricated bridge structures, prefabricated bridges are gradually widely used in the field of municipal engineering. In many projects, full prefabrication and assembly of piers, bent caps and main beams have been achieved, such as Jiamin Elevated Road and S7 Highway Project in Shanghai, Chengpeng Elevated Road in Chengdu, Xiangfu Road in Changsha, Fengxiang Road Rapid Reconstruction in Wuxi, etc. [3–5].

In the currently completed prefabricated bridge projects, the precast piers and bent caps are mostly installed by crawler cranes. This method has the defects of scattered working surfaces, large temporary land occupation for construction, and large traffic interference under the bridge, which cannot give full play to the advantages of prefabricated bridges. Therefore, the integrated installation method and equipment have started to be popularized and applied in domestic and foreign prefabricated bridge projects [6, 7]. The coastal viaduct project in Cartagena, Colombia needs to cross the environmental protection zone. In order to reduce environmental interference, the project adopts a fully prefabricated prefabricated bridge structure from pile foundation to main beam, and uses a cantilever integrated bridge erection machine to install all prefabricated components, realizing the integrated flow installation of fully prefabricated prefabricated bridges. In addition, there is no need to set up access roads along the project, which realizes zero interference to the environment during the construction period.. Subsequently, the Washington Elevations in the United States adopted the same structural form and construction technology. In the Yangang East Flyover Project in Shenzhen, China, in order to reduce the interference to the surrounding road traffic during the construction process, a new type of bridge girder integrated erection machine with landing front leg is used to erect concrete segmental beams and precast pier columns, and prefabricated components are transported and delivered on the completed bridge deck to the tail of the bridge erection machine, and the temporary land occupation for construction is small, and the traffic recovery is fast [8].

The integrated erection method allocates different types of prefabricated components to each working face of the bridge girder integrated erection machine for installation, and matches the installation efficiency of each working face, thus realizing the flow process of the entire installation. However, for rigid frame bridges with pier-beam consolidation, the installation of pier columns and the consolidation of pier and beams take a long time (10 days in total), which is lower than the installation efficiency of main beams (7 days). Therefore, it is difficult to achieve the efficiency matching of each working face by using conventional bridge girder integrated erection machines.

In view of the above problems, taking the reconstruction and expansion project of an expressway as the background, the conventional bridge girder integrated erection machine is upgraded to bridge girder integrated erection machines with dual landing front leg, and a rapid construction technology based on this new integrated bridge erection machine is proposed to realize the integrated flow construction of fabricated rigid frame bridge. Further more, the integrated installation efficiency was further improved by optimizing the conventional cantilever assembly method of segmental beam to the full cantilever assembly method.

2 Engineering Overview

The full rigid frame system is adopted for the main bridge of an expressway extension project, that is, the bearing less system with pier-beam consolidated is adopted for the middle pier and the transition pier. The span layout of this bridge is 45 m + 2 × 50 m + 45 m, as shown in Fig. 1.

The bridge is a fully prefabricated structure. The superstructure is single-box double-cell segment prefabricated box girder with a width of 20 m, which is erected by cantilever assembly method, as shown in Fig. 2. The middle pier and transition pier of the substructure are prefabricated curved column piers. Along the bridge, the middle pier is single column pier, and the transition pier is double column pier, as shown in Figs. 3 and 4

Most sections of the project are located in hills and water source protection areas. It is difficult to transport and deliver prefabricated components under the bridge. Therefore, prefabricated components need to be transported on the bridge deck and delivered from the tail of the bridge girder erection machine. Considering the construction conditions and structural characteristics of the supporting project, the integrated erection method is a reasonable process to realize the rapid construction of the project.

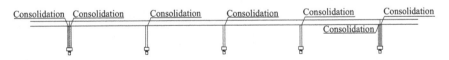

Fig. 1 Layout of full rigid frame bridge

Fig. 2 Sectional view of precast segmental box girder

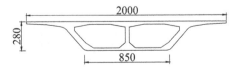

Fig. 3 Structural diagram of
middle pier

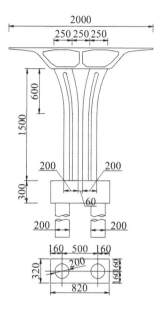

Fig. 4 Structural diagram of
middle pie

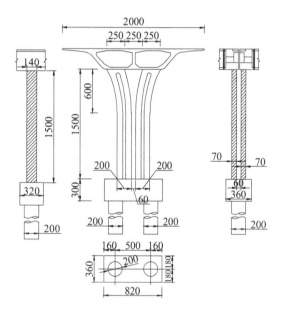

3 Rapid Integrated Construction Method

3.1 New Bridge Girder Integrated Erection Machine

The conventional bridge girder integrated erection machine consists of three spans. The front span is used to install the pier column and pier top block, and the middle span is used to install the main beam, and the tail span is used to lift the components, as shown in Fig. 5. The installation efficiency of each working face of the bridge girder erection machine is matched to realize the integrated flow construction. However, for the fabricated rigid frame bridge with pier-beam consolidation, the installation of pier column takes 3 days, and the consolidation of pier and beam takes 7 days, which takes 10 days in total, lower than the installation efficiency of main beam (7 days). Therefore, if the conventional integrated bridge erection machine is used, it is difficult to match the installation efficiency of the front span and the middle span, which will affect the efficiency of the construction.

In order to solve the above problems, it is considered to separate the working face of pier column and pier top block, and the bridge girder integrated erection machine with double landing front legs is proposed, as shown in Fig. 6. The main beam of the bridge girder erection machine adopts a double triangular truss structure, which is arranged from left to right in the order of component lifting section, main beam installation section, pier top block installation section and pier shaft installation section. The outrigger structure consists of 1# front leg, 2# front leg, 1# middle leg, 2# middle leg and the rear leg. During construction, the front leg is supported on the corbel, and the corbel structure is temporarily anchored to the cushion cap. The front leg of the bridge girder erection machine is equipped with a hydraulic pin system, which can adjust the height and fold. The front and rear lifting crane can rotate 360 degrees. The front lifting crane is specially used for the installation of pier column and pier top block, and the rear lifting crane is specially used for the installation of main beam.

Compared with the conventional bridge girder integrated erection machine, the supporting structure of the new bridge girder erection machine has an additional

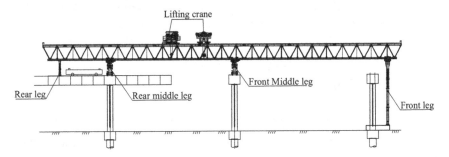

Fig. 5 Schematic diagram of common integrated bridge erecting machine

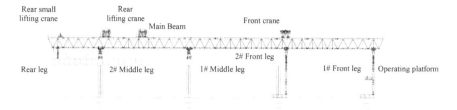

Fig. 6 Schematic diagram of new integrated bridge erecting machine

landing front leg. This additional span of bridge girder erection machine is used to erect the pier column of the previous span of the bridge, so as to separate the working face of pier column and pier top block, thus solving the problem of mismatched work efficiency.

3.2 Full Cantilever Assembly Method of T-Frame of Transition Pier

During the erection of the main beam of the rigid frame bridge, the segmental beam near the middle pier is generally installed by conventional cantilever assembly method, while the segmental beam near the transition pier is usually installed by half span suspension assembly method, which means that after the completion of the T-frame of the middle pier, the half span of the transition pier is installed by overall suspension, and then the bridge is completed after tensioning the prestress, as shown in Fig. 7. This construction method is the most widely used and mature construction method in balanced cantilever construction in China [9, 10]. However, the half span suspension assembly method for the side span of the transition pier will significantly increase the construction period. At the same time, this method has higher requirements on the lifting weight of the bridge girder erection machine.

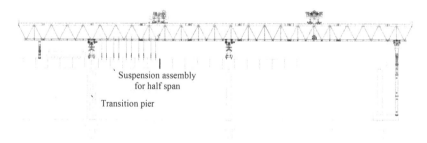

Fig. 7 Schematic diagram of integral suspension assembly for half span of side span at transition pier

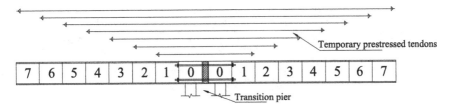

Fig. 8 Schematic diagram of T-structure formation of transition pier

In order to further improve the installation efficiency of segmental beam by using cantilever assembly method, a full cantilever assembly method is proposed to realize the cantilever assembly of segmental beam near the transition pier. In this method, the pier top block is temporarily fixed, and the cantilever assembled side span is formed by arranging temporary prestressed tendons, as shown in Fig. 8; When the side span is closed and the permanent prestress is tensioned, the temporary prestress and temporary fixation can be removed and the system transformation can be completed. This construction method can realize the full cantilever construction of rigid frame bridge and effectively shorten the construction period.

3.2.1 Temporary Fixation of Pier Top Block of the Transition Pier.

In order to ensure that the segmental beam near the transition pier form a T-frame structure and then adopt the cantilever assembly method, the pier top block of the transition pier shall be temporarily fixed. This pier top block is a prefabricated shell structure. Only one part of this pier top block is consolidated with the pier top, and the other part is suspended from the pier and supported on the bracket installed on the top of the transition pier. Cushion blocks shall be set in the intersection joints of transition piers. In order to facilitate removal in the future, hardwood blocks or steel plates that meeting the strength requirements can be filled in the intersection joints. At the same time, in order to reserve a certain amount of compressive stress at the joint after the installation of the pier top block, the fining twisted steel bar should be tensioned at the diaphragm of the transition pier. The details of temporary fixation is shown in Fig. 9.

3.2.2 Layout of Temporary Prestressed Tendons of T-Frame of the Transition Pier

The first type is the form of the top plate being grooved. In the process of cantilever assembly, the temporary prestressed tendons can be arranged in a straight line. The temporary prestressed tendons shall be arranged within the thickness range of the top plate according to the specification, and temporary prestressed ducts shall be reserved during the prefabrication of segmental beams. At the same time, reserved slots shall

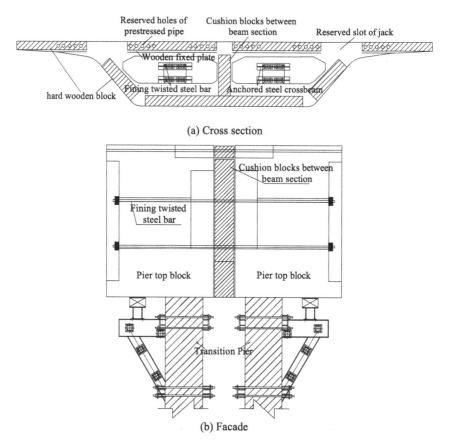

Reserved holes of prestressed pipe | Cushion blocks between beam section | Reserved slot of jack

Wooden fixed plate

hard wooden block | Fining twisted steel bar | Anchored steel crossbeam

(a) Cross section

Cushion blocks between beam section

Fining twisted steel bar

Pier top block | Pier top block

Transition Pier

(b) Facade

Fig. 9 Temporary anchoring measures for top block of transition pier

be set within a certain length range at the back end of the top plate to leave space for anchoring the temporary prestressed tendons and facilitate subsequent releasing and removal of the temporary prestressed tendons. Since the elongation of the prestressed tendon after tensioning is 30 cm, in order to facilitate extension of connector and releasing of prestressed tendons, the longitudinal size of the reserved slot should be greater than the length of the releasing of prestressed tendons, so it is advisable to reserve more than 50 cm. The structure and process are shown in Fig. 10.

The second type is the Form of tooth block. The anchorage end of the temporary prestressed tendon can also adopt form of tooth block, as shown in Fig. 11. At this time, the tensioning and removal operations are carried out inside the box room, and the temporary structure increases the self weight of the segment beam to a certain extent.

When the form of tooth block is adopted, the prefabricated members need to add reinforcement of anchor block and formwork of anchor block, which will increase the process. When the form of the top plate being grooved is adopted, there is no

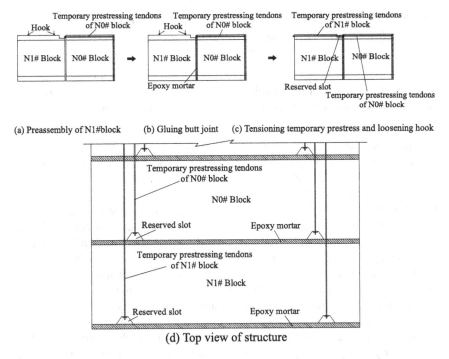

(a) Preassembly of N1#block   (b) Gluing butt joint   (c) Tensioning temporary prestress and loosening hook

(d) Top view of structure

Fig. 10 Schematic diagram of slotted temporary prestressed anchorage at the top plate

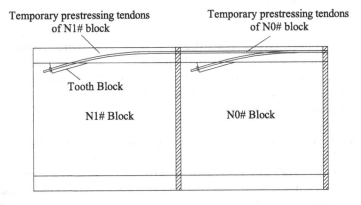

Fig. 11 Schematic diagram of temporary prestressed anchorage of tooth block form

need to add temporary anchor tooth blocks, and the temporary prestressed tendons can be tensioned directly at the roof. At this time, the construction efficiency is good. Therefore, it is recommended to adopt the form of the top plate being grooved for the temporary prestressed tendons of T-frame of the transition pier.

3.3 Rapid Integrated Construction Method Based on New Bridge Girder Integrated Erection Machine

According to the above discussion and the background project, a rapid integrated construction method with full cantilever assembly of rigid frame bridge based on bridge girder integrated erection machine with double landing front legs is proposed. The overall process is as follows:

(1) The N3 # precast pier column that has been temporarily stored at the platform of N1 # pier is lifted to the forward most span by two cranes, and the pier column is turned over and installed in place, and then the grouting connection between the pier column and the platform is started.

(2) While the first two cranes install the N3# pier column and conduct grouting connection between the pier column and the platform, the rear cranes lifts the N2# pier top block that has been temporarily stored at the platform of N1# pier to the top of the N2# pier, and then installs the pier top block of N2# pier. When the pier top block is aligned, the pier and beam is consolidated.

(3) While connecting the pier column with the platform and the pier top block with the pier column and waiting for the strength to meet the requirements, the T-frame of N1# pier is assembled with a bridge girder erection machine by symmetrical cantilever assembly method, and the shear cone is used to temporarily fix the segmental beam. After that, the permanent prestressed tendons are tensioned, and then the temporary prestressed tendons are removed, and the bridge girder erection machine loosens the hook.

(4) According to the previous step, the T-frame of N1 # pier is symmetrically assembled to the maximum cantilever position. After that, the cantilever of T-frame of N0# pier and N1 # pier is closed, and the wet joint is poured and cured. After the strength reaches the standard, the permanent prestressed tendons will be tensioned, and the construction for the T-frame of N1# pier will be completed. Then the temporary prestressed tendons of T-frame of N0# transition pier and the temporary fixing measures between the two top blocks of N0# transition pier are removed.

(5) 2# middle leg is moved forward to the top block of N2# pier, the rear leg is moved forward to the top block of N # transition pier, and 2# front leg is moved forward to the corbel of platform of N3 # pier and temporarily connected with 1# front legs. Then the top block of N3# pier and N4 # pier column are lifted from the tail of the bridge girder erection machine and transported on the bridge deck, and they are finally temporarily stored at the platform of N2# pier. After that, the main beam and rear leg of the bridge girder erection machine are moved forward, and then 1# front leg is moved forward to the corbel of platform of N4# pier, and the 2# front leg is folded up, moved forward across N3# pier, and supported on the corbel of platform of N3# pier. So the crossing span of the bridge girder erection machine is completed.

(6) Repeat steps 1 to 5 until the construction of T-frame of N4# transition pier is completed, that is, the construction of one-coupling bridge (N0# ~ N4#) is

completed. (No temporary fixing measures are required for the top block of middle pier).

The schematic diagram of the above construction process is shown in Table 1.

Table 1 Schematic diagram of rapid integrated installation process

Order number	Schematic diagram
1	
2	
3	
4	
5	
6	

4 Comparison and Analysis of Work Efficiency

Taking the four-span one-coupling prefabricated rigid frame bridge of the background project as the object, one side of the T-frame of a main beam is divided into seven sections, and three construction schemes are considered respectively. First, cantilever assembly of middle pier + half span suspension assembly of side span of transition pier + common integrated bridge girder erection machine; Second, cantilever assembly of middle pier + half span suspension assembly of side span of transition pier + new integrated bridge girder erection machine; Third, full cantilever assembly + new integrated bridge girder erection machine. The construction efficiency analysis of the above schemes is shown in Table 2–4. The brackets in the table indicate that the process time does not occupy the key path.

The following contents can be analyzed from Tables 2, 3, 4. First, the new integrated bridge girder erection machine separates the working face of installation of pier column and pier top block, which realizes the matching of the installation efficiency of the substructure and superstructure, and shortens the time that the installation of pier shaft and pier top block occupies the key path of flow construction from 10 days to 7.5 days, effectively improving the construction efficiency. Second, when the half span suspension assembly method is adopted for the segmental beam near the transition pier of side span, the construction period is significantly increased, and the integrated flow construction is discontinuous, which significantly reduces the construction efficiency. Third, based on the new bridge girder integrated erection machine and the full cantilever assembly method, the continuity of integrated flow construction is ensured, which effectively shortens the installation period and improves the installation efficiency from 39.5 days/unit to 32 days/unit.

Table 2 Construction efficiency analysis (Cantilever assembly of middle pier + Half span suspension assembly of side span of transition pier + common integrated bridge erecting machine)

Order number	Main procedure	Duration(day)	Remarks
1	Install N2# pier column and wait for the connection strength to meet the requirements	3	
2	Install N2# pier top block and wait for the connection strength to meet the requirements	7	
3	Install segmental beam of T-frame of N1# pier by cantilever assembly method	(7)	Start synchronously with step 1

(continued)

Table 2 (continued)

Order number	Main procedure	Duration(day)	Remarks
4	Install segmental beam of side span of N0# transition pier by half span suspension assembly method	4(7)	Immediately after the step 3, occupy the critical route for 4 days
5	Crossing span of the bridge girder erection machine	0.5	
6	Close T-frame cantilever, pour wet joint and maintain	(1)	Start synchronously with step 5
7	Repeat steps 1 ~ 3 and 5 ~ 6, continue to complete the construction of T-frame of N2# and N3# middle pier		
8	Repeat steps 4 ~ 6, complete the construction of side span of N4# transition pier by half span suspension method, that is, the construction of one-coupling bridge is completed		
Total		14.5 + 10.5 + 14.5 = 39.5	

Table 3 Construction efficiency analysis (Cantilever assembly of middle pier + Half span suspension assembly of side span of transition pier + new integrated bridge erecting machine)

Order number	Main procedure	Duration(day)	Remarks
1	Install N3# pier column	0.5	
2	Wait for the connection strength of N3# pier column and platform to meet the requirements	(2.5)	
3	Install N2# pier top block and wait for the connection strength to meet the requirements	7	Start synchronously with step 2
4	Install segmental beam of T-frame of N1# pier by cantilever assembly method	(7)	Start synchronously with step 3

(continued)

Table 3 (continued)

Order number	Main procedure	Duration(day)	Remarks
5	Install segmental beam of side span of N0# transition pier by half span suspension assembly method	7	Immediately after step 4
6	Crossing span of the bridge girder erection machine	0.5	
7	Close T-frame cantilever, pour wet joint and maintain	(1)	Start synchronously with step 6
8	Repeat steps 1 ~ 4 and 6 ~ 7, continue to complete the construction of T-frame of N2# and N3# middle pier		
9	Repeat steps 4 ~ 6, complete the construction of side span of N4# transition pier by half span suspension method, that is, the construction of one-coupling bridge is completed		
Total		$15 + 8 + 15 = 38$	

Table 4 Construction efficiency analysis (Full cantilever assembly + new integrated bridge erecting machine)

Order Number	Main procedure	Duration(day)	Remarks
1	Install N3# pier column	0.5	
2	Wait for the connection strength of N3# pier column and platform to meet the requirements	(2.5)	
3	Install N2# pier top block and wait for the connection strength to meet the requirements	7	Start synchronously with step 2
4	Install segmental beam of T-frame of N1# pier by cantilever assembly method	(7)	Start synchronously with step 3
5	Crossing span of the bridge girder erection machine	0.5	
6	Close T-frame cantilever, pour wet joint and maintain	(1)	Start synchronously with step 5
7	The next T-frame shall be constructed according to steps 1 to 6 until the completion of T-frame of N4# transition pier,that is,the construction of one-coupling bridge is completed		
	Total	$4 \times 8 = 32$	

5 Conclusion

At present, the conventional bridge girder integrated erection machine is not suitable for fully prefabricated rigid frame bridges. Due to the mismatch of the installation efficiency of the substructure and superstructure, enforced idleness may occur on the working face.

Aiming at the fully prefabricated rigid frame bridges in highly urbanized areas, a rapid construction technology of this type bridges based on the new bridge girder integrated erection machine with double landing front legs is proposed, which realizes the matching of the installation efficiency of the substructure and superstructure, and shortens the time that the installation of pier column and pier top blocks occupies the key path of flow construction from 10 days to 7.5 days, which effectively improves the construction efficiency and reduces the interference to the surrounding environment and traffic during the construction period.

For the precast segmental beam near the transition pier of side span, the conventional half span suspension assembly method is optimized to full cantilever assembly method, which ensures the continuity of the integrated flow construction, further improves the installation efficiency from 39.5d/unit to 32d/unit.

Acknowledgements Foundation Items: National Key Research and Development Program (2021YFF0500904).

References

1. Sun C (2021) Utilization and development of prefabrication and modular green construction for urban bridges. World Bridges 49(01):39–43
2. Yang W, Tsay J, Lau J, Huang J, He Y (2019) Development and application of technologies on prefabricated bridges. Guangdong Highway Commun. 45(05):67–73
3. Zhang H (2016) Field Istallation & Positioning technique of prefabricated column in Shanghai Jiamin Viaduct Project China Municipal Eng. 18(04):59–61
4. Zhang W (2016) Research on design characteristics of Xiangjiang River Bridge crossing Beijing Guangzhou railway in Xiangfu Road. Changsha Highway. 61(03):77–81
5. Li W, Ren C (2019) Application of prefabricated bent caps in Fengxiang Road Overpass in Wuxi Urban Roads Bridges & Flood Control. 11:121–124.
6. Zhang H, Zhang Y, Wang M, Xia H, Tian F (2018) Integrated construction method and equipment of assembled composite beam bridge. J China Foreign Highway. 38(06):140–143
7. Chen J (2018) Integrated construction technology of prefabricated bridge beam and pile column. Western China Commun Sci Technol 11:151–153(2018)
8. Cui C, Wu W (2020) Construction technology of prefabricated integrated bridge erecting machine. Constr Mach 11:96–98
9. Luo C, Zhou H, Zhu W (2020) Application of cantilever construction technology in bridge construction. Intell City. 06(11):206–207
10. Duan Z, Liang F (2020) Key techniques for assembly construction of segmental beam of approach bridge of Brunei. PMB Bridge Western Special Equip. 01:48–52

Study on Ground Deformation Law of Small Spacing Shield Tunneling in Sand and Rock Composite Stratum

Liwen Deng, Jianting Pan, Haolin Ye, Jinming Zeng, Hao Lu, and Hong Pan

Abstract The construction of shield tunnel often occurs in composite stratum, so it is of great significance to study the law of surface deformation under the working condition of composite stratum in order to ensure the safety of shield construction. Based on the actual tunnel engineering, the law of ground surface soil deformation caused by shield construction in the composite stratum of upper sand and lower rock is studied by using the method of field measurement. The results show that: The width coefficient of the surface settlement trough basically remains stable with the shield construction, and the stratum loss rate changes exponentially with the shield construction; Affected by the disturbance of the previous shield construction, the surface settlement trough caused by the subsequent shield construction is no longer on the top of the tunnel axis, and will move with the construction process, which is greatly affected by the shield attitude parameters; Taking into account the conditions of the center movement of the settlement trough of the subsequent tunnel, the total surface settlement caused by the successive shield constructions adopts the modified Mark-bolt formula to make the fitting parameters of the settlement trough more suitable for reality.

Keywords Surface Subsidence · Field Test · Composite Stratum · Double Tunnel · Fitting

L. Deng · J. Pan · J. Zeng · H. Lu
China Southern Power Grid, Guangzhou 510660, China

H. Ye
Guangzhou Mass Transit Engineering Consultant Co., Ltd., Guangzhou 510010, China

H. Pan (✉)
South China University of Technology, Guangzhou 510641, China
e-mail: hpan@scut.edu.cn

© The Author(s) 2023
G. Feng (ed.), *Proceedings of the 9th International Conference on Civil Engineering*,
Lecture Notes in Civil Engineering 327,
https://doi.org/10.1007/978-981-99-2532-2_20

239

1 Introduction

With the rapid development of domestic urban rail transit construction, the stratum of tunnel crossing is complex and diverse, and the construction frequency of shield tunnel in composite stratum is getting higher and higher. Composite stratum is composed of two or more strata with different physical and mechanical parameters within the range of tunnel excavation surface and in the direction of tunnel excavation.

Due to the differences in the properties of different strata, the primary problems to be solved in the construction of shield tunnel in composite strata are mainly focused on shield type selection design, construction technical measures, construction parameter optimization, shield adaptability [1–3]. There is also a high degree of attention to the study of surrounding stratum disturbance, but the current research methods are mostly numerical simulation [4, 5] and theoretical analysis [6, 7]. Among them, the numerical simulation is generally based on the finite element calculation of the continuum [8, 9], the analysis of the field measured data is less, and there is little research on the sand layer [10].

The upper soft strata and lower hard strata are the more common composite strata. In the current field measurements, the upper soft and lower hard strata mostly refer to the upper clay and lower rock strata [11, 12], but there is less research and analysis on the field monitoring of the surrounding strata disturbance caused by the shield construction in the upper sand and lower rock strata, and there is even less research and analysis on the secondary disturbance of the soil caused by the construction of two tunnels one after another.

The analysis and prediction of surface settlement due to shield construction is mostly done using the Peck formula [13], as follows:

$$S(x) = \frac{V_S}{\sqrt{2\pi} i} \cdot e^{\left(-\frac{x^2}{2i^2}\right)} \tag{1}$$

where $S(x)$ is the surface settlement value at distance x from the tunnel axis, i is the width factor of the settlement trough, V_S is the amount of ground loss per unit length, $V_S = \pi R^2 V_L$, V_L is the rate of ground loss per unit length, and R is the radius of cutter excavation. Peck assumes that the soil does not drain during construction and that the ground loss is uniformly distributed along the entire tunnel length.

O'reilly [14] based on the actual monitoring data in London, it is concluded that there is a linear relationship between i and tunnel depth z_0:

$$i = K \cdot z_0 \tag{2}$$

where K is the width parameter of settlement trough and is dimensionless.

Peck formula is often used to predict and evaluate the surface subsidence caused by shield construction, but two key parameters i and V_S are generally difficult to be determined in advance and have low versatility due to the influence of stratum conditions, tunnel parameters and construction quality. Therefore, dimensionless parameters K and V_L are usually used for reference between surface subsidence caused by shield engineering in different areas, which is practical and empirical [15, 16].

According to the assumption of Peck formula, V_L (or V_S) has only one value in the average sense in the whole process of shield construction, and K (or i) has only one value in most shield projects. At present, there are the following deficiencies in the research of K and V_L: (1) the timing of inversion of K and V_L parameters in different projects is not uniform and the explanation is not clear, which leads to a wide range of empirical values, which brings difficulties for prediction and calculation; (2) limited by the monitoring frequency, the actual change of the shape characteristics of surface settlement trough in the process of shield construction is not clear.

In order to find out the variation law of the characteristic parameters of the surface settlement trough with the shield construction and its performance in the upper sand and lower rock composite strata, this paper relies on the actual tunnel engineering, through on-site intensive monitoring, this paper analyzes the deformation law of the surface soil caused by the shield construction and the morphological characteristics of the final settlement trough after the construction.

2 Background

2.1 Tunnel Condition

220 kV Shijing-Huanxi Electric Power Tunnel (Xiwan Road ~ Shisha Road Section) is located in Baiyun District, Guangzhou City. It is constructed from north to south along Shikuan Road. Parallel to the power tunnel is the north extension of Guangzhou Metro Line 8, which is constructed from south to north.

The length of electric power tunnel shield machine is 9.62 m, the inner diameter of shield segment is 3.6 m, the outer diameter is 4.1 m, the diameter of cutter head is 4.35 m, and the length of segment is 1 m. The length of shield machine in subway tunnel is 10.29 m, the inner diameter of shield machine segment is 5.4 m, the outer diameter is 6.0 m, the diameter of cutter head is 6.28 m, and the length of segment is 1.5 m. The net distance between the two tunnels is about 3 m. In the monitoring area, the tunnel constructed first is subway, and the tunnel constructed later is electric power.

2.2 Survey Point Layout and Deformation Summary

The section where the monitoring section is located belongs to the upper sand and lower rock composite strata with karst development, and the stratigraphic distribution

Fig. 1 Monitoring section stratigraphic distribution and measuring point arrangement

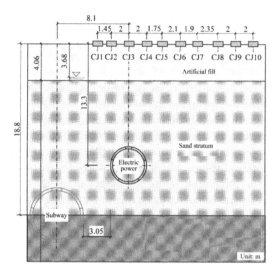

at the section is shown in Fig. 1. At the location of the monitoring section, nearly half of the sand layer is exposed on the construction surface of the subway tunnel, and the construction surface of the electric power tunnel is basically a full-section sand layer. The measuring points of surface deformation are set in the cross section, and the subway tunnel is located under the Shikuan Road, so the measuring points can not be arranged, and the positions of the measuring points are shown in Fig. 1. The front and rear 10 rings of the monitoring section belong to intensive monitoring, with monitoring points in each ring, and the other 5 rings are non-intensive monitoring, with one monitoring point in every other ring.

The surface deformation caused by subway and electric power tunnels construction is shown in Fig. 2 and Fig. 3 respectively. The more close to the tunnel, the greater the deformation. The maximum deformation caused by subway and electric power construction is CJ1 and CJ3, respectively, and the corresponding settlement is 15.24 and 23.31 mm. In addition, the settlement of each measuring point is stable with the shield construction.

3   Monitoring and Analysis of Surface Subsidence

Considering the disturbance of the soil near the electric power tunnel caused by the subway tunnel construction, the disturbance degree of the soil on both sides of the electric power tunnel is different, and the excavation causes the center of the horizontal surface settlement trough to shift the tunnel axis, so it is necessary to modify the Peck formula. As follows:

Fig. 2 Ground surface deformation of subway shield construction

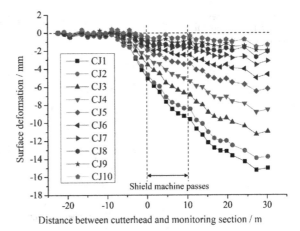

Fig. 3 Ground surface settlement of power shield construction

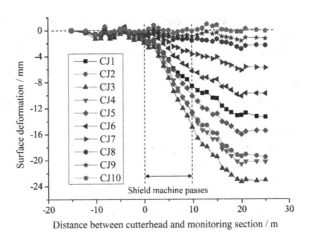

$$S(x) = \frac{V_S}{\sqrt{2\pi}i} \cdot e^{\left(-\frac{(x-x_c)^2}{2i^2}\right)} \tag{3}$$

Among them, x_c is the distance of the moving axis in the center of the settlement trough.

3.1 The Variation Process of Settlement Trough Parameters with Shield Construction

The curve fitting degree of settlement trough caused by subway and electric power tunnel construction is shown in Fig. 4. *adj-R²* reflects the curve fitting degree. The closer the value is, the higher the fitting degree is. It can be seen that the curve fitting

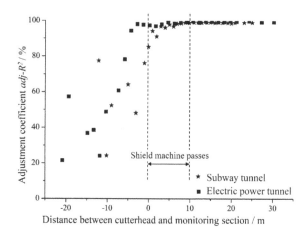

Fig. 4 The change process of settlement trough fitting degree with shield construction

degree is low when the cutterhead is at a certain distance from the monitoring section (all less than 80%). This is because the maximum surface fitting settlement values of subway and electric power tunnel cutterhead in front of the monitoring fault are 4.63 and 1.62 mm respectively, and the disturbance degree of soil mass is low at this stage. At the same time, non-construction factors such as monitoring errors account for a small proportion of the monitoring data in this stage. Therefore, the corresponding fitting parameters may not accurately reflect the shape of the settlement trough. According to the monitoring data of electric power tunnel, compared with the Peck formula, the curve fitting degree of the modified Peck formula can be increased by about 1.5 times when the cutter head reaches the monitoring surface, and the curve fitting degree can be increased by about 1.1 times when the shield tail comes off the monitoring surface. It can be seen that the modified Peck formula is more suitable for the situation of different degree of soil disturbance on the left and right sides of the shield tunnel.

On the whole, the Peck formula is suitable for the upper sand and lower rock composite strata after karst cave reinforcement, and the fitting degree can basically reach 95%, which can meet the engineering requirements.

Since the Peck formula is based on the undrained assumption, we can refer to the method of Fang [17] to take the tail of the shield machine about 2–3 days after passing through the monitoring section as the boundary between drained and undrained soil. Therefore, for subway and electric power shield, it can be considered that the surface measuring points after the tail of the shield machine leaves the monitoring section of 10.94 and 15.09 m respectively are basically not affected by the disturbance of shield construction.

Based on this, the curve fitting parameters which are directly affected by the disturbance of shield construction and the degree of fitting is more than 95% are analyzed. The change of K during the construction of subway and electric power tunnels is shown in Fig. 5. It can be seen that K fluctuates near a certain value during normal construction and remains stable with the tunneling process. The linear fitting

Fig. 5 Change process of K with shield construction

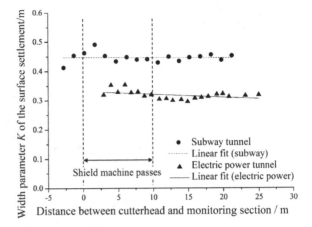

equations are respectively.

$$K = -1.07 \times 10^{-4} \cdot x + 0.45 \tag{4}$$

$$K = 1.05 \times 10^{-3} \cdot x + 0.33 \tag{5}$$

The fitting line is basically a horizontal line. Therefore, the K values of shield construction in the exposed rock layer and the single sand layer are 0.45 and 0.33 respectively.

It can be seen that K is affected by formation conditions, and can not be excluded from the influence of tunnel conditions. Therefore, when the strata and tunnel conditions are known, when the shield construction produces a certain degree of disturbance to the survey point, the K (or i) obtained by monitoring and fitting by Peck formula can be used to predict the surface lateral influence range in the whole process of shield construction, and can also be used to approximately calculate the surface lateral influence range caused by shield tunnel construction with similar strata and tunnel conditions. And assess the safety of important structures within the scope of influence.

The change of V_L during subway and electric power tunnels construction is shown in Fig. 6. V_L increases with shield construction, but the growth rate gradually decreases, and finally tends to remain the same. The V_L caused by subway and electric power tunnels construction is about 1.1% and 1.6%.

In addition, there is an exponential relationship between V_L and the monitoring section distance from the cutter head to the measuring point, and the fitting equations are respectively.

$$V_L = 1.830 - 1.449 \cdot e^{(-0.032)} \tag{6}$$

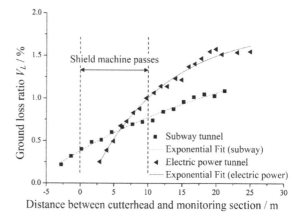

Fig. 6 Change process of V_L with shield construction

$$V_L = 1.821 - 2.060 \cdot e^{(-0.092)} \qquad (7)$$

The corresponding adj-R^2 is 98.9% and 98.8%, indicating that the index relationship can well reflect the changes of VL with the construction process. Therefore, the final stratum loss caused by shield construction under normal working conditions can be estimated by obtaining three data points through pre-monitoring.

3.2 Settlement trough Moves with the Construction of Electric Power Tunnel

Taking the monitoring data when the monitoring section is disturbed at the beginning of the electric power tunnel construction as the initial value, the translation of the center of the horizontal surface settlement trough during the construction is shown in Fig. 7. Combined with Fig. 4, Fig. 7 and Fig. 8, we can see that although the curve fitting degree of the settlement trough before the cutter head reaches the cross section is low, resulting in a certain discreteness of x_c, from the overall change trend, taking the axis of the electric power tunnel as the center, the center of the settlement tank moves from the left side to the right side (that is, from side A to side B), and finally remains stable, and the moving process mainly occurs during the shield machine passing through the monitoring section. Taking the average x_c of all the translation values after the tail of the shield machine is separated from the monitoring section, it can be known that the center of the surface settlement trough deviates from the axis about 0.5 m after the completion of the electric power tunnel construction.

As can be seen from Fig. 7 and Fig. 8, the center of the settlement trough is located on side A when the shield of the electric power tunnel is about to reach the monitoring section, because after the completion of the subway tunnel construction, the degree of disturbance of the soil is greater on the A side than on the B side. Subsequently, when the electric power tunnel gradually disturbs the monitoring point, the soil on

Fig. 7 The change process of settlement trough center with power shield construction

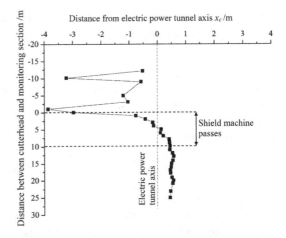

Fig. 8 Diagram of center movement of settlement trough

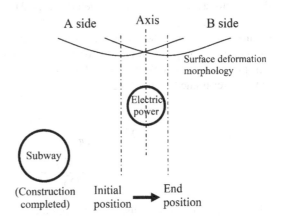

the A side is located in the superimposed area of the disturbance range of the subway and power shield construction, so that the center of the settlement trough is located on the A side at the beginning. After the subway construction, the subway tunnel construction grouting and karst cave grouting strengthen the soil layer, which makes the structure with high stiffness buried in the soil on the A side, and then the sand layer becomes denser after the disturbance of the electric power tunnel construction. The degree of disturbance of the soil is that the B side is greater than the A side, so that the center of the settlement trough gradually moves from side A to side B, after the tail of the shield machine leaves the monitoring section. The monitoring point is gradually out of the disturbance range, and the central position of the settlement trough remains stable. In addition, compared with the main construction parameters such as cutter head pressure and grouting pressure, the attitude parameters of shield machine have more influence on the center position of settlement trough.

3.3 Settlement trough Shape Caused by Shield Construction of Two Tunnels Successively

As mentioned earlier, after the tail of the shield machine leaves the monitoring Section 10.94 m, it is considered that the soil deformation is basically not affected by the shield disturbance, and the initial value of the shield monitoring of the electric power tunnel can be used to calculate the soil deformation value caused by the non-shield disturbance between the two monitoring. During this period, karst cave grouting treatment occurred near the monitoring section, because it only aims at the deep rock and soil, so the surface monitoring point mainly measures the consolidation deformation of the soil itself.

For double-line parallel tunnels, Liu Bo et al. [18] assume that the shield construction of two successive tunnels will not affect each other, and the surface settlement caused by the two tunnels is the same. With the help of superposition principle, two Peck formulas with the same parameters are used to calculate the final surface settlement. Ma Ke-Shuan [19] (Abbreviated as MKS) put forward the hypergeometric method, considering the influence of the later tunnel on the first tunnel, the Peck formula for superposition calculation has two different sets of calculation parameters, the calculation formula is as follows, but the parameters of the settlement trough of the latter tunnel are not specific.

$$S(x) = \frac{V_{S1}}{\sqrt{2\pi} \cdot i_1} \cdot e^{\left(-\frac{(x+L/2)^2}{2i_1^2}\right)}$$
$$+ \frac{V_{S2}}{\sqrt{2\pi} \cdot i_2} \cdot e^{\left(-\frac{(x-L/2)^2}{2i_2^2}\right)} \tag{8}$$

where L is the horizontal distance between the axes of the two tunnels.

Considering the translation of the center of the settlement trough caused by the rear tunnel shield, the MKS formula is modified as follows:

$$S(x) = \frac{V_{S1}}{\sqrt{2\pi} \cdot i_1} \cdot e^{\left(-\frac{(x+L/2)^2}{2i_1^2}\right)}$$
$$+ \frac{V_{S2}}{\sqrt{2\pi} \cdot i_2} \cdot e^{\left(-\frac{(x-L/2-x_C)^2}{2i_2^2}\right)} \tag{9}$$

For this project, subscript 1 and 2 represent subway tunnel and electric power tunnel respectively.

Only consider the soil deformation caused by the shield construction of subway tunnel and electric power tunnel, and the total amount of surface settlement during the monitoring period. If the MKS formula and the modified MKS formula are used to fit the total settlement caused by successive shield construction, the unknown parameters are 4 and 5 respectively, and there are many possibilities of permutation and combination of parameters that satisfy a certain degree of fitting. and the physical meaning of the fitting parameters may not necessarily accord with the reality.

As mentioned earlier, the parameters K and V_L can be obtained by fitting the surface subsidence with Peck formula. Therefore, it is known that the settlement trough parameters caused by single-line shield construction, K or V_L in the MKS formula is artificially set (modified) to be equal to the single-line fitting value, the deviation degree between the other fitting parameters and the known values is observed, and the engineering practicability of (modified) MKS formula is evaluated.

For the modified MKS formula, the moving distance of the center of the settlement trough in the power tunnel is known, and the x_c value is temporarily set to 0.44 m.

The K values of single and double lines are equal to the known values, and the V_L values of subway tunnel and electric power tunnel and their standard deviation (degree of deviation) are obtained by fitting. The V_L control value corresponding to subway tunnel and electric power tunnel construction is 1.09% and 1.55% respectively. Without considering different control conditions, the percentage of V_L value deviated from the control value obtained by different formulas shows that there is little difference in the standard deviation of VL value between the two shield tunnels under different control conditions, and the VL fitted by the modified MKS formula is closer to the control value.

The V_L value of the single line is equal to the known value, and the K value and its standard deviation of subway tunnel and electric power tunnel are obtained by fitting. The K control value for subway tunnel and electric power tunnel construction is 0.45 and 0.32 respectively. Without considering different control conditions, the percentage of K value deviated from the control value obtained by different formulas can be known. The standard deviation of K value caused by shield construction of subway tunnel and electric power tunnel is similar, the former is larger than the latter, and the K fitted by the modified MKS formula is closer to the control value.

From the analysis, it can be concluded that whether or not to control the K or V_L of single or double lines has little influence on the results of other fitting parameters, and the modified MKS formula is better to fit the total surface settlement caused by shield construction of two successive tunnels, but its use premise is that the center displacement of settlement trough caused by shield construction of the second tunnel is known. If the total surface settlement value and single-line settlement tank parameters caused by shield construction of two tunnels are known, the settlement trough parameters caused by another shield construction can be inverted by (modified) MKS formula.

The K or V_L of the double line is set to a known value to evaluate the influence of different degrees of x_c deviation on the fitting parameters. When the percentage of x_c deviation is-100%, the modified MKS formula is changed into the MKS formula, that is, the center of the settlement trough is located above the axis of the electric power tunnel. Therefore, when it is not clear which side of the tunnel the center of the settlement trough is caused by the backward shield construction, the MKS formula can be used. Although the fitting parameter value deviates from the reality to a certain extent, it will not cause too much error.

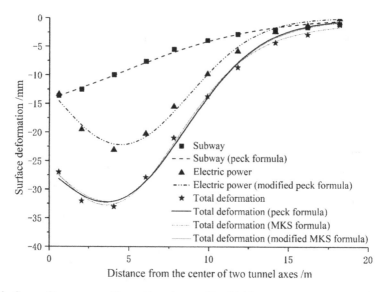

Fig. 9 Curve of transverse settlement trough caused by shield

The fitting curve of the (modified) MKS formula for controlling V_{LI} is drawn as shown in Fig. 9, which shows that the two settlement curves are highly consistent with the measured data, but there are differences in the physical meanings of the two corresponding parameters. As shown in Fig. 9, when the surface settlement values caused by subway tunnel and electric power tunnel construction are known, they can be fitted by Peck formula and modified Peck formula respectively, and then the final surface settlement curve can be obtained according to the linear superposition principle. At this time, the superposition curve can also fit the measured values well, but its use premise is that the measured data of surface subsidence caused by the construction of two tunnels are known.

4 Conclusions

Through the field measurement method, this paper studied the deformation law of surface soil mass caused by the construction of two tunnels successively in the upper sand and lower rock composite strata, and drew the following conclusions:

(1) The width coefficient (parameter) of the surface settlement trough is basically stable with the shield construction, and the lateral influence range of the surface does not change much with the construction, which can be used to evaluate the safety degree of the important structures in the influence area.

(2) The formation loss rate changes exponentially with the shield construction.

(3) Affected by the disturbance of the shield construction of the first tunnel, the surface settlement trough caused by the shield construction of the second tunnel is no longer located at the top of the tunnel axis, but will move with the construction process, which is greatly affected by the attitude parameters of the shield.

(4) The total surface settlement caused by the successive construction of the two tunnels is well fitted by the modified MKS formula considering the movement of the center of the settlement trough.

(5) for the total surface settlement caused by shield construction of two successive tunnels, under the condition that the surface settlement trough parameters V_L or K of the first or second tunnel are known, the surface settlement trough parameters of the second or first tunnel can be obtained by using MKS formula. Considering the movement of the center of the settlement trough of the back tunnel, the fitted settlement trough parameters will be more practical.

References

1. Jianguo L (2010) Countermeasures for shield tunneling in hard and soft mixed strata of Shenzhen Subway. Mod Tunnel Technol 47:79–84
2. Kaiwen H, Wang C, Jiang J, et al (2017) Type selection design and adaptability analysis of shield used in line No. 2 of Nanning metro. Tunnel Constr 37:1037–1045
3. Yazhou Z, Zhuyin W, Guangming Y et al (2019) Difficulties and countermeasures in design and construction of shield tunnels in upper-soft and lower-hard stratum. Tunnel Constr 39:669–676
4. Li Z, Wei L, Peng F (2014) Construction control and calculation of settlement induced by mixed ground shield undercrossing airport. J Railway Sci Eng 11:131–138
5. Xiao M, Gong Y, Zhou K, et al (2017) Shield tunnel settlement in complex strata based on strata loss theory. Constr Technol 20:113–117+174
6. Osman AS, Mair RJ, Bolton MD (2006) On the kinematics of 2D tunnel collapse in undrained clay. Géotechnique 56(9):585–595
7. Osman AS, Bolton MD, Mair RJ (2006) Predicting 2D ground movements around tunnels in undrained clay. Geotechnique 56(9):597–604
8. Dai X, Guo W, Cheng XS, et al (2021) Field Measurement and numerical analysis for evaluating longitudinal settlement induced by shield tunneling parallel to building. Yantu Lixue/Rock and Soil Mech 42(1):233–244
9. Avgerinos V, Potts DM, Standing JR (2016) The use of kinematic hardening models for predicting tunnelling-induced ground movements in london clay. Geotechnique 66(2):106–120
10. Zhu J-F, Xu R-Q, Liu G-B. Analytical prediction for tunnelling-induced ground movements in clays. J Geotech Geoenviron Eng 124
11. Meng F, Chen R-P, Kang K (2014) Effects of tunneling-induced soil disturbance on the post-construction settlement in structured soft soils. Tunnel Undergr Space Technol 41:165–175
12. Zhen Z (2015) Measurement and analysis on surface subsidence induced by shield tunneling in complex stratum of Nanjing. Constr Technol 44:83–86
13. Peck RB (1969) Deep excavations and tunneling in soft ground. In: Proceedings of 7th ICSMFE, pp 225–290
14. O'reilly M, New B (1982) Settlements above tunnels in the United Kingdom-their magnitude and prediction
15. Han X, Li N, Jamie R (2007) STANDING. Study on subsurface ground movement caused by urban tunneling. Rock Soil Mech 14:609–613

16. Gang W (2010) Selection and distribution of ground loss ratio induced by shield tunnel construction. Chin J Geotech Eng 32:1354–1361
17. Fang Y, Lin S, Lin J (1993) Time and settlement in EPB shield tunnelling. Tunnel Tunnel 25(11):27–28
18. Liu B, Tao L, Ding C, et al (2006) Prediction for ground subsidence induced by subway double tube tunneling. J China Univ Min Technol: 356–361
19. Ma K-S (2008) Research on the ground settlement caused by the shield construction and the protection of the adjacent buildings. Huazhong University of Science and Technology

Study on Soil Disturbance Caused by Shield Tunneling in Sandy Strata

Weiyu Zou, Shuyi Luo, Yingcheng Liu, and Hong Pan

Abstract Shield tunneling and construction in sandy soil have great safety risks, so it is necessary to study the soil disturbance caused by tunneling in sandy soil. However, the law of soil disturbance is still relatively vague, so this paper relies on the actual tunnel project, through high-density, close monitoring; The settlement law of soil around shield tunnel is studied and analyzed. The research shows that the settlement of soil disturbance is basically stable after the shield tail leaves the monitoring section about twice the length of the shield machine; During shield tunneling in sandy strata, the sand on the upper part of the tunnel is very easy to collapse, thus forming a "collapse arch". The formation of this feature will be one of the reasons for the construction accidents of large diameter shield machines; The remarkable feature of soil disturbance around the tunnel in the water rich sand layer is that the maximum horizontal displacement of the measuring points at different locations occurs at the same depth, and does not develop from the tunnel center to the 45-φ/2 diffusion angle; The maximum horizontal deformation during tunnel construction is far less than the vertical deformation, so the horizontal displacement can be used to better analyze the actual impact of shield construction disturbance.

Keywords Shield Construction · Sandy Strata · Layered Settlement · Collapsed Arch

W. Zou · Y. Liu
China Southern Power Grid, Guangzhou 510660, Guangdong, China

S. Luo
Guangzhou Mass Transit Engineering Consultant Co., Ltd., Guangzhou 510660, Guangdong, China

H. Pan (✉)
South China University of Technology, Guangzhou 510641, Guangdong, China
e-mail: hpan@scut.edu.cn

© The Author(s) 2023
G. Feng (ed.), *Proceedings of the 9th International Conference on Civil Engineering*,
Lecture Notes in Civil Engineering 327,
https://doi.org/10.1007/978-981-99-2532-2_21

1   Introduction

Nowadays, the scale of the city is increasing rapidly, and the development of underground space is also developing rapidly. with the improvement of the degree of development, the relationship between the projects of underground space is more complex, and the mutual influence can not be ignored. Shield construction as an important way of underground space development, it is very important to study the disturbance law of shield construction to the surrounding soil, so as to reduce stratum deformation and ensure the safety of the surrounding buildings and tunnels.

In the research on the disturbance of surrounding soil caused by tunnel shield construction, numerical simulation [1, 2] and theoretical analysis [3–6] are more common, in which the numerical simulation is generally based on the finite element calculation of continuum [7–9], while the theoretical analysis is basically aimed at saturated clay, and the undrained assumption [10, 11] is adopted at the same time, and there is little research on sand layer [12]. The research on the field measured data based on the actual project is less, and it is not systematic enough. At present, most of the existing studies on the displacement of the surrounding soil measured in the field only focus on a single deformation of the soil around the tunnel in the shield, such as surface settlement [13, 14], deep horizontal displacement [15, 16]. There is a lack of comprehensive research on the short-range and high-density soil displacement monitoring data for the same shield tunnel project.

Relying on the actual engineering, this paper carries out high-density monitoring of the surface settlement, layered settlement and the horizontal displacement of the deep soil around the shield tunnel, and studies the disturbance law of the surrounding soil caused by the shield construction in the upper soft and lower hard composite stratum. Combined with the synchronous grouting pressure, the fluctuation of the horizontal displacement of the deep soil when the tail of the shield machine leaves the monitoring section is discussed, and the disturbance law of the shield construction to the soil is studied.

2   Background

Shijing ~ Huanxi Power Tunnel (Xiwan Road ~ Shisha Road Section) is located in Liwan District and Baiyun District of Guangzhou City. The type of shield machine is mud-water balance shield, the length of shield machine is 9.62 m, the inner diameter of shield segment is 3.6 m, the thickness is 0.25 m, the outer diameter is 4.1 m, the diameter of cutter head is 4.35 m, and the buried depth is 415 m.

The monitoring section arranged in this project is located on the south side of Xiwan Road Crossing Zengpohe Highway Bridge in Guangzhou City. The strata of the tunnel are sand layer and silty clay layer. The buried depth of the tunnel axis is 14 m. The geological profile of monitoring section is shown in Fig. 1.

Fig. 1 Geological Profile

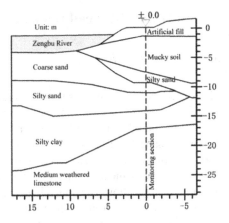

The monitoring contents include surface subsidence, layered settlement and deep horizontal displacement. The distribution of each measuring point on the monitoring section is shown in Fig. 2. The surface subsidence measuring points are arranged perpendicular to the tunnel axis and are located on the same tunnel section. The layered settlement measuring points and deep horizontal displacement measuring points are basically symmetrically arranged along the tunnel axis. Due to the influence of the actual construction level and complex strata, the actual measuring points can not be strictly symmetrically arranged on both sides of the tunnel.

Fig. 2 Layout of Monitoring Points

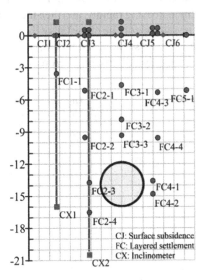

3   Study on Disturbed Deformation of Soil Mass

3.1   Surface Subsidence

The surface subsidence shows the surface subsidence of the monitoring section. There are six surface subsidence monitoring points in this project, numbered as CJ1, CJ2, CJ3, CJ4, CJ5 and CJ6,6. The measured surface subsidence is shown in Fig. 3. The Abscissa is the distance between the cutterhead and the monitoring section. The two red dotted lines in the Fig. divide the whole monitoring process into three stages. The dotted line on the left indicates that the cutterhead of the shield machine reaches the monitoring section. The dotted line on the right indicates that the tail of the shield machine is separated from the monitoring section.

As can be seen from Fig. 2. and Fig. 3, the settlement of the CJ4 measuring point with the smallest horizontal distance from the tunnel axis is the largest, and the settlement values of the other measuring points decrease with the increase of the horizontal distance between the location and the axis, which is consistent with the transverse surface subsidence curve proposed by Peck [17].

Before the cutterhead of the shield machine reaches the monitoring section, the surface settlement and settlement rate are very small, so it can be seen that the shield tunneling has little influence on the front surface subsidence; during the period when the shield machine passes through the monitoring section, the surface subsidence increases and the growth rate is larger, at this time, the surface is disturbed by construction; within 4 m from the monitoring section, the surface settlement value is still increasing rapidly, and then the growth rate slows down until it tends to be stable. The settlement rate of the tail of the shield machine is larger before 4 m away from the monitoring section, and then becomes smaller and more stable, the reason is mainly affected by tail grouting before 4 m, during this stage, a ring of segments forgot to

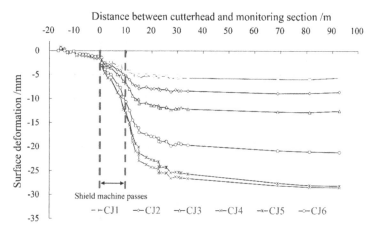

Fig. 3   Surface Subsidence

grouting (see Fig. 8.), resulting in increased settlement; after grouting, the deformation is mainly consolidation settlement, so the performance is relatively stable. The overall performance of the soil disturbance is that the settlement is basically stable after the tail of the shield machine is separated from the monitoring section about 2 times the length of the shield. And Lei Huayang [18], Bian [19] and Jiang [20]have come to the same conclusion when the diameter of the cutter head is close to the buried depth of the shield machine and the soil conditions are quite different. It can be seen that this conclusion can be used as a reference for the research of relevant shield engineering and the construction of projects with similar buried depth and cutter head diameter.

As shown in Fig. 4, the settlement of each measuring point varies with the distance between the cutter head and the monitoring section. The settlement law of the stratified settlement measuring point on both sides of the tunnel axis is the same as the surface settlement law mentioned above, but the three layered settlement measuring points directly above the axis have undergone a large displacement, in which the FC3-3 measuring point, which is only 2.6 m away from the top of the tunnel, sinks 0.6 m when the cutter head reaches the monitoring section, and then the measuring point can not continue to work. The FC3-2 measuring point also has a large settlement when the cutter head reaches the monitoring section, and the settlement is 20.24 mm, and then the settlement develops slowly when the shield machine passes through the monitoring cross-section. when the tail of the shield machine leaves the monitoring section, FC3-2 changes greatly again, develops to 98.94 mm, and then continues to increase rapidly to about 140 mm before it is basically stable. The variation trend of FC3-1, which is located in the lower buried depth and farther from the tunnel, is close to that of FC3-2. Because of the distance, the settlement is relatively small, and the maximum settlement is 67.34 mm.

3.2 Layered Settlement

The three measuring points directly above the axis of the tunnel have a large displacement, because the upper part of the tunnel and FC3-2 and FC3-3 are located in the sand layer, the bonding between the soil particles in the sand layer is small, and the sand layer is easy to collapse in a small range in the process of tunnel excavation, so the settlement caused by construction is small and large, so the settlement increases sharply when the cutter head arrives and the tail of the shield machine detaches. This phenomenon should be paid attention to in shield construction in similar strata, especially when there are important buildings above the tunnel axis in the sand layer to avoid excessive settlement.

As shown in Fig. 5, the layered settlement monitoring point FC1 series and the surface settlement monitoring point CJ2 are on the same side of the tunnel and are almost the same horizontal distance from the tunnel axis. Similarly, the FC2 series and FC3 series correspond to the surface settlement monitoring points CJ3 and CJ4 respectively. The settlement difference $\triangle S$ is obtained by subtracting the measured

Fig. 4 Layered Settlement
Curve on Both Sides of
Tunnel Axis

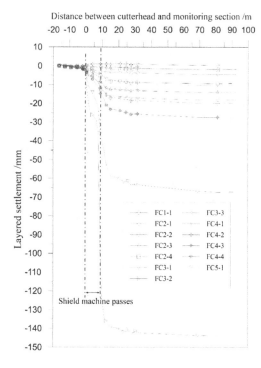

values of the upper and lower adjacent points in each settlement measuring point, as follows:

$$\Delta S = S_D - S_S$$

S_D is the settlement of the deeper measuring points in the two adjacent measuring points, and S_S is the settlement of the shallower measuring points in the adjacent two measuring points. A positive value of ΔS indicates that the soil between the two measuring points is compressed, while a negative value indicates that the distance between the two measuring points increases. Figure 5. shows the variation curve of the settlement difference between the upper and lower adjacent measuring points with the shield machine driving. Above the tunnel axis, the settlement difference of the CJ4 and FC3 series measuring points is relatively large. Therefore, the Y axis of the settlement difference curve of this series is the right axis. The Y axis of other curves is the left axis.

Because the settlement difference of CJ4 and FC3 series is larger than that of other measuring points above the tunnel axis, the Y axis of the settlement difference curve of this series is the right axis. The Y axis of other curves is the left axis. It can be seen from the diagram that most of the settlement differences of all measuring points occur in the interval between the arrival of the cutter head and the 3 m away from the monitoring section of the tail of the shield machine. The biggest difference

Fig. 5 Settlement
Difference Between
Monitoring Points

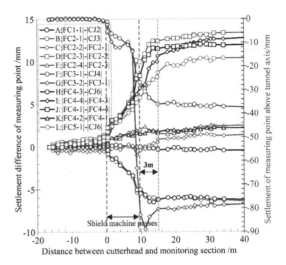

in the Fig. is about-90 mm, which occurs between FC3-2 and FC3-1 above the axis of the tunnel. Because FC3-3 collapses when the cutter head arrives, resulting in about 600 mm settlement, it is inferred that the settlement difference between FC3-3 and FC3-2 is as high as-450 mm. The settlement difference of each measuring point above the axis of the tunnel is negative, indicating that the distance between the measuring points increases, forming a tensile loosening zone.

The final settlement difference distribution in the monitoring section is shown in Fig. 6. In the Fig., the red indicates that the vertical distance between the two connected measuring points increases, showing a stretched state, while the blue indicates that the vertical distance between the two connected measuring points decreases, showing a compressed state. The number at the segment represents the vertical displacement at the measuring point at both ends of the segment, " + " indicates compression, and "-" indicates tension. From this picture, it can be seen that the deformation of the upper soil caused by the tunnel shows obvious characteristics of "collapse arch": the settlement and settlement difference of each measuring point (including CJ4 and FC3 series measuring points) on the tunnel axis (vault) is the biggest, and the settlement of the measuring point closer to the top of the tunnel is larger, and the whole is in a state of tension. It is noted that the settlement difference between FC3-3 and FC3-2 is about 0.45 m, and it can be inferred that there is a collapse crack between the two measuring points. At the measuring points on both sides of the tunnel, with the increase of the distance from the tunnel, the difference between settlement and settlement becomes smaller, so the arch effect is formed. Moreover, it can be seen that the range of the arch increases with the increase of the distance from the tunnel, but the smaller the settlement difference between the adjacent measuring points is, it tends to decrease gradually. It can also be seen from Fig. 6. that the settlement and differential settlement of the measuring points below the tunnel axis are very small.

Fig. 6 Settlement
Difference Between
Adjacent Monitoring Points

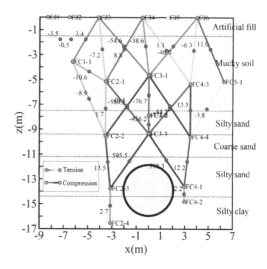

3.3 Deep Horizontal Displacement

The deep horizontal displacement of this project is the radial horizontal displacement
relative to the tunnel. The deformation curve of each inclinometer tube in the process
of shield machine tunneling is shown in Fig. 7. The swing amplitude of the curve
is: CX2 > CX1. It can be seen from Fig. 7. that the horizontal displacement mainly
moves towards the tunnel, and changes significantly when the distance between the
cutter head and the monitoring section is (-0.8 m ~ 0.4 m) and (6.4 m ~ 7.6 m).
Among them, -0.8 m and 0.4 m correspond to the arrival and passage of the cutter
head, and 6.4 m and 7.6 m correspond to the ring at the end of the shield that is about
to arrive and is not grouted.

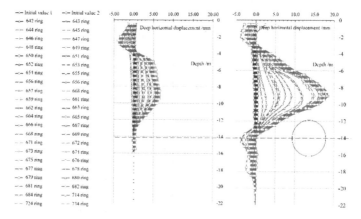

Fig. 7 Horizontal Displacement (CX1, CX2) Change Diagram of Deep Soil

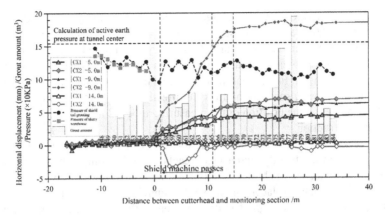

Fig. 8 Comparison of Horizontal Displacement of Characteristic Points and Changes of Shield Construction Parameters

The distance relationship between the two inclinometer tubes and the tunnel is: CX1 > CX2. The largest horizontal displacement occurs near -9.0 m, rather than 45-φ/2 angles upward from the center of the tunnel. The maximum horizontal displacement of CX2 is about 18 mm and the maximum horizontal displacement of CX1 is about 6 mm. There is a positive correlation between the distance between CX1 and CX2 from the axis of subway tunnel and the degree of disturbed deformation, that is, compared with CX2, the deformation degree of inclination curve of CX1 which is farther from the tunnel is smaller, which is consistent with the law of actual shield construction.

The maximum horizontal displacement of the two measuring points occurs near-9.0 m, and the soil layer in this depth is a silt layer with poor stability, and up to the middle line of the tunnel is a sand layer, which shows that the horizontal deformation of this depth is mainly the horizontal loss of the sand layer to the tunnel direction. The maximum horizontal displacement does not occur in the range of tunnel depth, but in the sand layer above the top of the tunnel, and the maximum horizontal increment caused by secondary under pressure of cutter head and shield tail grouting, and the maximum horizontal displacement in the monitoring process all occur at the same depth. This shows that the horizontal deformation of the soil around the tunnel in the water-rich sand layer may be caused by the flow of groundwater to the shield caused by the under pressure of the cutter head mud water bunker and shield tail grouting, resulting in the horizontal loss of the sand layer with poor self-stability. Therefore, when the top of the shield is dug in the sand layer, it may not only produce a large vertical displacement in the sand layer, but also lead to abnormal horizontal displacement. On the one hand, this abnormal horizontal displacement may have an adverse impact on the facilities adjacent to the shield.

Figure 8. shows the relationship between the measured values of deep horizontal displacement at three elevations in CX1 and CX2 and shield construction parameters.

The three elevations are the tunnel center elevation -14.0 m, the maximum displacement elevation -9.0 m and the silt soft soil center -5.0 m. The pressure and amount of shield tail grouting and the pressure of the slurry warehouse before the arrival of the cutter head are also given in the picture. It can be seen from the plane layout of the measuring points in Fig. 2. that the position of the inclinometer tube is 1.2 m in front of the monitoring section (the forward direction of the tunnel). Therefore, the cutter head arrives at the position of the measuring tube 1.2 m later than it arrives at the monitoring section.

In the process of shield machine tunneling, the deep horizontal displacement is roughly as follows: when the cutter head reaches 2 m in front of the inclinometer section, the displacement is basically zero, and then it deforms rapidly to the inside of the tunnel, and the displacement above the tunnel increases gradually when the shield machine passes through. The deformation of the tail of the shield machine gradually slows down and tends to be stable after it is separated from about 4 m. On the other hand, near the central side wall of the tunnel (|CX2|-14.0 m|), due to being squeezed by the cutter head, the deformation occurs to the outside of the tunnel, and the displacement increment is 4 mm. From the point of view of the displacement direction, whether it is horizontal displacement (Fig. 7.) or vertical displacement (Fig. 3 and Fig. 4), except for a small amount of outward extrusion deformation near the tunnel axis during the passage of the cutter head, the soil above the tunnel is displaced to the tunnel direction. This shows that the soil deformation in sandy stratum is mainly affected by slurry warehouse pressure or grouting pressure, and a large amount of grouting does not necessarily lead to soil uplift or compression. This may be related to the easy escape of grouting in sandy water-rich strata.

4 Conclusions

Based on the actual project, combined with the shield construction parameters, the surface settlement, layered settlement and deep horizontal displacement of the soil around the shield tunnel are analyzed and studied, and the following conclusions are drawn:

(1) Whether vertical displacement or horizontal displacement, the influence area of shield construction is mainly concentrated in the soil above the top of the tunnel, while the soil deformation in the range of shield is small. The overall performance of the soil disturbance is that the settlement of the tail of the shield machine is basically stable after the tail of the shield machine is separated from the monitoring section about twice the length of the shield machine.

(2) In the shield construction of sandy stratum, the soil deformation is mainly affected by slurry warehouse pressure or grouting pressure. Although the amount of grouting is large, it does not necessarily lead to soil uplift or compression. And the sand in the upper part of the tunnel is easy to collapse in a small area. Due to the small diameter and large buried depth of the tunnel in this project,

the formation of the collapsed arch does not have a great impact on the tunnel construction. Considering the large diameter tunnel engineering, this feature will be one of the causes of shield construction accidents.

(3) A remarkable feature of soil disturbance around the tunnel in the water-rich sand layer is that the maximum horizontal displacement of the inclinometer tube with different distance from the tunnel occurs at the same depth, not from the upward 45-$\varphi/2$ diffusion angle of the tunnel center. It may be that the insufficient pressure of slurry warehouse or the insufficient pressure of shield tail grouting cause the flow of groundwater to the shield, which leads to the horizontal loss of the sand layer with poor self-stability.

(4) When the pressure of slurry warehouse or shield tail grouting is insufficient in tunnel construction in water-rich sand layer, it will affect not only the vertical displacement of soil, but also the horizontal displacement of the surrounding soil. Compared with the settlement where the vertical displacement is greater than the order of 60 cm, the maximum horizontal displacement near the tunnel is only 5 ~ 18 mm, which is much smaller than the vertical displacement. Therefore, the actual influence distance of shield disturbance can be better analyzed by using horizontal displacement.

References

1. Zhou J, Meng G, Yan Q et al (2015) A study on the ground and building settlement caused by shield tunnelling in a peat soil stratum. Mod Tunnell Technol 3:160–167
2. Li Y, He P, Qin D et al (2012) Simulation of shield tunnel construction process base on 3D discontinuous geometry model. J Beijing Jiaotong Univ 4:38–43
3. Ye F, Gou C, Mao J et al (2015) Calculation of critical grouting pressure during shield tunneling in clay stratum and analysis of the influencing factors. Rock Soil Mech 4:937–945
4. Zhu Q, Ye G, Jianhua W et al (2010) Long-term settlement and construction disturbance during shield tunnelling in soft ground. Chinese J Geotechn Eng S2:509–512
5. Osman AS, Bolton MD, Mair RJ (2006) Predicting 2d ground movements around tunnels in undrained clay. Geotechnique 56(9):597–604
6. Osman AS, Mair RJ, Bolton MD (2006) On the kinematics of 2d tunnel collapse in undrained clay[J]. Géotechnique 56(9):585–595
7. Dai X, Guo W, Cheng XS et al (2021) Field measurement and numerical analysis for evaluating longitudinal settlement induced by shield tunneling parallel to building. Yantu Lixue/Rock Soil Mech. 42(1):233–244
8. Kasper T, Meschke G (2006) On the influence of face pressure, grouting pressure and TBM design in soft ground tunnelling. Tunn Undergr. Space Technol. 21(2):160–171
9. Avgerinos V, Potts DM, Standing JR (2016) The use of kinematic hardening models for predicting tunnelling-induced ground movements in London clay. Geotechnique 66(2):106–120
10. Shi C, Cao C, Lei M (2016) An analysis of the ground deformation caused by shield tunnel construction combining an elastic half-space model and stochastic medium theory. KSCE J Civ Eng 2016:1–12
11. Klar A, Klein B (2014) Energy-based volume loss prediction for tunnel face advancement in clays. Géotechnique 64(10):776–786

12. Zhu J-F, Xu R-Q, Liu G-B (2014) Analytical prediction for tunnelling-induced ground movements in sands considering disturbance. Tunn Undergr Space Technol 41:165–175
13. Chen S (2018) The Study of Surface Subsidence Caused by Metro Shield Construction. An Hui University of Science and Technology
14. Hao L, Chihao C, Shaoming L et al (2013) Measured deformation of dredger fill during slurry shield tunneling. Chinese J Geotechn Eng 35(S2):848–852
15. Yang J, Liu B (1998) Ground Surface Movement and Deformation Due to Tunnel Construction by Squeezing Shield. Rock Soil Mech. 03:10–13
16. Jiang X, Li L, Jie Y et al (2011) Dynamic analysis of strata horizontal displacements induced by shield construction of deep tunnel. Rock Soil Mech 32(04):1186–1192
17. Peck RB (1969) Deep Excavations and Tunneling in Soft Ground. In: Proceedings 7th ICSMFE, pp. 225–290
18. Lei H, Qiu W, Lv Q et al (2015) Study on the impact of the grouting factors on surface subsidence in the process of shield construction. Chinese J Undergr Space Eng 11(05):1303–1309
19. Bian J, Tao L, Guo J (2005) The ground settlement monitoring of a shield tunnel. Chinese J Undergr Space Eng. 02:247–249+254
20. Jiang A (2015) Analysis of the influential factors of and 3D analytical solution for ground deformation induced by shield tunnelling. Modern Tunnel Technol 52(01):127–135+142

Analysis of Blasting Vibration Signals at Different Initiation Positions of Tunnel Blastholes

Zhihao Yan, Genzhong Wang, Wenxue Gao, Min Zhou, Wangjing Hu,
Jin Ye, Yanping Li, Yuelan Xie, Yu Hu, and Xiaojun Zhang

Abstract In order to study the blasting vibration effect of different blasting points of the tunnel bore, the pilot tunnel in multi-arch tunnel was used to carry out the test of different blasting points of the bore. The results show that: 1) the peak velocity generated by blasting at different initiation points of the hole is different, and the magnitude of the peak velocity is reverse initiation > Intermediate initiation > Forward initiation; 2) The frequency distribution range of blasting vibration signal under the three initiation positions is: forward initiation > Reverse initiation > Intermediate initiation; At the same time, forward initiation is more conducive to the main frequency of blasting vibration signal moving to high frequency; 3) Compared with reverse initiation and intermediate initiation, forward initiation can better disperse energy, so that the energy is transferred from low frequency to high frequency, so that the blasting vibration is relatively small, which is more conducive to the safety of tunnel lining structure.

Keywords Multi-Arch Tunnel · Different Initiation Point of the Gun Hole · Wavelet Packet · Blasting Vibration · Signal Analysis

1 Introduction

In the construction process of tunnel drilling and blasting method, the change of different blasting positions of the hole will have a certain impact on the blasting vibration and blasting effect, mainly because the blasting position will determine the direction of the explosion energy propagation of the cylindrical charge pack and the time effect of the superposition of the blasting stress field of each charge pack [1, 2].

Z. Yan · W. Gao (✉) · Y. Hu · X. Zhang
Faculty of Architecture, Civil and Transportation Engineering, Beijing University of Technology, Beijing 100124, China
e-mail: wxgao@bjut.edu.cn

G. Wang · M. Zhou · W. Hu · J. Ye · Y. Li · Y. Xie
Zhejiang Lihua Blasting Engineering Co., Ltd., Suichang 323300, China

© The Author(s) 2023
G. Feng (ed.), *Proceedings of the 9th International Conference on Civil Engineering*,
Lecture Notes in Civil Engineering 327,
https://doi.org/10.1007/978-981-99-2532-2_22

Therefore, it is necessary to study the blasting vibration effect caused by blasting at different blasting positions.

In view of blasting at different initiation positions of the gun hole, relevant scholars have carried out corresponding research from theoretical analysis, field test and numerical simulation. Based on the Heelan short column theory, Gao Qidong et al. [3] discussed the influence law of different blasting positions of the hole on blasting vibration, and obtained the calculation formula of blasting vibration field generated by blasting at different blasting positions. Leng Zhendong et al. [4] found that double point initiation was conducive to reducing rock size and improving blasting effect by comparing the experiments of double point initiation with hole bottom initiation. Liu Liang [5] and Wang Yapeng [6] simulated blasting at different initiation positions based on LS_DYNA, and the research results showed that positive initiation was helpful to rock bottom fragmentation and reduce blasting vibration. At the same time, Gao Qidong [7] also used LS_DYNA to study the impact of blasting at different initiation positions of cut holes on the tunnel support structure. The research results showed that the forward initiation was helpful to reduce the vibration of the structure, while the reverse initiation was helpful to rock fragmentation.

The above scholars conducted a detailed study on the blasting effect of different initiation points of the blasthole from theory and simulation experiments. However, what was relatively lacking was related research on different initiation points of the full-section blasthole on site, and the relationship between blasting vibration signal frequency and energy under different initiation positions of the blasthole on site. Therefore, based on the reconstruction project of G528 National Road Suichang Xinluwan to Shilian section, this paper conducts blasting test research on the full-section blasting holes of the tunnel at different initiation positions in the guided tunnel, and studies the relationship between the frequency and energy of each signal.

2   Field Blasting Test

2.1   Project Background

This paper takes the dam tunnel in the reconstruction project of G528 National highway from Suichang Xinluwan to Shilian section as the background. The length of the tunnel is K4 + 928~K5 + 025, the total length is 97 m, and the maximum buried depth is 25 m. The surface of the tunnel is composed of residual slope silt clay containing gravel, the thickness is 1–2 m, and the underside is strongly to moderately weathered tuff, the rock mass is broken and the integrity is poor, which belongs to the IV grade surrounding rock. The whole section of the tunnel is excavated, and the area of the excavation section is 31.5 m^2.

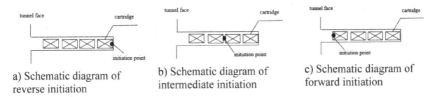

a) Schematic diagram of
reverse initiation

b) Schematic diagram of
intermediate initiation

c) Schematic diagram of
forward initiation

Fig. 1 Schematic diagram of different initiation points of the hole

2.2 Field Experiment Scheme

In view of the factors such as broken surrounding rock and poor stability of the tunnel from the dam, and considering that the vibration generated by the blasting of the main tunnel is likely to have a greater impact on the middle partition wall, the blasting vibration reduction experiment is carried out in advance in the middle pilot tunnel of the tunnel, that is, reverse blasting, intermediate blasting and positive blasting are respectively adopted for all blasting holes in the full section. The blasting vibration effect under each blasting mode is obtained by conducting experiments on different blasting positions of the blast hole, which provides guidance for the subsequent main tunnel to select the appropriate blasting position of the blast hole. The schematic diagram of the positions of different initiation points of the blast hole is shown in Fig. 1 below. Detonation is conducted through the detonating tube, and the TC-6850 blasting vibration monitor is used to monitor the ground vibration. The measuring point is arranged 20 m from the ground to the working face.

2.3 Blasting Parameter Design of Middle Guide Tunnel

Based on the site engineering geological conditions and considering the use of the full-section method in the blasting construction of the guide tunnel, the layout of the shot holes in this experiment is designed, as shown in Fig. 2 below, with a total of 6 sections. Its hole parameters are shown in Table 1 below.

3 Analysis of Blasting Vibration Signal at Different Positions of Blasting Holes

3.1 Time History Curve Analysis of Blasting Vibration Signal

In the process of blasting vibration monitoring, the peak value of blasting vibration velocity collected in the X-axis direction (tunnel longitudinal direction) is the largest, so only wavelet packet decomposition of signals in this direction is carried out later

Fig. 2 Layout of gun holes
(Unit:m)

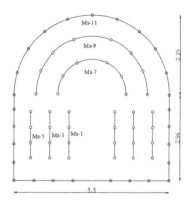

Table 1 Charge amount of each section for each hole

Hole category	Hole depth (m)	Number of holes (each)	Single well dosage (kg)	Total drug dose (kg)	Detonator period of don't
Cut hole	3.2	8	2.25	18	Ms-1
Cut spreader hole	3	8	2.1	16.8	Ms-3
Via hole	3	8	0.9	7.2	Ms-5
Via hole	3	6	1.5	9	Ms-7
Via hole	3	8	1.2	9.6	Ms-9
Contour hole	3	27	0.6	16.2	Ms-11
Amount	-	65	-	76.8	-

in this paper to study the relationship between frequency and energy. The blasting vibration signals of three kinds of holes at different initiation positions measured in the field are shown in Fig. 3 below. It can also be seen from Fig. 3. that the three initiation modes all reach the maximum vibration velocity during the first cut hole blasting, in which the peak velocity of reverse initiation is 7.15 cm.s^{-1} and the occurrence time is 0.033 s. The peak velocity of intermediate initiation is 5.84 cm.s^{-1}, and the occurrence time is 0.031 s. The peak velocity of forward initiation is 5.58 cm.s^{-1}, and the occurrence time is 0.029 s, the peak vibration velocity: reverse initiation > intermediate initiation > forward initiation, and peak velocity time point: forward initiation < intermediate initiation < reverse initiation. Investigate its reason, mainly when columnar cartridge blasting, energy spread along the hole, priority of reverse initiation energy spread along the hole in the direction of constraints, initiation energy transmission along the hole to both sides, among which are initiating explosive energy along the gun Kong Chao spread of surrounding rock, which causes reverse initiation at the ground vibration caused by tunnel is the largest. Further observation of the waveforms in Fig. 3 shows that, with the increase of the detonator stage, the waveforms of reverse initiation and forward initiation are

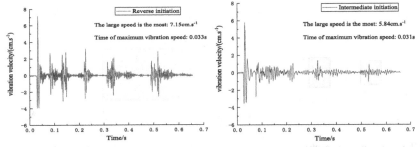

a) Reverse initiation blasting vibration signal b) Intermediate initiation blasting vibration signal

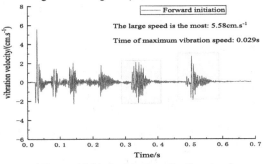

c) Forward initiation blasting vibration signal

Fig. 3 Blasting vibration signals at different initiation positions of the hole

wide in MS9 and MS11 sections, while the waveforms of intermediate initiation are narrow in MS9 and MS11 sections, indicating that when the intermediate initiation of the detonator is in the middle of the hole, all the holes in the two sections of the detonator can easily achieve simultaneous initiation. However, the reverse and forward initiations are easy to cause separate initiation of each hole in the high stage detonators.

3.2 Frequency Analysis of Blasting Vibration Signal

In order to further analyze the frequency of blasting vibration signals under different initiation positions of the three holes, Fourier transform is carried out for the blasting vibration signals in Sect. 3.1 above and normalization is carried out to obtain the spectrum diagrams of different initiation positions of the three holes, as shown in Fig. 4 below. It can be found from Fig. 4 that the frequency distribution of forward initiation is wider than that of reverse initiation and intermediate initiation. The frequency of forward initiation is mainly distributed in the range of 0–400 Hz, the frequency of reverse initiation is mainly distributed in the range of 0–300 Hz, and the frequency of intermediate initiation is mainly distributed in the range of 0–200 Hz. At

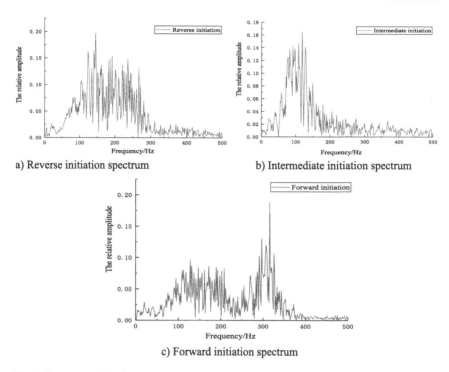

a) Reverse initiation spectrum   b) Intermediate initiation spectrum

c) Forward initiation spectrum

Fig. 4 Spectrum of blasting vibration signal at different initiation positions

the same time, the main frequency range of reverse initiation is 100–200 Hz, the main frequency of intermediate initiation is 50–150 Hz, and the main frequency of forward initiation has two peaks, one is 100–200 Hz, and the other is 300–400 Hz. This indicates that compared with reverse initiation and intermediate initiation, forward initiation is more conducive to energy transfer and high frequency dispersion.

4 Wavelet Packet Analysis of Blasting Vibration Signal

According to the blasting vibration signals at different initiation positions of the hole in Sect. 3.1 above, wavelet packet decomposition is carried out to study the energy distribution of each signal in different frequency bands. The wavelet basis function and the number of corresponding decomposition layers should be chosen reasonably when wavelet packet decomposition is carried out. The reasonable wavelet basis function can effectively remove the noise generated in the tunnel construction process and extract the real blasting vibration signal. However, the number of decomposition layers will directly affect the denoising effect of the signal. Too low decomposition layers will easily lead to mixed useful information in low frequency band, while too high decomposition layers will easily lead to deviation of the results. Therefore, this

Different hole initiation points	Main frequency band/Hz	The energy of/%
Reverse initiation	87.5–100 Hz	12.25
	137.5–150 Hz	13.53
	200–212.5 Hz	12.82
Intermediate initiation	100–112.5 Hz	57.25
	112.5–125 Hz	10.03
Forward initiation	87.5–100 Hz	10.17
	287.5–300 Hz	10.73

Table 2 Main frequency bands of blasting vibration signals at different initiation points of the hole

paper adopts db8 basis function in db series, which is similar to the tunnel blasting vibration signal waveform, to decompose the blasting vibration signals at different initiation positions of the three holes in eight layers [8].

TC-6850 blasting vibration vibrometer produced by Chinese Academy of Sciences was used in this tunnel blasting test. The sampling frequency was set as 6400 Hz, and the Nyquist sampling frequency was 3200 Hz. A total of 256 frequency bands were generated after 8-layer decomposition of wavelet packet. The size of each frequency band is 12.5 Hz, that is, the minimum frequency band after decomposition is 0–12.5 Hz. The energy of each frequency band is obtained through the above wavelet packet decomposition, and the frequency band interval of the main frequency band is defined in this paper for each sub-band whose energy ratio exceeds 10% [8]. The energy proportion of each main frequency band of blasting vibration signals at different initiation positions of the three holes is statistically shown in Table 2 below.

As can be seen from Table 2, the blasting vibration energy of intermediate initiation is mainly concentrated at 100–125 Hz, accounting for 67.28% of the total energy, and the energy is relatively concentrated. Compared with intermediate initiation, the main frequency band of reverse initiation is below 100 Hz and above 125 Hz, and the energy ratio of the three main frequency bands is 38.6%, indicating that the energy is relatively scattered. However, the frequency range of the two main frequency bands for forward initiation is quite different, one of which is in 87.5–100 Hz and the other is in 287.5 Hz–300 Hz. The energy ratio of the two main frequency bands is not very different, occupying 20.9% of the total energy, indicating that forward initiation can better disperse energy compared with reverse initiation and intermediate initiation. And make the energy transfer from low frequency to high frequency, so that the blasting vibration is relatively small, more conducive to the safety of tunnel lining structure.

5   Conclusions

To study tunnel hole blasting vibration effect different detonation point, with national highway G528 local practice period of reconstruction engineering of the new road bay to stone dam to tunnel as the background, elaborated to carry out the whole section in tunnel hole, experimental study on different detonation position of hole respectively take the reverse initiation, intermediate initiation, forward initiation, the results show that the:

(1) The three initiation positions all reach the maximum vibration velocity during the first cut hole blasting, but there are differences in the occurrence time of the peak velocity. The time points of the peak velocity are: forward initiation > Intermediate initiation > Reverse initiation; The peak velocity is: reverse initiation > intermediate initiation > Forward initiation; At the same time in the high stage detonator, the intermediate initiation is more conducive to the simultaneous initiation of the hole.

(2) The frequency ranges of blasting vibration signals at different initiation positions of the three holes are different, and the frequency distribution ranges are as follows: forward initiation > Reverse initiation > Intermediate initiation; At the same time, compared with reverse initiation and intermediate initiation, forward initiation is more conducive to the movement of blasting vibration signal frequency to high frequency.

(3) Compared with reverse initiation and intermediate initiation, forward initiation can better disperse energy and transfer energy from low frequency to high frequency, so as to make blasting vibration relatively small, which is more conducive to the safety of tunnel lining structure.

References

1. Gao QD, Jin J, Wang YQ et al (2022) Study on influence law of initiation position on transmission of explosion energy and its comparision and selection in tunnel cutting blasting. China J Highw Transport 35(05):140–152
2. Gao QD, Lu WB, Leng ZD et al (2020) Regulating effect of detonator location in blast-holes on transmission of explosion energy in rock blasting. Chin J Geotech Eng 42(11):2050–2058
3. Gao QD, Jin G, Wang YQ et al (2021) Acting law of in-hole initiation position on distribution of blast vibration field. Explosion Shock Waves 41(10):138–152
4. Leng ZD, Fan Y, Lu WB et al (2019) Explosion energy transmission and rock-breaking effect of in-hole dual initiation. Chin J Geotech Eng 38(12):2451–2462
5. Liu L, Zheng BX, Chen M, et al (2015) Numerical simulation analysis of influence of different detonation methods on bedrock in bench blasting. BLASTING 32(03):49–54+78
6. Wang YP, Wang X, Wang HL et al (2016) Influence analysis of continuous detonation mode on blasting effect. Mining Res Dev 36(12):101–103
7. Gao QD, Lu WB, Leng ZD, et al (2018) Optimization of cut-hole's detonating position in tunnel excavation. J Vib Shock 37(09):8–16
8. Chen JH, Qiu WG, Zhao XW, et al (2022) Vibration characteristics analysis of the metro tunnel subarea blasting based on wavelet packet technique. J Vib Shock 41(06):222–228+255

DSM Based Optimization to the Design Process of Prefabricated Construction

Xinlong Yang

Abstract The linear way of prefabricated construction (PC) design can induce extensive occurrence of reworks and increase costs and worktime. To optimize a reasonable and concurrent PC design workflow, a system of detailed PC design processes was built based on literature review and expert investigation. Based on graphic theory and DSM, the interdependent relationships between the processes were confirmed, and clusters of the coupled processes were identified. The paper proposed a model with an improved genetic algorithm to optimize the workflow to reduce rework costs and worktime. Then, a case study is used to estimate the model. The result shows that the optimization can reduce 12% costs and 12.43% worktime, and suggests that the cooperation between work of different design stages needs to be strengthened for a better product and less resource usage. The proposed model can effectively reduce the impact of rework. It improves the optimization methods of PC processes, and is valuable to PC planning and control.

Keywords Prefabricated Construction · Concurrent Engineering · Process Optimization · DSM · Genetic Algorithm

1 Introduction

China's construction industry is undergoing huge changes. Traditional construction methods in China consume high energy and produce large amounts of waste and carbon emissions. In 2020, the construction industry contributes 46.5% of total energy consumption and 51.3% of the total carbon emissions of the country [1], and the number is still rising as the housing demand grows. Considering the sustainable

X. Yang (✉)
Department of Civil Engineering, Faculty of Civil Engineering, Taiyuan University of Technology, No. 79 West Street Yingze, Taiyuan, Shanxi, China
e-mail: wy15123251601@163.com

© The Author(s) 2023   275
G. Feng (ed.), *Proceedings of the 9th International Conference on Civil Engineering*,
Lecture Notes in Civil Engineering 327,
https://doi.org/10.1007/978-981-99-2532-2_23

development of the country, China has announced its goal of cutting carbon emissions from construction and providing higher housing quality, which requires new construction methods for the industry.

Prefabricated construction (PC) differs from the traditional construction methods. It transfers a part of cast-in-situ work (mainly component manufacturing) into factories, which is known as off-site work. Therefore, the PC is more efficient, clean and has better quality. Compared with traditional construction methods, PC can reduce construction wastes, shorten project periods, reduce carbon emissions, save energy, and raise productivity [2–6].

Recently, China has introduced a series of policies including standards, initiatives, guidance and incentives to develop PC [7]. Yet the Chinese PC is still facing many challenges, including the immaturity of technology, the lack of experienced workers and the unfamiliarity of working process, etc. [8–10]. These problems can lead to reworks, which increases costs and worktime. Some other countries also faced the problems before, North America [11] once encountered the transportation issues of PC components; PC in Australia [12] development was hindered by unfriendly research environment, deficient planning and marketing; PC in South Korea [13] was subjected to the high construction costs, due to the immature design, workforce related problems and insufficient market size. China is still facing these problems in varying degrees.

Rework [14] (probably due to the errors or the nature of the process) is one of the major reasons for the high cost and prolonged working time in the Chinese practice of prefabricated construction. The ideal way of PC is closer to a concurrent process [15], but the common practice adopted in China is a linear process, which leads to reworks. DSM is an optimization method suitable for the manufacturing industry [16], and it mainly focuses on dealing with the coupled processes which cause wide ranges of reworks, and can make the linear process concurrent from a macro perspective. As PC has similar characteristics as manufacture in production methods, DSM can be applied to optimize PC process. Some researchers found that the design stage is a critical part, and it can cause a wide range of reworks when things go wrong in this phase [14].Therefore, it is necessary to confirm the design work and optimize the process through DSM. This paper proposes an optimization model with DSM to reduce costs and worktime, and the model can improve the optimization methods of PC processes, and be valuable to PC planning and control.

The structure of the contents is as follows: Sect. 2 briefly reviews the relevant research; Sect. 3 introduces DSM method and the DSM model building; Sect. 4 shows the optimization process and discusses the results of the work with a case study, and Sect. 5 summarizes the work.

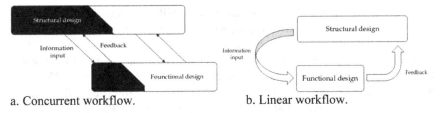

a. Concurrent workflow.   b. Linear workflow.

Fig. 1 The comparison of PC design process and traditional design process

2   Literature Review

2.1   Concurrent Engineering

Traditional Chinese building design is a linear process, while PC design is more suitable for a concurrent one [15, 17]. The comparison is shown in the Fig. 1.

The PC design needs to be integrated [18]. The PC component manufacturing and construction can be ideally implemented as early as in the design stage [19]. However, the PC design process now adopted in most projects in China is the linear type [17], which means that the work teams of different phases have little communication at early stages. The communication gap exists through the PC designing [15], causing increasing chance of errors. Hence, reworks can occur widely and increase costs and worktime.

Concurrent Engineering (CE) [20] is an idea proposed by IDA (Institute for Defense Analysis), widely adopted in manufacturing industry. CE emphasizes the idea of customer-oriented design (integrated design) and whole lifecycle design and communication through whole design stages [15]. It rearranges the process of product design and integrates the processes with strong interactions, in clusters [21]. The information exchanges can be intensive, but in smaller groups, because the information exchanges occur in different clusters concurrently [22]. This enhances the communication between different work teams within the clusters, which avoids a wide range of reworks. Therefore, it can provide higher working efficiency and better products. The current studies [23, 24] indicate that the concurrent way of construction integrates project teams, improves the quality of buildings and reduces time and costs of the projects, which gives the company competitive edge. Therefore, to reduce rework, optimization of design process needs to be carried out to make the process concurrent.

2.2   Optimization Methods

There are a lot of studies on the optimization methods of complex processes, e.g., Critical path method (CPM), IDEF method and DSM method.

CPM [25] is usually used in combination with PERT as evaluations. It optimizes the process by judging the critical path within the process system and optimizing the work on the path to minimize the overall worktime and resource usage. CPM is widely used in project management. Sroka, et al. [26] linearized CPM-Cost model for construction projects and tested it with an example. Mazlum, et al. [27] used the fuzzy CPM and PERT method to plan and improve a project of online internet branch. These studies show the utility and compatibility of CPM. However, CPM does not focus on the interactions between the work. Therefore, it is not an ideal way to solve the process coupling problems.

IDEF method [28] can represent the process systems by images. It is a comprehensive modeling method, including 16 sets of specific methods to represent the system in terms of functions, information exchanges and process obtaining, etc. In a product research, Li, et al. [29] used three methods of the IDEF to build the functional model, the message model and the model of its semantic relationships. However, IDEF method cannot show the interaction strength between processes in an explicit way. It is not an ideal way to deal with the coupling processes, either.

DSM method can show the interdependent strength between the processes in a matrix. It provides a simple but effective tool to model the interactions in a system and evaluate the coupling strength of the system. The DSM model is compatible with multiple approaches of optimization to improve the coupling relations. Therefore, it can be an ideal way to address the coupling problems. PEI, et al. [30] used DSM model and a genetic algorithm (GA) to cluster the process of complex product. The model can find the optimum results with the least coupling strength. It also can evolve with iterations to ensure the credibility of the optimal results. The DSM is often used in manufacturing industry [31] for project management and it is compatible with software simulation [32].The PC shares some similar features with manufacturing, which includes component production and shipment, and multiple involvers [14]. Unlike the traditional construction method, component production and erecting should also be considered. That means an integrated design [33] for PC is needed to coordinate different project teams to fulfill the needs of involvers and to manage this concurrent process. Therefore, DSM is suitable for PC processes, because the method has shown its utility in the integration of manufacturing projects and in the concurrent processes [34].

Some research used DSM for the PC process optimization in various degrees of depth. Chen [35], et al. built a model of QFD (Quality Function Deployment) to form the DSM matrix of the process in the PC component design, and an evaluation system was built. With the model, GA can be used to find the most reasonable working order. WANG [34], et al. and WANG [22] decomposed the PC design process, and set up a process system. These studies introduced the combination of DSM method and GAs to cluster process. Shen, et al. [14] proposed an optimizing model and an evaluation system of PC processes based on the DSM method, and built a risk management system for dealing with rework risks. In summary, the research shows the utility of the combination of the DSM optimization with the GA, and found that design stage is the most influential part of work [14, 33]. However, the current research only clusters the process, which is not practical for the PC design process. It is still necessary to optimize the sequence of the detailed design process considering rework impacts like costs and time.

3   Methodology

This paper aims at optimizing the ideal sequence of the design work. First, a questionnaire survey was implemented to verify the system of PC design processes, which was based on literature review [17, 33, 36, 37]. Then, a matrix representing the process system was built and the clustering was carried out based on graphic theory. A DSM model with an improved GA was used for process optimization. The optimization model considers the optimum sequence of work within the clusters and evaluates the optimization effects.

3.1   DSM Model

DSM (Design Structure Matrix) [38] is a method using matrices to optimize projects. There are two types of DSMs. One type is the Boolean type, or binary matrix, which contains only "0" and "1" as elements in the matrix. And the "0" indicates that the subjects do not have any relationships, while the "1" indicates that they do. The other is called Numeric Design Structure Matrix (NDSM), using numbers to show the strength of the relationships. Based on the directed graphs (digraphs) of the process chains, the DSM can be used to represent a complex system and show the interactions between the processes. The digraphs can be represented by binary DSM or NDSM, as in the Fig. 2. (digraph) and in the Fig. 3. (the DSM based on the digraph), since the DSM and the adjacency matrices of digraphs have the same structures [39].

Fig. 2 Digraph

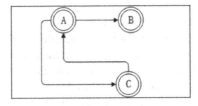

Fig. 3 DSM based on the Fig. 5

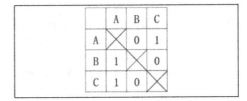

3.2   Questionnaire Survey

This study investigated some of the experienced experts in the PC industry to get a general evaluation of the relationships between the work for this optimization.

There are two questionnaires designed to identify the design processes and to build the DSM model of the PC design process. One questionnaire contains the background information of the experts and a five-point scale of recognition, and are sent to confirm the process system. The results of 11 questionnaires are shown in Fig. 4. The experts were from different cities such as Beijing, Shanghai, Chongqing, etc., and these cities can represent the development of PC in China. The scores of their recognition for the process system are shown in the Table 2. The average score of that is 4.27/5, which presents positive feedback for the process system. Some modifications were also mentioned in the questionnaires, including "The deepening design requires the complete involvement of the general contractor", "The component design will lag behind", "In the scheme stage, some contents were suggested should be advanced, and local product research can also be added". Then, the confirmed process system is shown in the Table 1. Another questionnaire used a four-point scale for the designers in the PC industry to determine the strength of relationships between the design processes. The scale looks into the information exchanges between the work, such as the frequency of communication between designers at different stages and the number of the transferred documents and files. The scores range from 0 to 3. Zero represents that there is no connection between the work, while 3 means a strong relationship between them, and 1 represents a mild relationship, and 2 represents a moderate relationship. Then, the NDSM was built based on the results of the four-point scale. A five-point scale was also used to evaluate the importance of the work as for rework impact.

3.3   Optimization Model Building

Two approaches were considered to optimize the process. One is based on graphic theory, the other is a simulation using GA. The first approach clusters the processes, so the work in different clusters can be done currently. The second approach finds the optimal workflow in the clusters.

Optimization Based on the Graphic Theory. In graphic theory, coupled processes have two-way links in digraph. Therefore, recognizing coupled processes can be done by finding the strongly connected paths in the digraph. As mentioned before, the digraph can be used to represent the DSM model of the designing process. Given that there is a path from work i to work j, and there is also a path from j to i, it is called a strongly connected path. The processes on the strongly connected paths are coupled processes (they have two-way links), as shown in Fig. 5.

An Accessibility Matrix (when there is a path from i to j, the element (i, j) is 1, otherwise it is 0) can be built from the digraph of the PC design process to find the

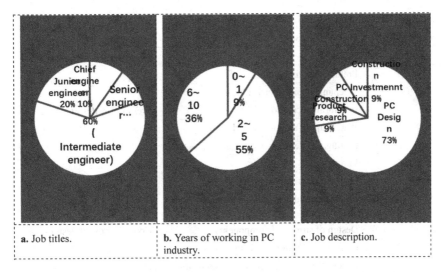

a. Job titles.

b. Years of working in PC industry.

c. Job description.

Fig. 4 Back ground information of the participants

strongly connected paths. If work i strongly connects work j, the multiplication of the element (i, j) and the element (j, i) is 1. Therefore, the matrix D of the Eq. (1) can determine the strongly connected paths, and so the clusters of coupled work.

$$D = P \cdot * P^T \tag{1}$$

In the equation, the P is the accessibility matrix of the NDSM (Fig. 6.), and the P^T is the transpose of the P.

Optimization through GA. Genetic algorithm [40] is an intelligent algorithm for solving multi-objective problems. Its idea originates from genic behaviors (inheritance, crossover, mutation) and the theory of "Survival of Fittest". It introduces the idea by coding the initial group, choosing the fittest groups by a function. And the winners inherit, crossover and mutate the codes. Then, it chooses the fittest and continues the operation again, until there is a final solution or it reaches the predetermined number of iterations. Yet, GA has the nature that it might premature and lead us to an unwanted result. An improved algorithm [14] can solve the problem, which changes the rate of crossover and mutation according to the convergence of the results. When the results of the offspring group are scattered, the crossover and mutation rate are low to accelerate the algorithm converging. When the results are highly converged, the rate goes high and the results can be better, as shown in the Eqs. (2) and (3). Then, the result of the GA can avoid the prematurity and the model can get the optimal result. The "roulette wheel selection" was used to select the fittest groups.

$$P_c = \begin{cases} Pc'(1/\beta), \alpha > a \text{ and } \beta > b \\ Pc', \text{ else} \end{cases} \tag{2}$$

Table 1 The confirmed system of the PC design processes

No	Design process	Stages
1	Building design	Conceptual design phase
2	Prefabrication planning	
3	Mechanical and electrical design	
4	structural calculation	Preliminary design phase
5	Water supply and drainage design	
6	heating and ventilation design	
7	Fire design	
8	Civil air defence design	
9	Environmental protection and energy saving design	
10	Design of structural components	Component split design
11	Connection Design of PC members	
12	Layout design of the prefabricated members	PC detail design
13	Node sample diagram	
14	Calculation book of prefabricated components	
15	Prefabricated (assembly) rate calculation	
16	Deepening the design of prefabricated components	Deepening design
17	Collision detection	Collision detection
18	Construction site layout drawing	Construction management plan
19	Selection of lifting machinery	
20	Transportation design	
21	PC moulds design	PC moulds design
22	Completion design	Completion design

Table 2 Recognition scores for the system

Recognition Scores	
3 points	1
4 points	6
5 points	4

Fig. 5 Strongly connected path between i and j

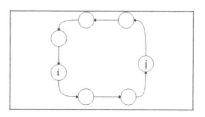

Fig. 6 The strength of relationships between PC design work (NDSM)

	1	2	3	4	5	6	7	8	9	10	11	12	13	14	15	16	17	18	19	20	21	22
1		3	0	2	0	0	1	0	1	1	0	0	0	0	0	0	0	0	0	0	0	0
2	3		1	2	1	1	0	0	0	1	0	1	0	0	0	0	0	0	0	0	0	0
3	3	1		1	3	3	2	2	3	0	0	0	0	0	0	0	0	0	0	0	0	0
4	2	2	0		0	0	0	0	2	1	3	1	2	2	1	0	0	0	0	0	0	0
5	1	1	2	0		2	1	1	3	0	0	0	0	0	0	3	0	0	0	0	0	0
6	1	1	2	0	1		1	1	3	0	0	0	0	0	0	3	0	0	0	0	0	0
7	3	2	0	1	1	1		0	1	0	0	0	0	0	0	1	0	0	0	0	0	0
8	1	0	1	1	1	1	0		0	0	0	0	0	0	0	0	0	0	0	0	0	0
9	3	1	2	0	1	1	0	0		0	0	0	0	0	0	0	0	0	0	0	0	0
10	3	2	0	1	0	0	0	0	0		3	3	2	3	3	3	3	3	2	2	0	0
11	1	1	0	1	0	0	0	0	0	3		3	2	3	3	3	2	0	0	0	0	0
12	1	1	0	1	0	0	0	0	0	3	3		2	3	3	3	2	3	3	2	0	0
13	2	1	0	0	0	0	0	0	0	3	3	3		1	1	1	0	0	0	0	0	0
14	0	0	0	0	0	0	0	0	0	3	3	3	3		3	3	0	0	0	0	0	0
15	0	0	0	0	0	0	0	0	0	3	3	3	3	3		3	0	0	0	0	0	0
16	0	0	0	0	0	0	0	0	0	3	3	3	3	3	3		0	1	1	2	0	0
17	3	3	3	2	2	2	1	1	0	2	2	2	1	0	0	0		0	0	0	0	0
18	1	0	0	0	0	0	0	0	0	0	0	0	0	0	0	0	0		3	3	0	0
19	1	0	0	0	0	0	0	0	0	3	0	0	0	0	0	0	0	0		2	0	0
20	2	3	0	0	0	0	0	0	0	3	0	0	0	0	0	0	0	0	0		0	0
21	0	1	0	0	0	0	0	0	0	3	1	0	0	0	0	0	0	0	0	0		0
22	0	0	0	0	0	0	0	0	0	0	0	0	0	0	0	0	0	0	0	0	0	

$$P_m = \begin{cases} Pm'(1/\beta), \alpha > a \text{ and } \beta > b \\ Pm', \text{ else} \end{cases} \tag{3}$$

Here, the P_c is the rate of the crossover, and the P_m is the rate of the mutation, in the improved algorithm. The Pc' and the Pm' are the presupposed values of the rate of the crossover and the mutation. The α equals the average of the fitness value divided by the greatest fitness value, in the groups, and the β equals the minimum fitness value divided by the greatest fitness value. The fitness value is the result of the Eq. (10) for every subject in the groups. The a and b, are presupposed numbers, ranging from 0.5 to 1 and 0 to 1 respectively. The lower of the value of the a and b, the easier the algorithm converges.

Based on DSM, real encoding is explicit to represent the sequence using sequential numbers. According to the project goals, the functions were designed to find results with lowest working time and costs. Based on the Iteration Risk matrix (**IR** matrix) [41, 42], the Rework Factor matrix (**RF** matrix), the Core Work evaluation (**CW(j)**), the functions are built for multi-objective optimization. Iteration Risk matrix (**IR** matrix) [43] is built to show the impact of the sequence of the process on reworks (errors that happen in the latter part will cause more reworks). The elements in **IR** matrix are calculated according to Eq. (4), in which the i and j represent the different processes and the n is the total number of the processes.

$$IR(i, j) = \exp(j/n - (i - j)/n) \tag{4}$$

Based on the research on rework propagation, a Rework Factor matrix (**RF** matrix) [44] shows the overall probability of the rework. In Eq. (5), the m represents the number of steps for every path available from process i to process j, in the digraph. Therefore, the value of "m" ranges from 1 to i-j-1. While the "C_{i-j-1}^{m-1}" is the total number of all the possible paths from i to j, and the "k" represents the specific path. The $PR (i, j)$ is the elements from the Rework Probability matrices [42] and it shows the probability of the rework when the information is delivered from work i to j. The $PR_k{}^{(m)}$ is the multiplication of all the $PR (i, j)$ on the path "k".

$$RF(i, j) = 1 - \prod_{m=1}^{i-j} \prod_{k=1}^{C_{i-j-1}^{m-1}} (1 - PR_k^{(m)}(i, j)) \tag{5}$$

The Core Work (CW) considers the importance of each work as for the rework impact. In Eq. (6), the $Wj1$ scores the probability of reworks, the $Wj2$ scores the costs to solve the reworks, and the $Wj3$ scores the extra worktime caused by rework.

$$w_j = w_{j1} \times \sum_{k=2}^{3} w_{jk} \tag{6}$$

$$CW(j) = \frac{w_j}{\sum_j^n wj} \tag{7}$$

Then, the multi-objective optimization model considers the minimum costs and time based on the aforementioned parameters and variables The main equation $f(x)$ and the relevant equations are shown as Eqs. (8), (9) and (10). Here, the ω_{ERC} and the ω_{ERT} are weight coefficients.

Main functions:

$$ERC = \sum_{i=1}^{n} \sum_{j=i+1}^{n} (RF(i, j) \times \sum_{u=i}^{n} (Cost_u \times RF(u, i)) \times CW(j) \times IR(i, j)) \tag{8}$$

$$ERT = \sum_{i=1}^{n} \sum_{j=i+1}^{n} (RF(i, j) \times \sum_{u=i}^{n} (Time_u \times RF(u, i)) \times CW(j) \times IR(i, j)) \tag{9}$$

$$f(x) = \min TEL = \omega_{ERC} \times ERC + \omega_{ERT} \times ERT \tag{10}$$

4 Optimization Results and Discussion

4.1 Cluster Result with Graphic Theory

A real case is used to estimate the model. According to the expert investigation, the NDSM of the design process is shown in the Fig. 6. Based on the graphic theory and the NDSM, the accessibility matrix is built, as in the Fig. 7, and the result of

Fig. 7 Accessibility Matrix

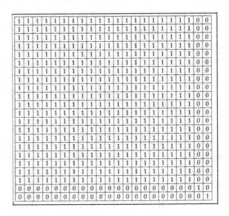

Fig. 8 The matrix *D*

the Eq. (1), which determine the clusters of the coupled processes, is shown in the Fig. 8.

The result *D* shows that the work in the PC design process is highly connected with each other. All the processes, except for the work No. 21 (PC molds design) and the work No.22 (Completion design), are coupled work and is put in a cluster.

4.2 Optimization Model Based on GA

By the given data of the time and costs of a practical case and the functions mentioned, we gained the optimal sequence of design processes through the GA. The result is shown in Fig. 9. The Fig. 11. shows that the maximum value of 1/TEL (minimum value of TEL) finally converged after 1200 times iterations, providing a credible answer. And the result also shows that the optimization reduces the cost (ERC) by 13.9% and the worktime (ERT) by 14.81%.

Building design	Prefabrication planning	Environmental protection and energy saving design	Fire design	heating and ventilation design	Mechanical and electrical design	Water supply and drainage design	Layout design of the prefabricated member	Design of structural	Collision detection	Connection Design of PC members	Structural calculation	Transportation design	Civil air defense design	Node sample diagram	Selection of lifting machine	Construction site layout	Deepening the design of prefabricated components	Calculation book of prefabricated components	Prefabricated (assembly) rate calculation	PC moulds design	Completion design
2	1	9	7	6	3	5	12	10	17	11	4	20	8	13	19	18	16	14	15	21	22

Fig. 9 Sequence of PC design process after optimization

The optimal result shows that the optimization cut down the cost and worktime by sequencing. However, there is a logical error in the sequencing manipulation, probably due to the subjectivity of the data. The work of Structural calculation should start before the Design of structural components, because the latter work relies on the information provided by Structural calculation. Therefore, rectification has to be done to the optimization result, as in the Fig. 10. And the calculation shows that the final result accomplished 12% decrease on cost (ERC) and 12.43% on working time (ERT). In the optimized work sequence, some processes which originally tended to be in the later part were put ahead, such as Environment protection and energy saving design, Fire design and Collision detection, because they are strong information-output processes. Accordingly, the strong information-receiving processes were put behind, as "Civil air defense design" and "Prefabricated rate calculation". The information-outputting processes start first and then the information receiving processes start later. Therefore, the information flow is strong forward rather than backward, and reduces the occurrence of reworks. Generally, the optimized sequence of the work is more reasonable and has positive impacts on costs and worktime reduction. According to the scores on the importance, the process 1 (Prefabrication planning) and the process 2 (Building design) are the most important work, with the highest costs and worktime increase if reworks occur. The process 3 (Mechanical and electrical design) was rated to the process with the biggest probability of rework. Therefore, these processes have to be carefully implemented, and the corresponding involvers should closely communicate with each other.

Building design	Prefabrication planning	Environmental protection and energy saving design	Fire design	heating and ventilation design	Mechanical and electrical design	Water supply and drainage design	Layout design of the prefabricated member	structural calculation	Design of structural component	Collision detection	Connection Design of PC member	Transportation design	Civil air defense design	Node sample diagram	Selection of lifting machine	Construction site layout	Deepening design of prefabricated components	Calculation book of prefabricated components	Prefabricated (assembly) rate calculation	PC moulds design	Completion design
2	1	9	7	6	3	5	12	4	10	17	11	20	8	13	19	18	16	14	15	21	22

Fig. 10 The optimization result after rectification (final result)

Fig. 11 The change of the minimum TEL in each population group with iterations

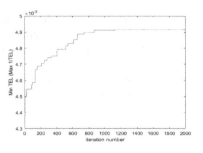

5 Conclusions

The PC design is a critical part in the prefabrication process. Reworks occurring in the linear design process causes the increase of costs and time. Therefore, it is necessary to optimize the process. A system of detailed PC design processes was confirmed through literature review and expert investigation. A matrix of interactions between processes was constructed, and clusters of coupled processes were found based on DSM and graphic theory. A DSM model based on the GA was built to find the optimal sequence of the PC design processes in the concurrent way. Then, a case study was used to demonstrate the optimization effect. Besides positive outcomes, it is found that PC design has concurrent features, but the majority of the processes cannot be done in a separate way, as indicated by the matrix D (the result of the dot product). Therefore, the optimized PC design is concurrent between clusters, and the processes are linear in the clusters. In the logical sequence of the detailed work in the clusters, the information flow is strong forward, and the work of the different phases can be done concurrently. Therefore, the communication is enhanced and the probabilities of errors are reduced, which improves the impacts of reworks.

In summary, the optimization model provides a perspective to view the PC design, and the results and the findings can offer some new insights of the development of the PC in China. The model extends existing process optimization methods and has value for planning and controlling PC management. The main contributions are as follows:

- A process system in the PC design stage was analysed, and the detailed processes were confirmed according to the literature review and expert investigation.
- Necessary data were measured, including the relationship strength between the processes, the probability of reworks, and the importance evaluation on each work in terms of rework prevention.
- Based on the DSM and GA, the entire optimization processes were simulated using GA by Matlab. The results reduced the costs and work time.

This paper also needs further research. In the future, more practical information and objective data should be collected to identify more detailed processes and improve the model. Some other optimization methods, such as Particle swarm optimization, Discrete event system simulation method, and Euler net, can be used to compare the results and find an optimal one.

Acknowledgements The authors would like to thank all the experts investigated for their assistance in data collection. The authors also would like to thank Mr. Teng Chen in China Architecture Design & Research Group for his assistance in theoretical support and data collection.

References

1. CABEE (2021) China building energy consumption annual report 2020. Building Energy Effi. 49(2):1–6
2. Kong A, Kang H, He S, Li N, Wang W (2020) Study on the carbon emissions in the whole construction process of prefabricated floor slab. Appl Sci 10(7):14
3. Song Nuo J Y-S W D-D (2017) The sustainability differences in the construction stage between prefabricated and conventional buildings: an empirical analysis. J Eng Manage (6):5
4. Liu Meixia WZ, Wang Jiening, et al (2015). Energy efficiency and carbon emissions evaluation of prefabricated construction in housing industrialization. Building Struct 45(12):5
5. Arditi D, Mochtar K (2000) Trends in productivity improvement in the US construction industry. Constr Manag Econ 18(1):15–27
6. Jaillon L, Poon CS (2009) The evolution of prefabricated residential building systems in Hong Kong: a review of the public and the private sector. Autom Constr 18(3):239–248
7. Luo T, Xue X, Wang Y, Xue W, Tan Y (2021) A systematic overview of prefabricated construction policies in China. J Clean Prod 280:17
8. Jiang L, Li Z, Li L, Gao Y et al (2018) Constraints on the promotion of prefabricated construction in China. Sustainability 10(7):17
9. Liu YC, Rui-Dong Z et al (2022) What leads to the high capital cost of prefabricated construction in China: perspectives of stakeholders. Eng Const Archit Manage 2022:28
10. Yuan Li XM, Eric Z, et al (2018) Unlocking the green opportunity for prefabricated buildings and construction in China. Resour Conser Recycl 139: 259–61
11. Rippon JA (2011) The benefits and limitations of prefabricated home manufacturing in North America. 2011:36
12. Navaratnam S, Ngo T, Gunawardena T, Henderson D et al (2019) performance review of prefabricated building systems and future research in Australia. Buildings 2019:14
13. Lee J-S, Kim Y-S (2017) Analysis of cost-increasing risk factors in modular construction in Korea using FMEA. KSCE J Civil Eng 21(6):1999–2010
14. Kaicheng S (2020) Research on Rework Risk of Prefabricated Construction Process. Tsinghua University (2020)
15. Qianhong D (2019) Research on Optimization of Assembly Architectural Design process based on parallel Theory. Chongqing University
16. Xuanzi PXZ, Xiuli W (2021) DSM optimal development module division of complex product based on improved genetic algorithm. Ind Eng Manage (026–006)
17. Chengwei L (2016) The prefabricated housing design process of parallel engineering. Tianjin Constr. Sci. Technol. 26(6):4
18. Duan YU et al (2018) Research on BIM collaborative application of the whole process of prefabricated building construction. China Stand 18:2
19. Yanna MHZHW (2020) Conflict recognition on working space in concurrent construction of prefabricated building based on BIM. J Saf Sci Technol 16(02):97–103
20. Umemoto K, Endo A, Machado M (2004) From to : the evolution of concurrent engineering at Fuji Xerox. J Knowl Manage 8(4):89–99

21. Xiong Guanglen ZH, Li B (2000) Research and application of parallel engineering in China. Comput Integr Manuf Syst. 6(2):7
22. Wang J (2021) Research on design process optimiaztion of prefabricated residential building based on design structure matrix. Zhengzhou University
23. Khalfan M, Raja N (2012) Improving construction process through integration and concurrent engineering. Australas J Constr Econ Build 5(1):58
24. Latortue X, Minel S (2013) Implementing concurrent engineering in the construction industry. 2013:655–659
25. Cai Chen WW (2003) Key chain management based on PERT/CPM. Chin J. Manag Sci 11(6):35–39
26. Radziszewska-Zielina E, Sroka B (2017) Linearised CPM-COST model in the planning of construction projects. Procedia Eng 208:129–135
27. Mazlum M, GüNERI AF (2015) CPM, PERT and project management with fuzzy logic technique and implementation on a business. Procedia Soc Behav Sci 210:348–357
28. Jeong KY, Wu L, Hong JD (2009) IDEF method-based simulation model design and development framework. J Ind Eng Manag 2(2):337–359
29. Al L et al (2000) The IDEF integrated modelling which supports concurrent engineering. Mach Des 17(4):3
30. Xiuli et al (2021) DSM optimal development module division of complex product based on improved genetic algorithm. Ind Eng Manage 26(6):11
31. Zhang H, Qiu W, Zhang H (2006) An approach to measuring coupled tasks strength and sequencing of coupled tasks in new product development. Concurrent Eng Res Appl 14(4):305–311
32. Chen Dong-Yue QW-H, Yang M et al (2008) DSM based product development process modeling and simulation optimization. Comput Integr Manuf Syst 14(4):6
33. Xu YBG, Guo J (2021) Analysis on whole process design of prefabricated building with concrete structure. Constr Sci Technol. (002):52–55
34. Wang QZL, Wang D et al (2019) Research on modeling and optimization of the design process of prefabricated housing based on DSM. Ind Constr 7:194–200
35. Chen Wei SX-J et al (2020) Design, management and control optimization of prefabricated components of prefabricated buildings based on QFD-DSM. J Civil Eng Manage 37(4):17
36. Zhang SKZ (2019) Study on the design process of prefabricated reinforced concrete structure. Fujian Archit Constr 256(10):4
37. Shaoqing Y (2021) Analysis on the architectural design management process of prefabricated finished residential buildings. Jiangxi Building Mater 2021(10):2
38. Steward DV (1981) The design structure system: A method for managing the design of complex systems. Eng Manage IEEE Trans 28(3):71–74
39. Prince AA, Jose I, Agrawal VP (2012) Concurrent design, modeling and analysis of microelectromechanical systems products - design for 'x' abilities. Micro Nanosyst 4(1):56–74
40. Yuan JCD (2002) Multicriteria optimal model for scheduling using genetic algorithms. Syst Eng 20(3):1–8
41. Browning TR, Eppinger SD (2002) Modeling impacts of process architecture on cost and schedule risk in product development. IEEE Trans Eng Manage 49(4):428–42
42. Cho S-H, Eppinger SD (2005) A simulation-based process model for managing complex design projects. IEEE Trans Eng Manage 52(3):316–28
43. Qing Y et al (2014) The impact of uncertainty and ambiguity related to iteration and overlapping on schedule of product development projects. Int J Project Manage 32(5):827–37
44. Yang Q et al (2017) The impact analysis and optimization of communication on rework risk between overlapped activities in R&D project. Syst Eng Theor Pract 37(9):10

Research on Vibration Signal Characteristics of Multilateral Boundary Deep Hole Blasting

Junkai Chen, Xiangjun Hao, Zheng Wei, Wenxue Gao, Xiaojun Zhang, and Zhaochen Liu

Abstract The essence of blasting action is the interaction between explosive energy and medium, and the blasting effect depends on the characteristics of explosive energy, medium and interaction law. The multi-boundary comprehensive blasting theory of stonework establishes the functional relationship among boundary conditions, throwing rate and explosive quantity. In this paper, a formula for calculating the explosive charge of multi-boundary deep-hole blasting is put forward, and the experimental study is carried out based on HHT analysis technology. Research shows that: (1) Based on HHT analysis technology, the main frequency of blasting vibration signal energy is in 0–40 Hz, and mainly in low frequency band, while the energy contained in high frequency component gradually decays. (2) Multi-boundary deep-hole controlled blasting has evenly broken rock blocks and all collapsed within the design range, which has achieved good blasting effect.

Keywords Multi-boundary · HHT · Deep Hole Blasting · Signal Characteristics · Blasting Effect

1 Introduction

Wang [1] takes micro-topography as a main condition and directly introduces it into the theory of multilateral boundary rock blasting. Zhang et al. [2] analyzed and processed the blasting vibration signal by HHT method, and obtained that EMD can decompose the signal well according to different time scales, and the decomposed intrinsic mode function can reflect the intrinsic characteristics of the signal itself. Chen et al. [3] studied the control blasting design of rock mass structure, put

J. Chen · W. Gao (✉) · X. Zhang · Z. Liu
College of Architecture and Civil Engineering, Beijing University of Technology, Beijing 100124, China
e-mail: wxgao@bjut.edu.cn

X. Hao · Z. Wei
Inner Mongolia Kinergy Blasting Co., Ltd., Erdos 017000, China

© The Author(s) 2023
G. Feng (ed.), *Proceedings of the 9th International Conference on Civil Engineering*,
Lecture Notes in Civil Engineering 327,
https://doi.org/10.1007/978-981-99-2532-2_24

forward the control theory of rock mass structure, and established the basic theory of rock mass blasting under geological boundary conditions. Guan et al. [4] uses the extreme point symmetric continuation method to eliminate the endpoint effect, which improves the computational efficiency and practicability of HHT. Zhang et al. [5] analyzed the empirical formula of blasting vibration prediction, and concluded that the nonlinear regression correction formula has the highest accuracy. Based on rock wave impedance and rock mass integrity coefficient, Hu et al. [6] constructs the expression of attenuation of peak vibration velocity of open pit bench blasting seismic wave with equivalent distance. Gao et al. [7] obtains the time-history relationship of seismic wave energy by wavelet transform, and deduces the calculation formula of vibration reduction delay of bench blasting interference. Zhang et al. [8] uses wavelet transform decomposition and response spectrum analysis to study the energy distribution characteristics of blasting vibration signal, and obtains that the main vibration frequency band of blasting vibration signal energy tends to low frequency band.

The above research results mainly focus on the horizontal boundary conditions, and it is not enough to consider the micro-topography change as an additional factor to the blasting theory and blasting method. Based on HHT analysis technique and multi-boundary rock comprehensive blasting theory, considering various terrain and geological conditions, a formula for calculating the charge of multi-boundary deep-hole blasting is put forward in this paper, which significantly improves the blasting effect.

2 Basic Principle of HHT Analysis Method

2.1 EMD Decomposition Principle

EMD decomposition is an important part of HHT transform, the core of which is to decompose vibration signal into a series of modal functions.

Assuming that the sampling sequence of slope blasting vibration signal is $S(t)$, the EMD decomposition process can be expressed as:

$$S(t) = \sum_{i=1}^{n} c_i(t) + R_n(t) \tag{1}$$

In Eq. (1), n is the order of signal decomposition. $c_i(t)$ is the i-order IMF component. $R_n(t)$ is the residual component after n-order decomposition.

The remaining part $R_1(t)$ of the signal can be obtained by the first IMF component $c_1(t)$:

$$R_1(t) = S(t) - c_1(t) \tag{2}$$

By analogy, the n-order IMF component $c_n(t)$ and residual component $R_n(t)$ of the original signal can be obtained. When both $c_n(t)$ and $R_n(t)$ are less than the threshold, the EMD decomposition process ends.

EMD decomposition is based on the local characteristics of the signal, and the operation process of the algorithm is adaptive. IMF component is stable, linear and symmetrical, and is suitable for processing blasting vibration signals.

2.2 Hilbert Transform and Spectrum

The combination of several IMF components of the original signal is obtained by EMD decomposition, and the Hilbert spectrum can be obtained by combining the instantaneous spectra of all IMF components by Hilbert transform. Hilbert transform is a kind of linear transformation, which has intuitive physical meaning and emphasizes the local properties of signals.

Hilbert transformation of IMF components:

$$H[c(t)] = \frac{1}{\pi} PV \int_{-\infty}^{\infty} \frac{c(t')}{t - t'} dt' \tag{3}$$

In Eq. (3), PV is Cauchy principal value.

Constructing the analytic signal $z(t)$:

$$z(t) = c(t) + jH[c(t)] = a(t)e^{j\Phi(t)} \tag{4}$$

In Eq. (4), $a(t)$ is the amplitude function.

$$a(t) = \sqrt{c^2(t) + H^2[S(t)]} \tag{5}$$

$\Phi(t)$ is the phase function:

$$\Phi(t) = \arctan \frac{H[c(t)]}{c(t)} \tag{6}$$

Define the instantaneous frequency as:

$$f(t) = \frac{d\Phi(t)}{dt} \tag{7}$$

The expression for the Hilbert instantaneous energy spectrum is:

$$H(\omega, t) = \mathrm{Re} \sum_{i=1}^{n} a_i(t) e^{\int \omega_i(t) dt} \tag{8}$$

$H(\omega,t)$ integrates time to get Hilbert marginal spectrum:

$$h(\omega) = \int_0^t H(\omega, t)dt \tag{9}$$

Hilbert marginal spectrum expresses the amplitude of each frequency in the whole world.

Hilbert energy spectrum expression:

$$E(\omega) = \int_0^t H^2(\omega, t)dt \tag{10}$$

Hilbert energy spectrum can express the energy accumulated at different frequencies in the whole time.

3 Calculation Principle of Multi-boundary Blasting Charge

3.1 Principle of Charge Layout for Multi-face Empty Blasting

(1) The upper charge adopts small hill bag and short hill bag, while the lower charge adopts long hill bag and all kinds of empty blasting overlap each other. When the lower charge has favorable terrain of lateral throwing, throwing blasting should be adopted.

(2) The upper cartridge is in favorable terrain, commanding and has large relative potential energy, so a larger cartridge should be arranged, and the design throwing rate can be relatively small.

(3) The upper charge should be arranged on the side near the high slope in the middle line, and it is strictly forbidden to over explode or damage the slope.

(4) At the foot of the low slope, in the section with poor or no accumulation conditions, the collapse of each layer of charge should be reduced as much as possible, and the charge should be arranged along the free surface in a plane charge-type multi-face empty blasting.

(5) The initiation sequence of the upper and lower charge is that the upper charge is initiated first, and the lower charge is initiated again after the upper rock mass collapses. Secondly, the charges on both sides of the lower layer or the lowest layer are detonated, and finally the intermediate charges in the lower layer are detonated again.

3.2 Calculation Formula of Multi-boundary Blasting Charge

With the development and application of blasting technology, it is more and more important to study the relationship and interaction between boundary conditions and explosive explosion energy. Considering the interaction between blasting effect and topographic and geological boundary conditions, the formula for calculating the explosive charge of multi-boundary deep-hole blasting is obtained as follows:

$$Q = K W^3 F(E, \alpha) = K F(E, \alpha) a W h \tag{11}$$

In Eq. (11), a is the hole distance (m). h is the drilling depth (m).

The multi-boundary blasting principle considers the stability and leakage of rock mass, the formation mechanism and development law of blasting cracks, and the formation mechanism of blasting engineering geological disasters. Combined with the classification of blasting rock mass, the classification of blasting rock mass structural plane and the design of controlled blasting of rock mass structure, the comprehensive blasting theory of medium potential energy and explosive blasting energy is put forward.

4 Project Example

4.1 Engineering Geological Conditions

Jinou Coal Mine is located in the west wing of Zhuozishan anticline, with scarce surface vegetation, fragile ecology, drought and little rain, and strong wind in winter and spring. The strata in the mining area and its vicinity are Carboniferous, Permian and Quaternary from old to new. Sinan, Cambrian and Ordovician limestone are distributed in turn from the axis of Zhuozishan anticline to the two wings, which constitute the sedimentary basement of coal-bearing strata and are in parallel and unintegrated contact with the overlying coal-bearing strata.

4.2 Design of Multi-boundary Deep Hole Blasting

Using hole packing or air interval charge to control explosive energy distribution. The millisecond delay reverse initiation technology in the hole is adopted to limit the maximum initiation charge in a single stage and control the blasting vibration and flying rocks. The anhydrous pore is made of ANFO explosive, and the water pore is made of emulsion explosive. Blasting design parameters are shown in Table 1.

Table 1 Table of blasting parameters

Number	Name	Unit	Design value
1	Step height	m	16
2	Step slope angle	°	80
3	Borehole diameter	mm	90
4	Unit consumption of explosives	kg/m^3	0.22
5	Chassis resistance line	m	4
6	Hole spacing	m	5
7	Row spacing	m	4
8	Hole depth	m	16.5
9	Ultra deep	m	1
10	Blocking length	m	2.8

4.3 Design of Accurate Delay Initiation Network

Based on the principle of stress wave superposition, free surface increase and rock block collision, the millisecond delay time is calculated by using the empirical formula of multi-boundary theory:

$$\Delta t = K_1 \cdot W (24 - f) \qquad (12)$$

In Eq. (12), Δt is the millisecond delay interval (ms). K_1 is the fracture coefficient of rocks, which is 0.5 for rocks with few fractures, 0.75 for rocks with medium fractures and 0.9 for rocks with developed fractures. W is the chassis resistance line (m). f is the rock firmness coefficient.

After calculation, the delay time is designed as 25 ms between holes and 60 ms between rows (Fig. 1).

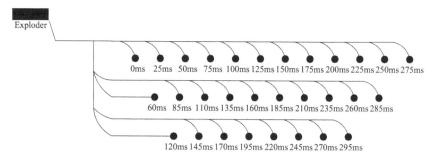

Fig. 1 Detonation network diagram of deep hole blasting

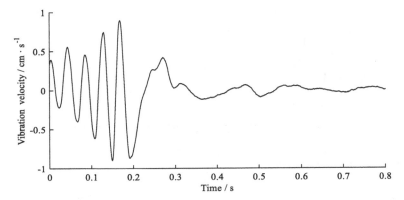

Fig. 2 Time history curve of vibration velocity

5 Characteristic Analysis of Blasting Vibration Signal

5.1 Vibration Velocity Analysis

Measuring points are arranged at the horizontal distance of 110 m, 130 m, 150 m and 170 m from the blasting source to sample the blasting vibration velocity. In this paper, the blasting vibration velocity at a horizontal distance of 110 m from the blasting source is taken as an example to analyze the characteristics of blasting vibration signals, and the sampling time is 0.8 s.

According to Fig. 2, the blasting vibration velocity increases first and then decreases, and reaches the maximum at 0.171 s. Then, with the passage of blasting load time, the particle vibration velocity decays continuously, and gradually approaches 0 at about 0.5 s.

5.2 HHT Analysis

As can be seen from Fig. 3, the frequency components contained in the original signal are very rich, but most of them are below 60 Hz. The dominant frequency of the signal is mainly between 0 and 40 Hz, and the main frequency of vibration is 24.43 Hz. The energy contained in the high frequency components gradually decays.

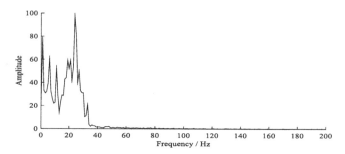

Fig. 3 Hilbert marginal spectrum

6 Conclusions

Multi-boundary blasting technology reduces the engineering cost, protects the rock mass outside the excavation boundary from damage as much as possible, accelerates the construction progress and improves the quality of blasting engineering.

(1) Based on HHT analysis technique, the main frequency of blasting vibration signal energy is in 0–40 Hz, and it is mainly in the low frequency band, and the energy contained in the high frequency component gradually decays.
(2) Multi-boundary deep-hole controlled blasting has uniform rock fragmentation, and all of them collapse within the design range, which has achieved good blasting effect.

References

1. Wang HQ (1992) On the calculating formula of top radius of blast funnel in a multi-boundary blasting system. China J Highway Transport 5(02):26–31
2. Zhang YP, Li XB (2005) Application of hilbert-huang transform in blasting vibration signal analysis. J Central South Univ (SciTechnol) 36(05): 68–173
3. Chen JP, Gao WX, Tao LJ (2006) Theory of rock blasting control in geology engineering. J Eng Geol 14(05):616–619
4. Guan XL, Yan JL (2012) The HHT time-frequency power spectrum analysis of the blasting vibration signal. Explosion Shock Waves 32(05):535–541
5. Zhang QB, Cheng GH, Xu ZH (2018) Study on propagation law of blasting vibration in open-pit mine based on regression. Anal Mining Res Dev 38(05):37–40
6. Hu XL, Qu SJ, Jiang WL et al (2017) Attenuation law of blasting induced ground vibrations based on equivalent path. Explosion Shock Waves 37(06):966–975
7. Gao FQ, Zhang GX, Yang J (2016) Mechanism of vibration reduction by waveform interference for bench blasting in open-pit mine and its application. Mining Res Dev 36(11):18–21
8. Zhang SH, Liu LS, Zhong QL et al (2019) Energy distribution characteristics of blast seismic wave on open pit slope. J Vib Shock 38(07):224–232

Experimental Investigation on the Factors Influencing the Shear Strength of Loess-Mudstone Composite Layer

Xiaodong Zhou, Tienan Wang, and Bowen Xu

Abstract Geological slopes are often disturbed by external engineering, leading changes of the upper moisture content and dry density in loess-mudstone composite layer. It directly affects the shear strength of the loess-mudstone composite layer interface. In order to explore the influence of the upper loess moisture content and dry density on the shear strength of the loess-mudstone composite layer, taking the cutting slope in a test base in Shaanxi Province as the engineering background, different loess moisture contents (10%, 13%, 16%, 19%) and dry densities (1.4, 1.45, 1.5, 1.55 g/cm^3) are employed to investigate the shear strength of loess-mudstone composite layers. Scanning electron microscopy (SEM) is used to observe the failure interface of the loess-mudstone composite layer to analyze the failure mechanism of the samples. The results show that: (1) The shear strength of the loess-mudstone composite layer is lower than those of pure loess and mudstone samples. (2) The moisture content of loess will deteriorate the shear strength of the composite layer, and its effect is greater than that of homogeneous loess; but the dry density of loess will enhance the shear strength of the composite layer, and its effect is less than that of homogeneous loess (3) The moisture content and dry density of loess will affect the distribution of pores in the composite layer interface, changing the shear strength of the composite layer. The research can provide certain data and theoretical basis for the prevention and control of landslides at the loess-mudstone interface.

Keywords Loess-mudstone composite layer · Shear strength · Moisture content · Dry density · Microstructure

X. Zhou
School of Geological Engineering and Surveying, Chang'an University, Xi'an 710054, China

T. Wang
School of Civil Engineering, Southeast University, Nanjing 210000, China

B. Xu (✉)
Shaanxi Energy Vocational and Technical College, Xi'an 710054, China
e-mail: 1440562313@qq.com

© The Author(s) 2023
G. Feng (ed.), *Proceedings of the 9th International Conference on Civil Engineering*,
Lecture Notes in Civil Engineering 327,
https://doi.org/10.1007/978-981-99-2532-2_25

1 Introduction

Loess-mudstone interface landslide refers to the loess landslide sliding along the interface between the overlying loess and the underlying mudstone, which is one of the common types of loess landslide in the Loess Plateau [1–3]. At present, the research results of loess landslide mainly focus on the movement characteristics of loess landslide [4, 5], sliding failure mechanism [6, 7], landslide susceptibility and risk assessment [8–10], landslide types and spatial–temporal distribution [11, 12]. There is still a lack of in-depth research on the mechanical properties and influencing factors of loess-mudstone and other double heterogeneous soils.

The shear strength of the contact surface of double heterogeneous soil is one of the main aspects to characterize the mechanical properties of the contact surface. In engineering practice, the contact surface of double-layer heterogeneous rock and soil structures is often regarded as the weak surface, and has attracted great attention from some scholars. Potyondy [13] was the first to study the mechanical properties of the contact surface between a variety of construction materials and rock and soil mass with the aid of strain-controlled direct shear instrument, and concluded through experimental research that the moisture content of rock and soil mass, the roughness of the structure and the vertical stress would all affect the shear strength of the contact surface. Yoshimi and Kishida [14] introduced X-ray photography technology into shear test of sand-steel contact surface and found that the shear strength was significantly affected by steel surface roughness. Hu et al. [15] applied CCD digital camera technology to the direct shear test of sand-steel contact surface, and pointed out that there is a critical contact surface roughness, which can be divided into shear failure modes of low roughness surface and high roughness surface, and the ideal elastic–plastic failure and strain localized failure with strong strain softening and dilatation respectively. Zhang Ga et al. [16, 17], based on the direct shear and single shear tests on the contact surface of coarse-grained soil and steel plate, pointed out that the shear strength is greatly affected by the properties of rock and soil, vertical stress and the roughness of the contact surface, and is independent of the shear direction. Cheng Hao et al. [18], based on large-scale direct shear tests of red clay-concrete contact surface, found that with the increase of contact surface roughness, the shear failure mode of contact surface gradually evolved from contact surface slip to internal soil failure. Although a series of in-depth studies have been carried out on double-layer heterogeneous rock and soil structures, there are still few experimental studies on the influence of moisture content and dry density of loess on the shear strength of loess-mudstone composite layer.

Based on this, taking a road cut slope in a test base in Shaanxi Province as the engineering background, loess-mudstone composite samples with different loess moisture content and dry density were prepared according to relevant test standards, and direct shear tests were carried out. The influence of loess moisture content and dry density on the shear strength of loess-mudstone composite layer was analyzed. Finally, microscopic failure mode of loess-mudstone interface were observed by

SEM to further reveal the mechanism of the influence of moisture content and dry density of loess on the shear strength of loess-mudstone composite layer.

2 Experimental Design

2.1 Soil Sample Preparation

The soil sample was taken from a road cut slope containing loess-mudstone contact surface in a test base in Shaanxi Province, as shown in Fig. 1. below. As can be seen from the Fig., the site loess-mudstone assemblage can be divided into three layers according to color. The uppermost layer is loess, the middle is mudstone with high moisture content, and the bottom is mudstone with low moisture content. The main reason for this phenomenon is that mudstone is more dense than loess, and it is difficult for rainwater to penetrate mudstone layer completely, resulting in the formation of aquifer between loess and mudstone layer.

According to the Standard of Geotechnical Test Method (GB/T 50,123–2019), the basic physical property parameters of the test soil samples were determined respectively. In view of the high heterogeneity of the soil, four sites were selected to collect soil, and three groups of samples were prepared at each location for repeated measurement and the average value was obtained, as shown in Table 1.

The remolded loess and fully weathered mudstone samples retrieved from the site after air drying were put into the grinding dish for moderate grinding. In order to obtain well-graded test soil samples, the screening method was used to pass the

Fig. 1 Distribution of soil layers in the cutting slope

Table 1 Basic physical parameters of samples

Sample	Moisture content/%	Density/g·cm³	Maximum dry density/g·cm³	Liquid limit /%	Plastic limit/%	Plastic index	Hydraulic conductivity cm/s
Loess	12.98	2.64	1.69	30.42	20.92	9.5	4.73×10^{-5}
Mudstone	15.31	2.73	1.78	38.51	23.19	15.32	5.85×10^{-7}

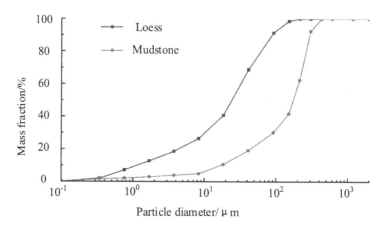

Fig. 2 Particle size gradation of loess and mudstone for testing

loess and mudstone through sieves with pore sizes of 2, 1.18, 0.6, 0.425, 0.3, 0.212 and 0.15 mm successively. The gradation composition of loess with fully weathered mudstone with particle size <0.15 mm was analyzed by laser particle size analyzer. Finally, the gradation parameters of the test loess and mudstone are shown in Fig. 2. In order to keep the basic physical property parameters of the control test soil samples unchanged, each group of soil samples was prepared according to this grading method.

Considering that the influence of rainfall and engineering disturbance on slope soil mostly stays in the loess layer. This experiment kept the physical parameters of the underlying mudstone unchanged, and mainly explored the influence of the loess moisture content and dry density on the shear strength of the loess-mudstone interface. The field loess moisture content ranges from 10.31% to 19.80% and the dry density ranges from 1.38 to 1.56 g/cm^3. Therefore, the design of the loess soil moisture content of 10, 13, 16 and 19% (at this time the dry density of loess is 1.5 g/cm^3), and the dry density of 1.4, 1.45, 1.5 and 1.55 g/cm^3 (at this time the moisture content of loess is 13%). Then, mudstone with moisture content of 15% and dry density of 1.7 g/cm^3 is configured as loess-mudstone composite layer. The vertical stress of 100, 200, 300 and 400 kPa is applied respectively, and the fast shear without consolidation and drainage is carried out. A total of 64 kinds of composite layer conditions are obtained. Meanwhile, parallel control tests of pure loess sample and mudstone group under vertical stress are designed. The shear rate of 0.8 mm/min and shear displacement of 6 mm were used in all tests. The test was repeated 3 times for each condition, and the results were averaged.

Mudstone compresses when vertical stress is applied. In order to avoid the actual shear position deviating from the preset contact surface, the loading mass of mudstone should be calculated according to the predetermined moisture content and dry density (accurate to 0.01 g when weighing). The pure mudstone ring cutter samples were prepared, and the consolidation and compression tests of the soil samples were

Table 2 Compression of mudstone samples under different normal pressures

Vertical stress/kPa	Moisture content/%	Control dry density/g·cm^{-3}	Initial sample density/g·cm^{-3}	Consolidation test compression/mm	Sample density after consolidation /g·cm^{-3}
100	15	1.7	1.955	0.235	1.978
200	15	1.7	1.955	0.363	1.991
300	15	1.7	1.955	0.494	2.005
400	15	1.7	1.955	0.617	2.017

conducted under normal pressures of 100, 200, 300 and 400 kPa respectively. The compression amounts of the pure mudstone samples under different normal pressures were recorded, and the density of the compressed samples was calculated according to the compression amounts. That is, the actual mudstone sample density when the sample with predetermined moisture content and dry density is prepared under the corresponding vertical stress. The test results are shown in Table 2.

2.2 Direct Shear Test Scheme

The direct shear test of loess-mudstone contact surface adopts strain direct shear instrument, whose upper and lower shear boxes have the same size and can accommodate samples with a cross-sectional area of 30 cm^2 and a height of 2 cm (as shown in Fig. 3.). The loess is located in the upper shear box, and the mudstone is located in the lower shear box. By applying horizontal shear stress to the lower shear box, the sample is shear at the contact point of the upper shear box, that is, the loess-mudstone contact surface.

Fig. 3 Strain-controlled direct shearing instrument

The direct shear instrument can apply shear forces under different normal pressure loads (100, 200, 300 and 400 kPa) and shear four samples simultaneously. However, in order to reduce instrument error, only one sample is placed at a fixed position for each shear. The shear displacement value of the display screen (accuracy 0.001 mm) and the deformation of the measuring ring measured by the electric dial indicator (accuracy 0.01 mm) were recorded by the software dynamometer system 3.9A, and the horizontal shear stress was obtained according to the coefficient of the measuring ring.

As the shear area gradually decreases during the test, the test result is smaller than the vertical stress applied by the actual instrument. In addition, if the set shear displacement and the actual shear displacement are not distinguished, the measured shear strength will be less than the actual shear strength of the soil. In this paper, the results of direct shear tests are modified from the aspects of effective shear area and effective vertical stress.

Due to the symmetry of the shear plane, 1/2 of the shear area is taken for area correction. The initial shear area A_0 is regarded as invalid parts A_1 and A_2, as shown in Fig. 4.

According to the geometric relation in Fig. 4., the expression of A_2 can be obtained:

$$A_2 = 2(S_{sectorO'AB} - S_{\triangle O'AB}) = \frac{D^2 \cos^{-1}\left(\frac{x}{D}\right)}{2} - x\sqrt{\frac{D^2}{4} - \frac{x^2}{4}} \qquad (1)$$

where, D is the diameter of the sample, which is 6.18 cm here. x is the shear displacement (cm). When x = 0, the initial shear area A_0 is 30 cm^2, and when x = 6 mm, the initial shear area A_2 is 26.29 cm^2. Let the area correction factor α satisfy:

Fig. 4 Schematic diagram of the effective shear area after the upper and lower shear boxes are displaced (the gray part is the actual shear area after displacement)

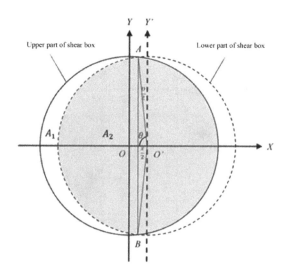

$$\alpha = \frac{A_0}{A_2} \tag{2}$$

there is:

$$\tau_1 = \alpha\tau_0 \tag{3}$$

where, τ_1 is the shear strength of the modified sample (kPa); τ_0 is the shear strength of sample without area correction (kPa).

The vertical stress on the effective shear area was corrected, and the force analysis of the upper shear soil sample was shown in Fig. 5. In Fig. 5, σ_1 is the uniform vertical stress (kPa) applied to the non-effective shear area A_1, σ_0 is the uniform vertical stress (kPa) applied to the top of the shear soil sample, σ_2 is the uniform vertical stress applied to the effective shear area A_2, and P is the applied shear force.

According to the moment equilibrium $M_{yQ} = 0$ of each stress component in the y direction of the central point Q of the effective shear area in Fig. 5, it satisfies [19]:

$$M_{\sigma 1} = M_{\sigma 0} + M_{\tau 1} \tag{4}$$

According to the geometric relationship in Fig. 4. and Fig. 5, it can be obtained:

$$\sigma_2 = \sigma_0 - \frac{\tau_1 l}{\pi x}[\pi - 2\theta + \sin(2\theta)] \tag{5}$$

where, I is the height of the soil sample in the upper shear box, which is 10 mm. Let the normal correction coefficient β satisfy:

$$\beta = \frac{\sigma_2}{\sigma_0} \tag{6}$$

The shear strength of the comprehensively modified sample τ_2 can be obtained:

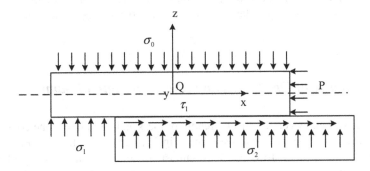

Fig. 5 The force diagram of the soil sample in the upper shear box after shear dislocation

$$\tau_2 = \beta\tau_1 \tag{7}$$

3   Shear Test Results and Analysis of Sample

3.1   Effect of Moisture Content on Shear Strength of Sample

The relationship between shear strength and vertical stress of loess, loess-mudstone composite layer and mudstone under different moisture content is shown in Fig. 6. As can be seen from the Fig. 6, under the same vertical stress, the loess-mudstone composite sample with double heterostructure has the lowest shear strength, followed by the loess sample on the upper layer, and the mudstone sample has the highest shear strength. This indicates that the contact surface is the weak strength of loess-mudstone composite layer. When the vertical stress was the same and the vertical stress was 400 MPa, for example, when the moisture content increased from 10 to 19%, the shear strength of loess continued to decrease by 3.71, 6.93 and 12.08% to 10.02, 18.72 and 32.62 kPa, respectively. The shear strength of loess-mudstone composite layer decreases by 11.67, 22.26 and 37.51 kPa, respectively, by 4.87, 9.29 and 15.65%. This indicates that moisture content has a more significant effect on the deterioration of shear strength at the loess-mudstone interface.

In to the pressure, the upper loess pore structure are pressure, saturation increases, the soil particles between weakly bound water and free moisture content increases, and the lower mudstone is more dense loess, poor permeability and a relative impermeable layer, makes the bottom of the north part of the water accumulation in the loess-mudstone contact area, contact area of soil In the overwatering soft plastic-saturated state, Thus, the shear strength of the contact surface is significantly reduced. This conclusion is consistent with the conclusion proposed by Xing Lin et al. [20] that the loess shear strength decreases with the increase of saturation, that is, the higher the moisture content of the upper loess, the more water accumulated at the loess-mudstone interface, and the higher the deterioration degree of the shear strength of the interface.

3.2   Influence of Dry Density on Shear Strength of Sample

The relationship between the shear strength of loess and loess-mudstone composite layer and the dry density of loess is shown in Fig. 7. It can be seen from the Fig. 7 that, under the same vertical stress, the loess-mudstone composite layer has the lowest shear strength in various working conditions that only change the dry density of loess, that is, the contact surface is still the weak strength surface of the whole loess-mudstone double-layer heterogeneous structure. However, when the vertical stress

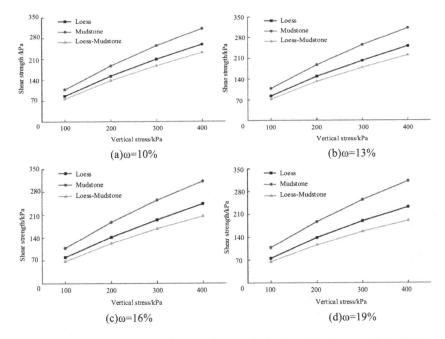

Fig. 6 Variation of shear strength of loess and composite layers with moisture content of loess

was the same and the vertical stress was 400 kPa for example, with the dry density of loess increasing from 1.4 g/cm^3 to 1.55 g/cm^3, the shear strength of loess increased by 14.92, 22.81 and 39.35 kPa, respectively, by 6.29, 9.61 and 16.58%. The shear strength of loess-mudstone composite layer increases by 11.70, 15.16, 24.99 kPa, 5.50, 7.12, 11.74%, respectively. It can be seen that the dry density of loess has greater influence on the shear strength of homogeneous loess than that of composite loess.

The relationship between the shear strength of loess and loess-mudstone composite layer and the dry density of loess is shown in Fig. 7. It can be seen from the Fig. 7 that, under the same vertical stress, the loess-mudstone composite layer has the lowest shear strength in various working conditions that only change the dry density of loess, that is, the contact surface is still the weak strength surface of the whole loess-mudstone double-layer heterogeneous structure. However, when the vertical stress was the same and the vertical stress was 400 kPa for example, with the dry density of loess increasing from 1.4 to 1.55 g/cm^3, the shear strength of loess increased by 14.92, 22.81 and 39.35 kPa, respectively, by 6.29, 9.61 and 16.58%. The shear strength of loess-mudstone composite layer increases by 11.70, 15.16, 24.99 kPa, 5.50, 7.12, 11.74%, respectively. It can be seen that the dry density of loess has greater influence on the shear strength of homogeneous loess than that of composite loess.

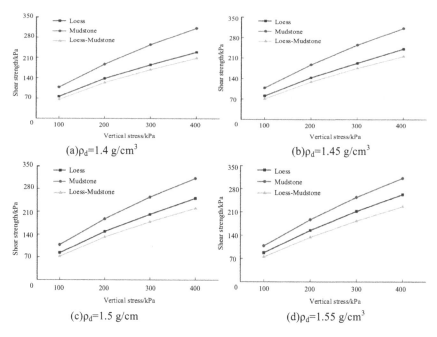

Fig. 7 Variation of shear strength of loess and composite layers on dry density of loess

4 Microstructure of Loess-Mudstone Interface after Shear Failure

In order to further reveal the influence of different loess moisture content and dry density on the shear strength of loess-mudstone composite layer, it is necessary to observe the structural changes of soil samples at the interface of the composite layer under different working conditions by microscopic means. The damaged loessie-mudstone interface block was observed by FEI Quanta 400 FEG environmental scanning electron microscope (SEM), State Key Laboratory of Continental Dynamics, Northwest University.

The microstructure of contact surface after shear failure of composite layers with different loess moisture content is shown in Fig. 8. As can be seen from the Fig. 8, when the moisture content is low, the soil particles are larger on the whole, with looser arrangement and more pores. In addition, it can be seen that small clay groups fill between the larger structural units and play an internal cementation role. Under this condition, the contact particles mainly have mechanical friction under shear. With the increase of moisture content, the amount of water accumulated on the contact surface increases, the amount of clastic material and clay material on the contact surface increases, and some large particles are aggregated into a whole state under the connection of clay particles, and the pore area decreases continuously. When the moisture content increased to 19%, the cement inside the clay mass was softened, dismembered into small clay particles and filled in the pores, and there were no obvious large pores in the visual field. The shear strength of the contact surface decreases due to the thicker bonded water film adsorbed by the disintegrated slime.

Figure 9. shows the microstructure scanning of the interface of composite layer with different dry density of loess after shear failure. Furthermore, PCAS software was used to binarize the image, and the results showed that the porosity of 1.4, 1.45, 1.5 and 1.55 (g/cm³) loess under dry density conditions were 8.52, 7.33, 6.17 and 5.31%, respectively. The average pore area is 228.88, 231.88, 185.94, and 183.65 (pixel). Therefore, it can be seen that with the increase of the dry density of the upper loess, the microscopic characteristics of the pores on the contact surface have a great change, and the pore number and average pore area on the whole decrease, while the porosity decreases. This is mainly because when the dry density of the upper loess increases, more soil particles will be embedded and adhered to the contact surface, filling the original pore structure on the contact surface. The large pores in the overhead are transformed into small and micro pores, and the particles are arranged from loose to tight, and the degree of occlusion is enhanced. Macroscopically, the shear strength and parameters of the contact surface increase to a certain extent with the increase of the dry density of the upper loess.

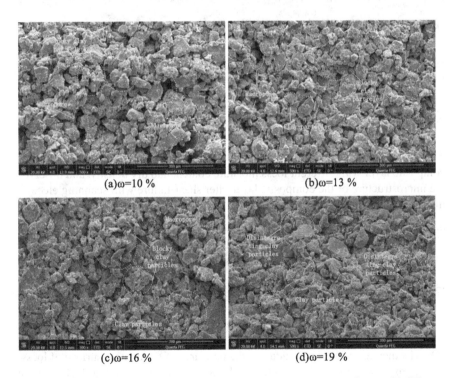

(a)ω=10 %   (b)ω=13 %

(c)ω=16 %   (d)ω=19 %

Fig. 8 Scanning images of the contact surface microstructure in the composite layers with different loess moisture contents

(a)ρ_d=1.4 g/cm^3 (b)ρ_d=1.45 g/cm^3

(c)ρ_d=1.5 g/cm^3 (d)ρ_d=1.55 g/cm^3

Fig. 9 Scanning images of the contact surface microstructure in the composite layers with different loess dry density

5 Conclusion

This paper conducted an experimental study on the shear strength of loess-mudstone composite layer with different loess moisture content and dry density, and observed the microstructure of the composite layer after shear failure with scanning electron microscopy. It can be concluded that:

(1) The shear strength of loess-mudstone composite layer is lower than that of pure loess and mudstone samples, indicating that loess-mudstone double-layer heterostructure is the weak surface of shear strength of the whole sample, which needs to be paid more attention to.

(2) The moisture content of loess will deteriorate the shear strength of composite layer, and its effect is greater than that of homogeneous loess. However, the dry density of loess will enhance the shear strength of composite layer, which is less than that of homogeneous loess.

(3) The increase of loess dry density, the pore area of the contact surface of loess-mudstone composite layer will be reduced, and the degree of interlock between the two layers of heterogeneous soil particles will be increased, leading to the increase of the shear strength of the composite layer. However, the increase of

loess moisture content will reduce the pore area of the contact surface of the composite layer, resulting in the accumulation of water at the contact surface, and reduce the shear strength of the loess-mudstone composite layer.

References

1. Wu W, Wang N (2002) Basic types and active features of loess landslide. Chin J Geolog Hazard Control 2:38–42
2. Tonglu L, Jianhui L, Xinsheng L (2007) Types of loess landslides and methods for their movement forecast. J Eng Geolog 4:500–505
3. Wu W, Su X, Liu W et al (2014) Loess-mudstone interface landslides: characteristics and causes. J Glaciol Geocryolog 36(05):1167–1175
4. Zhang D, Wang G (2007). Study of the 1920 Haiyuan earthquake-induced landslides in loess (China). Eng Geol 94(1–2):76–88
5. Zhou Q, Xu Q, Zhou S et al (2019) Movement process of abrupt loess flowslide based on numerical simulation—a case study of chenjia 8# on the heifangtai terrace. Mount Res 37(04):528–537
6. Wen BP, Yan YJ (2014) Influence of structure on shear characteristics of the unsaturated loess in Lanzhou, China. Eng Geol 168:46–58
7. Kong JX et al (2021) A landslide in Heifangtai, Northwest of the Chinese loess plateau: triggered factors, movement characteristics, and failure mechanism. Landslides 18(10):3407–3419
8. Zhang M, Liu J (2010) Controlling factors of loess landslides in western China. Environ Earth Sci 59(8):1671–1680
9. Bai SB, Lu P, Wang J (2015) Landslide susceptibility assessment of the Youfang catchment using logistic regression. J Mount Sci. 12:816–827
10. Li B, Xu Q, Cheng Q et al (2020) Characteristics of discontinuities in Heifangtai landslide area in Gansu. China. Appl Geophys 17:857–869
11. Peng J, Wang S, Wang Q et al (2019) Distribution and genetic types of loess landslides in China. J Asian Earth Sci 170:329–350
12. Qiu H, Hu S, Wang X et al (2020) Size and spatial distribution of loess slides on the Chinese Loess Plateau. Phys Geogr 41(2):126–144
13. Potyondy JG (1961) Skin friction between various soils and construction materials. Geotechnique 11(4):339–353
14. Yoshimi Y, Kishida T (1981) A ring torsion apparatus for evaluating friction between soil and metal surfaces. Geotech Test J 4(4):145–152
15. Hu L, Pu J (2004) Testing and modeling of soil-structure interface. J Geotech Geoenviron Eng 130(8):851–860
16. Zhang G, Zhang J (2004) Experimental research on cyclic behavior of interface between soil and structure. Chin J Geotech Eng 2004(02):254–258
17. Zhang G, Zhang J (2005) Reversible and irreversible dilatancy of soil-structure interface. Rock Soil Mech 2005(05):699–704
18. Cheng H, Chen X, Zhang J et al (2017) Experimental research on residual shear strength of red clay-concrete structure interface. J Central South Univ (Sci Technol) 48(09):2458–2464
19. Kai Y, Xin Y, Yongshuang Z et al (2014) Analysis of direct shear test data based on area and stress correction. Chin J Rock Mech Eng 33(01):118–124
20. Xing L, Ren Y, Shen X (2020) Effect of saturation on the shear strength of unsaturated loess. J Shandong Agric Univ (Nat Sci Ed) 51(06):1074–1079

Compaction Effect Due to Single Pile Driving in PHC Pile Treated Soft Clayey Deposit

Guowei Li, Ruyi Liu, Chao Zhao, Yang Zhou, and Li Xiong

Abstract The compaction effect of extra-long prestressed high-strength concrete (PHC) piles in deep soft soil foundation was studied by field test. The pore water pressure gauge, inclinometer were embedded in different plane positions or different depths of the foundation to monitor the pore pressure and deformation of the foundation when driving pile. The research shows that the magnitude of excess pore water pressure caused by single pile installation is mainly related to buried depth of the measuring point and the linear distance between the pile tip and the measuring point. The shorter the distance or the deeper the depth is, the greater the excess pore pressure caused by pile installation. The horizontal influence radius of pile compacting on the pore water pressure is about 10.7 m. The excess pore pressure induced by pile installation increases with depth, and is obviously affected by stratum properties. In the vicinity of soil with high permeability coefficient, such as thin sand layer or silty fine sand layer, the excess pore pressure cannot be accumulated in a large amount. The existing subgrade obviously restricts the lateral deformation of soil between piles and PHC piles. The pile deformation is small at the top and bottom, and large in the middle. The inflection point of the deformation curve appears at the pile connection position. The relationship the excess pore pressure of the measuring point with the depth and distance of the measuring point is given.

Keywords PHC pile · Compaction effect · Pore water pressure · Lateral displacement

G. Li · R. Liu · L. Xiong
Hohai University, Key Laboratory of Ministry of Education for Geomechanics and Embankment Engineering, Nanjing 210024, China

C. Zhao
Guangdong Road and Bridge Construction Development Co., Ltd, Guangzhou 510510, China

Y. Zhou (✉)
Henan University of Technology, Civil Engineering School, Zhengzhou 450001, China
e-mail: robertzhouy@163.com

G. Feng (ed.), *Proceedings of the 9th International Conference on Civil Engineering*,
Lecture Notes in Civil Engineering 327,
https://doi.org/10.1007/978-981-99-2532-2_26

315

1 Introduction

PHC pile has reliable quality and is widely used in soft foundation treatment. And the compaction effect of PHC pile during installation is significant. The pile penetration makes the soil around the pile compact which will induce lateral extrusion and vertical uplift of the soil and in turn causes tilting or floating of the adjacent pile, even crack and overturning of the adjacent buildings [1–3] (a. Zhang et al.2006 b. Liu et al. and c. Wang et al.). In addition, compaction effect leads to the increase of excess pore pressure, and the dissipation of the induced excess pore pressure, i.e., consolidation, results in considerable soil settlement and negative surface friction around the pile. Field test is a commonly used research method by many scholars because of its authenticity and reflecting the engineering practice. Li et al. (2011) [4] and Liu et al. (2012) [5] studied the influence of pile installation process on foundation deformation, excess pore pressure and plugging effect through field tests. Xing et al. (2009) [6] found that the lateral displacement of the soil mass at the distance of 16 d (d is the pile diameter) from the edge of the pile group edge was the largest, equal to15 mm. Wei et al. (2020) [7] investigated the effect of the pre-drilled hole and isolation effect in PHC pile installation; Wan et al. (2020) [8] studied the effect of soil properties and pile driving speed on soil compaction during pile installation. In summary, the PHC piles has been widely researched and lots of useful results were obtained.

However, the existing research on PHC pile driving is basically within 10 m long, lacking of research on the extra-long PHC pile which usually exceeds 40 m and requires pile extensions more than three times. For extra-long PHC pile, the pile installation process has a more s13ignificant impact on the lateral deformation than the normal pile [9–12]. In the southeast regions of China, the thickness of the soft foundation often reaches tens of meters, even nearly 100 m which acquires longer pile length. The existing research showed that the excess pore pressure is positively related to the penetration depth. Thus, it is necessary to study the compaction effect of extra-long PHC piles. In addition, previous studies mostly focused on the excess pore pressure and lateral deformation of soil caused by driving piles, and there has no research about the compaction effect on the lateral deformation of adjacent piles.

In this paper, field test was conducted to study the compaction effect of extra-long PHC piles. Thorough this test, excess pore pressure changes, lateral displacements of the soil and pile shaft during the pile installation were monitored. And the compaction effect caused by the penetration of extra-long PHC piles was analyzed assisted with the measured data.

Table 1 Physical–mechanical indexes of soil

soil category	H(m)	γ(kN/m^3)	w (%)	c(kPa)	φ (°)
Filling	2.5	17.12	20.2	—	—
Silty clay	3	16.38	31.2	4.20	15.8
Shell layer	2	—	—	—	—
Silty	29	18.10	43.5	6.13	25.0
Fine sand	8	19.20	27.6	—	—

2 Field Test

2.1 Test Site

The field test was carried out in an expansion project of express way in Guangdong province in China. The test site is relatively flat but with very deep soft clay deposits. At the top, there is a layer of sand filling with a thickness of 2.5 m, and the followed by 34 m – thick clay layer, including silty clay, a shell layer and a layer of silty, the bottom of a 8 m silty sand located on the moderately weathered rock stratum. The main physical parameters are shown in Table 1.

2.2 Tested Program

The moderately weathered rock stratum below the fine sand was designed as the bearing stratum of the piles. The design length of the pile is ranging from 45 to 150 m, the pile has an outer diameter of 400 mm and an inner diameter of 310 mm, thus the thickness of the annular of the PHC pile is 90 mm. The material of the pile was C80 concrete. A total of 79 piles were installed. The piles were installed in a rectangular shape. The pile space is 3 × 2 m along the longitudinal and transversal direction. The pile location layout and installation route procedure are illustrated in Fig. 1. The PHC pile installation in the test area started on A-1 (5 May) and ended on V-16 (4 June), and lasted for 31 days. The layout of the test instruments is shown in Fig. 1. Two inclinometer tubes were installed, designed as CX1 and CX2 with a depth of 42 m and 44 m, respectively. CX1 was used for lateral deformation measurement of the soil within piles. And CX2 was used for the pile shaft and placed inside the hollow of the pile I-5. In addition two of the piezometers were set at K1 points with a buried depth of 12 m and 18 m respectively. As shown in Fig. 1, K1 has an equal radial distance of 3.5 m from the surrounding 4 piles.

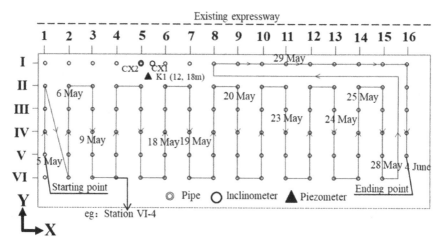

Fig. 1 Instrument layout and construction sequence diagram

3 Test Results and Analysis from Single Pile Driving

3.1 Pore Water Pressure

In the section, the compaction effect of three piles with different radial distance on the target pile I-5 was investigated. In this paper, pile II-4, pile III-4 and pile VI-4 are selected and these piles are representative of the pile installation at close distance, middle distance and far distance. The excess pore pressure variation with pile installation time was shown in Fig. 2(a)–(c) for the above three piles, respectively. Note the excess pore pressure plotted in Fig. 2. are the incremental pressure induced by the penetration from an initial time. Figure 2(a) shows the relationship between the excess pore water pressure in K1 and the driving time of pile II-4. It can be seen that when the pile tip depth is less than 13 m, the pore water pressure of K1–12 increases rapidly. When the pile depth exceeded 13 m, the excess pore water pressure of K1-12 basically remains unchanged, keeps at 6 kPa. For K1-18, the excess pore pressure is zero when the pile depth is within 13 m. When the pile depth ranges within 13 –37 m, the excess pore pressure rises rapidly, with the maximum value of 15.8 kPa. It is interesting to find that the excess pore water pressure starts to slowly decrease as the pile tip depth is beyond 37 m, and the excess pore water pressure of K1-18 is 13.3 kPa after pile installation.

Figure 2. (b) shows the relationship between the excess pore pressure and the driving time for pile III-4. The total penetration depth of the PHC pile is 45 m. Basically, the trend of pore pressure with time is consistent with pile II-4. For both piles, i.e., II-4 and III-4, the excess pore pressure rises rapidly when the pile tip reaches near the measuring point. And the induced excess pore pressure measured at the deep point (18 m) is larger than that at the shallow point (12 m). For K1-12, the

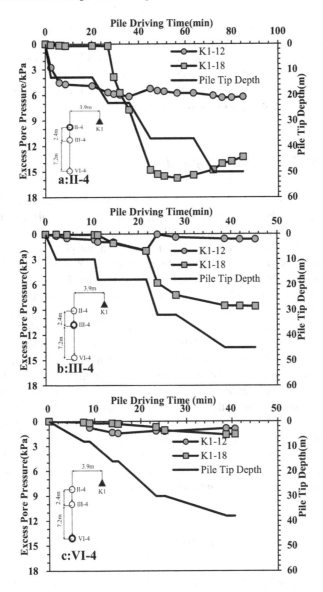

Fig. 2 Curve of excess pore water pressure generated by single pile installation

peak value of the induced excess pore pressure is 2.0 kPa, and the residual value after installation is 0.6 kPa. For K1-18, the peak value is 8.5 kPa, and the residual value is 8.5 kPa. Figure 2. (c) shows relationship between the excess pore water pressure and the driving time for pile VI-4. The total penetration depth is 38 m with a peak excess pore pressure 1.4 kPa and a residual value of 0.9 kPa at K1-12, and a peak value of 1.6 kPa and a residual value of 1.54 kPa at K1-18. Clearly, as the distance of increasing between the PHC pile and the measuring point, the excess pore water pressure caused by the single pile installation decreases gradually. For example, the excess pore water pressure during the penetration of pile VI-4 is very small generally indicating no significant compaction effect due to pile driving at farther distance. Therefore, the horizontal distance between pile VI-4 and the measuring point, e.g., K1, can be taken as the impacting radius, which is 10.7 m in this study.

3.2 Lateral Displacement of Soil

The pile II-6 and II-8 were selected to study the impact of penetration on the lateral deformation of soil between piles. Figure 3. shows the lateral displacement with depth in X and Y directions after each pile's installation. The lateral displacement were obtained from the inclinometer at CX1. In general, the pile installation pushing the soil mass away results in positive displacement in Y direction and negative displacement in X direction. In Y direction, the maximum lateral displacement of pile II-6 is 3.8 mm at the depth of about 24 m and the value induced by pile II-8 is basically close to 0. And the maximum lateral displacement caused by pile II-6 and II-8 in X direction is 4.8 mm and 2.5 mm, respectively. Definitely, the lateral displacement of soil induced by the PHC pile installation decreases with the distance of the driven pile. It also can be seen from Fig. 3. that the lateral displacement of soil in the Y direction is significantly smaller than that in the X direction. Due to the existence of the in-service expressway, there is a constraint effect on soil deformation induced by pile installation in the Y direction. In addition, the lateral displacement at the upper and lower parts in Y direction are small, while the displacement in middle height is relatively larger. Contrast to Y direction, the soil displacement in X direction shows a continuous increase from bottom to the top. As mentioned above, the existing embankment has significant resistant effect on the soil deformation in Y direction, but it is not true in X direction.

3.3 Lateral Deformation of PHC Pile

The influence of pile installation on the lateral deformation of adjacent piles mostly researched by numerical simulation or model test [13–16] (a. Yao et al. b. Zhan et al. and c. Wei et al. d Luo et al.) at present. Rare reports about field tests on this topic were found up to date. In this paper, by setting inclinometer in the pile the lateral deformation of the pile was directly measured. Thus the impact of the compaction effect of subsequent pile installation on the existing pile can be directly estimated. As shown in Fig. 1, pile I-5 is selected to be the target pile with a 44 m-long inclinometer. First, four consecutive piles (II-5, III-5, IV-5 and V-5) are selected as the research object along the Y direction. Figure 4. shows the curve of lateral displacement of pile I-5 with depth as the four piles (II-5, III-5, IV-5 and V-5) were penetrated. Among

Fig. 3 Lateral displacement
curve of soil caused by
single pile installation

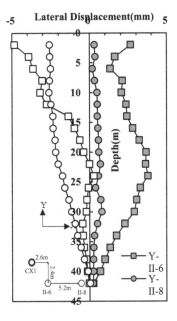

them, the penetration of the nearest PHC pile II-5 has the greatest impact on the lateral displacement of the pile shaft, and the maximum lateral displacement is 4.3 mm at the depth of 22 m. In addition, the pile's displacement mainly occurs at 1/3–2/3 of the pile length. Considering the connection position of the prefabricated pile shaft, the lateral displacement turning points (1/3 and 2/3 depth) exists consistent with the connection points. It is possible that the joint of the pile shaft is a relatively weak point resulting in the bending and deformation of the whole pile due to installation of near pile. Therefore, special attention should be paid to the joint of shaft of a PHC pile during driving. The results also show that the influence of driving for pile V-5 on the deformation of pile I-5 can be ignored. The influence range of pile installation on the deformation of adjacent piles is about 4 times the pile spacing, generally equal to 10.7 m.

3.4 Influencing Radius

Figure 5. shows excess pore pressure with horizontal distance from pile-center to the K1. Note that the excess pore water pressure in Fig. 5. (a) is the maximum value after penetration of each pile. It can be seen from Fig. 5. (a) that the maximum excess pore water pressure caused by single pile installation at a depth of 12 m is 26 kPa, and the value at 18 m is 80 kPa at horizontal distance of 1.8 m. As mentioned above, deeper soil yields larger pore water pressure. As the horizontal distance increases, the pile driving induced excess pore water pressure caused by

Fig. 4 Lateral displacement
of pile shaft caused by single
pile installation

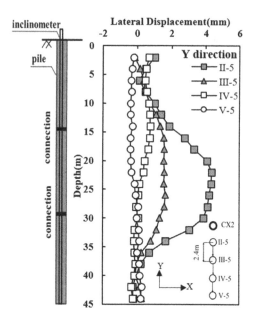

single pile installation gradually decreases. When the horizontal distance exceeds
10.7 m, the PHC pile installation will not produce any excess pore water pressure.
In this case, the influence range of excess pore pressure generated by single pile
installation on soil mass is about 10.7 m, that is, 27 times the pile diameter. In Fig. 5.
(b), pile driving induced pore water pressure was normalized by the effective stress.
The effective stress at 12 and 18 m was estimated respectively as 120 and 180 kPa by
assuming a fully saturated condition and a unit weight of 20 kN/m^3. It was found the
normalized results for pore pressure at both 12 and 18 m can be identically expressed
as Eq. (1):

$$\frac{u}{\sigma_v} = 0.8R^{-1.6} \tag{1}$$

u: excess pore water pressure (kPa)

σ_v: effective stress (kPa)

R: distance between measuring point and pile axis (m)

By using Eq. (1), the pile driving induced pore water pressure can be estimated
with known radial distance and depth.

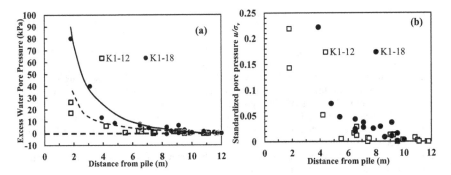

Fig. 5 Curve of induced excess pore water pressure changing with horizontal distance between pile and measuring point

4 Conclusions

In this paper, based on a high-speed reconstruction and expansion project, a field test study on the compaction effect of PHC piles in super deep soft foundation is carried out. According to the test results, the following conclusions are drawn:

1. The value of excess pore water pressure caused by single pile installation is mainly related to the linear distance between the pile tip and the measuring point. The shorter the distance is, the greater the excess pore water pressure induced by the pile installation process is. When the pile tip crosses the measuring point, the excess pore pressure will not increase. In this test, the horizontal influence radius of PHC pile soil compacting on excess pore pressure is about 10.7 m.
2. The induced pore water pressure increases with the depth and is obviously affected by the stratum property. In this study, the peak value of the induced pore water pressure is 165 kPa. The pore water pressure cannot be accumulated in the vicinity of soil with large permeability coefficient, such as thin sand layer or silty fine sand layer.
3. The existing subgrade has obvious restraint effect on the soil between piles and the lateral deformation of PHC piles. The deformation of PHC piles is small at the top and bottom, and large in the middle. The inflection point of deformation occurs at the pile connection position.
4. The relationship between excess pore water pressure and horizontal distance can be obtained by normalization. Pile driving induced pore water pressure can be estimated with known radial distance and depth.

Acknowledgements The authors wish to express their gratitude and sincere appreciation for the financial support received from the National Natural Science Foundation of China (Grant No. 42177126).

References

1. Zhang Z, Xin G, Yu H, Xiong L (2006) Study on floating PHC piles and disposal measures in soft soil foundation. Chin J Geotech Eng 05:549–552
2. Liu J, Yang Q, Yu D, Zhang S (2019) Case study and analysis of the squeezing soil effect of preformed pile on the settlement and deformation of pile foundation. China Civil Eng J (S2):95–101. https://doi.org/10.15951/j.tmgcxb.2019.s2.014
3. Wang X, Chen L, Dou R (2003) Research on soil deformation caused by pile-driving and the formula calculating soil deformation. Rock Soil Mech.S2:175–179. https://doi.org/10.16285/j.rsm.2003.s2.040
4. Li Z, Liu Z, Zhao Y, Zhang H (2011) File test study of new-type soil displacement screw pile. Chin J Rock Mech Eng 30(02):411–417
5. Liu H, Jin H, Ding X, Li J (2012) Field test research on squeezing effects of X-section cast-in-place concrete pile. Rock Soil Mech 33.S2:219–223+228. https://doi.org/10.16285/j.rsm.2012.s2.062
6. Xing H, Zhao H, Xu C, Ye G (2009) Driving effect of PHC piles. Chin J Geotech Eng 31(08):1208–1212
7. Wei L, Du M, He Q, Liao P, Li S (2020) Filed test of static pressure pile group based on high precision automatic monitoring system. Chin J Rock Mech Eng 39.S2:3627–3635. https://doi.org/10.13722/j.cnki.jrme.2019.0928
8. Wan X, Ding J, Huang C, Ding C (2020) Field test research on compaction effects during installation of group piles in layered soils. J Southeast Univ (Nat Sci Ed) 50(06):1090–1096
9. Zhao C, Du X, Zhao C, Xie X (2013). Squeezing effect of inner-digging prestressed piles. Chin J Geotech Eng 35.03:415–421
10. Lei H, Li X, Lu P, Huo H. (2020) Field test and numerical simulation of squeezing effect of PHC pile. Rock Soil Mech 33.04:1006–1012. https://doi.org/10.16285/j.rsm.2012.04.039
11. Li G, Bian S., Lu X, Yang T, Lei G (2013). Field test on extruding soil caused of PHC pile driving by static pressure for improving soft foundation of widened embankment. Rock Soil Mech 34.04:1089–1096. https://doi.org/10.16285/j.rsm.2013.04.018
12. Lei G, Ai Y, He Z, Shi J (2012) Driving response of an open-ended PHC pile group in silty fine sands. Chin J Geotech Eng 34(02):294–302
13. Yao X, Hu Z (1997) Estimation of pore water pressure caused by pile sinking in saturated soft soil. Rock Soil Mech 04:30–35. https://doi.org/10.16285/j.rsm.1997.04.006
14. Zhan L, Li J, Rao P (2010) Model tests on compacting effect of jacked group piles near slope top. Chin J Geotech Eng 32(S2):150–153
15. Wei L, Li S, Du M, Zhang H, He Q (2021) Numerical analysis of soil extrusion effect of hydrostatic PHC pile based on CEL method. J South China Univ Technol (Nat Sci Ed) 49(04):28–38
16. Luo Z, Gong X, Zhu X (2008). Soil displacements around jacked group piles based on construction sequence and compacting effects. Chin J Geotech Eng 06:824–829

Study on Design and Application of Ultra-Thin Cover for Noise Reduction and Skid Resistance of Old Cement Concrete Pavement

Hui Huang, Jie Chen, and Honggang Zhang

Abstract According to the comparative analysis from the whole life cycle economic cost accounting, the capital investment of preventive maintenance of old cement concrete pavement is less than that of corrective maintenance. Based on the old cement concrete pavement, the problem such as anti-sliding noise, from how to improve the driving comfort and improve the stress of the cement road panel considering structure of old cement concrete pavement preventive maintenance are discussed in this paper, recommend a kind of cement concrete pavement noise reduction against sliding type ultra-thin rubber asphalt surface preventive maintenance structure, It can provide favorable technical reference for the applicability and rational selection of preventive maintenance structure of old cement concrete pavement.

Keywords Cement concrete pavement · Noise reduction and skid resistance · Ultra-thin · Rubber asphalt · Preventive maintenance

1 Introduction

Ultra-thin cover is a common technical measure in the preventive maintenance technology of old cement concrete pavement [1]. The so-called preventive maintenance is a cost–benefit maintenance strategy for existing pavement under the condition of good pavement condition. Preventive maintenance is carried out before obvious damage to the pavement. The benefit cost ratio of the pavement with preventive maintenance is 6−10 times that of the pavement with any maintenance measures.But from the national scope, the highway and national provincial highway have just entered

H. Huang · J. Chen (✉) · H. Zhang
Guangxi Transportation Science and Technology Group Co., Ltd., Guangxi 530007, Nanning, China
e-mail: 1157537047@qq.com

Guangxi Key Lab of Road Structure and Materials, Guangxi 530007, Nanning, China

Research and Development Center on Technologies, Materials and Equipment of High Grade Highway Construction and Maintenance Ministry of Transport, Guangxi 530007, Nanning, China

© The Author(s) 2023
G. Feng (ed.), *Proceedings of the 9th International Conference on Civil Engineering*,
Lecture Notes in Civil Engineering 327,
https://doi.org/10.1007/978-981-99-2532-2_27

a new stage from construction mainly to construction and maintenance. The traditional road maintenance consciousness is still the mainstream, and is still the "correct maintenance" of "road not bad (repair)". It has not yet entered the preventive maintenance stage. Prevent and delay the pavement diseases occur, however, focus on maintenance, the preventive maintenance technology can continue to keep the pavement performance, prolong DaZhongXiu maintenance cycle, reduce the cost of the whole life cycle economic investment, prolong the service life of road surface integral, and the comparison, from the total life cycle cost accounting preventive maintenance money less than corrective maintenance maintenance [1, 2].

At present, the preventive maintenance measures for the old cement concrete pavement at home and abroad are to directly overlay the asphalt surface cover, make full use of the stiffness and strength of the cement pavement, improve the driving comfort, but the reflection cracks and interlayer bonding problems, and the noise reduction of the protection board are still urgent problems to be solved [2, 3].

According to the present situation of and the demand, consider the caking property of rubber asphalt mixture, flexibility, crack resistance, at the same time, considering the economic costs and the status quo of traffic paving layer, function orientation and durability problems, this paper tested after evaluation of pavement and so good, and poor board edge deflection of old cement concrete pavement, A preventive maintenance structure with noise reduction and anti-skid type of ultra-thin rubber asphalt overlay for old cement concrete pavement is proposed, which can effectively reduce the impact load of driving load on the old cement concrete pavement panel, improve the driving comfort, and extend the service life of the old cement pavement and the whole pavement after paving. From top to bottom, it is as follows: 3.0 cm optimized ARAC-10 rubber-asphalt concrete ultra-thin cladding, rubber-asphalt adhesive layer, old cement concrete surface layer, noise reduction and anti-slip type ultra-thin rubber-asphalt cladding preventive curing structure schematic diagram is shown in Fig. 1.

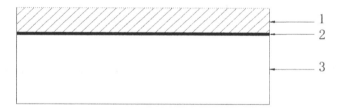

Fig. 1 Noise reduction anti slip type ultra thin rubber asphalt cover preventive maintenance structure diagram

Table 1 Test results of main performance indexes of rubber asphalt

Performance index	Value	Methods method
Rotary viscosity(180°C)/Pa.s	3.2	T0625
Penetration(25°C, 100 g, 5 s)/0.1 mm	42.2	T0604
Softening point/°C	68.6	T0606
°Celastic recovery(25)/%	78.0	T0662
Ductility (15°C, 5 mm/min)/cm	9.5	T0605
Relative density(25°C)	1.022	T0603

2 Raw Materials and Experimental Study

2.1 Rubber Asphalt

In this paper, the experimental study shows that the matrix asphalt is No. 70 Grade A road petroleum asphalt from Maoming, with rubber powder (30−80 mesh).And the stirring and shearing processing technology was adopted, the processing temperature was (175−185) °C, the stirring time was 1 h, and the shearing time was 30 min. Rubber asphalt was prepared indoors. The performance test results of rubber asphalt were shown in Table 1, in which the content of rubber powder in rubber asphalt was 21% (i.e. the quality of rubber powder: asphalt quality = 21:100).

2.2 Aggregate and Packing

In this paper, the coarse aggregate is 1# (6−11), 2# (3−6) mm limestone gravel, fine aggregate is 3#(0−3) mm limestone rock debris, the filler is limestone mineral powder. The screening test results of aggregate and packing used in this test are shown in Table 2. below.

Table 2 Test results of aggregate and packing screening

Mineral aggregate	Mass percentage (%) through the following sieve (mm)										
	19.0	16.0	13.2	9.5	4.75	2.36	1.18	0.6	0.3	0.15	0.075
1#coarse aggregate	100	99.8	96.6	68.4	1.2	0.5	0.5	0.5	0.5	0.5	0.5
2#coarse aggregate	100	100	99.9	99.0	54.9	1.0	0.8	0.8	0.8	0.8	0.8
3#fine aggregate	100	100	100	100	99.3	70.0	57.2	42.6	31.6	25.1	13.3
Mineral powder	100	100	100	100	100	100	100	100	100	100	91.1

2.3 Optimized Ore Grading Design of ARAC-10 Rubberized Asphalt Mixture

Considering the characteristics of rubber asphalt slurry, the structure of multi-gravel small skeleton ore is adopted. The aggregate dosage range of 4.75−9.5 mm is 60−65%, the aggregate dosage range of 2.3−4.75 mm is 15−20%, and the aggregate dosage range of 0.075−2.36 mm is 20−25%. The key screen pass rate of 9.5 mm is 85−90%, 4.75 mm passes 35−40%, 2.36 mm passes 23 − 28%, 0.075 mm passes 4−6%, resulting in the formation of dense and stable skeleton structure between aggregates, so as to improve its high temperature, fatigue, deformation and cracking resistance.

According to the screening test results of aggregates and fillers, combined with existing engineering practice and field practice experience, ARAC-10 rubber-asphalt mixture ore grading design is selected and determined, as shown in Fig. 2.The water stability test results and high temperature stability test results of rubberized asphalt mixture are shown in Table 3.The optimized ARAC-10 rubber asphalt mixture has excellent performance, water stability and high temperature stability far exceed the technical requirements of the specification, and has good high and low temperature, fatigue resistance, deformation resistance, cracking resistance, and noise absorption. The function is very suitable for hot and humid high temperature areas.

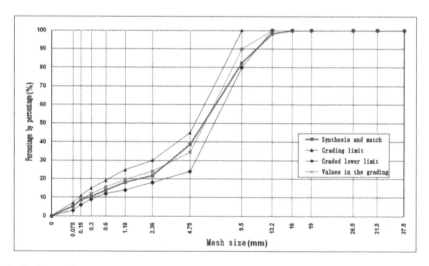

Fig. 2 Ore grading curve

Table 3 Water stability and dynamic stability performance test results

Performance index		Water immersion residual stability $MS_0(\%)$	Freeze–thaw splitting tensile strength ratio(%)	Dynamic stability (times/mm)
Graduation type	Asphalt-stone ratio(%)	123.4	85.7	8124
ARAC-10	6.0			

3 Study on Mechanical Response of Ultrathin Overlay Structure on Old Cement Concrete Pavement

3.1 Construction of Finite Element Calculation Model of Ultra-Thin Overlay on Old Cement Concrete Pavement

In order to grasp the mechanical properties of the ultra-thin overlay structure of cement concrete pavement, this paper uses the finite element analysis software to build the overlay structure finite element model with overlay thickness of 3, 4, 5, 6 cm, as shown in Fig. 3.Because the numerical model of asphalt pavement structure is symmetrical and the stress state of the structure is axisymmetric, the axisymmetric model is used to calculate the stress characteristics of the pavement structure.The numerical model is calculated by using a single garden load pattern, the uniform pressure is 0.7 MPa, the equivalent garden diameter is 30.4 cm, and the thickness and elastic modulus of each structural layer of the old cement concrete pavement are selected as shown in Table 4.The numerical simulation model is shown in Fig. 3 [4, 5]. The numerical simulation model uses a four-node quadrilateral element, the length of the element is divided according to the principle of dense on the top and sparse on the bottom. The boundary conditions of the model are set as follows: the bottom surface of a certain depth of the subgrade is a fixed surface, the left and right surfaces parallel to the Y axis have no displacement in the X direction, and the structural layers of the pavement are completely continuous contact [5, 6].

Fig. 3 Numerical simulation model diagram

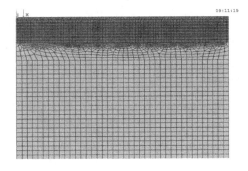

Table 4 Parameters of ultrathin overlay structure model for old cement concrete pavement

Structural layer	Layer thickness(cm)	Elastic modulus (MPa)	Poisson ratio
Asphalt surface	(3, 4, 5, 6 cm)	1200	0.25
Cement concrete surface	24 cm	30,000	0.20
Base course	20	1800	0.25
Subbase	20	1200	0.25
Cushion	20	600	0.30
Earth base	400	35	0.35

3.2 Finite Element Calculation Analysis of Ultra-thin Overlay on Old Cement Concrete Pavement

In this paper, only considering the traffic load, the finite element analysis software is used to simulate the changes of horizontal compressive stress at the bottom of asphalt layer, shear stress and shear strain in the depth direction with the thickness of asphalt layer under different overlay thickness (3, 4, 5, 6 cm), the calculation results are shown in Figs. 4, 5 and 6.

The results show that when the thickness of asphalt layer increases from 2 to 5 cm, the mechanical response index of ultra-thin overlay structure on old cement concrete pavement does not change obviously, and the horizontal compressive stress, shear stress and shear strain at the bottom of asphalt layer change greatly.

The relationship between the change of mechanical response index and the change of asphalt layer thickness in the ultra-thin overlay structure layer of old cement concrete pavement is shown in Fig. 7. The road surface displacement and the shear strain in the structure layer are positively correlated with the change of layer thickness, that is, with the increase of asphalt layer thickness, the peak value of road surface displacement and shear strain in the structure layer increases. The other indexes are

Fig. 4 Variation law of horizontal compressive stress distribution at asphalt layer bottom with asphalt layer thickness

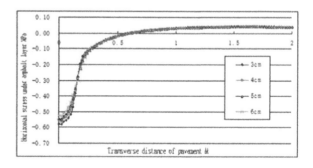

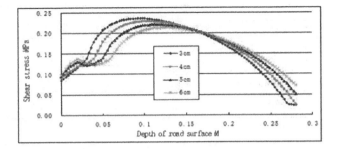

Fig. 5 Variation of shear stress distribution in depth direction of structural layer with asphalt layer thickness

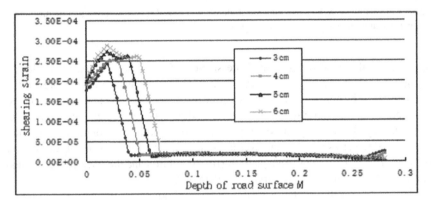

Fig. 6 Variation of shear strain distribution in depth direction of structural layer with asphalt layer thickness

negatively correlated with the change of layer thickness, but the change range of each index is not large. Considering the economic applicability and construction workability, it is recommended that the thickness of 3 cm should be used for the overlay of old cement concrete pavement.

4   Study on Application of Ultrathin Overlay Structure on Old Cement Concrete Pavement

A section of the old cement concrete pavement has excellent road conditions, basically no broken plates and no obvious dislocation. Therefore, it is considered to protect the force of the existing cement plate, reduce the direct impact of heavy load traffic on the cement plate, prolong the service life, etc. At the same time, considering the economic cost input and the current road conditions, the function orientation and

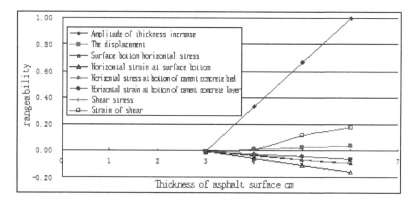

Fig. 7 Correlation between variation amplitude of mechanical response index in structural layer and increase amplitude of asphalt layer thickness

durability of the overlay layer, etc., the section adopts 3.0 cm optimized ARAC-10 rubber asphalt concrete ultra-thin overlay for transformation during preventive maintenance. The comparison diagram of the test section of the old cement concrete pavement overlaid with ultra-thin rubber asphalt overlay before and after paving is shown in Fig. 8. After 3 years of follow-up investigation, it is found that the road condition of the test section of the overlay is good, and there is no obvious disease on the road surface. The overall change of the road condition and the tracking survey test data is not large. The tracking survey test data of the old cement concrete pavement overlaid with ultra-thin rubber asphalt overlay are shown in Table 5.

After the old cement concrete pavement is paved with 3cmARAC-10 rubber asphalt concrete ultra-thin overlay, the pavement condition is good, which effectively solves the noise and anti-skid problems of the cement pavement, improves the driving comfort, and reduces the driving load on the old cement pavement panel. The load effect, especially the impact vibration load, effectively solves the problem of increased plate breakage and performance attenuation in the later period, thereby

Fig. 8 Comparison diagram of old cement concrete pavement with ultra - thin overlay test section before and after construction

Table 5 Parameters of ultrathin overlay structure model for old cement concrete pavement

serial	Overlay type	Overlay operation years	PQI value (分)	Asphalt pavement core sample test data			Splitting tensile strength of old concrete face slab (Mpa)	Compressive strength of cement stabilized gravel base (Mpa)
				deflection value(0.01 mm)	porosity y(%)	splitting tensile strength(Mpa)		
1	3cm Rubber Asphalt Concrete Overlay+24cm Old cement concrete	1年	97.79	44.62	4.2	1.16	5.06	25.8
2	surface+1 cm lower seal coat+18cm cement stabilized	2年	95.25	45.28	5.5	1.02	5.21	26.2
3	macadam base+18cm gravel-sorted subbase	3年	93.42	43.92	4.7	1.12	5.15	24.3

prolonging the service life of the cement concrete panel. Compared with ordinary asphalt overlay, rubber asphalt ultra-thin overlay can slow down the expansion rate of reflection cracks and prolong the service life of old cement concrete pavement and pavement structure after overlay.

5 Conclusion and Suggestion

(1) Aiming at the old cement concrete pavement with good road conditions such as no void, broken plate and poor deflection of plate edge after detection and evaluation, considering the adhesion, flexibility and crack resistance of rubber asphalt mixture, economic cost investment and current road conditions, functional orientation and durability of overlay, this paper proposes a noise reduction and anti-sliding type ultra-thin rubber asphalt overlay preventive maintenance structure for old cement concrete pavement, which can effectively reduce the impact load of driving load on the old cement concrete pavement panel, improve driving comfort, and prolong the service life of old cement pavement and the whole pavement after overlay.

(2) The experimental results of water stability and high temperature stability of rubber asphalt mixture show that the ARAC-10 rubber asphalt mixture proposed in this paper has good high and low temperature, fatigue resistance, deformation

resistance, cracking resistance and noise absorption function, which is very suitable for hot and humid high temperature areas.

(3)   In this paper, through the finite element simulation analysis, it is known that the surface displacement and the internal shear strain of the structural layer of the old cement concrete pavement overlay ultra-thin overlay structure are positively correlated with the change of layer thickness, and the remaining indicators are negatively correlated with the change of layer thickness, but the change range of each index is not large. Comprehensively considering the economic applicability and construction workability, it is recommended that the thickness of 3 cm should be adopted for the ultra-thin overlay of the old cement concrete pavement.

(4)   After paving 3 cm ARAC-10 rubber asphalt concrete ultra-thin overlay on the old cement concrete pavement, the pavement condition is good, which effectively solves the noise and anti-sliding problems of the cement pavement, improves the driving comfort, and reduces the load of the driving load on the old cement pavement panel, especially the impact vibration load. Compared with ordinary asphalt overlay, rubber asphalt ultra-thin overlay can slow down the expansion rate of reflection cracks and prolong the service life of old cement pavement and the whole pavement structure after overlay.

Acknowledgements The research of this article is supported by Guangxi's major science and technology project (Guike AA18242032),and Guangxi's key research and development plan (Guike AB19245019).

Declaration of Competing Interest The authors declare that they have no known competing financial interests or personal relationships that could have appeared to influence the work reported in this paper.

References

1. Sun L et al (2003) Behavior theory of asphalt pavement structure. Tongji University Press
2. Sha Q (2008) Early Damage phenomenon and prevention of expressway asphalt pavement. People 's transportation publishing house
3. JTG D50—2017 (2017) Specification for design of highway asphalt pavement
4. Zhang H, Huang H, Aijun Y (2013) Study on the mechanical response of asphalt overlay structure of different structural types of old road reconstruction and upgrading. Highw Eng 37(1):129–132
5. Cao Z Zhang Q (2010) Mechanical response analysis of thin asphalt overlay on old cement concrete pavement. Highw Eng 35(4):42–46
6. Xu X et al (2009) Analysis of mechanical properties of asphalt overlay on old cement pavement. Highw Motor Transp (3):55–58

Seismic Data Denoising Analysis Based on Monte Carlo Block Theory

Hongliang Jiang, Chaobo Lu, Chunfa Xiong, and Mengkun Ran

Abstract Denoising of seismic data has always been an important focus in the field of seismic exploration, which is very important for the processing and interpretation of seismic data. With the increasing complexity of seismic exploration environment and target, seismic data containing strong noise and weak amplitude seismic in-phase axis often contain many weak feature signals. However, weak amplitude phase axis characteristics are highly susceptible to noise and useful signal often submerged by background noise, seriously affected the precision of seismic data interpretation, dictionary based on the theory of the monte carlo study seismic data denoising method, selecting expect more blocks of data, for more accurate MOD dictionary, to gain a higher quality of denoising of seismic data. Monte carlo block theory in this paper, the dictionary learning dictionary, rules, block theory and random block theory is example analysis test, the dictionary learning algorithm based on the results of three methods to deal with, and the numerical results show that the monte carlo theory has better denoising ability, the denoising results have higher SNR, and effectively keep the weak signal characteristics of the data; In terms of computational efficiency, the proposed method requires less time and has higher computational efficiency, thus verifying the feasibility and effectiveness of the proposed method.

Keywords Seismic data denoising · Monte Carlo theory · Block theory

H. Jiang · C. Lu · C. Xiong · M. Ran (✉)
Guangxi Transportation Science and Technology Group Co., Ltd., Guangxi 530007, Nanning, China
e-mail: 851067167@qq.com

H. Jiang · C. Lu · C. Xiong
Guangxi Key Lab of Road Structure and Materials, Guangxi, Nanning 530007, China

Guangxi highway tunnel Safety warning Engineering Research Center, Guangxi 530007, Nanning, China

© The Author(s) 2023   339
G. Feng (ed.), *Proceedings of the 9th International Conference on Civil Engineering*,
Lecture Notes in Civil Engineering 327,
https://doi.org/10.1007/978-981-99-2532-2_28

1 Introduction

Sparse representation of signals has attracted wide attention in recent years. The principle is to use a small number of atoms in a given supercomplete dictionary to represent seismic signals, and the original signals are expressed through a linear combination based on a small number of basic signals [1]. Sparse coding is also known as dictionary learning. This method seeks appropriate dictionaries for samples with common dense expression and converts the samples into appropriate sparse expression forms, thus simplifying the learning task and reducing the complexity of the model. The two main tasks of signal sparse representation are dictionary generation and signal sparse decomposition. The selection of dictionaries generally falls into two categories: dictionary analysis and dictionary learning.

In essence, Monte Carlo method is a very important numerical calculation method guided by the theory of probability and statistics. Monte Carlo method has been widely used in architecture, automation, computer software and other fields [2]. The Monte Carlo method is an effective method to find the numerical solution for the complicated problem. For large-scale image filtering, Chan et al. proposed a Monte Carlo local nonmean method (MCNLM) [3], which speeds up the classical nonlocal mean algorithm by calculating the subset of image block distance and randomly selecting the position of image block according to the preset sampling mode.

Monte Carlo thought has been widely applied. For example, in mechanical engineering, Rodgers et al. built a micro-structure prediction tool based on Monte Carlo thought, and explored the correlation between the average number of remelting cycles during construction and the generated cylindrical crystals by setting the correlation coefficient of the model [4].

2 Dictionary Learning Theory

In theory, sparse dictionary learning includes two stages: dictionary construction stage and sparse dictionary representation sample stage. Learning every dimension of dictionary atoms requires training, and sparse representation is to represent the signal with as few atoms as possible in the over-complete dictionary. The trained dictionary is too large to be used in practice. The construction and training of the dictionary require a large number of training samples. The over-complete dictionary contains too many dimensions and is not practical and operable. The block theory can overcome this limitation and establish a training set. Data blocks can be extracted from noisy data or similar but non-noisy data as training sets.

The data is divided into small blocks and denoised separately, so the learning problem becomes [5, 6]:

$$\left(\hat{A}, \ \hat{D} \right) \in \arg \min_{A, D} \sum_{i, j} 0.5 \left\| R_{(i, j)} (X) - Da_{i,j} \right\|^2 + \lambda \left\| a_{i,j} \right\|_1 \tag{1}$$

Is the sparse decomposition of data blocks in the dictionary and is a parameter used to balance the fidelity of the data with the sparsity of the data representation in the dictionary. And are regarded as known quantities, and the variables in formula 1 correspond to the minimization of a single variable.

The update of sparse coding can be completed by the sparse matrix solver, which can find and solve the sparse representation of each training sample data in a fixed dictionary [7]:

$$\hat{\alpha} \in \arg \min_{\hat{\alpha}} \frac{1}{2} \|x - D\alpha\|^2 + \lambda \|\alpha\|_1 \tag{2}$$

There are many algorithms to solve the problem of sparse representation of signal and decomposition in complete dictionary. In this paper, orthogonal matching tracking (OMP) is used.

Dictionary updates include minimizing data representation errors, i.e.

$$\hat{D} \in \arg \min_{\widehat{D} \in \mathbf{D}} \|P - D\alpha\|_F^2 \tag{3}$$

$$\mathbf{D} = \left\{ d : \|d_j\|^2 = \sum_i |d_j[i]|^2 \leq 1 \right\} \tag{4}$$

When the dictionary is retrieved in the set, the norm of the atom can become arbitrarily large, and the degree of variation of the coefficient can be arbitrarily small. Artificial reduction of energy minimizes the possibility of dictionary return [8].

The method adopted in this article is the MOD method, which essentially updates the dictionary by minimizing the mean residual. In the training with noisy data, the noise variance control parameters, first,

$$\|x - \tilde{x}\|_2 \leq C\sigma \sqrt{n} \tag{5}$$

Formula 5 represents the level of noise. When the error between the vector and its sparse approximation meets the conditions, OMP calculation will be stopped. Among them, the gain factor C is used to balance the fidelity and sparsity, and the gain factor makes the sparse coding step more flexible. The denoising gain factor balances data fidelity and sparsity, and a larger gain factor will give a sparse and less accurate patch representation in the dictionary [9]. When learning on noiseless samples, a fixed number of coefficients estimated by OMP must be selected in advance.

After the dictionary is trained, the standard of formula 5 is used to calculate the sparse approximation in the dictionary, denoise each data block, and then calculate the average of the denoised data blocks to reconstruct the whole data.

3 Block Theory

Chunking is the theory that data is broken up into small chunks and processed before the dictionary is built. There are certain similarities in the data blocks, which may appear as coming from the same in-phase axis or signal edge region in the seismic data. When the similarity is not obvious, the whole data can be sparsely represented by a set of basis functions, and a new matrix is constructed after vectorization of the selected data blocks. The stacking of data blocks is called the overlap degree. In general, if the upper limit of the number of selected blocks is not set, the data size is fixed, the higher the overlap degree is, the more selected blocks are, the more samples are selected, and the more beneficial it is for dictionary construction and update. However, a large number of sample selection increases the calculation amount.

3.1 Regular and Random Chunking

Regular chunking and random chunking are both chunking methods that limit the upper limit of selected blocks and the degree of overlap. The selection of these two methods has nothing to do with the characteristics of the data itself.

The theory of regular block selection specifies the step size and direction of block selection in the process of selection. Formula 6 represents the regular block partitioning process of a 3×3 matrix with a block size of 2×2:

$$
\begin{bmatrix} x_{11} x_{12} x_{13} \\ x_{21} x_{22} x_{23} \\ x_{31} x_{32} x_{33} \end{bmatrix} \rightarrow \begin{bmatrix} x_{11} & x_{12} & x_{12} & x_{13} \\ x_{21} & x_{22} & x_{22} & x_{23} \\ x_{21} & x_{22} & x_{22} & x_{23} \\ x_{31} & x_{32} & x_{32} & x_{33} \end{bmatrix} \tag{6}
$$

In seismic data, the process of regular block selection is uniform throughout the data. The difference between the random block theory and the regular block theory lies in that the block selection position of the random block is completely random, so the situation of data block cluster stacking may occur:

$$
\begin{bmatrix} x_{11} & x_{12} & x_{13} & x_{14} \\ x_{21} & x_{22} & x_{23} & x_{24} \\ x_{31} & x_{32} & x_{33} & x_{34} \\ x_{41} & x_{42} & x_{43} & x_{44} \end{bmatrix} \rightarrow \begin{bmatrix} x_{11} & x_{12} & x_{12} & x_{13} \\ x_{21} & x_{22} & x_{22} & x_{23} \\ x_{21} & x_{22} & x_{33} & x_{34} \\ x_{31} & x_{32} & x_{43} & x_{44} \end{bmatrix} \tag{7}
$$

Therefore, in seismic data, random blocks may be repeated selection of a large number of areas, obvious aggregation of data blocks, or no selection of data in some areas.

3.2 *Monte Carlo Data Chunking*

In dictionary learning, dictionary directly determines the quality and efficiency of denoising. In traditional data block and sample selection method is mostly random and selection rules, each sample is there, and we hope the dictionary thin enough, namely the limited samples can contain more useful information, can be enough to show the characteristics of complex data, so in the process of selecting data contains information should choose more blocks of data. If the variance of the set data block is associated with the contained valid information, set the standard we expect. When the variance is greater than the expected value, the corresponding data block will be retained, and then all the retained data blocks will undergo the next step of dictionary training. This selection method is called the Monte Carlo sample selection method.

Randomness to itself quality problem, usually need to probability of correct description and simulation process, for the deterministic randomness of itself is not quality problem, you need to build the probability of an artificial process, to a certain parameter set to solution of the problem, will not be the randomness of quality problem into a random. For example, using Monte Carlo method to estimate PI is a very classic problem. As shown in Fig. 1, the number of selected points n is 1000, and the calculated PI is 3.1568, with an error of about 0.048.

In seismic data, we want the data block to contain information about the same phase axis as much as possible. Therefore, variance can be used to determine whether the selected block region contains in-phase axis information or whether it contains enough effective information, which is defined as the complexity of the data block. If the amplitude of the in-phase axis does not change in the region selected from the noise-free data, the variance is 0. If the amplitude of the in-phase axis changes in the region, the variance will increase [10].

Fig. 1 Calculation of π using Monte Carlo idea

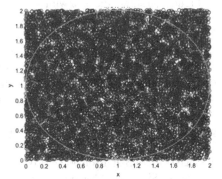

4 Examples of Measured Seismic Data

In order to explore the practical effect of Monte Carlo block theory dictionary learning, the dictionary learning method based on Monte Carlo block theory is applied to the processing of measured data. The data will be processed by random block theory dictionary learning, regular block theory dictionary learning and Monte Carlo block respectively. Moreover, the denoising effect is studied by marking the location of different block theory dictionary learning. Wherein, the iteration times of the dictionary learning algorithm is 50, the coincidence degree is 4, the noise variance is 0.2, and the size of a single data block selected is 9 × 9.

Figure 2. shows the data block selection diagram of the original data and its different block theory dictionary learning. Figure 2. a is the original data diagram of the data. There are a large number of weak amplitude signals, which are in obvious contrast with the in-phase axis of the main strong amplitude. From monte carlo block selection is shown in Fig. 2. b, data block is focused on the overall data of the upper signal strong area, compared with the signal is weak regional selection of data block is less, and most of the weak signal region selection data block in the subdivided by multiple axis gathered area, has the obvious and primary and secondary data without fine feature selection. As shown in Fig. 2. c, the data blocks selected by the rule are evenly spread over the entire data, which is close to the whole selection. However, due to the different proportions of data blocks in signal areas with different intensifies, it needs to be verified by the denoising process whether the dictionary can represent the data sparsely and accurately enough. In contrast, as shown in Fig. 2. d, the randomly selected data blocks are very messy, because randomness leads to a large number of selected data blocks in the blank area and a single area of the signal, and a small number of areas containing a lot of data feature information are not selected, which has an impact on later sparse representation and dictionary update.

Figure 3. a shows the data with 20% noise added to the data, mainly strong signals with high amplitude. When 20% noise is added, most weak signals are submerged. Figure 3. b shows the Monte Carlo block denoising diagram, the SNR is 16.4 dB, the running time is 5.0 s, and the main horizontal in-phase axis is reconstructed with relatively complete denoising. Use a dictionary to learn the data sparse representation to rebuild and remove most of the noise, but some detail feature does not show it, the data distortion caused by Fig. 3. c denoising Fig. for the dictionary selection rules theory learning, denoising SNR is 14.9 dB, running time of 5.1 s, in Fig. 3. d random graph theory dictionary learning method, The signal-to-noise ratio of denoising is 15.5 dB and the running time is 5.5 s. The two methods have the most serious damage to the signal. From the perspective of running time, the previous speculation is verified. Since the Monte Carlo block selects the least data blocks, the running time is the shortest, and the processing efficiency of data denoising is relatively high.

Figure 4. is the residuals of the data after denoising, and Fig. 4. a is the residuals of the Monte Carlo selection method. In the Fig. 4., except for some strong amplitude signals, there is basically no feature of weak signals, and they are mostly noise,

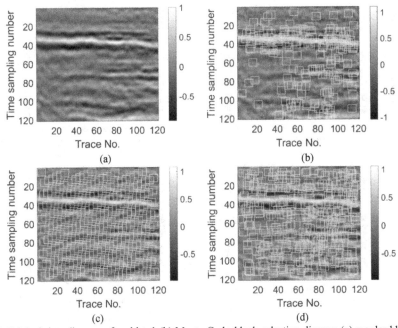

(a) Original data diagram of realdata1 (b) Monte Carlo block selection diagram (c) regular block selection diagram (d) random block selection diagram

Fig. 2　Theoretical diagram of different blocks of realdata1 measured seismic data

indicating that the denoising is more thorough. Figure 4. b shows the residual signal-to-noise ratio selected by the rule, and weak signal features can be distinguished in the lower left and lower right parts. The random residual Fig. shown in Fig. 4 c shows that the signal features in the lower left part are more obvious, which indicates that in the original denoising process, the retention effect of data detail features in this area is poor. However, the residual graphs of the three methods can all see some features of the original data and have a certain degree of damage, while the Monte Carlo block theory dictionary learning denoising achieves better denoising results.

Through this dictionary learning method based on monte carlo theory demonstrates how to applied to seismic data denoising, and the results with the stochastic block theory dictionary learning dictionary and rules of block theory, from the perspective of the denoising results of measured data, the dictionary to study seismic data denoising still has an obvious effect in the treatment of the complex seismic data. Comparing three methods, monte carlo block theory dictionary learning can be relatively better reconstructed seismic data, suppression of noise at the same time can effectively preserve the weak signal characteristics of seismic data, in smaller seismic data, the monte carlo block will priority selection signal strong area, weak signal area of less data, lead to appear weak in phase axis, the condition of the damaged However, when the data is large enough, the overall data de-noising and

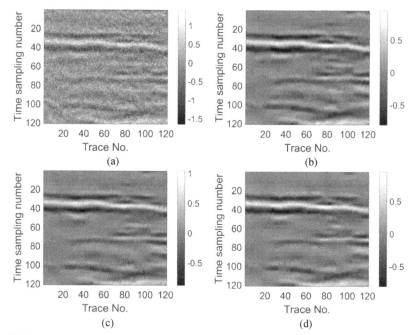

(a) realdata1 denoised data map (SNR=4.6dB) (b) Monte Carlo block denoising map
(SNR=16.4dB) (c) Regular block denoising map (SNR=14.9dB) (d) Random block denoising map
(SNR=15.5dB)

Fig. 3 Denoising results of realdata1 data dictionary learning with different block theory

reconstruction is relatively complete and accurate, and the processing efficiency is
fast, and a high signal-to-noise ratio is obtained. Random block theory dictionary
learning and regular block theory dictionary learning are obviously inferior to Monte
Carlo block theory dictionary learning in detail feature protection, and the processing
efficiency is relatively poor, exposing a lot of disadvantages.

On the basis of data 1, we have conducted 6 comparative studies of measured data
and 7 comparative studies and analyses to obtain the following tables and figures
below.

Figure 5. for the three methods all data denoising SNR contrast Fig., it is not hard
to see the monte carlo block theory in dealing with different types of data dictionary
learning can obtain good denoising effect, is the most excellent in three ways. Table
1 is the running time of the three methods in seven data, Fig. 6. for three methods run
time contrast Fig., it is not hard to see, Monte Carlo block theory dictionary learning
has a higher signal-to-noise ratio and the shortest time. Due to its self-adaptability,
it can use fewer data blocks to contain enough effective information and meet the
requirements of dictionary construction denoising. Therefore, it is a more efficient
denoising method.

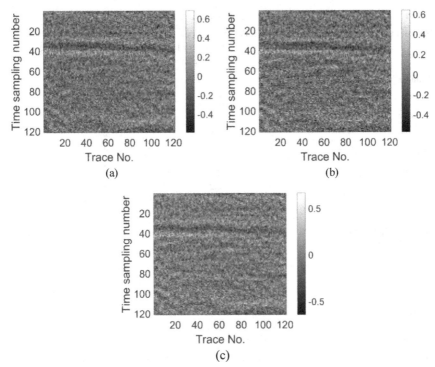

(a)   (b)

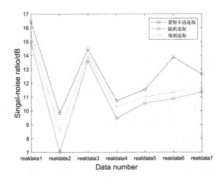

(c)

(a) Monte Carlo block residuals (b) regular block residuals (c) random block residuals

Fig. 4 De-noising residuals of realdata1 data dictionary learning of different blocks theory

Fig. 5 Comparison of SNR
of different denoising
methods

Table 1 List of running time of different methods of measured data

The data type	Monte Carlo block theory dictionary learning running time/s	Rule block theory dictionary learning running time/s	Random block theory dictionary learning running time/s
readata1	5.0	5.1	5.5
readata2	6.1	8.1	6.8
readata3	5.6	6.6	7.2
readata4	4.6	5.4	5.2
readata5	5.0	11.4	8.6
readata6	5.5	6.3	6.4
readata7	8.7	12.2	10.8

Fig. 6 Compares the running time of different methods

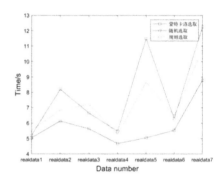

5 Conclusion

The data processing ability of dictionary learning and Monte Carlo block theory dictionary learning in complex cases is verified by the processing of actual data. By comparing the three methods, the method proposed in this paper achieves the highest signal-to-noise ratio and the highest processing efficiency, especially the weak features of the data are preserved. The results show that Monte Carlo block theory dictionary learning has a good performance of denoising under different conditions.

Because the dictionary directly determines the quality of the denoising result. Afterwards monte carlo block theory dictionary learning dictionary, selection rules of block theory and random selection rules of block theory dictionary learning, although three approaches to sparse coding, dictionary update and a series of core algorithm is consistent, but as a result of the selected data block, a higher effective information block of data of different, caused the denoising results of three methods there is a certain gap.

In this paper, dictionary learning based on a block theory provides a new attempt for seismic data of dictionary learning, and makes a certain breakthrough compared with the traditional methods. The dictionary learning algorithm itself is optimized to a certain extent. The advantage of dictionary learning lies in the extraction of the most essential features of data, which is of great significance in the research of data processing in the direction of geophysics.

Acknowledgements The research of this paper is supported by Guangxi Key Research and Development Project (Guike AB21220069) and Guangxi Communications Investment Group Co., LTD., independent project: Safety Analysis, management technology and long-term performance observation of Goaf under highways.

References

1. Zhou Y, Gao J, Chen W (2015) Research on separation method of multi-source mixed seismic records based on sparse representation. In: Proceedings of geophysical prospecting technology symposium 2015, china petroleum society, pp 165–168
2. Werner MJ, Ide K, Sornette D (2011) 2011, Earthquake forecasting based on data assimilation: sequential Monte Carlo methods for renewal point processes. Nonlinear Process Geophys 18(124):49–70
3. Chan SH, Zickler T, Lu YM (2014) Monte Carlo non-local means: random sampling for large-scale image filtering. IEEE Trans Image Process 23(8):3711–3725
4. Abrol V, Sharma P, Sao AK (2016) Greedy double sparse dictionary learning for sparse representation of speech signals. Speech Commun 85:71–82
5. Guo L, Yao L, Gao H, et al (2015) Vibration Signal denoising technology based on dictionary learning and sparse coding. Vibr Test Diagn 35(04):752–756+802
6. Wang AQ, Xu K, Song AM (2017) Image denoising method based on local self-similar dictionary learning. J Dalian Jiaotong Univ 38(04):192–195
7. Li Q-Q, Wang H (2019) Review of optimization algorithms based on sparse representation theory. Surveying Mapp Geogr Inf 44(04):1–9
8. Tang Z, Tang G, Liu X et al (2019) A new dictionary update and atom optimization image denoising algorithm. Comput Technol Dev 29(04):33–37
9. Bao C, Ji H, Quan Y et al (2016) Dictionary learning for sparse coding: algorithms and convergence analysis. IEEE Trans Pattern Anal Mach Intell 38(7):1356–1369
10. Yu SW (2017) Seismic data reconstruction based on adaptive sparse inversion. Heilongjiang: Harbin Institute of Technology, (in Chinese), pp 37–47

Influence of Anti-Mud Agent on the Performance of Gangue Backfilling Paste

Wei Zhou, Zhaoyang Guo, Kangkang Wang, Haibo Zhang, and Xuemao Guan

Abstract Anti-mud agents could improve the efficiency of the action of water reducers in concrete by preferentially. The anti-mud agent was preferentially adsorbed on the clay surface, which reduces the ineffective adsorption of the water reducing agent to the paste, thereby improving the water reduction efficiency. However, its application in high-sediment content coal gangue gypsum backfill materials had not been reported. In this paper, The competitive adsorption mechanism echanism of anti-mud agent was first described. Tested its competitive adsorption with water reducer molecules on the surface of gangue powder. The influences of anti-mud agent on the slump, coagulation time and compressive strength of the paste at different ages were studied. The results showed that: as the dosage of anti-mud agent increased, the amount of desorption of the water reducer from the surface of gangue powder increased. When adding the same extra amount of water reducer, the slump of the backfill paste material increased with prolonged the coagulation time. The strength of the paste decreased at 3 d, and the strengths of 7 d and 28 d were not significantly deteriorated. A small amount of anti-mud agent could greatly improve the fluidity of the paste. This study provides a scientific basis for the pumping of pure solid waste paste.

Keywords Backfilling Paste · Water Reducer · Anti-Mud Agent · Slump · Strength

1 Introduction

The massive mining of coal resources in China had led to various social and environmental problems such as surface subsidence and gangue accumulation. As one of the important contents of green mining of coal resources, paste backfill technology had developed rapidly in China in recent years [1–4]. Backfill paste was a Bingham

---

W. Zhou · Z. Guo · K. Wang · H. Zhang (✉) · X. Guan
Henan Key Laboratory of Materials On Deep-Earth Engineering, School of Materials Science and Engineering, Henan Polytechnic University, Jiaozuo 454003, China
e-mail: zzhb@hpu.edu.cn

© The Author(s) 2023
G. Feng (ed.), *Proceedings of the 9th International Conference on Civil Engineering*,
Lecture Notes in Civil Engineering 327,
https://doi.org/10.1007/978-981-99-2532-2_29

fluid made of gangue, fly ash, cement and admixture mixed with water. Fluidity and strength were important indicators to characterize the performance of backfill paste. The use of water reducing agent could greatly improve the paste concentration, increase the strength of the backfill body [5–7]. And achieve step by step improvement of the performance of the backfill paste while ensuring the paste had good fluidity.

The action mechanism of the water reducer was to reduce the adsorption of water molecules by the particles of the cementitious material by adsorbing on the surface of the paste, increase the dispersion, and release a large amount of free water [8–10]. So as to reduce the water consumption of the paste and ensure the fluidity of the paste. However, the gangue aggregate in the paste usually had a high mud content [11]. And the gangue powder with montmorillonite and kaolin as the main components will preferentially adsorb the water reducer, greatly affecting the effect of the water reducer [12–14]. A small amount of anti-mud agent could effectively increase the action efficiency of water reducing agent [15–17], which had been applied in the field of building materials [18, 19], but its effect on backfill paste had not been reported.

In order to study the influence of anti-mud agent on the performance of backfill paste, the competitive adsorption between anti mud-agent and water reducer (FDN) and its mechanism were analyzed, and the influence of anti-mud agent content on the performance of backfill paste was tested. The addition of a small amount of anti-mud agent could greatly reduce the amount of water reducing agent, which provides a new way to reduce the cost of coal gangue gypsum body backfill materials.

2   Analysis of Action Mechanism of Anti-Mud Agent

Because coal gangue powder contains a large amount of montmorillonite, the action mechanism of anti-mud agent was illustrated by taking montmorillonite as an example. The action mechanism of anti-mud agent was shown in Fig. 1. In the cement paste mixed with montmorillonite and water reducing agent, montmorillonite will absorb more water reducing agent. The water reducing agent could not fully combine with the cement particles, which causes the cement particles to absorb a lot of water molecules and agglomerate, affecting the performance of the cement paste [20, 21]. Anti-mud agent was more sensitive to montmorillonite. After adding anti-mud agent, it could replace part of water reducer and montmorillonite for adsorption, releasing a large number of water molecules and water reducer. A large number of water reducer could act on the surface of cement particles, increasing the dispersion of cement particles and releasing water molecules, improving the performance of cement paste [22].

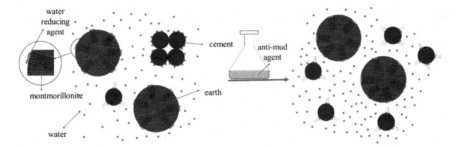

Fig. 1 Mechanism of action of anti-mud agent

3 Experiment

3.1 Experimental Materials

Gangue: after secondary crushing of the washed gangue from Changcun Coal Mine of Shanxi Lu'an Environmental Protection and Energy Development Co., Ltd., 4.75 ~ 16 mm gangue was used as coarse aggregate and 0 ~ 4.75 mm gangue was used as fine aggregate. The silt content of fine gangue was up to 5.78%. Fly ash: Grade II ash discharged by Shanxi Lu'an Environmental Protection Energy Development Co., Ltd. Cement: Grade 42.5 ordinary Portland cement produced by Jiaozuo Qianye New Materials Co., Ltd. Water reducer (FDN): naphthalene series superplasticizer produced by Shanxi Yonghong Building Materials Chemical Co., Ltd., with the molecular formula of $(C_{21}H_{14}Na_2O_6S_2)$ n. Anti-mud agent (KN): self-made soluble powder composed of sodium tripolyphosphate and citric acid.

3.2 Experimental Proportion and Method

Competitive Adsorption of FDN and KN on the Surface of Gangue Powder. Pass the gangue aggregate through a 0.075 mm square sieve, take part of the gangue powder under the sieve, weigh the raw materials according to Table 1, stir and mix them for 30 min, so that FDN and KN could be fully dissolved in water and adsorbed on the soil surface, and then stand for 24 h, after the soil was fully precipitated (Fig. 2.), filter out the supernatant of the precipitation by vacuum suction filtration method, dilute it 10 times with distilled water, and measure the concentration of S element in the diluted solution by inductively coupled plasma emission spectrometer (ICP-OES). The matching numbers in Table 1. were recorded as J1, J2, J3, J4, J5 and J6 respectively. Since KN did not contain S element, while FDN contains S element, the greater the content of S element in the clear night, indicating that the less FDN the soil absorbs, the more significant the effect of KN.

Table 1 Mixing ratios for competitive adsorption experiment between anti-mud agent and water reducer

number	Gangue earth/ g	FDN/ g	KN/ g	water/ g
J1	33	0	0	315
J2	33	2	0	315
J3	33	2	0.225	315
J4	33	2	0.45	315
J5	33	2	0.675	315
J6	33	2	0.90	315

Fig. 2 Solution precipitation

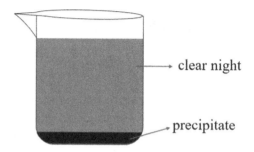

→ clear night

→ precipitate

Paste Preparation and Performance Test. The mass concentration of 85% remains unchanged. Weigh the raw materials with the ratio in Table 2, fully mix them evenly, and then test the paste slump and coagulation time according to NB/T 51,070–2017 Test Method for Paste Backfill Materials in Coal Mines. Pour 100 mm × 100 mm × In 100 mm mold, demould after curing for 24 h under standard conditions (20 °C, 95% humidity), and test the paste strength after curing to the specified age under standard conditions. The matching numbers were recorded as F, F0, F5, F10, F15 and F20 respectively.

Table 2 Experimental ratios.$(kg \cdot m^{-3})$

number	cement	fly ash	coarse gangue	fine gangue	FDN	KN	water
F	100	350	792	528	3	0	315
F0	100	350	792	528	1	0	315
F5	100	350	792	528	1	0.05	315
F10	100	350	792	528	1	0.10	315
F15	100	350	792	528	1	0.15	315
F20	100	350	792	528	1	0.20	315

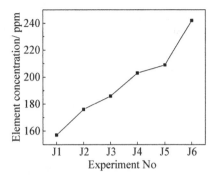

Fig. 3 Effect of KN on the concentration of element S in solution

4   Experimental Result

4.1   Results of Competitive Adsorption Experiment

The influence of KN on the concentration of element S in the solution was shown in Fig. 3. It could be seen that when FDN and KN were not added, the concentration of S element in the solution was 157 ppm, which was the S element dissolved from the coal gangue powder. After FDN was added, the concentration of S element in the solution reaches 176 ppm, an increase of 12.1%, which was caused by FDN molecules not adsorbed on the surface of coal gangue powder. With the increase of the content of KN, the concentration of S element in the solution shows a linear growth trend. Compared with J1, the concentration of S element in J6 solution increases from 176 to 242 ppm, an increase of 37.5%. This was because KN could preferentially adsorb with montmorillonite [8], releasing FDN. FDN dissolution in the solution increases the concentration of S element in the solution.

4.2   Paste Fluidity

The influence of KN content on the paste slump was shown in Fig. 4. It could be seen that the paste slump first increases rapidly with the KN content. When the KN content exceeds $0.10\ kg \cdot m^{-3}$, the growth rate of the slump slows down. The slump of F0 sample was 135 mm, and the slump of F10 fresh paste was increased to 232 mm, with an increase of 71.9%, which meets the pumping requirements [21] and was equivalent to the slump of F sample, indicating that a small amount of KN could greatly improve the fluidity of paste and reduce the amount of FDN. This was because the incorporation of KN releases a large number of FDN molecules adsorbed by coal gangue powder, which improves the water reduction efficiency.

Fig. 4 Effect of KN dosage
on the slump of paste

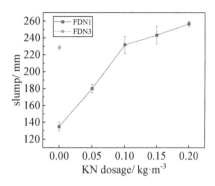

4.3 Paste Coagulation Time

The influence of KN content on the coagulation time of the paste was shown in Fig. 5. It could be seen that with the increase of KN content, the coagulation time of the paste was gradually extended. Before the content was 0.05 kg·m^{-3}, the coagulation time increases rapidly, and then the increase rate slows down. The coagulation time of sample F0 was 9.0 h, and that of sample F10 increases to 13.1 h, with an increase of 45.6%. The coagulation time of sample F was equivalent to that of sample F10. The extension of coagulation time was due to the increase of free water content in the paste. The incorporation of KN releases FDN molecules adsorbed by coal gangue powder. FDN could disperse agglomerated cementitious materials and release adsorbed free water, which was conducive to improving the controllability of paste pumping. The amount of FDN in F sample without KN should reach 3 kg·m^{-3}, because its FDN molecules were largely adsorbed on the surface of coal gangue powder and could not play a role in reducing water dispersion. In addition, the KN used in this study was a mixture of sodium tripolyphosphate and citric acid, which had a retarding effect [22], and will also extend the coagulation time of the paste.

Fig. 5 Effect of KN dosage
on the coagulation time of
paste

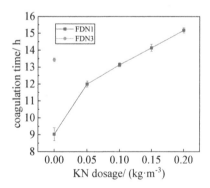

Fig. 6 Effect of KN dosage on compressive strength of paste at different ages

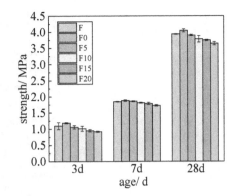

4.4 Compressive Strength of Paste

The influence of KN on the compressive strength of paste at 3 d, 7 d and 28 d was shown in Fig. 6. It could be seen that the strength of paste at different ages decreases with the increase of KN content. When the content of KN increases from 0 to 0.20 kg·m^{-3}KN, the paste strength decreases from 1.19 MPa to 0.92 MPa at 3d, by 22.7%, from 1.88 MPa to 1.73 MPa at 7d, by 8.7%, and from 4.05 MPa to 3.65 MPa at 28d, by 9.9%. The strength of sample F and sample F10 at the same age had little change. It shows that KN had a great influence on the strength of the backfill paste at 3 d, and had a little influence on the strength at 7 d and 28 d. The KN used in this study was a mixture of sodium tripolyphosphate and citric acid. With the increase of the dosage, the KN not adsorbed on the surface of coal gangue powder will be adsorbed on the surface of the cementitious material, which had an inhibitory effect on its hydration reaction [23, 24]. Therefore, the strength of the paste decreases significantly in 3 days, the inhibition disappears when the age increases, and the strength decreases in the later period.

The SEM structure observation of F0 sample and F10 sample at the age of 3 d and 28 d was shown in Fig. 7. It could be seen that with the increase of age, the hydration product C-S-H gel of the two groups of samples increases, and the structural pore decreases. Therefore, the strength of the paste increases with the increase of age. Compared with F0 sample, the structural pores of F10 sample at 3 d and 28 d age were larger, and the amount of hydrated product C-S-H gel was less. Therefore, the strength of F10 sample was less than that of F0 sample. The difference between the micro pores of the two groups of samples was no longer obvious at 28 d age, so the strength decline of F10 sample at 28 d age was reduced, which indicates that the KN had little effect on the later strength of the paste.

Fig. 7 Observation of SEM structure of F0 and F10 specimens at different ages

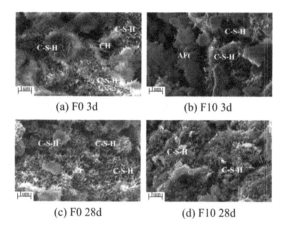

(a) F0 3d   (b) F10 3d

(c) F0 28d   (d) F10 28d

5 Conclusion

(1) With the increase of KN content, the adsorption amount of water reducer on the surface of coal gangue powder decreases, the slump of coal gangue gypsum body backfill material increases, and the coagulation time extends.

(2) A small amount of KN could greatly improve the fluidity of the paste and reduce the amount of water reducing agent. The slump of coal gangue gypsum body backfill material with 0.10 kg·m^{-3} KN and 1 kg·m^{-3} water reducing agent was similar to that without KN and 3 kg·m^{-3} water reducing agent.

(3) With the increase of the dosage of KN, the strength of coal gangue gypsum backfill material decreases in 3 days, but it did not decrease significantly in 7 days and 28 days.

Acknowledgements The authors appreciate the support from the National Natural Science Foundation of China (U1905216), and the Natural Science Foundation of Henan Province (182300410207).

References

1. Yang K, Zhao XK, Wei Z, Zhang JQ (2021) Development overview of paste backfill technology in China's coal mines: a review. Environ Sci Pollut Res 28(48):67957–67969
2. Sun Q, Zhang JX, Zhou N (2018) Study and discussion of short- strip coal pillar recovery with cemented paste backfill. Int J Rock Mech Min Sci 104:147–155
3. Behera SK, Mishra DP, Singh P et al (2021) Utilization of mill tailings, fly ash and slag as mine paste backfill material: Review and future perspective. Constr Build Mater 309:125120
4. Qi CC, Fourie A (2019) Cemented paste backfill for mineral tailings management: review and future perspectives. Miner Eng 144:106025
5. Huang CL, Cheng ZR, Zhao JH, et al (2021) The influence of water reducing agents on early hydration property of ferrite aluminate cement paste. Crystals 11(7): 731

6. Han Z, Zhang YS, Qiao HX et al (2022) Study on axial compressive behavior and damage constitutive model of manufactured sand concrete based on fluidity optimization. Constr Build Mater 345:128176
7. Guo YX, Liu GY, Feng GR, et al (2020) Performance of coal gangue-based cemented backfill material modified by water-reducing agents. Adv Mater Sci Eng 1–11
8. Li FX, Chen YZ, Long SZ et al (2013) The retardation effect of super-retarding polycarboxylate-type superplasticizer on cement hydration. Arab J Sci Eng 38(3):571–577
9. Liu Y, Li H, Wang H et al (2020) Effects of accelerator–water reducer admixture on performance of cemented paste backfill. Constr Build Mater 242:118187
10. Dalas F, Nonat A, Pourchet S et al (2015) Tailoring the anionic function and the side chains of comb-like superplasticizers to improve their adsorption. Cem Concr Re 67:21–30
11. Li JY, Wang JM (2019) Comprehensive utilization and environmental risks of coal gangue: a review. J Clean Prod 239:117946
12. Pourchet S, Liautaud S, Rinaldi D et al (2012) Effect of the repartition of the PEG side chains on the adsorption and dispersion behaviors of PCP in presence of sulfate. Cem Concr Res 42(2):431–439
13. Ma YH, Shi CJ, Lei L et al (2020) Research progress on polycarboxylate based superplasticizers with tolerance to clays - a review. Constr Build Mater 255:119386
14. Lei L, Plank J (2014) A study on the impact of different clay minerals on the dispersing force of conventional and modified vinyl ether based polycarboxylate superplasticizers. Cem Concr Res 60:1–10
15. Haijun X, Shenmei S, Jiangxiong W et al (2015) β-Cyclodextrin as pendant groups of a polycarboxylate superplasticizer for enhancing clay tolerance. Ind Eng Chem Res 54(37):9081–9088
16. Lei L, Plank J (2012) A concept for a polycarboxylate superplasticizer possessing enhanced clay tolerance. Cem Concr Res 42(10):1299–1306
17. Tang XD, Zhao CL, Yang YQ et al (2020) Amphoteric polycarboxylate superplasticizers with enhanced clay tolerance: preparation, performance and mechanism. Constr Build Mater 252:119052
18. Qian SS, Yao Y, Wang ZM et al (2018) Synthesis, characterization and working mechanism of a novel polycarboxylate superplasticizer for concrete possessing reduced viscosity. Constr Build Mater 169:452–461
19. Gang C, Jiaheng L, Yong D et al (2018) Synthesis of a novel polycarboxylate superplasticizer with carboxyl group as side chain terminal group to enhance its clay tolerance. J Wuhan Univ Technol-Mater Sci Ed 33(1):226–232
20. Song YM, Guo CC, Qian JS et al (2014) Adsorption mechanism of polycarboxylate-based superplasticizer in CFBC ash-Portland cement paste. J Wuhan Univ Technol-Mater Sci Ed 29(5):945–949
21. Abile R, Russo A, Limone C et al (2018) Impact of the charge density on the behaviour of polycarboxylate ethers as cement dispersants. Constr Build Mater 180:477–490
22. Zhong DM, Liu QD, Zheng DF. (2022) Synthesis of lignin-grafted polycarboxylate superplasticizer and the dispersion performance in the cement paste. Colloids Surf. A: Physicochem Eng Aspects 642: 128689
23. Tan HB, Guo YL, Ma BG et al (2018) Effect of sodium tripolyphosphate on clay tolerance of polycarboxylate superplasticizer. KSCE J Civ Eng 22(8):2934–2941
24. Möschner G, Lothenbach B, Figi R et al (2009) Influence of citric acid on the hydration of Portland cement. Cem Concr Res 39(4):275–282

Application of Ununiform Arrangement Piling in Reinforcement and Correction of Existing Structures

Shengbin Zhou, Jianxing Tong, Xu Li, Xunhai Sun, Yahui Wang, Jialu Li, Guoqiang Cao, and Zhao Li

Abstract Due to the construction problem of the original concrete mixing pile, the inclined settlement of a building occurred. After analysis and calculation, the uneven pile distribution in the north and south region is adopted to reduce the subsidence and correct the deviation. According to the comparison between the calculated results and the measured data of the project, the two are basically consistent, which indicates that the method of uneven pile distribution can successfully adjust the distribution of the reaction force of the base and achieve the effect of subsidence reduction and deviation correction.

Keywords Inclined settlement · Ununiform arrangement piling · Reaction force of the base

1 Introduction

Adopting new pile foundation reinforcement scheme for existing buildings foundation reinforcement and correction is more common reinforcement scheme, and each part of the existing buildings foundation quality accident is often accompanied by foundation uneven settlement, and the problem of excess tilt, therefore, in the process of strengthening full tilt, with uneven sheet pile can reduce the deformation, optimization of basal counterforce distribution difference.

S. Zhou (✉) · J. Tong · X. Sun · Y. Wang · J. Li · G. Cao · Z. Li
State Key Laboratory of Building Safety and Built Environment, Beijing 100013, China
e-mail: 67139960@qq.com

Institute of Foundation Engineering,
China Academy of Building Research, Beijing 100013, China

Beijing Engineering Technology Research Center of Foundation and City Underground Space Development and Utilization, Beijing 100013, China

X. Li
China Building Technique Group Co., Ltd, Beijing, China

© The Author(s) 2023
G. Feng (ed.), *Proceedings of the 9th International Conference on Civil Engineering*,
Lecture Notes in Civil Engineering 327,
https://doi.org/10.1007/978-981-99-2532-2_30

2 Project Overview

The main buildings of this project are shear wall structure, raft foundation, foundation buried depth of about 2.5–3.5 m. Among them, Building B has 15 floors above ground and 1 floor underground, and the basement is connected to the underground garage. The surface layer of the site is plain fill soil, and the underlying layer is Quaternary and Tertiary Marine sedimentary soil. From top to bottom, it is divided into 5 geotechnical engineering layers. The measured buried depth of stable water level is about 3 m. The typical geological profile is shown in Fig. 1, and the physical and mechanical parameters of the soil layer are shown in Table 1.

The basement soil of Building B is layer ② which is Medium sand with a bearing capacity characteristic value of 120 kPa. The superstructure design requires that the bearing capacity after foundation treatment is 300 kPa and 250 kPa, respectively. The bearing capacity and deformation of natural foundation can not meet the requirements of superstructure. The original foundation treatment scheme uses cement soil mixing pile composite foundation.

The design parameters of composite foundation of raw soil–cement mixing pile in Building B are as follows: The characteristic value of bearing capacity of single pile is 300kN, the characteristic value of bearing capacity of composite foundation is 250 kPa, the pile length is 15−18 m, the pile end falls in the fourth layer gravel sand layer, the pile spacing is 1 m and 1.2 m, the pile diameter is 500 mm, the strength grade of cement body of the pile is not less than 5.1 MPa, the thickness of mattress is 250 mm.

Since the construction of the main structure in July 2012, up to October 2013 before the foundation reinforcement, the settlement of the two buildings has been too large, with the settlement of building A reaching 30 cm and the settlement of Building B approaching 20 cm, and the inclination reaching 4.76‰. The settlement

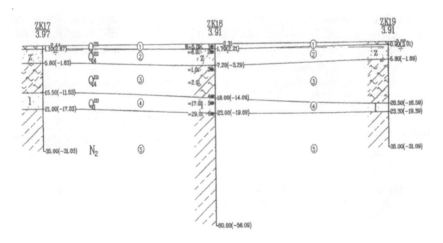

Fig. 1 Typical geological profile

Table 1 Physical and mechanical parameters of soil layer

Soil layer number	Soil layer name	Mean layer thickness(m)	Dry density ρ_d (g/cm^3)	Natural water content ω (%)	Natural porosity ratioe	Liquidity index I_l	Plasticity I_p	Standard value of lateral resistance of limit pile q_{sk} (kPa)	Standard value of ultimate pile tip resistance q_{pk} (kPa)	Modulus of compression$E_{s(1-2)}$ (MPa)	characteristic value of subsoilfak (kPa)
①	Plain fill soil	1.44	1.41	27.7	0.933	0.29	13.93			4.74	70
②	Medium sand	4.45	1.60	23.4	0.681			30		12.48	130
②$_1$	silt	1.39	1.59	20.7	0.672			18		11.29	120
③	Mucky silty clay	13.15	1.19	46.5	1.320	1.26	16.60	19		2.37	80
④	Gravelly sand	5.39	1.82	16.1	0.446			116	2500	13.92	260
⑤	Silty clay	unexposed	1.49	28.66	0.825	0.23	14.61	85	1200	8.66	280

and differential settlement have no convergence trend. Therefore, it can be judged that there are defects in the construction quality of soil–cement mixing pile of this project, which has not reached the expected purpose of design.

3 Calculation Method and Results

When the existing building is reinforced and the new piles under the base pass through the soft soil layer and enter the distributed friction piles of relatively good soil layer, the settlement of the middle point of the reduced-sink composite pile foundation can be calculated according to the following formula [1, 2].

$$S = \psi \left(S_s + S_{sp} \right)$$

$$s_s = 4 p_0 \sum_{i=1}^{m} \frac{z_i \overline{\alpha}_i - z_{(i-1)} \overline{\alpha}_{(i-1)}}{E_{sic}}$$

$$s_{sp} = 280 \frac{\overline{q}_{su}}{\overline{E}_s} \cdot \frac{d}{(s_a/d)^2}$$

$$p_o = \eta_p \frac{F - n R_a}{A_c}$$

s- settlement amount at the center of pile foundation;

s_s- midpoint settlement caused by additional pressure of foundation soil at the bottom of cap (Fig. 2.);

Fig. 2 Stratification diagram of settlement calculation of reduced sink composite pile foundation

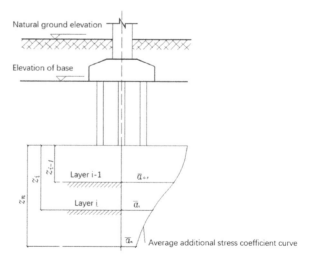

s_{sp}– settlement caused by pile-soil interaction;

p_o– the hypothetical average additional pressure of natural foundation calculated by the combination of quasi-permanent values of load effects (kPa);

E_{sic}– The compression modulus of the soil under the bottom of the cap shall be the modulus of the soil loaded after unloading;

m– The number of soil layers in the depth range of foundation settlement calculation; The calculated depth of settlement can be determined by stress ratio or deformation ratio.

q_{su} 、 E_s– the average thickness-weighted ultimate friction and average compression modulus of the pile side within the range of the pile;

d– Pile diameter, when it is a square pile, d = 1.27b(b is the length of the section side of the square pile);

s_a/d– equivalent distance to diameter ratio;

z_i, z_{i-1}– the distance between the bottom of the cap and the bottom surface of the i and i-1 layers;

α_i, α_{i-1}– the average additional stress coefficient of corner points from the bottom of cap to the bottom of the i and i-1 soil layers; Rectangle aspect ratio a/b and depth aspect ratio zi/b = 2zi/Bc were calculated according to equivalent cap area, which were determined by Appendix D of Technical Code for Building Pile Foundation (JGJ 94–2008). The equivalent width of the cap: $B_c = B \sqrt{A_C/L}, B$ 、 L is the width and length of the outer edge plane of the building foundation;

F– Total additional load (kN) acting on the bottom of the bearing platform under the combination of quasi-permanent value of load effect;

η_p– Influence coefficient of foundation pile penetration deformation; According to the soil quality of the bearing layer of pile end, the sand is 1.0, the silt is 1.15, and the viscous soil is 1.30.

Ψ– The experience coefficient of settlement calculation is 1.0 if there is no local experience.

Building B has 14 floors above ground and 1 floor underground. The settlement amount at the maximum point (north side of the building) is 299.96 mm, at the minimum point (south side of the building) is 240.88 mm, and the maximum differential settlement amount from south to north is 59.08 mm. Therefore, uneven pile distribution is adopted to reduce subsidence and correct deviation in the north and south areas of the project. The steel pipe pile is φ299 mm/φ245 mm directly, the pile length is about 18 m, and the pile end falls on the gravel sand layer in layer 4. The bearing capacity of φ245 mm single pile is 250 kN, and that of φ299 mm single pile is 300 kN. The pile replacement ratio on the north side of the foundation is about 62%, and the pile replacement ratio on the south side is about 35% (Fig. 3.).

According to the results, the reaction force of the base on the south side of the foundation is about 50 kPa−200 kPa, and the reaction force of the base on the north side of the foundation is about 50 kPa−110 kPa, and the stress concentration state appears at the base side (Fig. 4, Fig. 5).

Through the adjustment of uneven pile distribution, the settlement gradually decreases from south to north. The new settlement in the south of the foundation is about 36 mm, and the north is about 21 mm, the settlement difference in the

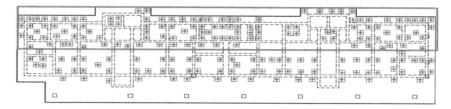

Fig. 3 Layout Plan of pile reduction for Building B (the area is the north side of the foundation)

Fig. 4 Reverse diagram of
basement soil of Building B

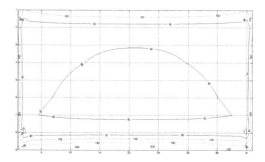

Fig. 5 Pressure curve of the
axial base of Building B
(vertical)

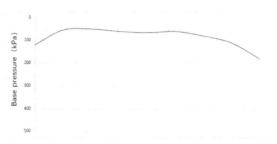

central axis is 15 mm. The foundation tilts to the south, and the foundation tilts back
about 0.0013L. It shows that the variable stiffness leveling design of uneven pile can
adjust the uneven settlement of foundation (Fig. 6, Fig. 7).

Fig. 6 New deformation of
Building B after
reinforcement

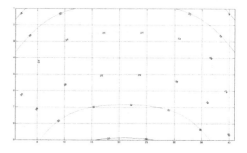

Fig. 7 New settlement curve
of the central axis of
Building B (vertical)

4 Comparative Analysis of Engineering Measurement and Design [3]

After the settlement of the building stabilized, the average measured soil stress between piles on the north side of the foundation of Building B was 45 kPa, on the south side was 85 kPa, and the average stress on the top of the mixing pile was 149 kPa. According to the results, the base reaction on the north side of the foundation was 80 kPa, and on the south side was 105 kPa (Fig. 8, Fig. 9).

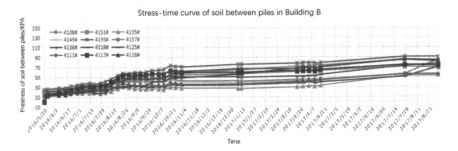

Fig. 8 In-situ stress-time curve of soil between piles in Building B

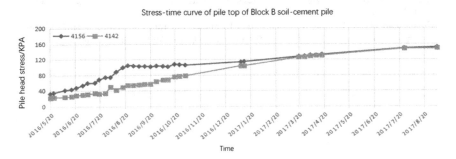

Fig. 9 Stress-time curve of pile top of cement soil pile in Building B

In November 2013, the north pile was connected with the foundation and sealed. By May 2017, the measured post-construction deformation was 32 mm−46 mm. Excluding the settlement generated according to the original settlement rate, the construction disturbance settlement was about 15 mm, and the new settlement was about 17 mm−31 mm.

5 Conclusion

After the uneven pile distribution method was adopted in this project, the average measured soil stress between piles on the north side of the foundation was 45 kPa, on the south side was 85 kPa, and the average stress on the top of the stirred pile was 149 kPa. The pile replacement ratio of the north side of the foundation (the side with large settlement) is 62%, the calculated base reaction value of the north side of the foundation is 80 kPa, the pile replacement ratio of the south side of the foundation (the side with small settlement) is 35%, the calculated base reaction value of the south side of the foundation is 105 kPa. The new settlement in the south of the foundation is about 36 mm, in the north is about 21 mm, and the foundation receding is about 0.0013L. The results show that the distribution of basal reaction can be adjusted successfully by the method of uneven pile distribution, and the effect of subsidence reduction and deviation correction can be achieved.

Acknowledgements This article was supported by the Science and Technology Program of the Ministry of Housing and Urban-Rural Development (2016-K5-055).

References

1. China Academy of Building Research (2011) Code for design of building foundation (GB50007–2011) (Beijing: China Architecture & Building Press) p 28
2. China Academy of Building Research (2008) Technical code for building pile foundations (JGJ94–2008) (Beijing: China Architecture & Building Press) p 57
3. China Academy of Building Research (2012) Technical code for improvement of soil and foundation of existing buildings (JGJ123–2012) (Beijing: China Architecture & Building Press) p 37

Research and Application of Anchorage Vertical Loading System in Building Reinforcement and Rectification

Zhao Li, Xunhai Sun, Ning Jia, Shengbin Zhou, Xinhui Yang, Jianxing Tong, and Haitao Yang

Abstract In this paper, the settlement and tilt of a building in north China are taken as an example. The differential settlement of the building is not stable, and the tilt continues to increase. After the rectification measures of forced landing by digging out the soil were taken, the expected effect was not achieved. Then the auxiliary rectification was carried out by means of centralized and cyclic loading through the anchorage vertical loading system, and the automatic dipmeter and manual monitoring are adopted in the process of deviation rectification to achieve information construction. At present, the construction of this project has been completed and the effect of reinforcement and rectification is good, which can provide reference for similar projects and has certain promotion and guidance significance.

Keywords Reinforcement and Rectification · Anchorage Vertical Loading System · Automatic Dipmeter · Information Construction

1 Introduction

For the reinforcement and rectification of existing buildings, the commonly used measures mainly include jacking method and crash landing method. Compared with the jacking method, which has a long construction period, high cost and difficulty, the crash landing method is relatively simple. The crash landing method mainly includes loading, precipitation, stress relief and excavation method, etc. Relatively speaking, excavation method is the most commonly used because of its simplicity,

Z. Li (✉) · X. Sun · N. Jia · S. Zhou · X. Yang · J. Tong · H. Yang
State Key Laboratory of Building Safety and Built Environment, Beijing 100013, China
e-mail: lz417130724@126.com

Institute of Foundation Engineering, China Academy of Building Research,
Beijing 100013, China

Beijing Engineering Technology Research Center of Foundation and City Underground Space
Development and Utilization, Beijing 100013, China

© The Author(s) 2023
G. Feng (ed.), *Proceedings of the 9th International Conference on Civil Engineering*,
Lecture Notes in Civil Engineering 327,
https://doi.org/10.1007/978-981-99-2532-2_31

371

short construction period and strong controllability. However, a single excavation method often has certain limitations. If the site soil layer is dense, or gravel, gravel and other inclusives are more, the excavation method is not easy to achieve the desired results. Excessive excavation may lead to the risk of sudden increase of building settlement. At the same time, in the process of grouting hole sealing after rectification, the amount of grouting is not easy to grasp due to the looseness of the soil, which is easy to lead to the tipping of the building in the process of grouting [1, 2].

This paper takes a building in the north as an example, the vertical loading system of anchor cable was adopted to assist the rectification of tilting on the basis that the forced landing failed to reach the expected effect. The system made use of the original raft of the building to cut holes in the raft for anchor cable construction, and the anchor end of the anchor cable entered a good soil layer, so that the anchor cable and the original raft formed a good vertical loading system. Through a number of jacks to carry out concentrated circulation loading to crash landing and prevent tipping. At the same time, automatic dip meter combined with manual monitoring is used in the process of rectification, so as to achieve information construction. The project has achieved good results and can provide some reference for similar projects.

2   Engineering project

2.1   Project Overview

The project is located in a city in the north of our country, the use functions above ground are houses, underground use functions for warehouse, equipment houses and parking, etc. There are 11 floors above ground and 1 floor underground. The width of the foundation is 15.95 m and the length of the foundation is 48.30 m. The structure type is shear wall structure, the foundation type is graded sand and stone replacement foundation, the foundation type is flat raft foundation, the thickness of the raft is 600 mm, and the design grade of the foundation is Class B.

The original foundation treatment scheme is as follows: remove all the soil above the layer of ③ silt (or ③1 gravel), then use graded sand and stone with a thickness of no more than 300 mm to backfill to the design elevation, the compaction coefficient is 0.97.

According to relevant construction and monitoring data, since the structure capping up to now, the overall relative settlement of the southwest corner is 5.18 cm, and the overall relative settlement of the southeast corner is 6.18 cm. The inclination rate of the relative settlement of the south and north sides exceeds the requirement of the design code by 0.3%. Field monitoring data show that the settlement of the tower is not stable and still presents a trend of continuing to increase.

2.2 Settlement Cause Analysis

The building site is generally high in the northwest and low in the southeast, showing a gentle slope shape. According to the survey report, the lithologic soil of the site consists of loess-like silt and silty soil on the upper part, with round gravel layer distributed among them, and pebble layer on the lower part, which can be divided into 5 layers from top to bottom, namely plain filled soil, loess-like silt, silt, gravel and pebble in turn, among which loess-like silt layer has collapseability. The typical geological profiles are shown in Figs. 1 and 2.

Fig. 1 North basement lithology map

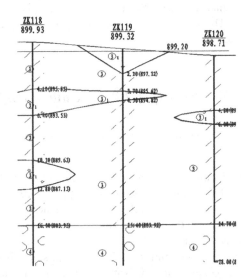

Fig. 2 South basement lithology map

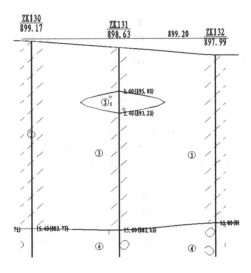

No groundwater was revealed within the scope of the survey. Considering the buried depth of the foundation of the proposed building, the adverse effects of the site's water on the construction of the project are not considered.

According to the survey report, combined with the calculation model and the actual situation of the site, the main reasons for the building tilt are analysed as follows.

According to the geological survey report, there are ③ silty soil and ③1 gravel layers under the replacement padding layer, and the compression modulus of the silty soil layer is 8 MPa, and the compression modulus of gravel layer is 30 MPa. The north side of the building has many interlayers of ③1 gravel, and the thickness is thicker. The south side of the building basically has no ③ gravel layer.

Through modeling analysis, the overall center of gravity is south, and the eccentricity ratio is 2.88.

2.3 The Original Rectification Plan for Reinforcement

(1) Reinforcement Scheme

The original reinforcement scheme adopts the form of external raft and micro pile to control the continued settlement of the building.

The design parameters of micro-pile are as follows, effective pile length is 19 m, the raft anchored in the upper part is no less than 0.5 m, ordinary welded pipe with DN80 wall thickness of 3 mm is selected for steel pipe, the total length of welded pipe is no less than 19.5 m, the diameter of micro-pile is 200 mm, the distance between piles is 1000 mm, PSA 32.5 cement slurry is injected into the hole and inside the welded pipe, and the water-cement ratio should be 0.5 ~ 0.55.

The width of the south external raft is 1.5 m, the concrete strength grade is C35. The rib plate is set on the new foundation to connect with the outer wall of the original basement, and the rib plate is set at the place of the original shear wall.

(2) Inclination correction plan

The original rectification plan adopted the form of forced landing.

Two small drill hammers with equal spacing from east to west are used to cut soil simultaneously. For the first time, dig a deep hole every 2 m with a depth of 8 m and a diameter of no more than 8 cm (the diameter of drill pipe should be selected according to the site test drilling). The second dig is the middle position of the first dig hole, and the dig depth is 5 m. The third excavation is the middle position of the excavation holes after the two sides are encrypted, that is, the spacing of the excavation holes is 0.5 m, and the depth of the holes is 8 m. The fourth excavation is in the middle of the above excavation holes, with a depth of 10 m; Adjust whether to encrypt and increase the depth of excavation hole according to the effect of excavation and building tipping. When encountering the location of elevator shaft and catchment hole, dig under the catchment hole for excavation construction, with the depth of excavation not less than 12 m.

2.4 Effect Evaluation of the Original Reinforcement and Rectification Scheme

According to the reinforcement and rectification plan, the construction of micro pile and external raft was carried out first. The monitoring data showed that the settlement of the south side of the building increased more than that of the north side at the initial stage, indicating that the uneven settlement and tilt were still intensifying and did not show a convergence trend. After the construction of micro pile and external raft, the differential settlement trend gradually slows down.

After the construction of the external raft is completed, the north side of the work begins to dig. The excavation work is carried out symmetrically and synchronously with the micro-drill hammer. The depth, speed and position of excavation are determined according to the requirements of the program and the actual situation on site. Monitoring data show that with the progress of excavation work, the north side of the settlement is obvious, showing a relatively obvious inclination trend on the whole. With the progress of excavation work, the inclination trend gradually becomes stable, and the maximum inclination rate is inclined from 5.2‰ to about 4.0‰, which is still a certain gap from the standard requirements.

3 Anchorage Vertical Loading System Scheme

In view of the dense soil layer in the site of this project, there are gravel and boulders in the original stratum, and the original foundation treatment plan adopts graded sand and gravel backfill, resulting in more gravel and boulders in the excavation are. The excavation method is not easy to achieve the desired results, and excessive excavation may lead to the risk of sudden decline of building settlement. Therefore, in order to further accelerate the settlement rate of the north side of the building and avoid the tilting rebound when the grouting hole is closed after the building is rectified, the anchorage vertical loading system is adopted to assist the tilting correction in the northeast and north side of the raft overhanging area.

In this project, 6 anchor cables are set at the northeast side of the anchor cable loading system, with a spacing of 1 m and a locking value of 500kN. On the north side, set 24 anchor cables with a spacing of 1.5 m and a locking value of 500kN. The diameter of the anchor cable is 150 mm, three bundles of 15.2 steel strands are used, and the depth of the anchor cable is 20 m. The cement is P.O. 42.5 cement, and the water-cement ratio is 0.5 ~ 0.55. When the construction age of the anchor cable is satisfied, the 6 cables shall be taken as a group for simultaneous grading tensioning.

The 6 cables on the east side shall be taken as a group, and the 6 cables on the north side (separated by 3 knots) shall be taken as a group. After the completion of the first group of tensioning, the next group of cables shall be taken until the tensioning of all the anchor cables is completed. In the tensioning process, automatic monitoring equipment is used for real-time monitoring, so as to achieve dynamic design and information construction. At the same time, the tensioning is carried out timely according to the slack situation of the anchor cable. After the correction is complete, the anchorage loading system will be locked as a permanent anchor cable.

4   Construction Process Control

The anchorage vertical loading system uses the original raft to make holes in the original raft, the construction equipment adopts the anchor rig, the construction technology adopts the double casing followed by hydraulic drilling into the hole, the pile end bearing layer into the pebble layer, so that the anchor cable and the original raft form a good vertical loading system, the raft is cyclically loaded by several jacks, thus increasing the overall settlement of the north side. At the same time, the anchor cable is tensioned twice a day, and the tensioning process is also carried out synchronously for multiple groups of anchor cables, so as to solve the situation of automatic unloading of the lock on the anchor cable after the raft settlement. After the rectification work is completed, in order to prevent the building from tilting back when grouting holes are sealed, each anchor cable would be locked.

5   Monitoring Data Analysis

In order to control the deformation of the building in the whole process of tilting correction, the monitoring means of automatic inclinometer combined with manual monitoring and dual control are adopted in the process of reinforcement and tilting correction. The two monitoring methods verify each other, complement each other with advantages, increase the credibility of monitoring data, and enable the monitoring to better serve the tilting correction construction and truly achieve dynamic design and information construction.

Manual tilt monitoring adopts the drop method for tilt measurement, and the measuring instrument is the total station. Before construction, the used instrument shall be inspected and calibrated according to the standard requirements. Each observation shall be carried out according to the same observation Angle and height interval, the same instrument shall be used, and the observation personnel shall be fixed.

Automatic incline monitoring is performed using a Geokon BGK6150-1 automatic incline meter, which uses a uniaxial sensor to accurately measure incline using a incline sensor attached to the monitored structure. A dip meter was set in the southwest corner and the northeast corner of the building respectively during the rectification process. The data collector adopted the BGK Logger data acquisition system, which was set to automatically collect data every 6 h for analysis. During the construction of the anchorage vertical loading system, real-time monitoring of the dip meter data with each level of load tension can guide the construction and control of vertical loading of the anchor cable, so that the tilting correction of the building can be controlled. The manual and automatic monitoring curves are shown in Fig. 3, 4, 5.

At present, the rectification work has been completed. The whole rectification process has realized safety and control, and the overall tilt of the floor meets the standard requirements. The above data are the curves of manual monitoring and automatic monitoring.

Figure 3 shows the tilting change curve of the main body of the building monitored manually. It can be seen from the figure that the tilting trend is obvious in stage 1–1, that is, from the excavation on July 12 to around August 10, and the tilting trend

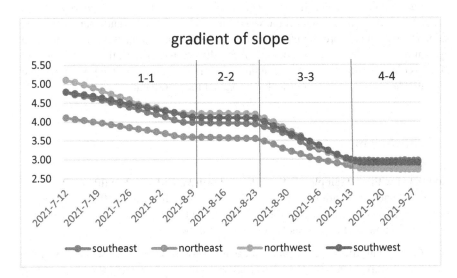

Fig. 3 Artificial tilt monitoring curve

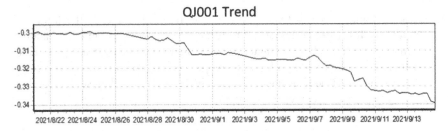

Fig. 4 Automatic tilt west side monitoring curve

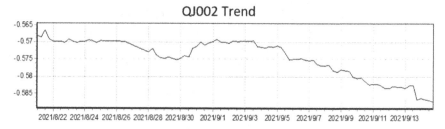

Fig. 5 Automatic tilt east side monitoring curve

gradually becomes gentle in stage 2–2 due to stratum conditions and other reasons. In stage 3–3, the tilting effect is obvious after the vertical anchor loading system combined with the crash landing. And the four corners all returned to within 0.3%, meeting the requirements of the specification. In Stage 4–4, the overall tilt trend tends to be stable after the grouting sealing in mid-September.

Figure 4 and Fig. 5 are the trend charts of automatic monitoring curves, among which Fig. 4 is the trend chart of the west side and Fig. 5 is the trend chart of the east side. The two curves are the trend graphs of the tilt curve after the vertical loading system of the anchor cable is adopted. According to the figures, the overall tipping effect is obvious after the vertical loading system of the anchor cable is adopted, and the trend tends to be consistent with the manual measurement.

6   Conclusion

Combined with an example of a building in northern China, when the original reinforcement and rectification measures failed to achieve the desired effect, the vertical loading system of anchor cable was adopted to rectify the tilting measures, and the effect was good. The main conclusions are as follows.

(1) Appropriate tilting rectification measures should be selected according to the soil condition. For the dense soil layer, or the stratum with more gravel and boulders, it is not easy to achieve the expected effect with a single excavation method, and the anchorage vertical loading system can be used to assist tilting rectification.

(2) The anchorage vertical loading system can be used to rectify the tilting of buildings. It uses the original raft to form a good loading system, and adopts several jacks to load the original raft in circulation, thus increasing the overall settlement of the building block. In complex geological conditions, excessive excavation may lead to a sudden increase in the risk of building settlement, is a very good measure to rectify the tipping.

(3) The vertical loading system locks each anchor cable as a permanent anchor cable after the end of the tilting correction, which can prevent the inclined rebound in the process of building grouting hole sealing and the subsequent settlement stabilization process of the building.

(4) Information construction should be carried out in the process of rectification and reinforcement, manual monitoring and automatic monitoring should be combined, and the two monitoring methods should verify each other, complement each other's advantages, increase the credibility of monitoring data, so that the monitoring can better serve the rectification and rectification construction, and truly achieve dynamic design and information construction.

Acknowledgements This article was supported by the Science and Technology Program of the Ministry of Housing and Urban-Rural Development. (2016-K5-055)

References

1. China Academy of Building Research (2012) Technical code for improvement of soil and foundation of existing buildings (JGJ123-2012) China Architecture & Building Press Beijing
2. China Academy of Building Research (2011) Code for design of building foundation (GB50007–2011) (Beijing: China Architecture & Building Press)

Stability Analysis of Earth and Rock Dams During the Construction Period of Soft Foundation Reinforcement

Jiahao Li, Jingnan Xu, Shiwen Zhao, and Junhua Wu

Abstract The consolidation of earth and rock dam foundations and its effect on stability is a key concern in water conservancy and water transportation projects. It is common to use the vacuum combined pre-pressure method to deal with soft ground foundation, where the reasonable selection of soil consolidation coefficient is one of the important factors to determine the success of the project. In this paper, numerical simulation is used to study the strength law of soil and rock dam foundation under different consolidation coefficients of soft ground, and the corresponding stability coefficients of soil and rock dam are obtained. On this basis, the reinforcement effect of an additional counter pressure platform is proposed and analyzed. The results show that the larger the consolidation coefficient of soft ground is, the more it is conducive to the growth of the strength of the soft soil layer and can effectively improve the stability coefficient of the earth and rock dam. At the same time, the additional counter-pressure platform has a certain effect on the overall stability enhancement of the earth and rock dam. The research results can provide a reference for the application of vacuum combined pre-pressure method to earth and rock dam projects.

Keywords Soft soil foundation · Coefficient of consolidation · Stability against sliding · Numerical analysis

J. Li · J. Xu · J. Wu (✉)
School of Civil Engineering and Architecture, Nanchang Hangkong University, Nanchang 330063, Jiangxi, China
e-mail: wjhnchu0791@126.com

Key Laboratory of Water Conservancy and Water Transport Engineering, Ministry of Education, Chongqing Jiaotong University, Chongqing 400074, China

Key Laboratory of Damage Mechanism and Prevention and Control Technology for Earth and Rock Dams, Ministry of Water Resources, Nanjing 210029, Jiangsu, China

S. Zhao
Nanjing Rui Di Construction Technology Co., Ltd, Nanjing 200120, Jiangsu, China

© The Author(s) 2023
G. Feng (ed.), *Proceedings of the 9th International Conference on Civil Engineering*,
Lecture Notes in Civil Engineering 327,
https://doi.org/10.1007/978-981-99-2532-2_32

1 Quotes

Earth and rock dams are widely used in China and are common building in water conservancy projects. However, due to long-term water storage and other reasons, over-saturated, making it a typical soft soil characteristics, resulting in poor foundation-bearing capacity, excessive foundation settlement and uneven settlement problems during construction. These problems have a great impact on the stability of earth and rock dams, and are prone to dam destabilization and damage, so the stability of earth and rock dam foundations is a key concern today.

Numerical simulation analysis has been widely used in engineering projects, such as slope, roadbed and dam construction. Yang Kun, Guo Qing et al.[1–3] conducted infiltration as well as stability analysis based on numerical simulation for various working conditions in the field. Zhang Songyun[4] explored the effect of flooding process on the stability of buildings using the embankment of Koster Power Station as an example. Li Sen, Zhang Zhaojun [5, 6] et al. Studied the effect of different parameters on dams based on ANSYS. Feng Wang7 Used the Kriging method to assess the slope safety of earth and rock dams with reliability as the index. Zhang Lijuan and Cheng Ping [8, 9] analyzed and evaluated the permeability and stability of earth-rock dams based on numerical analysis and limit equilibrium theory.

In this paper, we will use a water storage project as the background, and used Geostudio [10] finite element software to numerically simulate the overall slope of the earth and rock dam, to investigate the influence of the growth of the strength of the soft soil layer on the stability of the earth and rock dam under different consolidation coefficients, and to improve the overall stability of the earth and rock dam by adjusting the range of the backpressure platform of the earth and rock dam.

2 Numerical Analysis

2.1 Initial Parameter Selection

In this paper, a typical section is used for analysis, and the section diagram is shown in Fig. 1. According to the field test, where the average value of the consolidation coefficient of layers 2–3 is $1.46 \times 10 \text{ cm}^{-32}/\text{s}$ and the average value of small value is $0.32 \times 10 \text{ cm}^{-32}/\text{s}$; the average value of the consolidation coefficient of layers 2–4 is $1.01 \times 10^{-3} \text{cm}^2/\text{s}$ and the average value of small value is $0.22 \times 10 \text{ cm}^{-32}/\text{s}$. Considering that the measured consolidation coefficient in the field is much larger than the measured value in the room, in order to compare the effect of consolidation coefficient on consolidation degree, the consolidation coefficient was enlarged by 5 times, and $0.22 \times 10 \text{ cm}^{-32}/\text{s}$ and $1.1 \times 10^{-3} \text{cm}^2/\text{s}$ were used to calculate the strength growth of 2–3 and 2–4 soil layers, respectively, and the parameters of each soil layer are shown in Table 1.

The strength growth is calculated using the following equation.

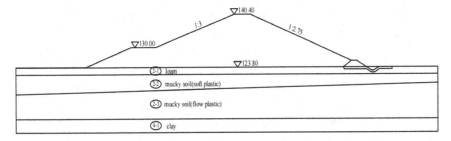

Fig. 1 Typical cross section of earth and rock dam

Table 1 Physical and mechanical properties of each layer of soil

Soil layer	Water content(%)	Saturated density (g/cm³)	Specific Gravity	Permeability coefficient (10⁻⁶ cm/s)	Compression modulus (MPa)	Cohesion (kPa)	Friction angle
2–1	19.5	1.94	2.69	9.21	6.02	19.8	16.3
2–3	52.2	1.69	2.70	1.21	2.94	10	6.5
2–4	65.3	1.72	2.69	Impervious to water	2.801	10	5.0
4–1	24.5	1.98	2.68	1.703	5.71	19.8	16.1

$$\Delta\tau_f = \eta U \Delta\sigma_c' \tan\varphi_{cq}$$

where. η is the discount factor, according to the building foundation treatment specification is generally $0.8 \sim 0.85$, this paper is taken as 0.825. U is the consolidation degree of the dam base soil. $\Delta\sigma_c'$ is the effective consolidation pressure before shear. φ_{cq} is the angle of internal friction measured by the consolidation fast shear test. The strength growth at different fill heights is shown in Table 2 and Table 3.

Table 2 Strength growth at different fill heights (Cv = 0.22×10^{-3} cm² /s)

Filling height(m)	Solidity	Strength increase value (kPa)		Strength after growth (kPa)	
		2–3 layers	2–4 floors	2–3 floors	2–4 floors
0	0	0	0	18.9	18.1
3	0.19	10.48	9.78	29.38	27.88
7	0.35	19.31	18.02	38.21	36.12
12	0.55	30.35	28.31	49.25	46.41
17	0.99	54.63	50.97	73.53	69.07

Table 3 Strength growth at different filling heights (Cv = 1.1 × 10-3cm2/s)

Filling height(m)	Solidity	Strength increase value (kPa)		Strength after growth (kPa)	
		2–3 floors	2–4 floors	2–3 floors	2–4 floors
0	0	0	0	18.9	18.1
3	0.35	19.31	18.02	38.21	36.12
7	0.56	30.90	28.83	49.8	46.93
12	0.84	46.35	43.25	65.25	61.35
17	1.16	64	59.72	82.9	77.82

2.2 Numerical Analysis Results

In this paper, the SLOPE module in Geo-Studio software is used for analysis, and the hazardous starting arc location is obtained by Bishop method, and the selected model is a homogeneous model, and the hazardous slip surface is determined by specifying the import and export range. The results are shown in Table 4, and the change law of safety coefficient is shown in Fig. 2.

The above calculation results show that the safety coefficient of the earth and rock dam decreases as the filling height rises, and the safety coefficient of the sea side and land side slope is still greater than 1 in a stable state when the filling height is

Table 4 Calculation results of safety factor (without reinforcement)

Filling height(m)	$Cv = 0.22 \times 10^{-3}$ cm^2 /s		$Cv = 1.1 \times 10^{-3}$ cm^2 /s		Normally allowed values
	Seaside	Landside	Seaside	Landside	
7	1.152	1.076	1.257	1.299	1.25
12	0.957	0.861	1.060	1.022	1.25
17	0.903	0.807	0.988	0.951	1.25

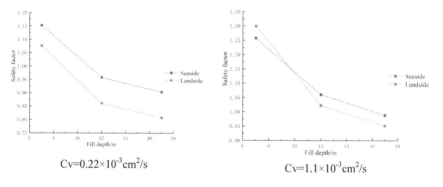

Cv=0.22×10^{-3}cm^2/s   Cv=1.1×10^{-3}cm^2/s

Fig. 2 Variation rule of safety factor

7 m. When the filling height reaches 12 m, the dam is already in the ultimate stable state and there is a possibility of instability, and the calculation results are in line with the actual situation. At the same time, with different consolidation factors, the safety factor of the dam slope with a consolidation factor Cv of $1.1 \times 10^{-3} cm^2/s$ is generally higher than that with a consolidation factor Cv of $0.22 \times 10^{-3} cm^2/s$. This paper concludes that without considering the occurrence of large lateral displacements, a larger consolidation factor will shorten the consolidation time of the foundation soil, increase the average consolidation degree of the foundation soil, and significantly increase the strength of the soft soil layer, thus This paper concludes that a larger consolidation coefficient will shorten the consolidation time, increase the average consolidation of the foundation soil, and increase the strength of the soft soil layer, thus improving the bearing capacity of the foundation and increasing the overall safety factor of the dam slope.

2.3 Program Adjustment

According to the above analysis, the dam body is in the ultimate stable state, the safety factor does not meet the code requirements, and there is a possibility of instability. For this reason, it is necessary to consider adjusting the reinforcement scheme and calculating the length of the additional counterpressure platform required on the original foundation according to the two different soil consolidation factors to make it meet the code requirements. The calculation results of the safety factor are shown in Table 5, and the law of change of the safety factor is shown in Fig. 3.

According to the above calculation results, when the consolidation factor is $0.22 \times 10^{-3} cm^2/s$, the length of the additional sea-side ballast platform is 43.3 m and the height is 5 m. The length of the additional land-side ballast is 40.8 m and the height of the counter-pressure is 4.4 m. When the consolidation factor is $1.1 \times 10^{-3} cm^2/s$, the length of the additional sea-side ballast is 37.4 m and the height is 5 m. The length of the additional land-side ballast is 35 m and the height is 4.4 m. From the results, it can be seen that the overall slope safety factor is significantly improved by the additional counterpressure platform.

Table 5 Calculation results of safety factor (with reinforcement)

Filling height(m)	Cv = 0.22 × 10⁻³ cm² /s		Cv = 1.1 × 10⁻³ cm² /s		Normally allowed values
	Seaside	Landside	Seaside	Landside	
7	1.842	2.283	1.610	2.062	1.25
12	1.441	1.389	1.379	1.435	1.25
17	1.255	1.256	1.250	1.251	1.25

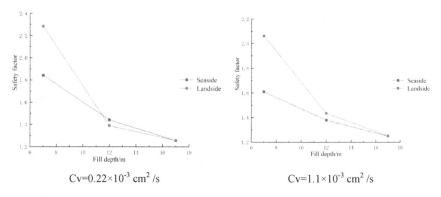

Cv=0.22×10⁻³ cm² /s   Cv=1.1×10⁻³ cm² /s

$Cv=0.22\times10^{-3}$ cm^2/s $Cv=1.1\times10^{-3}$ cm^2/s

Fig. 3 Change law of safety factor after extension

3 Conclusions

(1) The pore water discharged is mainly from the soft soil layer due to the poor permeability and high water content of the soil in the soft soil layer, so the consolidation of the soft soil layer is the main reason for the consolidation of the foundation soil. Therefore, the selection of the consolidation coefficient of soft soil layer in the numerical simulation analysis will affect the bearing capacity of the consolidated foundation to a certain extent.

(2) According to the comparison of the above calculation results, the foundation strength increases significantly in the calculation process for a larger consolidation coefficient. When the soil consolidation coefficient is larger, the consolidation effect is more obvious, and the safety factor of earth-rock dam is significantly improved. Therefore, it is considered in this paper that when the consolidation coefficient is set as $1.1 \times 10\text{-}3\text{cm}2/\text{s}$, it is more conducive to the overall stability of earth-rock dam.

(3) The overall stability of an earth and rock dam is not only related to the bearing capacity of the foundation, but also to the structural dimensions of the dam itself. The calculation results show that with a significant increase in the bearing capacity of the foundation, the increasing fill height will still reduce the overall safety factor to less than 1. In addition, the ultimate state of stability occurs at the filling height of 12 m. Therefore, additional counter-pressure platforms need to be installed on the seaside and landside. The results show that the additional counter-pressure platform has significantly improved the overall slope safety factor and significantly improved the overall stability of the earth and rock dam.

Acknowledgements This work is supported by Project (51869013) of the National Natural Science Foundation of China, Project (YK321013) of Open Research Fund of Key Laboratory of Failure Mechanism and Safety Control Techniques of Earth-Rock Dam of the Ministry of Water Resources, and Project (SLK2021A05) of Key Laboratory of Hydraulic and Waterway Engineering of the

Ministry of Education. All opinions, findings and conclusions in this work represent the views of the authors only.

References

1. Kun Y (2021) Slip stability analysis of overflow dam section of a reservoir slurry masonry gravity dam. Water Sci Technol Econ 27(06):72–78
2. Qing G (2021) Stability review of a reservoir dam. Shanxi Water Conserv Sci Technol 2021(01):26–29
3. Lina G, Wang L (2021) Application and development of finite element slip surface stress method in slope stability analysis. Water Resour Hydropower Technol (in English) 52(S2):416–420
4. Zhang S, Hu M (2020) Seepage and slope stability analysis of Kausw power station embankment in the Philippines during flooding_Zhang Songyun. Water Transp Eng 2020(5):55–60
5. Zhao L (2021) Stability analysis of Diaokou reservoir dam in Pu County. Dams Safety 2021(02):37–41
6. Sen L, Zhang Z, Ren Z (2021) Analysis of the influence of foundation parameters on the safety of dams based on ANSYS. Northeast Water Resour Hydropower 2021(02):13–15
7. Wang F (2016) Study on the application of Kriging method in the analysis of stability reliability of saturated-unsaturated earth and rock dam slope. Journal of Nanchang University
8. Zhang L (2017) Numerical Analysis of Seepage and Stability and Safety Evaluation of Chaohe Main Dam of Miyun Reservoir. Journal of Tsinghua University
9. Cheng P, Wang LF, Ren QY, Zeng T, Li LG (2020) Analysis of stability and reliability of rocky slopes in the abatement zone of Three Gorges Reservoir. People's Changjiang. 51(03):113–118
10. Yuan H (2020) Application of geostudio-based 2d geological modeling and numerical simulation. In: Proceedings of the 2020 Annual Academic Conference of the Chinese Society of Civil Engineering, pp. 68–75

Seepage Analysis of Drainage Decompression in Sloping Basement

Shuijiang Li, Quanbin Wan, Haorong Yan, and Hong Pan

Abstract The accidents of bottom floor cracking occur on underground structure due to uplift pressure. Especially in the southern rainy area, super large buildings on slope with high underground water level are in a complex seepage field and it is hard to deal with. In order to solve this problem, a residential district in Guangzhou as an example is analyzed through multiple stratigraphic level of 2D finite element method to the most dangerous conditions of the seepage field. The more reasonable and economic drainage decompression measures, " relief well + sand cushion", are put forward, which can meet the requirement of decompression. Finally, the monitoring data further validate the rationality of the method.

Keywords Drainage decompression · Uplift pressure · Relief well

1 Introduction

A large amount of groundwater is stored in the soil voids and rock fissures, and the groundwater will produce uplift pressure on the underground structures buried in or above the rock and soil bodies. The problem of excessive uplift pressure of underground structures is often not given enough attention to cause accidents. The reason for this is that the design does not consider the most unfavorable situation in the construction conditions, and the second is that the groundwater level exceeds the design water level (such as during heavy rainfall). Insufficient decompression will eventually lead to pit instability, pit surge, basement floor cracking, basement overall uplift and other problems, and may also affect the surrounding pipelines or neighboring buildings [1].

S. Li · Q. Wan
Guangzhou Environment Protection Investigation Co.,Ltd, Guangdong 510030, China

H. Yan · H. Pan (✉)
South China University of Technology, Guangdong 510641, China
e-mail: hpan@scut.edu.cn

© The Author(s) 2023
G. Feng (ed.), *Proceedings of the 9th International Conference on Civil Engineering*,
Lecture Notes in Civil Engineering 327,
https://doi.org/10.1007/978-981-99-2532-2_33

Especially for mega buildings on slopes in the southern rainy areas, the groundwater level is high, and the original slope seepage field forms a new seepage condition under the interference of mega buildings, and the influence on the basement becomes very complicated. If the value of basement water uplift pressure calculation is small, it will cause low safety reserve or even accidents, and if the value is conservative, it will lead to high construction cost [2].

Since the damage occurring in the uplift pressure problem is mainly infiltration damage of the soil, the terminology used in the literature of previous studies on infiltration damage is more varied and has changed considerably over time. Terzaghi in doing sheet pile weir impermeable sand trough model experiments, the downstream sand surface is floated by the top of the water flow phenomenon called pipe surge, indicating the general term for the infiltration deformation of the foundation soil [3, 4]. USACE used internal stability to define the ability of the filter layer to resist separation of coarse and fine particles and the formation of tubular channels [5]. Subsequently, Kenney and Lau defined the internal stability of the soil as the ability of the granular material to resist the loss of fine particles due to a disturbance, which can be seepage or vibration [6]. Kezdi first used the term suffusion to define this phenomenon: when seepage flows through the pores of a medium, the water carries away the fine particles of the medium but does not destroy the structure of the medium [7]. Mansur let piping is defined as: in the dike base permeable layer, due to water pressure and local channel concentration seepage resulting in the bottom of the dike or the overlying soil layer of sand particles or other soil particles are gradually eroded phenomenon [8]. Moffat and Fannin gradually unified the understanding that when fine particles in cohesionless soils are lost in the pores of coarse particles without changing the total volume of the soil, the phenomenon is called "suffusion"; and as the number of lost fine particles increases, it causes a reduction in the soil skeleton, which is called "suffosion" [9, 10].

For the calculation of depressurized wells, the steps for solving the multi-well system given by S.Y. Wu are: superimposing the solution of a single well and making it conform to the boundary conditions, listing the equations for the coefficients to be determined, and then solving for the coefficients to be determined to obtain the solution of the multi-well system. Based on the above method, for wells under general double-layered foundations, S.Y. Wu gave approximate and rigorous solutions for single-row wells [11]. Mao Changxi, on the other hand, gave the solution for a cluster of pressurized wells with arbitrary arrangement [12]. If the well cluster is arranged along a closed boundary, the precipitation funnel in the area of the well cluster is nearly circular, the radius of influence of each well is nearly equal, and the flow rate of each individual well is approximately the same, a simplified calculation of the well cluster equivalent to a circular distribution can be obtained [13]. In the design of the pit precipitation, the water barrier of the water stop curtain was considered by Lingao Wu, who treated the water stop curtain as a complete water barrier boundary and solved the unsteady seepage process from the beginning of pumping to stabilization in the pit [14].

In this paper, for a sloping basement residential district basement construction project in Guangzhou, the drainage decompression method based on the combination of water stop curtain and pressure reducing well is adopted, and the regional multi-layer horizontal surface seepage finite element calculation is used to simulate the seepage field, the monitoring of basement floor uplift pressure is carried out during the construction period, the drainage decompression scheme is developed, The applicability of this method to the problem of excessive uplift pressure is analyzed, which will provide useful experience for solving engineering problems.

2 Project Overview

The project is located in the eastern part of Luogang District, Guangzhou, and the site is surrounded by mountains on three sides, with the elevation of the top of the mountains being about 80 ~ 100 m, and located in a valley, the elevation of its low point is about 34.5 m, which is basically in the south west - north east direction. Due to the mountains on three sides, abundant groundwater converges to the site, causing surface overflow and forming fish ponds and ditches. There are roads around the site, the elevation of the southeast road is 39 ~ 45 m, the elevation of the northeast road is about 37 m, and the elevation of the west road is about 38 ~ 45 m. Figure 1 shows a typical geological profile.

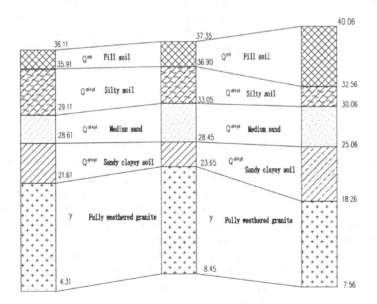

Fig. 1 Typical geological profile

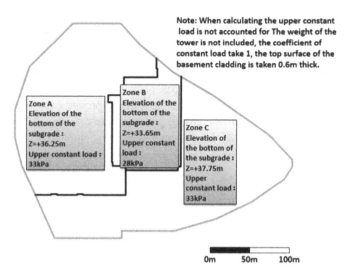

Fig. 2 Basement floor bottom elevation and upper constant load

2.1 Basement Burial Depth

Several buildings are proposed to be built in the site area, mostly 12-storey high, with single-storey integral basement in the lower part, the area of basement is about 56,000 m². the ± 0.00 elevation of the building is + 42.15 m. The elevation of basement floor is not uniform and there are leap levels, the main elevations are as follows: + 36.25 m (Zone A), + 33.65 (Zone B) and + 37.75 (Zone C), as shown in Fig. 2.

2.2 Hydrological Conditions

The groundwater types include upper layer stagnant water, pore water in the sand layer and fracture water in the rock layer: upper layer stagnant water is mainly stored in the fill, and the source of recharge is mainly atmospheric precipitation, which is obviously influenced by the season; diving is mainly stored in the sand layer, and the source of recharge is mainly atmospheric precipitation and fish pond recharge, which is obviously influenced by the season and locally has micro-pressure; fracture water in the rock layer is related to the fracture development of the bedrock and its connectivity. The fracture water in the rock layer is related to the fracture development of bedrock and its connectivity, and the main source of recharge is fish pond recharge, and the amount of recharge is obviously influenced by the degree of rock fragmentation and the degree of topographic relief.

2.3 Problems

The ground elevation, groundwater level, top elevation of the sand layer and bottom elevation of the sand layer in the analysis area were interpolated by the natural proximity method based on 140 geological drilling data holes in the site. From the interpolation results of the ground and water levels, the analysis of several typical sections shows that the permeable layer presents a pressurized nature and the basement is in the overburden layer, which does not penetrate deeply into the permeable layer and does not interfere with the hydraulic conductivity of the permeable layer. Comparing the burial depth of the basement and the water table, it can be seen that the basement is subjected to an average uplift pressure of 30 ~ 40 kPa, and the local foot of the hill is as high as 50 kPa. The water table was measured under the condition that it did not rain during the exploration period, and the problem of floating stability of the basement will be more prominent in the rainy season. Especially near the foot of the hill (south, southwest and northwest of the site) or at the lower basement floor, the basement is more threatened by the uplift pressure.

The preliminary analysis knows that the following two main problems should be solved: first, under unfavorable (heavy rainfall) conditions, seepage analysis of the site to derive the uplift pressure on the groundwater chamber; second, for the characteristics of the project, to propose effective, safe and economic drainage decompression measures.

3 Seepage Analysis

The seepage analysis is calculated by the regional multilayer horizontal surface seepage finite element method[15], and the calculation program is developed by Professor Cao Hong of South China University of Technology, which has been applied to many important water conservation projects, foundation excavation projects and urban groundwater environment research projects.

3.1 Calculation Area and Boundary Conditions

The boundary of the calculation area should be theoretically divided to the watershed. Since the road is opened at the side of the cell, the seepage water of the mountain ridge will overflow at the roadside during heavy rain, so the road is taken as the calculation boundary. The northeastern direction extends along the rill towards Nangang River, taking about 180 m, and the area of the calculation area is $130,000 m^2$.

The sand layer is only exposed in the middle of the site, and the sand layer is missing at both sides of the site, i.e. at the foot of the hill. There is no obvious demarcation line between the area with sand layer disclosure and the area with sand layer disappearance, so a transition area is set between the two areas in the calculation.

Considering that the overflow will occur at the foot of the mountain when the rainfall is high, the calculation boundary of various working conditions is taken as the head boundary under the most unfavorable conditions when calculating with the finite element model, i.e. the head value is equal to the boundary elevation.

3.2 Calculate the Model Soil Layer

In order to properly reflect the interrelationship of seepage characteristics of each soil layer, the site soil layers are simplified when establishing the calculation model. In the missing sand layer, the sandy clayey soil is treated as a strong permeable layer, and the silt, clay and fill soil are treated as a water barrier. In the transition zone, the sand layer and sandy clayey soil are treated as strongly permeable layers, and the silt, clay and fill soil are treated as water barrier layers. In the area with sand layer exposure, because the permeability coefficient of sand layer is much larger than that of sandy clay, the sand layer is regarded as a strong permeable layer, and the sandy clay, silt, clay and fill are regarded as a water barrier. The bedding layer is made of gravel and sand with better permeability, which is a strong permeable layer.

From the finite element calculation model, it can be seen that the distribution of the strongly permeable layer is in the shape of an obvious washout, and the strongly permeable layer at the foot of the hill is thin and located in the shallow part of the stratum, while the sand layer in the middle of the site is unevenly distributed with a large variation in thickness. The maximum thickness of the strongly permeable layer is 12.4 m, the minimum value is 0.5 m, and the average thickness is 5.5 m.

3.3 Selection of Calculation Parameters

The permeability coefficient of the sand layer in the finite element model calculation was selected based on the results of the field pumping test; the permeability coefficients of other soil layers were selected based on engineering experience; the fine sand was exposed in only four holes in the site and was not considered in the calculation. The values of permeability coefficient of each soil layer are shown in Table 1.

3.4 Original Working Condition Seepage Field Analysis

After finite element calculation, the seepage field of the original landform of the site during the storm period and after the basement was completed were obtained. For the

Table 1 Calculated parameters

Soil type	Permeability coefficient(cm/s)
Fill Soil	1.00×10^{-5}
Silt	1.00×10^{-5}
Clay	1.00×10^{-5}
Medium sand	1.30×10^{-3}
Sandy clayey soil	1.00×10^{-4}
Concrete substrate	1.00×10^{-10}

original geomorphology, the results of the distribution of equal head lines show that the groundwater seepage in the site is the seepage field of the pressurized aquifer, and the head is higher than the ground, and the head value is higher at the foot of the hill. The maximum head value of the proposed basement site is 40.5 m; there is a sand layer in the middle of the site to expose the area with a small hydraulic slope; the groundwater of the site flows from the foot of the hill in the southwest along the gully to the Nangang River in the northeast.

After the basement was completed, the basement played a role in suppressing the overflow, and the head congestion was about 1 m high; the head value of the basement floor bottom and the strong permeable layer was basically the same. The whole basement floor is subject to high head pressure, and the maximum head pressure is about 50 kPa in area B (see Fig. 2 for partition). In order to reflect the head distribution in the basement more visually. As shown in Fig. 3, four typical profiles were selected to analyze the results. the head values of all four profiles were higher than that of the basement floor. The profiles AB and CD are subjected to a maximum uplift pressure of about 50 kPa, and the head is higher at the foot of the hill as shown by the profile GH.

The comprehensive seepage field calculation results show that the bottom of the basement is subjected to a large uplift pressure, and the uplift pressure in most areas of the basement is greater than the upper load of 30 kPa, so drainage decompression measures must be taken. In the area without sand layer exposure, due to the small

Fig. 3 Typical profile

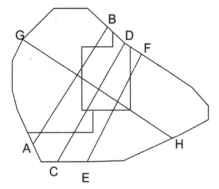

permeability coefficient, it is suitable to use anti-drawing piles. In the area of basement B, sand layer is exposed and the uplift pressure is up to 50 kPa, so it is not economical to use anti-drawing piles. It is recommended to use the combination of pressure reducing well and anti-drawing pile for anti-floating measures.

3.5 Recommended Scheme and its Seepage Field Analysis

In this project, the groundwater is mainly stored in the sand layer, and the head is controlled by the topography around the area, so it is not easy to drain by simple precipitation; the basement floor is above the sand layer, and the uplift pressure on the floor mainly depends on the head of the sand layer; the sand layer contains a lot of clay and powder particles, and the permeability is low, and it is missing at the foot of the hill, which is slightly unfavorable to the drainage and pressure reduction; the stratum at the floor is silt, silty soil or clayey soil, and the permeability is small. Based on the above characteristics, the drainage decompression scheme of "pressure reduction well + sand bedding" is recommended, as shown in Fig. 4.

The decompression wells were arranged in the B area where the elevation of the basement floor is low. The radius of the well is 0.6 m, and the bottom of the well is 1 m into the residual soil, which is 1 m lower than the bottom elevation of the basement floor in this area. to enhance the decompression effect of the decompression well, a 0.5 m thick bedding layer is arranged at the bottom of the basement floor in area B. Area B extends 50 m to area A and C, and a 0.15 m thick sand bedding layer is arranged. the elevation of the floor in areas A, B and C are different, and the

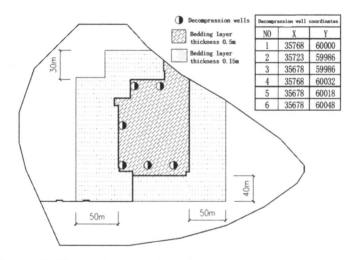

Decompression wells		
Bedding layer thickness 0.5m		
Bedding layer thickness 0.15m		

Decompression well coordinates		
NO	X	Y
1	35768	60000
2	35723	59986
3	35678	59986
4	35768	60032
5	35678	60018
6	35678	60048

Fig. 4 Layout of bedding and depressurization wells

sand should be backfilled outside the intersection side wall of each area to make it connected, so that the water can flow to the decompression well in area B.

The "decompression well + sand bedding" scheme was calculated by finite element to obtain the iso-head line as Fig. 5, and the pumping flow rate of the decompression well was 60 m³ /d. The head values in the figure are equal to the bottom elevation of the base plate in each area, and the contour lines are shown by thick dashed lines, from this it is clear that: 1)The bottom elevation of the bottom plate in area B is + 33.65 m, the calculated head value is 34.5 ~ 33.8 m, the uplift pressure is 8.5 ~ 1.5 kPa, the average uplift pressure is 5 kPa; 2)The bottom elevation of A area is 36.25 m, the calculated head value is 41 ~ 35 m, the uplift pressure is 48 k ~ 0 kPa, except for the local uplift pressure at the foot of the west side of the hill is big, most of the area uplift pressure is less than 20 kPa, and the uplift pressure of a small half of the area is 0 kPa. 3)The elevation of the bottom slab of area C is 37.75 m, the calculated head value is between 41 ~ 35 m, the uplift pressure is 33 ~ 0 kPa, except for the uplift pressure is big at a very small part at the foot of the southwest side of the mountain, the uplift pressure is less than 13 kPa in most of the area, and the uplift pressure is 0 kPa in a large part of the area.

It can be obtained that "pressure reducing well + sand layer" can effectively solve the anti-floating problem of basement slab, so that the uplift pressure of basement slab is mostly 0 or much smaller than the overlying load, and there is no need to set additional anti-drawing piles or anti-drawing anchors, which can effectively solve the problem and also save cost and improve the construction efficiency.

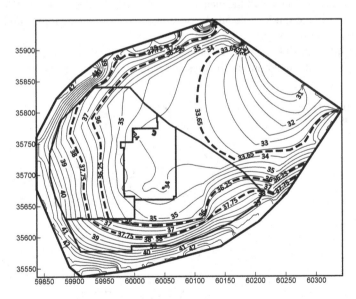

Fig. 5 Isohydraulic head line of strongly permeable layer after deployment of depressurization wells

4 Basement Floor Uplift Pressure Monitoring

For special reasons, the project could not implement drainage decompression measures in accordance with the recommended scheme in the previous section, and only raised the elevation of the basement slab to avoid uplift pressure damage. In order to ensure the safety of the project, the construction period to monitor the project basement floor uplift pressure.

Figure 6 shows the layout of 13 measurement points, half black circle marked as deep observation wells, full black circle for general observation wells. The water level is measured by the water level meter, which is converted into the head value of the basement floor, and then the uplift pressure of the basement floor is obtained.

From the completion of the basement floor pouring observation, a total of 25 times for a year. The frequency is about once every half month, and the observation frequency will be increased in case of heavy rain.

From the monitoring results of the pressure change law of the orifice of the pressure measurement well with time, we know that the groundwater head pressure is obviously affected by the season, and the head value of the orifice is not high from February to June, while it is higher from August to November. In addition, according to the method of the above section, the finite element analysis of the actual construction conditions of the anti-floating seepage, the head calculation value of the observation points and the actual monitoring results are compared and analyzed as shown in Fig. 7, it can be seen that the predicted head values of L1 and L2 sections are accurate. L3 section part of the measurement points exceeded the predicted value, from the site situation, it may be due to the southeast of the site near the application of mixing pile caused by the water level rise.

As a result, the results obtained from the simulations are more accurate and can correctly predict the head and uplift pressure values at the basement floor, but the

Fig. 6 Plan layout of
pressure measurement wells

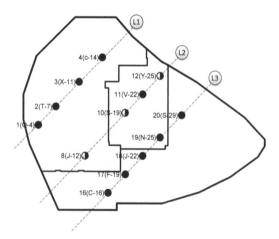

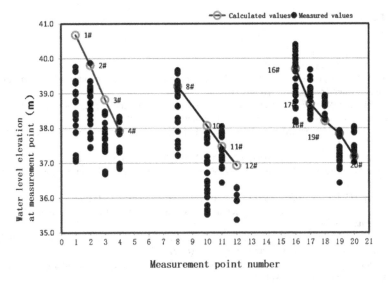

Fig. 7 Comparison of calculated head values and measured values

actual situation will also be affected by the surrounding environment and construction conditions, so the real construction should be analyzed in specific problems.

5 Conclusion

Based on the actual project, the head and uplift pressure values at the basement floor were obtained through numerical simulation, compared with the actual monitoring data, and the site seepage conditions and seepage field were analyzed, and the following conclusions were obtained.

(1) The uplift pressure checking calculation of the basement on the slope should analyze the seepage characteristics in detail according to the geological conditions of the site, and highlight its seepage characteristics through stratigraphic simplification. Using multi-layer horizontal surface finite element method for seepage field analysis, the results are effective and reliable.

(2) "Pressure reduction well + sand layer" measures have been shown by finite element calculations to be effective in uplift pressure reduction, which can make the basement floor uplift pressure mostly zero or much smaller than the overlying load.

(3) Monitoring results show that the calculation and analysis method to predict the head value results are more accurate, the specific analysis also need to consider the actual construction situation and the impact of the surrounding environment.

References

1. Dixuan Z (2007) Experimental study of underground structure floatation model. Shanghai Jiaotong University, Shanghai
2. Guocai Z (2010) Research on the calculation of buoyancy of underground structures on sloping ground. Water Transp. Eng. 10:88–91
3. Terzaghi K (1996) Failuare of dam foundations by piping and means for preventing it (in German). Die Wasserkraft, Special Forchheimer. 17:445–449
4. Changxi M (2005) Study of tube surge and filter layer: The part of tube surge. Geotechnics 26(2):209–215
5. U.S. Army Corps of Engineers, USACE (1953) Investigation of Filter Requirements for Underdrains. Technical Memo., No.3-360, U.S. Waterways Experiment Station, Vicksburg, Mississippi
6. Kenney TC, Lau D (1985) Internal stability of granular filters. Can Geotech J 22:215–225
7. Kezdi A (1979) Soil physics-selected topics. Elsevier Scientific Publishing Company, Amsterdam, 160p.
8. Mansur CI, Postol G, Sally JR (2000) Performance of relief well systems along Mississippi River levees. J Geotechn Geoenviron Eng 126(8):727–738
9. Moffat R, Fannin RJ (2011) A hydromechanical relation governing internal stability of granular soil. Can Geotech J 48(3):413–424
10. Moffat R, Fannin RJ (2011) Spatial and temporal progression of internal erosion in granular soil. Can Geotech J 48(3):399–412
11. Wu SY (1980) (Anhui Institute of Water Resources Science). Theory of seepage calculation for multi-layered foundations and decompression trench wells. Water Resources Press, Beijing
12. Mao C (2002) Computational analysis and control of seepage flow. China Water Conservancy and Hydropower Publishing House, Beijin, 2nd edn.
13. U.S. Army Corps of Engineers, USACE (1992) Design, construction and maintenance of relief wells. EM 1110-2-1914, Washington, D.C. 20314–1000
14. Wang R-D (1957) Slightly removing the effect of flow bed sheet pile thickness on the calculation of filtration flow (seepage). J Zhejiang Univ 3:9–27
15. Cao H, Zhang T et al (2003) Research on the application of seepage calculation of multi-layered strongly permeable foundation. J Rock Mech Eng 22(7):1185–1190

Model Testing Technique for Piles in Soft Rock Considering the Overlying Layers

Meiqi Liu, Guirong Li, Kunming Wu, Yuheng Wang, Xiaosen Zhang, and Bin Huang

Abstract Model test is a common method to study the bearing peculiarity of pile foundation. The influence of overlying soil thickness and overburden pressure on the bearing capacity of soft rock-socketed pile should be considered in the physical model test of mini piles in soft rock. In this paper, the influence of coverage on the bearing characteristics of rock-socketed sections is studied by finite element analysis, and the modelling method of equivalent overburden pressure is proposed. This method can be used to study the carrying peculiarity of soft rock-socketed pile and reveal the failure mechanism of pile tip. The development of pile model test technology considering overburden pressure promotes more scientific design methods for pile foundation.

Keywords Soft rock socketed pile · Model experiment · Coverage · Coverage pressure

1 Introduction

Rock-socketed pile is widely used in high-rise buildings and bridge engineering with high load and high settlement requirement. Soft rock socketed pile is a common type in China. When some scholars study the model test of overlying soil thickness and earth pressure on rock-socking piles in soft rock, the 1 g model test used in the laboratory is generally scaled down, which leads to the reduction of soil layer thickness, so the stress field of the model soil is difficult to reflect the in-situ dead weight stress. In order to reflect the in-situ stress field, it can be realized by the centrifuge model with high cost. When this test condition is not available, some scholars use air bag, pile load or rigid support to load the rock soil around the pile. Airbag load pressure range is limited, more than 400 kPa pressure air easy to stress concentration and damaged, hard to ensure the success rate. Pile load test need

M. Liu · G. Li · K. Wu · Y. Wang · X. Zhang · B. Huang (✉)
School of Architecture and Civil Engineering, Huizhou University, Huizhou, China
e-mail: huangbin@hzu.edu.cn

© The Author(s) 2023
G. Feng (ed.), *Proceedings of the 9th International Conference on Civil Engineering*,
Lecture Notes in Civil Engineering 327,
https://doi.org/10.1007/978-981-99-2532-2_34

space is large, is not convenient operation, low pressure range. Rigid support load applied load can be larger, but exist eccentric load, stress non-uniform problem in geotechnical engineering.

In this paper, the effect of coverage on the bearing capacity of rock-socketed part is discussed by using finite element method, and the equivalent simulation method of overburden pressure is proposed, the data of rock-socketed section is extracted and analysed. The aim is to simulate the in-situ stress field without limiting the free deformation of the soft rock around the pile and to ensure uniform loading. Therefore, how to simulate the in-situ stress field without limiting the free deformation of the soft rock around the pile and ensuring uniform loading is of great significance to reveal the bearing mechanism of the soft rock-socketed pile under different stress states.

2 The effect of Coverage On Bearing Capacity of Soft Rock Socketed Part and Model Testing Method

2.1 FEM Analysis of Effect of Coverage Layer on Standing Capacity of Soft Rock-Socketed Pile

The soft rock socketed piles with different burden thickness and pressure are simulated by finite element method, and the overburden is clay. By comparing the influence of different overburden thickness and overburden pressure on the carrying peculiarity of rock-socketed section, we are able to judge whether the impact of overburden and equivalent overburden pressure on the carrying peculiarity of rock-socketed section is consistent (Table 1).

Using axisymmetric model, 15-node triangular elements are adopted for pile [1], soft rock and clay overburden, and the thickness of soft rock below pile bottom is 10 m. The calculation model is shown in Fig. 1. The bottom of the model is a fixed

Table 1 FEM Analysis Scheme of Soft Rock Rock-socketed Piles

Test number	Diameter of pile (mm)	Rock-socketed length	Overburden pressure (kPa)	Thickness of cover (m)
1	800	6d	10	-
2	800	6d	100	-
3	800	6d	300	-
4	800	6d	500	-
5	800	6d	-	0.5
6	800	6d	-	5
7	800	6d	-	15
8	800	6d	-	25

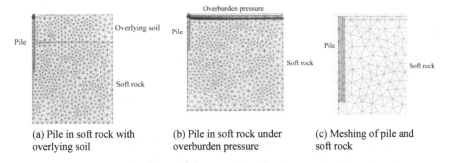

(a) Pile in soft rock with overlying soil

(b) Pile in soft rock under overburden pressure

(c) Meshing of pile and soft rock

Fig. 1 FEM Model and Meshing of Pile & Rock/Soil

boundary, both sides are fixed in the horizontal direction, and can move freely in the vertical direction.

In this example, the parameters of soft rock are selected with reference to the soft rock simulation sample, which is still in the linear elastic stage when its peak strength is 50%, so the deformation modulus is equal to the initial tangent modulus, instead of the confining pressure in the Janbu formula with the mean stress, the formula for calculating the deformation modulus E is as follows:

$$E = K p_a \left(\frac{p}{p_a} \right)^n \tag{1}$$

where K and n are the test parameters, for this test, 1959.3 and 0.1066, respectively; p is the mean stress; p_a is atmospheric pressure, 0.1 MPa.

For the case of overburden pressure or overburden self-weight stress is $\sum \gamma h$, the average stress P of soft rock can be calculated by the formula below:

$$p = \frac{1 + 2K_0}{3} \sum \gamma h \tag{2}$$

where K_0 is the static lateral pressure coefficient, $K_0 = 1\text{-sin } 43.4° = 0.313$; and $\sum \gamma h$ is the overlying pressure or the self-weight stress of the overlying layer.

When the overburden pressure is 10,100,300.500 kPa or the overburden thickness is 0.5,5,15,25 m respectively, the deformation modulus of soft rock is 144,184,206 and 218 MPa respectively. The calculation parameters of the soft rock rock-socketed pile model are exhibited in Table 2.

Table 2 Model Parameters

Materials	Constitutive model	γ (kN/m^3)	E (MPa)	μ	c (kPa)	φ (°)
Soft Rock	Mohr–coulomb	23	144,184,206,218	0.24	343	43.4
Pile	Linear elasticity	25	30,000	0.18	-	-
Clay	Mohr–coulomb	20	5	0.40	60	20.0

2.2 Effects of Different Overlying Soil Thickness and Different Overlying Pressure on Carrying Characteristics of Rock-Socketed Section

The calculation results of load-settlement on the top of rock-socketed section of pile foundation with different overburden thickness are exhibited in Fig. 2. The top load of the rock-socketed section is the axial force of the pile, and the top settlement of the rock-socketed part is the pile shaft subsidence of the buried depth. The larger the thickness of overburden, the smaller the settlement under the same load (axial force) on the top of rock-socketed section, and the load corresponding to the yield point of the load-settlement curve increases with the thickness of overburden, and the corresponding settlement decreases. The calculation results of load-settlement relationship on the top of rock-socketed sections with different overburden pressures are exhibited in Fig. 3. The pile top load is the upper load of the rock-socketed part, and the pile top settlement is the top subsidence of the rock-socketed section. With the increase of the overburden pressure, the settlement becomes smaller and smaller under the top load of the same rock-socketed part. And the greater the overlying pressure, the greater the load corresponding to the yield point of the load-settlement curve, and the corresponding settlement is smaller.

Fig. 2 Load-settlement Curve of Soft Rock-Socketed Pile with Different Overlying Soil Thickness

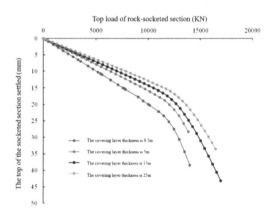

Fig. 3 Load-settlement
Curve of Soft Rock-Socketed
Pile with Different
Overburden Pressure

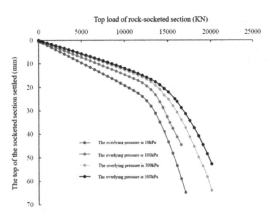

2.3 Equivalent Modelling of Overlying Soil Using Overburden Pressure for Piles in Soft Rock

Figure 4 shows the contrast of load-settlement curves at the top of rock-socketed segment of pile foundation with different covering thickness and equivalent overlying pressure. It can be seen that the top load and settlement relationship curve of rock socketed pile with upper coverage is basically the same as that of rock socketed pile without overburden after equivalent coverage pressure, and the load settlement curve after equivalent overburden pressure is slightly higher. This is because the distribution of dead weight load of the overlying soil layer on the rock-socketed top surface under the impact of pile is uneven in a small range, and the equivalent coverage stress homogenizes the upper load, which caused settlement is slightly smaller, but this discrepancy does not exceed 10%. This fully indicates that the effect of coverage on the bearing capacities of rock-socketed part can be equivalent to that of coverage pressure without the effect of deadweight stress of overburden.

3 Model Test Device of Micro Pile in Soft Rock Considering the Influence of Coverage Pressure

In order to directly understand the bearing mechanism of rock-socketed piles in soft rock, it is necessary to develop a physical model test device for the bearing mechanism of single pile which can be used for CT scanning. In order to simulate the soil stress field around a pile and visualize the deformation distribution, development process and failure mode of a single pile under vertical load, the bearing mechanism of a single pile is studied (Fig. 5). And the vertical bearing characteristics of the rock-socketed pile in soft rock are comprehensively interpreted [5].

Fig. 4 Comparison of top
Load-Subsidence
Relationship Curves of Soft
Rock-Socketed Pile with
Distinct Coverage Thickness
and Equivalent Overlying
Pressure

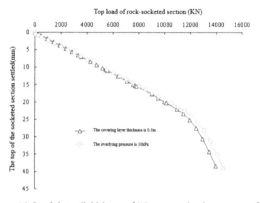

(a) Overlying soil thickness of 0.5m vs. overburden pressure of 10 kPa

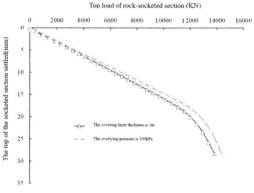

(b) Overlying soil thickness of 5m vs. overburden pressure of 100 kPa

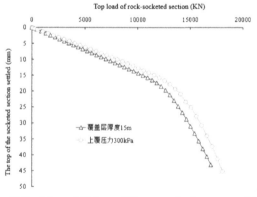

(c) Overlying soil thickness of 15m vs. overburden pressure of 300 kPa

Fig. 4 (continued)

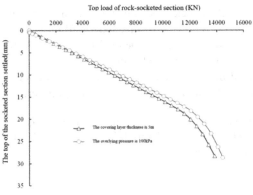

(d) Overlying soil thickness of 5m vs. overburden pressure of 100 kPa

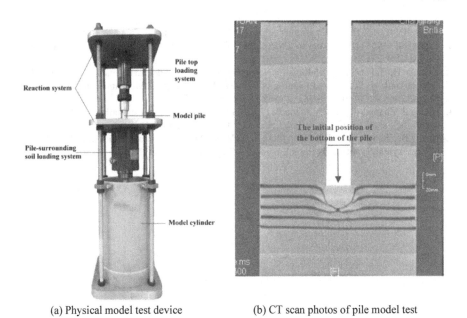

| (a) Physical model test device | (b) CT scan photos of pile model test |

Fig. 5 Application of Model Testing Technique for Piles in Soft Rock considering Overburden Pressure (Huang et al. 2019 [4])

4 Conclusions

When the overburden layer is equivalent to the overlying pressure of the relevant self-weight, due to the equivalent of the uniform load, the vertical displacement field and the vertical stress field of the rock-socketed depth of 1/4 ~ 1/3 pile length of the soft rock socketed pile will be different, while the vertical displacement field in the rock-socketed section and the lower part is almost the same as the vertical stress field. For the whole rock-socketed member, it is reasonable that the overburden is equivalent to the uniform overburden pressure, and the difference of the top settlement of the rock-socketed member caused by the uniform load is less than 10%. After the inner wall of the model barrel is treated with butter and foil to reduce friction, there is basically no effect of the edge wall effect. which provides a reliable technical guarantee for carrying out the model test of the carrying peculiarity of the model pile.

Acknowledgements This research was supported by the Technology Innovation Cultivation Special Fund (Grant No. pdjh2023b0497) and National undergraduate innovation and entrepreneurship training program (202210577015).

References

1. Plaxis BV (2018) Plaxis 2D reference manual. The Netherlands
2. Ladanyi B (1961) Discussion. In: Proceedings of 5th International Conference Soil Mechanics, Paris 3, 270–271
3. Kishida H, Takano A, Yasutomi Y, Nagatsura Y (1973) End bearing capacity of piles in sand ground. In: 8th Annual meeting of Japanese Society of Soil Mechanics and Foundation Engineering, vol 1, pp. 479–482
4. Huang B, Zhang YT, Fu XD, Zhang BJ (2019) Study on visualization and failure mode of model test of rock-socketed pile in soft rock. Geotech Test J 42(6):1624–1639
5. Huang B, Zhang YT, Fu XD, Zhang BJ (2021) Vertical bearing characteristics of rock socketed pile in a synthetic soft rock. Eur J Environ Civ Eng 25(1):132–151

Effect of Autoclaved Aerated Concrete on Dynamic Response of Concrete Gravity Dam Under Earthquakes

Fei Zhao, Shaoyu Zhao, and Shuli Fan

Abstract Autoclaved Aerated Concrete (AAC) is commonly used in lower floors buildings in low seismicity areas due to its lightweight property and high energy absorption capacity. This paper proposes a novel application of AAC as an effective seismic countermeasure in the reduction of vibrational energy for concrete gravity dam. According the vibrating characteristics and failure modes of gravity dam under earthquake excitation, AAC was placed in the upper zone of a gravity dam to reduce the seismic inertia force and consequently to increase the seismic safety of the dam. Dynamic responses of two non-overflow sections of a gravity dam were analyzed through finite element analysis utilizing a damaged plasticity constitutive model. The anti-seismic effect of using AAC in gravity dams is researched by inputting different kind of ground motion records. The comparison of the natural vibration characteristics, dam crest displacement, and dynamic damage of the dam were investigated. The results show that, AAC effectively improves seismic resistance of concrete gravity dams, particularly eliminating cracks in the concrete along reduced damage zones, through inertial force reduction and energy dissipation. The results warrant further considerations for applying AAC to gravity dams.

Keywords Autoclaved Aerated Concrete (AAC) · Concrete Gravity Dam · Earthquake · Mitigation of Earthquake Damage · Seismic Analysis

F. Zhao
National Disaster Reduction Center of China, Beijing 100124, China

S. Zhao
School of Civil Engineering, The University of Queensland, St. Lucia, QLD 4072, Australia

S. Zhao · S. Fan (✉)
State Key Laboratory of Coastal and Offshore Engineering, Dalian University of Technology, Dalian 116024, Liaoning, China
e-mail: shuli@dlut.edu.cn

© The Author(s) 2023   409
G. Feng (ed.), *Proceedings of the 9th International Conference on Civil Engineering*,
Lecture Notes in Civil Engineering 327,
https://doi.org/10.1007/978-981-99-2532-2_35

1 Introduction

Nearly 70% of the national total hydropower potential is concentrated in the southwest region of China which region is well known for its high seismic intensity. With strong earthquakes predicted to occur in the coming decades, enhancing the seismic capacity of a high dam is of great concern. For concrete gravity dams or Roller Compacted Concrete (RCC) gravity dams higher than 100 m, real accidents, shaking table experiments and numerical simulation results indicate that the downstream face with an abrupt slope change is the weakest point during a strong earthquake. The best known case is the 103 m high Koyna dam in India which was seriously damaged during the 1967 earthquake [1]. The largest damages were the horizontal cracks on both the upstream and the downstream surfaces of several monoliths. Other gravity dams that experienced strong earthquakes, such as the Hsingfengkiang dam, China, and Sefid-Rud dam, Iran, also experienced the similar failure modes. Experiments on small scale models [2–4] and numerical simulations [5–7] have effectively rendered the dynamic responses and failure mechanism of concrete gravity dams during seismic excitation [8]. All these researches illustrate that the upper region with abrupt slope changes will be damaged seriously during earthquake and the neck position of a concrete gravity dam is the most vulnerable to a strong earthquake.

Adding steel reinforcement in potential cracking zones is a common remedial measure to enhance the seismic performance of gravity dams. Long et al. [9] performed numerical analyses to evaluate the effect of the reinforcement on the seismic damage of a 160 m high gravity dam using a modified embedded-steel model. Wang et al. [10] conducted a series of experiments on a shaking table to investigate the dynamic damage reaction of RCC gravity dam with and without reinforcements. These works indicate that the steel reinforcement in dam improves the seismic resistance of gravity dam and limits the possible rapid cracks expansion throughout a dam structure. Installing post-tensioned anchors in dams is another widely-used method to enhance their stability and to control prospective failure paths. Leger and Mahyari [11] carried out the numerical studies of post-tensioning pre-stressed cable seismic strengthening of concrete gravity dams. Morin et al. [12] studied the global seismic response of a post-tensioned gravity dam model. Ghaemmaghami and Ghaemian [13], with the help of a shaking table, tested a model dam after being repaired and strengthened using epoxy grouting of cracks and the installation of post-tensioned anchors. However, these seismic measures require a great deal of steel and are cost-intensive to prevent seismic cracking in gravity dams for the reason that a large amount of steel bars will be consumed in mass concrete dam section to satisfy the needs of the seismic performance [14]. In the meantime, the application of reinforced steel bar will result in long construction period and notable difficulty in the construction and quality control of roller compared concrete. Other methods, such as placing air-cushion insulation [15] and bonding fiber-reinforced polymer materials [16] to dam surface to increase energy dissipation, are also costly.

Autoclaved aerated concrete (AAC) is a mixture cementitious product of quick-lime, water and sand, whose low density is formed by the existence of air bubbles in the mixture. The prominent advantages of AAC are its lightweight and high energy absorption capacity, which economize the design of bracing structures including the foundation and walls of lower floors and reduce the seismic inertial force of structures [17]. Experiments were carried out in Europe and the U.S. to evaluate the seismic behavior of AAC type of masonry [18–21]. Varela et al. [22] made a suite of seismic tests of AAC shearwall specimens and developed a rational procedure to determine seismic performance factors for the design of AAC structures. Based on the experimentally observed responses, the hysteretic characteristics and the maximum drift ratio of 1% were proposed to avoid collapse of AAC structures. Imran and Aryanto [23] observed that AAC blocks exhibited better seismic performance than conventional clay-brick masonry under cyclic loading. Costa et al. [24] carried out experiments and nonlinear analyses of AAC infill structures to obtain a reliable and authentic depiction of the lateral cyclic behavior of AAC masonry buildings. The AAC masonry panel displays the apparent displacement ductility capacity and the maximum ultimate drift for panels failing in flexure is up to 0.5%. Ferretti et al. [25, 26] performed several tests on AAC masonry and beams to obtain a complete and integrated characteristic of AAC masonry behavior with attention to the softening regime and to describe AAC masonry behavior under different loading regime, determining some useful and essential parameters and values of AAC masonry. Bose and Rai [27] tested the resistance properties and hysteretic behavior of a single-reinforced concrete frame with AAC infill masonry. The test results showed that AAC infill at lower story drifts undertook the majority of the lateral load. The frame with AAC infill was found the first crack in the infill panel. At same time boundary separation cracking appeared between the AAC infill and top beam. The significant damage induced in AAC infill contributed to the energy dissipation and hysteretic damping. These researches reveal that AAC has the characteristic of a good combination of mechanical strength and energy dissipation, making it applicable for the implement of shear walls in seismic zones. It is beneficial for reinforced frame buildings to improve the seismic response by infilling AAC masonry in the upper stories.

In this paper, considering the advantages of light weight and strong energy dissipation capacity of AAC, we conducted a series detailed studies on the feasibility and suitability of AAC used in gravity dam as a shock and vibration mitigation measure. AAC infills were implemented in a gravity dam to replace a part of RCC to decrease the weight of the upper dam and to reduce its earthquake inertial forces and vibration energy. Nonlinear numerical analyses were carried out to investigate the dynamic damage of two non-overflow sections with heights 102 m and 189 m with the consideration of vibrational energy dissipation brought by AAC under different earthquake excitation.

2 Materials and Methods

2.1 Plastic-Damage Constitutive Model

The concrete damaged plasticity model implemented in the software ABAQUS was based on the constitutive relationships proposed by Lublinear et al. [28]. To simulate cracking and crushing of concrete under dynamic or cyclic loadings, this model was later amended by Lee and Fenves [6]. Based on basic damage mechanics, the effective stress $\overline{\sigma}_{ij}$ is defined as:

$$\overline{\sigma}_{ij} = E^0_{ijkl} : \left(\varepsilon_{kl} - \varepsilon^p_{kl} \right) \tag{1}$$

where E^0_{ijkl} is the initial elastic stiffness tensor, ε_{kl} and ε^p_{kl} are the elastic strain tensor and plastic strain tensor, respectively.

Concrete experiences damage when the stress level reaches the failure stress. The damage causes the softening of concrete material stiffness. Considering that concrete destruction is mainly caused by tensile damage, only the role of tensile damage was considered in this paper. The damage can be expressed by the damage factor d. Accordingly, the relationship of Cauchy stress σ_{ij} and effective stress $\overline{\sigma}_{ij}$ is defined by:

$$\sigma_{ij} = (1 - d)\overline{\sigma}_{ij} = (1 - d)E^0_{ijkl} : \left(\varepsilon_{kl} - \varepsilon^p_{kl} \right) \tag{2}$$

The plastic flow potential function and yield surface make use of two stress invariants of the effective stress tensor, namely the hydrostatic pressure stress $\overline{p}$,

$$\overline{p} = -\frac{1}{3\overline{\sigma}_{ij}} : I \tag{3}$$

and the Mises equivalent effective stress $\overline{q}$,

$$\overline{q} = \sqrt{3/2 \left(\overline{S}_{ij} : \overline{S}_{ij} \right)} \tag{4}$$

where $\overline{S}_{ij}$ is the effective stress deviator, defined as

$$\overline{S}_{ij} = \overline{\sigma}_{ij} + \overline{p}I \tag{5}$$

As for as effective stresses, the yield function equation takes the following form:

$$F = \frac{1}{1 - \alpha} \left(\overline{q} - 3\alpha\overline{p} + \beta(\varepsilon^p)\langle \hat{\overline{\sigma}}_{max} \rangle - \gamma \langle -\hat{\overline{\sigma}}_{max} \rangle \right) - \overline{\sigma}_c \left(\varepsilon^p_c \right) = 0 \tag{6}$$

$$\gamma = \frac{3(1 - K_c)}{2K_c - 1} \tag{7}$$

where α and β are dimensionless material constants, $\widehat{\overline{\sigma}}_{max}$ is the maximum principal effective stress, K_c is the strength ratio of concrete under equal biaxial compression to triaxial compression, and $\overline{\sigma}_c\left(\varepsilon_c^p\right)$ is the effective compressive cohesion stress.

The concrete damage plasticity model presumes non-associated potential plastic flow,

$$\dot{\varepsilon}_{ij}^p = \dot{\lambda}\frac{\partial G\left(\overline{\sigma}_{ij}\right)}{\partial \overline{\sigma}_{ij}} \tag{8}$$

The flow potential G used for this model is the Drucker-Prager hyperbolic function:

$$G = \sqrt{(\in \sigma_{t0}\tan\psi)^2 + \overline{q}^2} - \overline{p}\tan\psi \tag{9}$$

where $\psi(\theta, f_i)$ is the dilation angle measured in the $p - q$ plane at high confining pressure, σ_{t0} is the uniaxial tensile stress at failure, and $\in$ is a parameter, referred to as the eccentricity, that defines the rate at which the function approaches the asymptote.

2.2 Stress–Strain Laws in Uniaxial Tension of AAC

Performance of AAC infill was gauged by Imran and Aryanto [23] from the strength and deformation characteristics, the observed hysteretic energy dissipation capacity and the measured ductility through a series tests. However, the investigation on the stress–strain relationship and on constitutive mode of AAC under axial load, simulating seismic forces, is still limited. Few research provided a complete description of AAC masonry nonlinear fracture behavior [25, 29, 30], including inelastic parameters governing plastic stress–strain relationship. The AAC used in this paper is supplied by Dalian Tangjia Modern Building Material Co., LTD. The density ρ is 550 kg/m^3, the average compressive strength f_c is 3.1 MPa, and the elastic modulus E is 1320 MPa. For the experimental condition limitation, tensile strength and Poisson's ratio were not obtained. The direct tensile strength $f_{ct} = 0.54$ MPa, and the Poisson's ratio $\gamma = 0.38$ were found in the reference [25] for the reason that the density, compressive strength and the modules of the AAC are similar to those in this reference. The tensile properties of AAC are shown in Fig. 1.

3 Model Description and Design for Reducing Vibration

The Huangdeng gravity dam was selected in this paper. The dam is located on the upstream of Lancang River in Yunnan Province of China. Two typical non-overflow sections were selected for analysis, as shown in Fig. 2: Section A is a typical right

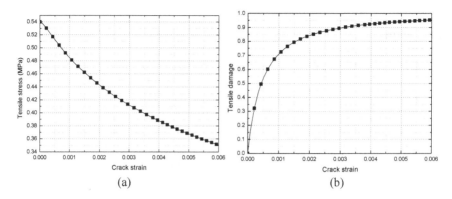

Fig. 1 AAC Tensile properties: **a** tensile stiffening; **b** tension damage

bank monolith 102 m tall with a 97 m deep reservoir, and Section B is a typical left bank monolith 189 m tall with a 184 m deep reservoir.

As mentioned above, cracks generally appear in the upper part of a gravity dam body during a seismic event, as it suffers from a whipping effect under earthquake excitation due to the sudden change of structural stiffness at the top of gravity dam. The acceleration at the crest of a gravity dam is 6 to 10 times that experienced at the dam heel [10, 31]. Thus, reducing the large inertial force in the upper part of a gravity dam is an effective method to decrease the damage sustained by a gravity dam during an earthquake. According the envelopes of maximum and minimum principal stresses for linear dynamic analysis of gravity dams [32], the middle region

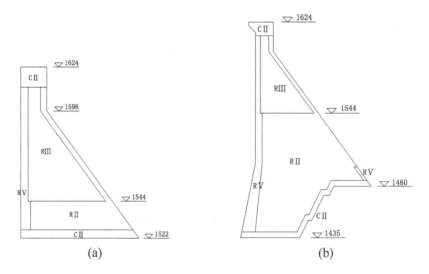

Fig. 2 Geometry and material partitions of two gravity dam sections: **a** Section A: the typical profile of right bank monolith; **b** Section B: the typical profile of left bank monolith

in the monolith is in low stress state, the larger tensile and compressive stresses just distribute at the surface of upstream and downstream. Therefore, low strength concrete can be used in the middle zones. And even these regions are designed to be empty, such as in the case of hollow gravity dam. Based on this characteristic, the authors first proposed the idea that AAC may be used in gravity dam as a seismic countermeasure to decrease the mass of the upper part. As a result, the inertial force of the upper part under earthquake excitation will be decreased. As shown in Fig. 2, the RIII partitions of the RCC dam were designed using low strength RCC. In this paper RCC in these zones were replaced by AAC in the new designed dam (named AAC dam) as a seismic countermeasure.

The plane stress conditions were assumed to conduct two-dimensional analyses of the non-overflow monolith. A finite foundation model was utilized and the finite foundations were extended three times the heights of the selected dam sections both in upstream, downstream and downwards directions. Considering energy dissipation resulted from the radiation damping at the far field and eliminating the effect of wave reflecting on artificial boundary, infinite elements were utilized at the boundaries of the finite foundation. The dam and the part of foundation which was in 20 m far away from the dam base were regarded as plastic-damage material. The partitions of dam concrete material were shown in Fig. 2. The mechanical parameters of materials were listed in Table 1. A dynamic amplification factor of 1.3 is took into account for the tensile and compressive strength to explain strain rate effects.

With the given dynamic tension softening curve and elastic properties of concrete and rock foundation, nonlinear dynamic analyses of damage development procedure in the gravity dam were conducted for different values of Peak Ground Acceleration (PGA). According the Specifications for Seismic Design of Hydraulic Structures in China (DL 5073-2000), three actual strong ground motion records at least should be used to research the failure mode of a hydraulic structure and evaluate its safety under earthquake excitation. In this paper, three typical ground motions recorded in the site of Class II were selected and applied in the stream and vertical direction of the dam-foundation system by taking account of duration, frequency characteristic and peak index of energy distribution. The typical strong ground motions are the records of Dec. 11, 1967 Koyna earthquake, the Taft record of the Kern County earthquake which was obtained on July 21, 1952 Kern county, California and the Qianan record which was obtained in Tangshan aftershock occurs on Aug. 9, 1976. The records

Table 1 The material mechanics parameters of dam and bedrock

Material partition	CII	RII, RV	RIII	AAC	Bedrock
Dynamic elastic modulus (MPa)	33,150	33,150	28,600	1716	8000
Poisson ratio	0.167	0.167	0.167	0.38	0.25
Density (kg/m^3)	2400	2400	2400	550	2610
Static compressive strength (MPa)	13.2	13.2	9.9	3.1	30
Dynamic compressive strength (MPa)	17.16	17.16	12.87	4.03	45
Dynamic tensile strength (MPa)	1.72	1.72	1.287	0.54	1.5

of these ground motions were given in Fig. 3. Keeping the same waveform, the amplitude of the acceleration time histories was increased step by step until the crack propagated through the dam. The PGA of Koyna horizontal record was adjusted to 502 gal for Section A as the exciting load, and 402 gal for Section B. The PGA of Taft record and Qianan record are 176 gal and 351 gal for Section A, respectively. The vertical earthquake acceleration representative value is two-thirds of the horizontal acceleration representative value according the provisions of DL 5073-2000. The dynamic analysis of the dam-foundation system was conducted considering the static analysis results as the initial conditions including gravity loads and hydrostatic loads.

The hydrodynamic effect of reservoir water was modelled using the method of additional mass [33]. The additional mass per unit area of the upstream wall was given in approximate form by the expression:

$$P_w(y) = \frac{7}{8}\rho_w\sqrt{h_w(h_w - y)}, \quad y \leq h_w \tag{10}$$

where $\rho_w = 1000$ kg/m^3 is the density of water, h_w is the depth of water, and y is the height from water surface to the location. It is assumed that the hydrodynamic pressures derived from the vertical ground acceleration component are neglected in all the numerical analysis for the reason that they are small.

4 Seismic Analysis

4.1 Dynamic Characteristics of Different Dam Sections

At first the changes on the dynamic characteristics of the selected dam sections were investigated. The first twenty natural frequencies of the AAC dam and RCC dam were shown in Fig. 4 and the first five frequencies were list in Table 2.

Since stiffness and mass changes simultaneously, the different order frequencies and different sections have an irregular rate of change. For section A, the first natural frequency of AAC dam is 1.56 Hz, decreasing 1.65% compared with that of RCC dam which is 1.58 Hz. However, the third natural frequency increases from 4.15 Hz to 4.21 Hz when the RCC in RIII district is replaced by AAC. As for Section B with 189 m height, the first natural vibration frequency of AAC dam is 1.03 Hz, which is 3% higher than that of RCC dam. All other frequencies decrease after the RCC is replaced by AAC in the upper part of Section B. The first three natural frequencies oscillate within a small amplitude that varies no more than 5% after RCC replaced by AAC in the RIII district of the selected sections. But for high frequencies, the vibration frequencies of AAC dams reduce quickly for the reason that the elastic modulus of AAC is extremely small. The rate of change is more than 20%, especially for Section A with 102 m height. The mass participation factor of AAC dam in vertical direction decreases in the second mode and increases in the

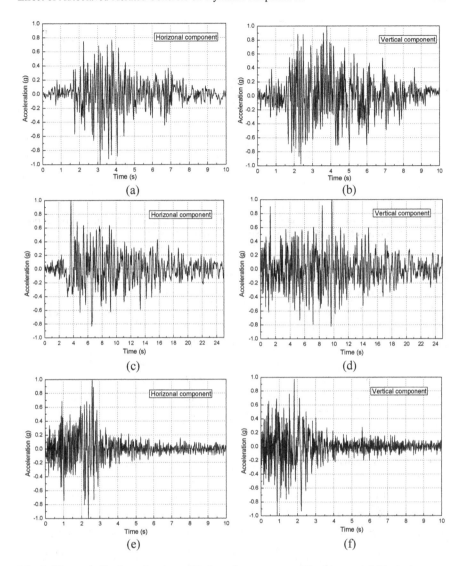

Fig. 3 The normalized acceleration: **a** Horizontal component of Koyna record; **b** Vertical component of Koyna record; **c** Horizontal component of Taft record; **d** Vertical component of Taft record; **e** Horizontal component of Qianan record; **f** Vertical component of Qianan record

third mode compared with the RCC dam, as shown in Table 3. Therefore, for the AAC dam, the first and second modes vibrate in the horizontal direction, and the third mode vibrates in the vertical direction.

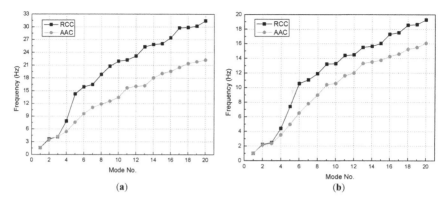

Fig. 4 Influence of AAC on the natural frequency of gravity dam: **a** Natural frequency of Section A: 102 m monolith; **b** Natural frequency of Section B: 189 m monolith

Table 2 The first five natural frequencies of gravity dam infilled with RCC and AAC

Mode No.	Section A			Section B		
	RCC dam (Hz)	AAC dam (Hz)	Change rate (%)	RCC dam (Hz)	AAC dam (Hz)	Change rate (%)
1	1.58	1.56	−1.65	1.00	1.03	3.05
2	3.65	3.49	−4.43	2.21	2.14	−3.13
3	4.15	4.21	1.34	2.47	2.36	−4.59
4	7.85	5.37	−31.53	4.40	3.52	−20.12
5	14.26	7.58	−46.81	7.41	4.96	−33.07

Table 3 The mass participation factors in x component and y component

Mode No.	Section A				Section B			
	RCC dam		AAC dam		RCC dam		AAC dam	
	X	Y	X	Y	X	Y	X	Y
1	0.75	0.02	0.66	0.01	0.76	0.02	0.72	0.02
2	0.08	0.85	0.21	0.0007	0.08	0.87	0.23	0.002
3	0.14	0.12	0.03	0.91	0.13	0.11	0.01	0.97
4	0.02	0.01	0.09	0.05	0.02	0.00	0.03	0.01
5	0.002	0.0005	0.01	0.0004	0.002	0.00001	0.01	0.0023

4.2 Time History Analysis of Displacement and Acceleration

The stream displacement time histories of the dam crest deviated from the values corresponding to static loads are shown in Fig. 5. The black line is the original dam response, whereas the red line is the response of the AAC dam. The positive

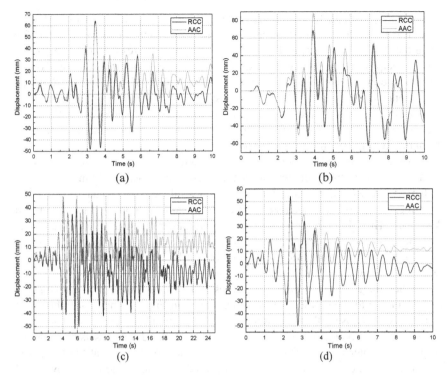

Fig. 5 Time history graphs of the horizontal displacement at the dam crest: **a** Displacement response of Section A (Koyna record); **b** Displacement response of Section B (Koyna record); **c** Displacement response of Section A (Taft record); **d** Displacement response of Section A (Qianan record)

direction of the horizontal displacement is in the stream direction. Comparing the curves, it can be found that the dynamic displacement response has less sensitivity on the replacing RCC by AAC. This means that the AAC infill has little effect on the dynamic deformation of the selected dam sections.

For the low gravity dam as Section A excited by Koyna record, the maximum dynamic displacement of AAC dam is 62.55 mm which occurred at 3.48 s, decreasing 1.64 mm compared with the results of the RCC dam, as shown in Fig. 5(a). A large deformation dam appears in AAC dam and the residual deformation at the dam crest is 20.36 mm at the end of the earthquake.

For Section B (height = 189 m), as shown in Fig. 5(b), the maximum displacement caused by the Koyna record increases from 68.62 mm to 88.32 mm as a result of the stiffness deterioration of structures when the RCC in RIII region is replaced by the AAC. Other than the increasing amplitudes at the peak points, there is only a slight difference in the horizontal displacement computed using AAC dam model and RCC dam model. The application of AAC has less effect on the dynamic deformation of a high gravity dam than that of a low gravity dam.

Comparing with the RCC dam, the first large dynamic displacement of AAC dam Section A in the stream direction increases from 43.93 mm to 48.33 mm at 4.13 s when the Taft record is applied in the structures. In Fig. 5(c) it can also be found that the reciprocating center axis of AAC dam is mainly concentrated in the 10 mm vicinity of downstream direction due to the long duration of the Taft record, which is different with the results of RCC dam. Residual deformation of AAC dam appears on the dam crest due to the light weight and low elastic modulus of AAC, which changes from -13.8 mm (upstream direction) to 10.48 mm (downstream direction) when the RCC in RIII region is replaced by AAC.

As shown in Fig. 5(d), the maximum displacement of AAC dam Section A in the stream direction is 49.35 mm which occurs at 2.40 s, decreasing 4.74 mm compared with that of the RCC dam when the dam section is subjected to Qianan record. After that time, the divarication emerges between the dynamic responses of RCC dam and AAC dam. The vibration range of AAC dam is smaller than that of RCC dam. The larger horizontal displacement appears the tendency to stream direction appears when the AAC is used in RIII partitions. At the end of earthquake, the residual deformation reaches 12.39 mm.

The horizontal acceleration time histories of dam crest for the RCC dam and the AAC dam due to the earthquake loads are shown in Fig. 6. When the lightweight material AAC is used, the amplitude of acceleration at dam crest oscillates with a slight change compared with that of RCC dam under different seismic motions despite the lighter upper part of the block. The most reason is that the three-first natural frequencies have small variation of amplitudes as mentioned above. And for a gravity dam, the first three natural frequencies play a dominant role in the dynamic response of gravity dam under seismic excitation. The replacement of RCC in RIII partitions by AAC does not change the acceleration response of gravity dam obviously. However, the density of concrete in upper part of gravity dam decreases from 2400 kg/m^3 to 550 kg/m^3. Therefore, the inertial force of the upper part will be reduced less than one quarter under the same earthquake excitation after RCC in RIII partitions is replaced by AAC.

4.3 Seismic Wave Effects on the Damage of the Dams

A strong earthquake often causes stiffness and strength deterioration of structures. In this section, the effects of AAC application on the dynamic damage response of the dam-reservoir-foundation system were investigated by comparison of the AAC dam and RCC dam dynamic solutions. The influence of upper block weight on the damage of gravity dam was researched to investigate the seismic performance of AAC with different spectral characteristics.

Figures 7 and 8 show the accumulated tensile damage profiles of the RCC dam and AAC dam Section A with different ground motions. A penetrating crack is found at the neck position of the RCC dam propagating from the downstream face to the upstream face at approximately a 45° angle to the vertical under the Koyna record

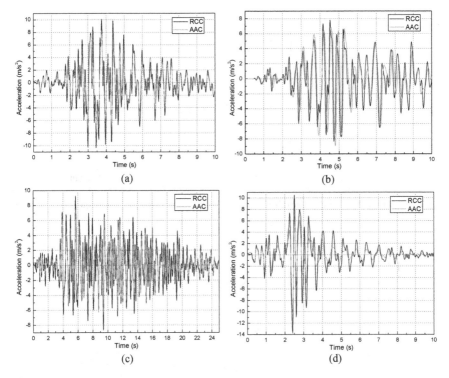

Fig. 6 Time history graphs of the horizontal acceleration at the dam crest: **a** Acceleration response of Section A (Koyna record); **b** Acceleration response of Section B (Koyna record); **c** Acceleration response of Section A (Taft record); **d** Acceleration response of Section A (Qianan record)

excitation, similar to the failure modes of the Koyna dam. At the same time, another crack appears at the heel of dam, however it propagates at approximately a 45° angle into the foundation, different with the damage profile reported in literatures [6, 32] which the foundation was assumed as a rigid body. As shown in Fig. 7(b) and (c), a similar damage profiles appear in the RCC dam Section A when the Taft record and Qianan record are applied in the low gravity dam section. However, the penetrating cracks at neck position appear a furcation in the interior of dam body, and more damage zones are found at the downstream surface under the Taft and Qianan records excitation.

For the AAC dam of Section A, damages are mainly concentrated in the dam heel and the contact layer of RCC and AAC at the upstream surface no matter which ground motion record is applied. The initial cracking profiles are nearly horizontal, and then extend into the foundation. The crack on the upstream surface extends into dam body about 4 m along the contact layer of RCC and AAC for the reason of stiffness deterioration. The cracking profiles of AAC dam in these two zones are more serious than those of RCC dam, as seen in the comparison of Fig. 7 and 8. The main reason is that the weight of the upper block of AAC dam is less than

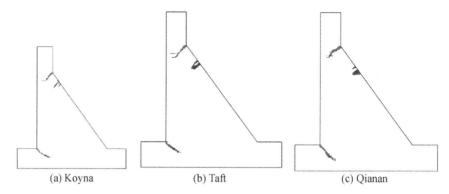

(a) Koyna   (b) Taft   (c) Qianan

Fig. 7 Evolution of tensile damage of RCC dam Section A under different ground motion excitation

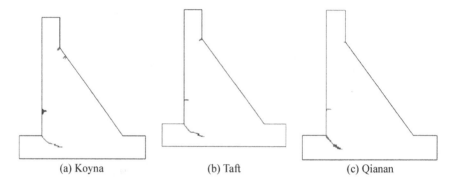

(a) Koyna   (b) Taft   (c) Qianan

Fig. 8 Evolution of tensile damage of AAC dam Section A under different ground motion excitation

that of RCC dam. However, the throughout crack at the slope change at downstream disappears with the application of AAC. Thus it can be inferred that AAC application is effective in decreasing the dynamic response of gravity dam under different ground motions excitation, consequently inhibiting damage propagation through the energy dissipation due to the plastic deformation of AAC.

4.4 Dam Height Effects on Damage of the Dams

In this section, the seismic damage analyses of the high dam Section B subjected to the Koyna record were performed. The dynamic responses of RCC dam and AAC dam were examined to evaluate the effects of dam height on AAC seismic performance.

For the RCC dam of Section B, damages are mainly concentrated in the vicinity of discontinuity at the upstream and downstream surface in addition to the dam heel, as shown in Fig. 9. There are two cracks at the downstream surface. One crack extends completely across the upper section. Due to the high compressive stress at the dam heel of the high dam section, there is less damage of Section B at dam heel than those of Section A. At the slope change of the upstream face, the crack almost propagates at a horizontal angle.

The damage propagations of AAC dam are different with those of RCC dam as shown in Fig. 9 and Fig. 10. No throughout crack occurs in the upper part of dam at the slope change for the AAC dam because of the light weight and the energy dissipation property of the AAC. However, as a tradeoff, the base crack propagates more in the case of AAC. There is another crack that appears at the upstream surface at the contact layer of AAC and RCC. These Figures show that the application of AAC in upper part helps to reduce the damage zone and retain the integrity of the dam's upper part, which is prone to cracking.

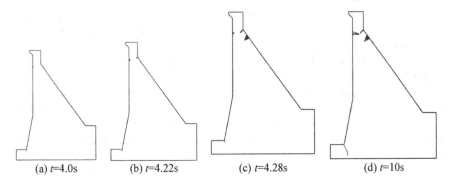

(a) *t*=4.0s   (b) *t*=4.22s   (c) *t*=4.28s   (d) *t*=10s

Fig. 9 Evolution of tensile damage of RCC dam Section B under Koyna record

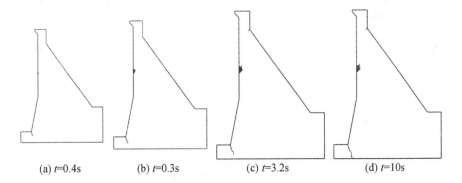

(a) *t*=0.4s   (b) *t*=0.3s   (c) *t*=3.2s   (d) *t*=10s

Fig. 10 Evolution of tensile damage of AAC dam Section B under Koyna record

4.5   Discussion of Results

The results mentioned above show that the horizontal displacement of RCC dam of section A oscillates with a shaft line. However, the displacement of AAC dam of section A basically increases and the oscillating axis shifts toward the downstream direction under different ground motions. Furthermore, earthquake duration has obvious influence on the displacement response of the gravity dam. Because of the high energy generated by the long duration of Taft record, displacement curves between AAC dam and RCC dam appear apparent separation at the time of approximately ten second of earthquake record.

In these numerical studies, the residual deformation at the crest of low gravity dam increases at the end of ground motions after the RCC in the upper part of gravity dam is replaced by AAC. There is a slight difference for high gravity dam in the dynamic deformation computed using AAC dam model and RCC dam model. Moreover, no matter how high the gravity dam is, the amplitude of acceleration at dam crest has a slight change between the RCC dam and AAC dam. So, it seems that AAC is more suitable to be used in high gravity dam to reduce the inertial force and vibrational energy.

The dynamic damage analyses of dams have shown that the weak parts of RCC gravity dam structures under seismic excitation is mainly concentrated in the vicinity of discontinuity at the upstream and downstream surface and the dam heel. However, the penetrating crack disappears at the neck position for AAC dam. It has been shown that the AAC plays a significant role of seismic response reduction and improves the seismic performance of the gravity dam body through energy dissipation. At same time the application of AAC leads to greater accumulate damage at the dam heel and the upstream surface at the contact layer of AAC and RCC, which calls for special aseismic design attention. Additional techniques should be used to guarantee sufficient strengthening for these locations. Optimizing studies on the location and region size of AAC should be further conducted to investigate the application of AAC in gravity dam as an aseismic measure for vibration reduction and damage mitigation.

5   Conclusions

This paper researched the feasibility of AAC as a seismic countermeasure to enhance the aseismic performance of gravity dams. The effect of AAC on the dynamic responses and failure modes of different height gravity dams under different ground motions was studied. Due to the lightweight and high energy absorption capacity of AAC material, the earthquake response of a gravity dam can be mitigated by replacing the RCC in upper low stress region to AAC. The application of AAC reduces the damage zone near the upper region with abrupt slope changes of the gravity dam sections. AAC is a suitable seismic countermeasure for improving the seismic performance of gravity dam. However, the weight decrease leads to greater accumulated damages at the dam heel, which calls for special aseismic design attention. In authors' future work, additional optimizing analyses on the location and region size of AAC will be performed.

Acknowledgements This work was partially supported by the National Key Research and Development Program of China (Grant No. 2017YFC1503002). The authors would like to thank them for their financial support. Thank Mrs. Ing Lim for professional editing and English corrections.

References

1. Chopra A, Chakrabarti P (1973) The Koyna earthquake and the damage to Koyna dam. Bull Seismol Soc Am 63(2):381–397
2. Donlon WP, Hall JF (1991) Shaking table study of concrete gravity dam monoliths. Earthq Eng Struct Dyn 20(8):769–786
3. Harris DW, Snorteland N, Dolen T, Travers F (2000) Shaking table 2-D models of a concrete gravity dam. Earthq Eng Struct Dyn 29(6):769–787
4. Mridha S, Maity D (2014) Experimental investigation on nonlinear dynamic response of concrete gravity dam-reservoir system. Eng Struct 80:289–297
5. Bhattacharjee SS, Leger P (1993) Seismic cracking and energy dissipation in concrete gravity dams. Earthq Eng Struct Dyn 22(11):991–1007
6. Lee J, Fenves GL (1998) A plastic-damage concrete model for earthquake analysis of dams. Earthq Eng Struct Dyn 27(9):937–956
7. Tinawi R, Leger P, Leclerc M, Cipolla G (2000) Seismic safety of gravity dams: from snake table experiments to numerical analyses. J Struct Eng-ASCE 126(4):518–529
8. Wang G, Wang Y, Lu W, Zhou C, Chen M, Yan P (2015) XFEM based seismic potential failure mode analysis of concrete gravity dam-water-foundation systems through incremental dynamic analysis. Eng Struct 98:81–94
9. Long Y, Zhang C, Xu Y (2009) Nonlinear seismic analyses of a high gravity dam with and without the presence of reinforcement. Eng Struct 31(10):2486–2494
10. Wang M, Chen J, Fan S, Lv S (2014) Experimental study on high gravity dam strengthened with reinforcement for seismic resistance on shaking table. Struct Eng Mech 51(4):663–683
11. Leger P, Mayahari A (1994) Finite element analysis of post-tensioned gravity dams for floods and earthquakes. Dam Eng 5:5–27
12. Morin PB, Leger P, Tinawi R (2002) Seismic behavior of post-tensioned gravity dams: shake table experiments and numerical simulations. J Struct Eng 128(2):140–152
13. Ghaemmaghami AR, Ghaemian M (2010) Shaking table test on small-scale retrofitted model of Sefid-rud concrete buttress dam. Earthq Eng Struct Dyn 39(1):109–118
14. Hall JF, Dowling MJ, El-Aidi B (1992) Defensive earthquake design of concrete gravity dams. Dam Eng 3(4):249–263
15. Liu H, Zhang S, Chen J, Sun M, Sun L, Li Y (2011) Simulation analysis theory and experimental verification of air-cushion isolation control of high concrete dams. Sci China Technol Sci 54(11):2854–2868
16. Zhong H, Wang N, Lin G (2013) Seismic response of concrete gravity dam reinforced with FRP sheets on dam surface. Water Sci Eng 6(4):409–422
17. Narayanan N, Ramamurthy K (2000) Structure and properties of aerated concrete: a review. Cem Concr Compos 22(5):321–329
18. Tanner JE, Varela JL, Klingner RE, Brightman MJ, Cancino U (2005) Seismic testing of autoclaved aerated concrete shearwalls: a comprehensive review. ACI Struct J 102(3):374–382

19. Penna A, Magenes G, Calvi GM, Costa AA (2008) Seismic performance of AAC infill and bearing walls with different reinforcement solutions. In: Proceeding of the 14th international brick and block masonry conference, Sydney, Australia
20. Siddiqui UA, Sucuoglu H, Yakut A (2015) Seismic performance of gravity-load designed concrete frames infilled with low-strength masonry. Earthq Struct 8(1):19–35
21. Rosti A, Penna A, Rota M, Magenes G (2016) In-plane cyclic response of low-density AAC URM walls. Mater Struct 49(11):4785–4798
22. Varela JL, Tanner JE, Klingner RE (2006) Development of seismic force reduction and displacement amplification factors for autoclaved aerated concrete structures. Earthq Spectra 22(1):267–286
23. Imran I, Aryanto A (2009) Behavior of reinforced concrete frames in-filled with lightweight materials under seismic loads. Civ Eng Dimension 11(2):69–77
24. Costa AA, Penna A, Magenes G (2011) Seismic performance of autoclaved aerated concrete (AAC) masonry: from experimental testing of the in-plane capacity of walls to building response simulation. J Earthq Eng 15(1):1–31
25. Ferretti D, Michelini E, Rosati G (2015) Cracking in autoclaved aerated concrete: experimental investigation and XFEM modeling. Cem Concr Res 67:156–167
26. Ferretti D, Michelini E, Rosati G (2015) Mechanical characterization of autoclaved aerated concrete masonry subjected to in-plane loading: experimental investigation and FE modeling. Constr Build Mater 98:353–365
27. Bose S, Rai DC (2016) Lateral load behavior of an open-ground-story RC building with AAC infills in upper stories. Earthq Spectra 32(3):1653–1674
28. Lubliner J, Oliver J, Oller S, Onate E (1989) A plastic-damage model for concrete. Int J Solids Struct 25(3):299–326
29. Trunk B, Schober G, Helbing AK, Wittmann FH (1999) Fracture mechanics parameters of autoclaved aerated concrete. Cem Concr Res 29(6):855–859
30. Yang K, Lee K (2015) Tests on high-performance aerated concrete with a lower density. Constr Build Mater 74:109–117
31. Chen J, Wang M, Fan S (2013) Experimental investigation of small-scaled model for powerhouse dam section on shaking table. Struct Control Health Monit 20(5):740–752
32. Calayir Y, Karaton M (2005) Seismic fracture analysis of concrete gravity dams including dam-reservoir interaction. Comput Struct 83(19):1595–1606
33. Westergaard H (1933) Water pressure on dams during earthquake. Trans ASCE 98(2):418–472

Application and Numerical Simulation of Key Technologies for Deep Foundation Pit Support

Jingsheng Tong, Hongmei Zhang, and Weiqiang Zhang

Abstract This paper adopts the key technology "tunnel and underground space support components", assembles it into a "honeycomb" structure type, and expands its application to deep foundation pit support engineering, which fully reflects the combination of "industry-university-research" theoretical research and engineering practical application. This key technology of support can optimize the setting of lateral support components in the deep foundation pit support process, optimize and improve the current conventional deep foundation pit support construction technology, and has good research and industrialization promotion and application value. In this paper, the deformation and stress–strain analysis of each working condition of excavation of the foundation pit are also carried out through numerical simulation, and the uplift displacement and stress change of the foundation pit bottom are within the safety range of stress control, and the supporting stress is far lower than the bearing capacity of the prefabricated components, which further verifies the practicability and reliability of the key innovative technology.

Keywords Deep Foundation Pit · Supporting Key Technology · Engineering Application · Numerical Simulation

J. Tong (✉) · W. Zhang
China Municipal Engineering Northwest Design and Research Institute Co., Ltd., Ürümqi, China
e-mail: 393064125@qq.com

H. Zhang
Lanzhou Institute of Technology, Lanzhou, China

J. Tong
Key Laboratory Mechanics on Disaster and Environment in Western China, Lanzhou University, Lanzhou, China

© The Author(s) 2023 427
G. Feng (ed.), *Proceedings of the 9th International Conference on Civil Engineering*,
Lecture Notes in Civil Engineering 327,
https://doi.org/10.1007/978-981-99-2532-2_36

1 Introduction

The research and application of the theory of underground space support (including tunnel lining support and deep foundation pit support) has a history of nearly 100 years [1]. Since the twentieth century, the development of support nursing theory has mainly gone through three stages: classical pressure theory, bulk pressure theory and modern theory of elastoplastic deformation pressure, and the more representative theories are: new Austrian support technology theory, axonal transformation theory three law theory, loose circle branch nursing theory and composite branch nursing theory and many other theories [2]. In recent years, China has made great achievements in underground space support technology and research theory, but due to the uncertainty and complexity of engineering geological conditions, along with the large-scale development and construction of urban underground space, various foundation pit projects continue to emerge, and the current foundation pit engineering support technology is still difficult to meet the actual needs, and further innovation and development are urgently needed.

This paper adopts the research patented technology "tunnel and underground space support components" [3], which belongs to the research and development of original basic theory and key technology application, which can be assembled into a "honeycomb" structure type and expanded to the application of deep foundation pit support engineering in underground space and the application of this technology fully reflects the combination of "industry-university-research" theoretical research and engineering practical application. This support technology can optimize the setting of lateral support components in the deep foundation pit support process of underground space, improve the current conventional deep foundation pit support construction method and technical problems, and has good research and industrialization promotion and application value. In this paper [4], the stress and strain analysis of each working condition of foundation pit excavation is carried out through numerical simulation, which further verifies the practicability and feasibility of this innovative technology.

2 Key Technologies and Applications of Deep Foundation Pit Support

2.1 Support Key Technical Principles

At present, the representative application technologies of deep foundation pit support mainly include new support technologies and groundwater control technologies such as composite soil nail support, cement-soil retaining wall, underground continuous wall and joint support [5]. According to the actual needs of deep foundation pit engineering and the application progress of new technologies, this paper summarizes the development trend of deep foundation pit engineering technology, studies the technology of combining supporting structure and main structure, and uses invention patent components to improve the construction technology of commonly used support structure, shorten the construction period and save resources according to the construction technology of partial or all permanent underground structure as temporary support structure of foundation pit.

The support technology process is to assemble the "honeycomb" prefabricated support members into a row of support structural units according to the requirements [6], and the two sides of the unit are fixed on the I-shaped steel sheet piles driven vertically into the deep foundation pit to form an overall supporting thin-walled wall structure. The earth pressure acting on the back of the supporting member on the component can be partially converted into horizontal thrust that can cancel each other between the components, so that the contact between the components is closer and the structural force mode is greatly optimized [7]. At the same time, the use of stiffener ribs increases the overall rigidity of the supporting wall and the ability to withstand loads. The support technology greatly optimizes the requirements for setting transverse support members usually required for deep foundation pit support measures, thereby solving the problem of affecting the overall construction due to the interference of transverse support members after deep foundation pit excavation.

The key technology of this support divides the length according to the standard required for the construction of deep foundation pits, assembles each unit separately, and then assembles into a whole structure. When assembling, it is necessary to first try to calculate the depth and width that can be safely carried in the longitudinal and horizontal directions of the excavation foundation pit, and then according to the calculation results, the assembled supporting members and vertical I-beam steel sheet piles are assembled into an integral wall structure to achieve the purpose of common force and support, and the supporting wall structure is shown in Fig. 1.

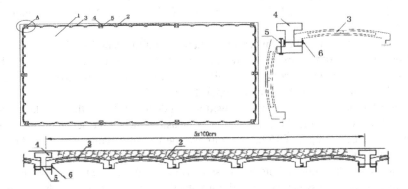

Fig. 1 Schematic diagram of the structure of the overall support wall of the deep foundation pit (In the picture: 1 - The inside of a deep foundation pit, 2 - poured filled waterproof concrete, 3 - arch support "member", 4 - I-beam steel sheet pile column, 5 - fixed bolt, 6 - fixed nut, 7 - bolt hole, 8 - stiffener "member", 9 - "member" reserved bolt hole)

2.2 Support the Application of Key Technologies

The key technology of deep foundation pit support in the underground space first needs to calculate the safety and stability coefficient of foundation excavation, lateral earth pressure, and the depth, length and stability time of each excavation according to the depth and width of deep foundation pit excavation, and carry out design and construction guidance according to the data, and support the lining in time within the effective time to achieve the purpose of guiding construction safety. The specific implementation steps combined with a practical application engineering implementation plan are detailed as follows:

Application example: A grade IV surrounding rock geological deep foundation pit needs to be excavated at a depth $H = 15.0$ m, a foundation pit length $L = 150.0$ m, width $B = 80.0$ m. The soil is medium groundwater, leakage, construction vibration is medium, cohesion reduction coefficient $K_c = 0.6$, pile height around the foundation pit $H_0 = 1.0$ m, soil severity $\gamma = 17.0$ KN/m^3, cohesion $c = 155 K P_a$, cohesion change coefficient 0.6, internal friction angle $\varphi = 23°$. According to the above conditions, the implementation plan of foundation pit excavation is designed.

According to the above data and requirements, the steps of excavation and support process of foundation pit construction are as follows:

Step 1: Release the line according to the length of the foundation pit L = 150 m and the width B = 80 m and determine the I-beam driving position, first drive the 17 m long I-beam sheet pile vertically at the four corner points, and then drive 31 and 14 I-beam sheet piles of the same specification along both sides of the length and width of the foundation pit at the determined position.

Step 2: Calculate and analyze a reasonable excavation plan, assuming that the excavation plan is as follows:

Scheme 1: Assume that the length of one excavation is L = 75 m, and the excavation depth is H1 = 7.5 m; The second excavation depth H2 = 7.5 m, the calculation results are shown in Table 1:

It can be seen from the comparison of Table 1 that the deep foundation pit excavation and support are carried out by scheme 1, the stability coefficient = 1.40, the total lateral horizontal force F = 34700KN, the maximum self-stabilizing depth Hz = 10.15 m, and the length without support Lmax = 75 m.

Calculation result: At the first excavation at a depth of 7.5 m, the foundation pit was stable. After taking the overall support measures, when the second excavation is carried out to a depth of 15 m, the stability coefficient is reduced to 1.13. At this time, the length of each excavation cannot exceed 20 m, and the foundation pit can be stable for 1 day. After taking timely support measures, excavation and support in the next cycle process can be carried out until the excavation reaches the required depth, which ensures construction safety.

Scheme 2: Assuming the excavation length L = 150 m, the excavation is divided into three depth directions: the first excavation depth H1 = 5 m, the second excavation depth H2 = 5 m, and the third excavation depth H3 = 5.0 m, the calculation results are shown in Table 2:

Table 1 Scheme 1 Calculation table of deep foundation pit excavation stability

Excavable length	The first excavation depth is 7.5 m and the length is 75 m			The second excavation was 7.5 m deep and 75 m long		
	stability Coefficient	stabilization time	Required completion time	stability Coefficient	stabilization time	Required completion time
1.0 m	6.79	stable	Completed within 1 year	7.42	stable	Completed within 1 year
5.0 m	2.41	Stable for several months	Completed within 6 months	2.12	Stable for 1 month	Within 3 weeks
10.0 m	1.87	Stable for several weeks	Within 4 weeks	1.46	Stable for several days	Within 3 days
20.0 m	1.59	Stable for several days	Within 5 days	1.13	**Stable for 1 day**	**Within five hours**
50.0 m	1.43	Stable for several days	Within 3 days			
75.0 m	1.40	Stable for several days	Within 2 days			
	The stability coefficient is 1.40, the total horizontal pressure is 17350KN, and the maximum self-stabilization depth is 10.15 m			The stability coefficient is 1.13, and the total horizontal pressure is 57300KN		

It can be seen from the comparison in Table 2 that the first excavation stability coefficient = 1.72, the total horizontal force F = 18100KN, the maximum self-stabilizing depth Hz = 10.15 m, and the length without support Lmax = 150 m.

Calculation result: At the first excavation of 5 m depth, the foundation pit was stable. After the completion of the support, the second excavation is carried out to a depth of 10.0 m, the safety factor is reduced to 1.2, and the total horizontal force is increased to 56300KN, at this time, the excavation length of each section cannot exceed 50 m, the foundation pit can be stable for 1 day, and the excavation of the next cycle section of 20 m length can be carried out after timely support measures; When the third excavation reaches a depth of 15 m, the safety factor is reduced to 1.13, and the total horizontal force is increased to 114600KN, at this time, the excavation length is required to not exceed 10 m long, and timely support is required after excavation before the excavation of the next cycle of construction section can be carried out until the excavation is completed.

Step 3: According to the comparative analysis of Table 1 and Table 2, according to the requirements of construction safety and time saving, it is recommended to adopt the excavation method of Option 2, which can meet the purpose of construction safety and investment saving.

As shown in Fig. 2, the first layer of soil in the deep foundation pit is excavated, the construction excavation depth is H1 = 5 m, the length is 150 m, and the excavation

Table 2 Scheme 2 Calculation table of deep foundation pit excavation stability

Excavable length	The length of the first excavation is 150 m and the depth is 5 m			The second excavation is 150 m long and 5 m deep			The length of the third excavation is 150 m and the depth is 5 m		
	stability Coefficient	stabilization time	Required completion time	stability Coefficient	stabilization time	Required completion time	stability Coefficient	stabilization time	Required completion time
1.0 m	6.35	stable	Completed within 1 year	7.07	stable	Completed within 1 year	7.42	stable	Completed within 1 year
5.0 m	2.62	Stable for several months	Completed within 6 months	2.28	Stable for several months	Completed within 6 months	2.12	Stable for 1 month	Within 3 weeks
10.0 m	2.15	Stable for several weeks	Within 4 weeks	1.68	Stable for several weeks	Within 4 weeks	1.46	Stable for several days	Within 3 days
20.0 m	1.92	Stable for several weeks	Within 2 weeks	1.38	Stable for several days	Within 3 days	1.13	Stable for 1 day	Within 5 h
50.0 m	1.78	Stable for several weeks	Within 10 days	1.20	Stable for 1 day	Within 3 h			
150.0 m	1.72	Stable for 1 week	Within 3 days						
	The stability coefficient is 1.72, the total horizontal pressure is 18100KN, and the maximum self-stabilization depth is 10.16 m			The stability coefficient is 1.2, and the total horizontal pressure is 56300KN			The stability coefficient is 1.13, and the total horizontal pressure is 114600KN		

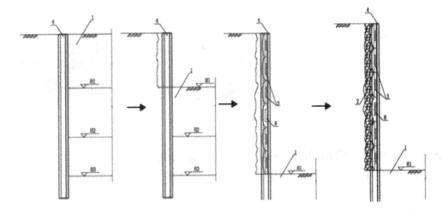

Fig. 2 Drawing of the excavation process of deep foundation pit

sequence is first the middle and then the four weeks. After the completion of each stage of construction and excavation, the excavated soil is cleared and transported in time, and the excavation of each layer of soil and the splicing and installation of the corresponding arch member support are completed within the required time to ensure that the foundation pit is stable and does not collapse.

Step 4, deep foundation pit support structure assembly: the supporting members are stitched horizontally according to one unit every 5 m, and the two ends of the unit are anchored and fixed on I-shaped steel sheet piles by bolts, from bottom to top, assembled row by row, until the first layer of soil is excavated and assembled. In the assembled components, the third row is the setting position of the stiffener component element layer. The stiffener member is set up to increase the overall bending stiffness of the supporting wall.

Step 5: The gap between the soil and the supporting structure after the support is completed is poured and compacted with 30 cm thick waterproof concrete, which plays the dual role of bearing earth pressure and stopping water.

Step 6: According to the requirements of the 2 plan, cycle the construction until the foundation pit is fully excavated to the required depth of construction and the support is completed.

After the construction of the deep foundation pit is completed, the prefabricated components can be used as permanent support walls of the foundation pit, or they can be dismantled and reused as wall support materials in underground parking lots or other projects to achieve the purpose of energy conservation and environmental protection.

Fig. 3 Schematic diagram of excavation and support model of foundation pit

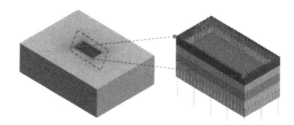

3 Numerical Simulation of Key Technologies for Deep Foundation Pit Support

3.1 Numerical Models and Parameters

The model simulates the working conditions of the construction stage of the excavation process of the extended application of deep foundation pit, adopts the MC (Moore-Coulomb) constitutive model, and the supporting structure adopts prefabricated components and vertical I-beam fixed support form. The calculation model is formulated according to the long-span deep foundation pit, the size is 40 m × 20 m × 15 m (length × width × depth), divided into three excavations, each excavation depth is about 2.5 m, the support adopts the studied "honeycomb" prefabricated support members, the model is shown in Fig. 3.

Initial Excavation. According to the analysis of the initial excavation ground stress cloud 4, the effective stress of earth pressure in the X, Y and Z directions of the three axes is small, which are:

S-XXmax = 209 kPa, S-XXmin = 30 kPa; S-YYmax = 209 kPa, S-YYmin = 30 kPa;
S-ZZmax = 480 kPa, S-ZZmin = 40 kPa, Effective range S-ZZ = 40~120 kPa.

It shows that the in-situ stress in the horizontal X and Y directions of deep foundation pit excavation is small and balanced, and the in-situ stress in the vertical direction gradually increases with depth. After the I-beam is driven, because the step-by-step excavation of the foundation pit has not yet been carried out, the initial displacement is basically 0, and the supporting stress is the same as the in-situ stress (Fig. 4).

Excavation −1~5 Stages. Only the displacement and effective stress cloud map in the Z-axis direction that has the main influence of the excavation −1~5 stage is analyzed, the step-by-step excavation depth is 2.5 m, and the prefabricated components are assembled and supported immediately after each excavation, and the displacement: effective stress in the Z-axis direction after excavation are analyzed as follows:

- Displacement: main Z direction

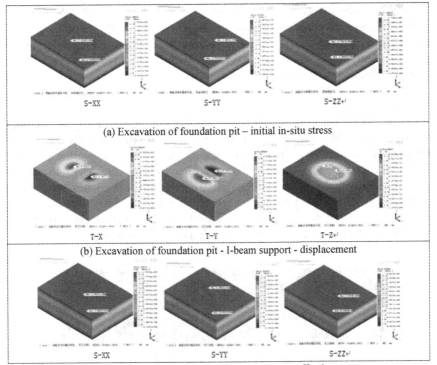

(a) Excavation of foundation pit – initial in-situ stress

(b) Excavation of foundation pit - I-beam support - displacement

(c) Excavation of foundation pit - I-beam support - effective stress

Fig. 4 Initial excavation stress cloud

excavation −1: T1-Zmax = 10.78 mm, T1-Zmin = −0.193 mm, effective range T1-Z = 0~2.5 mm.

Excavation-2: T2-Zmax = 20.80 mm, T2-Zmin = -0.43 mm, effective range T2-Z = 1.34~6.65 mm. Excavation-3: T3-Zmax = 29.40 mm, T3-Zmin = −0.67 mm, effective range T3-Z = 1.84~9.85 mm. Excavation-4: T4-Zmax = 37.08 mm, T4-Zmin = −4.5 mm, effective range T4-Z = 2.42~12.82 mm. Excavation −5: T5-Zmax = 43.8 mm, T5-Zmin = −15.94 mm, effective range T5-Z = −6.0~8.96 mm.

- Effective range of support stress

Support-1: S1-ZZ effective range = −480~0.25 kPa; Support-2, S2-ZZ effective range = −280~20 kPa; Support-3, S3-ZZ effective range = −340~50 kPa; Support-4, S4-ZZ effective range = −370~70 kPa; Support-5, S5-ZZ effective range = −400~120 kPa.

From the above data, it can be seen that after step-by-step excavation −1~5 stage, after step-by-step excavation at a depth of 2.5 m (total excavation depth 12.50 m, I-beam anchoring depth 5 m), the displacement deformation in the X and Y directions is small, and the maximum uplift displacement of the first excavation at the bottom of the vertical Z pit is 10.78 mm. After the timely support of the prefabricated supporting members is adopted, the prefabricated components can play the function of timely bearing and supporting, and the stress is 0.25~−480 kPa in the vertical change range and 0.45~−208 kPa in the horizontal direction, and the deformation and stress control in the excavation process are within the safe range.

By the completion of excavation, the maximum displacement of the excavation-5 pit bottom uplift reaches 43.8 mm, and the minimum displacement settlement is about 15.94 mm.

Figure 5 Cloud diagram shows that the stress of the excavation support member increases with the increase of excavation depth, and the vertical effective stress also gradually increases. After adopting prefabricated "honeycomb" supporting members in the direction of X and Y, the overall stress is effectively reduced. The Z-axis stress increases more with the increase of excavation depth, and the maximum stress range reaches −480~120 kPa, but it is far lower than the bearing capacity of prefabricated components, so the deformation and stress of the deep foundation pit of the entire excavation section are controlled within a safe and reasonable range

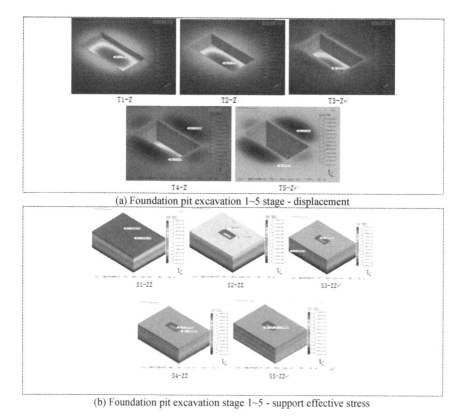

(a) Foundation pit excavation 1~5 stage - displacement

(b) Foundation pit excavation stage 1~5 - support effective stress

Fig. 5 Cloud diagram of excavation −1 excavation excavation and supporting stress

4   Conclusions

Combined with the key support technologies of the research topic, this paper introduces a new construction technology and construction method for the innovation of deep foundation pit support technology based on the application of engineering calculation examples and numerical simulation methods. The technology is green, low-carbon and environmentally friendly, using recyclable prefabricated support technology, which provides new ideas for solving the problems of high energy consumption, low utilization rate and waste of resources in conventional foundation pit support technology, has good research and industrialization promotion and application value, improves the commonly used support construction technology, broadens the new construction method of using prefabricated components in deep foundation pit, and concludes:

This paper adopts the key support technology, and provides a new construction idea and process method

(1) The key technology of support of this research topic, using arched prefabricated support components, assembled into the best force-bearing support structure shaped like "honeycomb", is applied to large-span deep foundation pit support engineering projects, simplifies the current commonly used deep foundation pit construction technology and technical treatment measures, solves the problem of common deep foundation pit excavation setting lateral support and affects a wide range of construction interference, and innovates the new construction method of using prefabricated assembly new material structure construction in deep and large foundation pit support.

(2) Using numerical simulation, the deformation and stress–strain analysis of each working condition of foundation pit excavation were carried out, and it was concluded that the maximum uplift displacement and stress change of the pit bottom after excavation and timely support of the deep foundation pit were within the safe range of support control. However, considering the safety and stability factors of construction, it is necessary to take static pressure treatment measures on the bottom of the pit to reduce the stress at the bottom of the foundation pit; In the process of deep excavation of the foundation pit, the supporting stress is far lower than the bearing capacity of the prefabricated components, and the deformation and stress of the foundation pit are controlled within the safe range, which further verifies the practicality and reliability of this key innovative technology.

Acknowledgements This research was funded by the research project of the China State Construction Engineering Corporation [No. CSCEC-2016-2017-(12)]. The author sincerely thanks Professor Zhouyouhe for his guidance and help.

References

1. Xu G (2003) Support Structure of Underground Engineering. China Water & Power Press. (in Chinese)
2. Tong J, Wang S, Han X, Chen S, Zhang W (2019) General Design and Calculation Method of Vertical Surrounding Rock Pressure in Deep and Shallow Tunnel and Underground Space, Patent No.: ZL201610824651.1. National Intellectual Property Administration, PRC, 2019.08.06. (in Chinese)
3. Tong J, Wang S, Han X, Zhang W (2016) Prefabricated Support Components for Construction and Assembly of Underground Space, Patent No.: ZL201521112761.2. National Intellectual Property Administration, PRC, 2016.05.25. (in Chinese)
4. The Professional Standards Compilation Group of People's Republic of China. Code for Design of Railway Tunnel TB 10003-2016 P J 449-2016. China Railway Publishing House, Beijing. (in Chinese)
5. The Professional Standards Compilation Group of People's Republic of China (2004) JGJ D70-2014 Code for design of road tunnel. China Communications Press, Beijing. (in Chinese)
6. Tong J (2020) General formulas for calculating surrounding rock pressure of tunnels and underground spaces. KSCE J Civ Eng 24(4):1348–1356
7. Tong J (2021) Research on the application of multifactor surrounding rock pressure calculation theory in engineering. KSCE J. Civil Eng. 25:2213–2224

Complex Frequency-Domain Oscillation Analysis of the Pumped-Storage Systems

Yang Zheng, Wushuang Liu, Xuan Zhou, Wanying Liu, and Qijuan Chen

Abstract Hydraulic impedance, as an efficient frequency-domain analysis alternative, has been widely utilized in hydraulic transient analysis for decades. Since the mathematical expressions of hydraulic systems with complicated pipe networks are usually rather complex, it is difficult for the traditional continuous impedance method to obtain the analytical solutions to the system's frequency responses directly. Therefore, an equivalent circuit modeling-based discrete impedance method is proposed to mathematically express the hydraulic systems of a pumped storage plant system with complicated pipe networks. Through drawing an analogy between various hydraulic facilities and different types of electrical circuits, the equivalent circuit topology of any hydraulic system can be obtained according to the circuit theory in electrical engineering. The oscillation analysis of the pumped-storage power plant is conducted, and influences of the pump-turbine impedance on the system's oscillation characteristics have been discussed.

Keywords Complex Frequency · Pumped-Storage System · Hydraulic Oscillation · Mathematical Model

1 Introduction

The pumped-storage power plant is known as a multi-physics coupling system [1] and usually has a very complicated flow passage layout, so the investigation of the hydraulic features of pumped-storage systems is of great importance. Over the past decades, researchers have proposed different mathematical models of pumped-storage systems for better hydraulic description, and the mathematical analysis schemes are divided into two categories, i.e., time-domain models and frequency-domain models [2].

Y. Zheng (✉) · W. Liu · X. Zhou · W. Liu · Q. Chen
School of Power and Mechanical Engineering, Wuhan University, 8 South Donghu Road, Wuchang District, Wuhan 430072, China
e-mail: zhengyang@whu.edu.cn

G. Feng (ed.), *Proceedings of the 9th International Conference on Civil Engineering*,
Lecture Notes in Civil Engineering 327,
https://doi.org/10.1007/978-981-99-2532-2_37

Time-domain models excel in simulating the time evolutions of the system states, e.g., the rotation speed, the head and the discharge [3]. Frequency-domain methods, such as the well-known transfer matrix [4] and hydraulic impedance [5], were extensively adopted in the stability criterion and the oscillation analysis of hydraulic systems. The hydraulic impedance [6] can explicitly express the relationship between head and discharge in complex domain. Existing literature has reported its applications in many practical water conveyance scenarios [7–9].

This paper introduces an equivalent circuit model-based discrete impedance method that can be applied to the complex frequency-domain analysis of pumped storage systems. The hydraulic transients in pipes are analogized with the electromagnetic characteristics in transmission lines. For a complex pumped storage plant system, the overall circuit topology can be obtained by combining the equivalent circuits of different hydraulic components according to the system structural layout. Oscillation analysis was performed in a pumped-storage power plant with a complex flow passage structure by applying the proposed discrete impedance model.

2 Methodology of the Equivalent Circuit Model

The Saint–Venant equations of the pressurized pipe [10, 11] are given as Eq. (1)

$$
\begin{cases}
\frac{\partial H}{\partial t} + v\frac{\partial H}{\partial x} + \frac{a^2}{gA}\frac{\partial Q}{\partial x} = 0 \\
g\frac{\partial H}{\partial x} + \frac{1}{A}\frac{\partial Q}{\partial t} + \frac{v}{A}\frac{\partial Q}{\partial x} = g\left(S_0 - \frac{n_c^2 Q|Q|}{R_a^{4/3}A^2}\right)
\end{cases}
\tag{1}
$$

where, H and Q denote the pressurized head and discharge, respectively. n_c, R_a and A represent the friction factor, the hydraulic radius and the cross-sectional flowing area of the pipe, respectively.

Since the value of wave propagation velocity a is bigger than that of flow velocity v, $v\frac{\partial H}{\partial x}$ is often ignored [11]. The hydraulic dynamics of the pipe can be modelled as a T-shaped equivalent circuit shown in Fig. 1, where R_e, L_e, C_e are the circuit resistance, inductance and capacitance per unit length, respectively.

The equivalent resistance, inductance, and capacitance parameters can be obtained as $R_e = \frac{n_c^2|Q|}{R^{4/3}A^2} - \frac{S_0}{Q} + \frac{1}{gA^2}\frac{\partial Q}{\partial x}$, $L_e = \frac{1}{gA}$, $C_e = \frac{gA}{a^2}$.

where, the partial differential term $\frac{1}{gA^2}\frac{\partial Q}{\partial x}$ corresponds to $v\frac{\partial h}{\partial x}$ in Eq. (2).

Fig. 1 The circuit of the transmission line per unit length

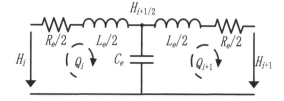

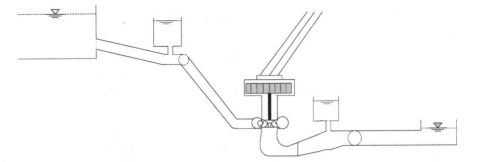

Fig. 2 Structural schematic of the pumped-storage power plant

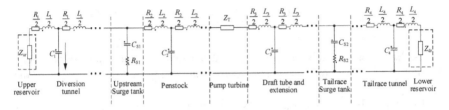

Fig. 3 Equivalent circuit topology of the hydraulic structure of a pumped-storage plant

3 System Plant Introduction and its Modelling

A pumped-storage power plant with a complex conduit system was taken as a case study. The flow passage comprises of a diversion tunnel, an upstream surge tank, a penstock, a volute, a draft tube and extension, a tailrace surge tank, and a tailrace tunnel. The structural schematic of the hydraulic system is illustrated in Fig. 2.

According to the equivalent circuit modeling theory, the overall equivalent circuit topology of the flow passage of the pumped-storage plant is shown in Fig. 3.

4 Simulation Result Analyses

The decay coefficients and eigen frequencies of the first ten orders of oscillation in free oscillation analysis with the proposed discrete model and the traditional continuous model are listed in Table 1. It shows that ECM can obtain very similar complex eigen frequency results to those of the traditional continuous impedance model. The modeling errors between the two models are quite small for low-frequency responses and gradually increase for higher frequency responses. The oscillation frequencies of the first two orders of oscillation are much lower than those of others. Compared with the classic theoretical oscillation frequencies of the surge tanks calculated by the equations in [10], the first two orders of oscillation are believed to coincide with

Table 1 The eigen mode analysis comparison of the two models

Order	Discrete ECM	
	Decay coefficient	Eigen Frequency
1	−0.0018	0.0656
2	−0.0021	0.1129
3	−0.0051	3.2430
4	−0.0053	6.4482
5	−0.0055	9.6120
6	−0.0057	12.7721
7	−0.0057	15.9697
8	−0.0056	19.2061
9	−0.0056	22.4475
10	−0.0055	25.6558

the oscillation modes of the tailrace surge tank and upstream surge tank, respectively. While the 3rd - 10th order oscillations correspond to the os cillation modes of the system pipelines.

The change regulations of the decay coefficients of the 3rd- to 10th-order oscillations are depicted in Fig. 4. The decay coefficients of eight oscillation orders share the same variation trends and decrease at different rates as pump-turbine impedance increases. The 3rd-order oscillation possesses the fastest decreasing rate, while the 5th-order oscillation possesses the slowest. The decay coefficients are positive when $Z_T < 0$ and negative when $Z_T > 0$. When $Z_T = 0$, the decay coefficient equals zero, which means the oscillation pattern at this condition is a constant amplitude oscillation.

To further investigate the variation trend of the eigen frequency in each order of oscillation, the precise variation trends of the 3rd- to 10th- order of oscillation are displayed in Fig. 5 (a)–(h), respectively. It shows that the eigen frequencies of

Fig. 4 Change regulations of the decay coefficients of the first eight orders that correspond to the pipelines

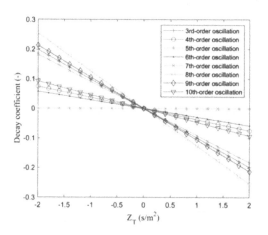

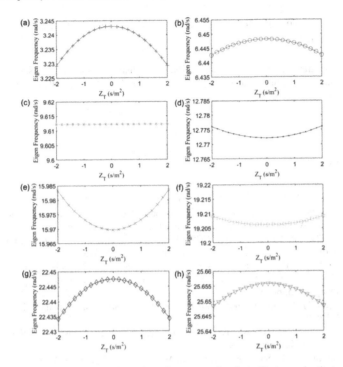

Fig. 5 The precise variation trend of the eigen frequency of each oscillation order that corresponds to the pipelines: (a) the 3rd- order; (b) the 4th- order; (c) the 5th- order; (d) the 6th- order; (e) the 7th- order; (f) the 8th- order; (g) the 9th- order; (h) the 10th- order

most oscillations vary along with the change of the pump-turbine impedance though the oscillation amplitudes are very small. Among the eight orders of oscillation, the frequencies of the 3rd-, 4th-, 9th-, and 10th- orders of oscillation decrease as the absolute value of the pump-turbine impedance increases, and the frequencies of the 6th-, 7th-, and 8th- orders of oscillation increase as the absolute value of the pump-turbine impedance increases. the frequency of the 5th- order oscillation remains constant as the pump-turbine impedance varies.

5 Conclusions

This work introduced an equivalent circuit modeling (ECM)-based discrete hydraulic impedance method in complex frequency-domain analysis of pumped storage systems. The proposed ECM-based impedance method can achieve satisfactory modeling accuracy. In the oscillation analysis of the pumped-storage power plant, numerical results showed that there exists a close relationship between the value of the pump-turbine impedance and the decay coefficients of the different orders of oscillation. However, the pump-turbine impedance can hardly affect the system's eigen frequencies.

Acknowledgements Authors wish to acknowledge the financial support from National Natural Science Foundation of China (grant number: 52009096) and China Postdoctoral Science Foundation (grant number: 2022T150498)

References

1. Feng C, Zheng Y, Li C, Mai Z, Wu W, Chen H (2021) Cost advantage of adjustable-speed pumped storage unit for daily operation in distributed hybrid system. Renew Energ 176:1–10
2. Yang J (2018) Applied fluid transients. Science Press, Beijing. (in Chinese)
3. Zheng Y, Chen Q, Yan D, Liu W (2020) A two-stage numerical simulation framework for pumped-storage energy system. Energy Conv Manag 210:112676
4. Vitkovsky JP, Lee PJ, Zecchin AC, Simpson AR, Lambert MF (2011) Head- and flow-based formulations for frequency domain analysis of fluid transients in arbitrary pipe networks. J Hydraul Eng 137:556–568
5. Kim SH (2011) Dynamic memory computation of impedance matrix method. J Hydraul Eng 137:122–128
6. Suo L, Wylie EB (1989) Impulse response method for frequency-dependent pipeline transients. J Fluids Eng ASME 111:478–483
7. Gong JZ, Lambert MF, Simpson AR, Zecchin AC (2013) Single-event leak detection in pipeline using first three resonant responses. J Hydraul Eng 139:645–655
8. Duan HF, Lee PJ, Ghidaoui MS, Tung YK (2011) Extended blockage detection in pipelines by using the system frequency response analysis. J Water Res Plan Man 138:55–62
9. Kim SH, Choi D (2022) Dimensionless impedance method for general design of surge tank in simple pipeline systems. Energies 15:3603
10. Chaudhry MH (2014) Applied hydraulic transients. Springer, London
11. Zhao Z, Yang J, Yang W, Hu J, Chen M (2019) A coordinated optimization framework for flexible operation of pumped storage hydropower system: nonlinear modeling, strategy optimization and decision making. Energy Conv Manag 194:75–93

Computer-Vision-Based Structure Shape Monitoring of Bridges Using Natural Texture Feature Tracking

Weizhu Zhu, Xi Chu, and Xin Duan

Abstract Structural health monitoring is carried out for a limited number of measuring points of the bridge. The root of the problem of bridge damage identification is that the mechanical equation inversion result is not unique due to the incomplete measured data. The full-field description ability of digital images to the structural shapes can effectively alleviate the problem of damage identification caused by incomplete measured data. This project aims to research the full-field shape monitoring method of the bridge. Firstly, the mathematical representation of the corresponding image points on the bridge surface is formed by using the image feature extraction method. Analyze the feature points position change mathematical model before and after deformation, and propose the structural full-field displacement vector calculation theory. Verify the full-field displacement calculation theory by a test beam. The results show that the maximum absolute error of the vector length is 0.24 mm and the relative error is less than 5%. The research realizes the structural full-field displacement monitoring under the natural texture condition for the first time. The results can promote the application and development of digital image processing technology in the field of structural health monitoring, improve the level of bridge safety evaluation, and realize the automation, intelligence and visualization of structural deformation monitoring.

Keywords Computer vision · Full-field displacement · Natural Texture Feature Tracking

W. Zhu · X. Chu (✉) · X. Duan
College of Civil and Transportation Engineering, Shenzhen University, Shenzhen 518060, China
e-mail: 2150471002@email.szu.edu.cn

Institute of Urban Smart Transportation and Safety Maintenance, Shenzhen University, Shenzhen 518060, China

© The Author(s) 2023
G. Feng (ed.), *Proceedings of the 9th International Conference on Civil Engineering*, Lecture Notes in Civil Engineering 327, https://doi.org/10.1007/978-981-99-2532-2_38

1 Introduction

When the sensors are arranged on the finite points of the structure to obtain the dynamic response information, the structural damage identification is carried out on the assumption that the structural model degree of freedom is consistent with the observation degree of freedom. However, in practical engineering, the observation data is often incomplete due to the influence of various conditions. In order to solve this problem, more sensors are needed, but the actual number of sensors is limited. For structural damage identification, if the monitoring data is incomplete, the useful information obtained is insufficient, making the damage identification problem only be solved under insufficient known information. Therefore, the damage identification method faces the problem of difficulty in damage identification due to insufficient test data.

In recent years, with the development of preinstalled cameras, unmanned aerial vehicles (UAV), wearable virtual reality devices and other hardware devices, damage detection and recognition technology based on machine vision have been applied to actual structures. Image processing technology has been widely used in the detection of local cracks in structural components, such as concrete cracks [1], concrete spalling [2], pavement cracks [3], underground concrete pipe cracks [4], asphalt pavement potholes [5] and so on. Prasanna [6] et al. researched the automatic detection of concrete bridge cracks. First, use the increasing structural elements to carry out alternating open-close filtering on the image, smooth the image and remove noise. Then the multi-scale morphological edge detector was used to accurately extract the edge of bridge cracks and track and locate the development of cracks. Sarvestani [7] et al. developed a vision-based robotic image acquisition device and proposed a more advanced automated visual surveillance system. The system adopts a remote control robot to acquire the image, and identifies the size of the acquired crack through digital image processing software, so that the detection process is more rapid, safe, reliable and low cost. Dyke [8] et al. proposed a vision-based bridge crack detection technology through automatic target detection and grouping processing. Yang [9] and Nagarajaiah [10] et al. used video to detect local structural damage in real-time by combining the low-rank sparse representation method.

These methods have no essential difference from the conventional sensor monitoring methods, and they are aimed at the local area of the structure, so the problem of damage identification caused by incomplete test data has not been fundamentally solved. With the development of digital image processing technology and image acquisition hardware, the holographic monitoring of bridge structures by the camera has become the future development direction of bridge structural safety monitoring. The ability of the image to describe the structure shape can effectively alleviate the problem of structural damage identification caused by incomplete test data. The surface of the bridge structure has many natural texture features. The feature extraction method can obtain the natural texture feature points on the surface of the bridge. Furthermore, analyze the mathematical model of the feature points' relative position change before and after deformation, propose the displacement field calculation theory of the feature points, and establish a full-field displacement monitoring method for the structure under natural texture conditions. The research results can conduct non-contact deformation monitoring on the bridge structure, significantly improving the completeness of the bridge structure monitoring data.

2 Method of Feature Point Tracking

The surface of the bridge structure is uneven, and some surface defects that are difficult to avoid, such as small holes, voids, pits, cracks, etc., constitute the natural texture of the structure surface. Can identify these natural texture features and track the positions of these natural textures before and after deformation. The displacement calculation theory of natural texture feature points can be analyzed, and the displacement of natural textures can be collected to form the displacement field of the structure surface.

In the case of small deformation, the natural texture of the structure surface does not change significantly. The image scale space extreme point detection method can extract the structure surface's natural texture features. Image scale-invariant feature transform algorithm (SIFT) [11, 12] can extract natural texture's extreme points in the scale space of the image through the method of image's gray difference, called feature points. The obtained feature points have scale, position, and orientation invariant features.

Use the camera calibration board to test the feasibility of natural texture feature extraction on the structure surface by SIFT algorithm. Place the calibration board in two different positions, and take images. Figure 1 shows the extracted feature points of the two images.

Match the feature points extracted in Fig. 1 as shown in Fig. 2. As the number of feature points extracted in Fig. 1 is up to 2117, to intuitively display the matching results of feature points, Fig. 2.only shows the matching effects of the first 40 feature points.

Fig. 1 Feature point extraction of two images

Fig. 2 Registration of
feature points

3 Image Acquisition Test of Bridge Structure

Figure 3 shows the structure and actual object of the steel truss-concrete composite beam specimen used in this test.

3.1 Method of Test Data Collection

The selection of the monitoring camera is based on the analysis of measurement accuracy, field angle, focal length and other aspects, and Fuji GFX 100 ordinary civilian camera is selected. According to the actual shooting distance to choose the appropriate focal length of the lens, this paper chooses Fuji GF 32-64/4 RLMWR lens. See Table 1 for the technical parameters of the camera and lens.

The camera was positioned 5 m from the center of the test beam. In the test, dial indicators are used to obtain the measured value of structural deformation. A total of 13 dial indicators are arranged at the bottom nodes of the test beam test. Figure 4

Fig. 3 Physical Drawing of
Steel Truss-Concrete
Composite Beam (mm)

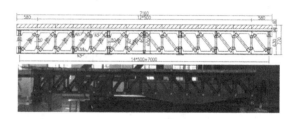

Table 1 Parameters of Camera and Lens

Number of pixels	Sensor size	Data interface	Aspect ratio	Photosensitive original
102 million	43.8 × 32.9 mm	USB3.0	4:3	CMOS
Lens model	Relative aperture	Focal length of lens	Diameter and length	Lens weight
GF 32-64/4 RLM WR	F4.0-F32	32–64 mm	φ92.6 × 116 mm	φ92.6 × 116 mm

Fig. 4. Measuring Camera and Dial Indicator Arrangement

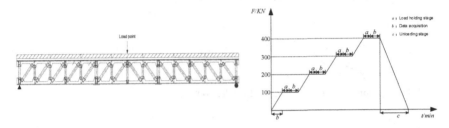

Fig. 5 Test beam's loading scheme

shows the camera placement position and percentile table arrangement are shown in Fig. 4.

3.2 Loading of Test Beam

The test beam was loaded at the midspan in 100kN increments from 0 to 400kN. After each loading is completed, the load is held for two minutes to capture images and collecte the dial indicator data. Figure 5 shows the loading scheme.

3.3 Acquiring the Image of Test Beam

As an example of 100kN working condition, the obtained test beam image is shown in Fig. 6(a). SIFT is used to extract the feature points of the images of the test beams under various working conditions. Figure 6(b) shows the feature extraction results of the test beams under 100kN working condition.

The main edge contains the shape information of the structure. It is the main area of the feature points distribution of the structure image, which constitutes an important carrier for extracting the holographic deformation information of the structure. Simplify the feature extraction results of each working condition, delete unnecessary environmental feature points, and highlight the main features of the structure, as shown in Fig. 7.

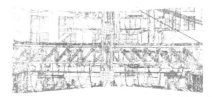

(a) Image of test beam at 100kN   (b) Feature extraction results of test beams under
                                               100kN

Fig. 6 Feature extraction results of the test beam under 100kN

Fig. 7 Distribution of test beam's main body under 100kN

4 Method of Extracting Full-Field Displacement Vector of Structural Surface Based on Feature Point Tracking

4.1 Structural Full-Field Displacement Vector Calculation

Feature point matching establishes the relationship between the natural texture of the structure before and after deformation, but the displacement is still uncertain. A fixed point can be arranged on the image measurement plane to constrain the position of each feature point, and then the displacement of the feature points before and after deformation can be calculated. Propose a mathematical model for the relative position change of feature points, as shown in Fig. 8.

Fig. 8 Mathematical model of feature points' position change

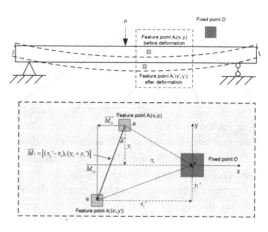

In Fig. 8, the red square denotes a fixed point. Theoretically, the fixed point O can use any feature point of the chessboard. Select the feature point at the center of the checkerboard as the fixed point. The full-field displacement vector algorithm of the structure is as follows:

(1) As shown in Fig. 8, displacement vector $\overrightarrow{A_i A_i'}$ is composed of the starting point A_i before deformation and the ending point A_i' after deformation. This displacement vector is the basic element of the structure's full-field displacement vector, as represented by the blue vector in Fig. 8, which M_i denotes.

(2) M_i is generated according to the following algorithm: Given the bridge images (i_1, i_2) before and after deformation, first, the feature points are extracted from the images, expressed as A_i and A_i', and the initial matching C, $C = \{(A_i, A_i') : i = 1, 2, \ldots n\}$ are obtained, where $A_i \in i_1$, $A_i' \in i_2$.

(3) Feature points A_i, A_i' are projected to the fixed point O to obtain the feature point coordinates $A_i(x_i, y_i)$, $A_i'(x_i', y_i')$, extract the coordinate values (x_i, y_i), (x_i', y_i') between the feature points A_i, A_i', and fixed point O, and obtain $\overrightarrow{M_i} = \left[(x_i' - x_i), (y_i + y_i')\right]$. M_i is a displacement vector with two parameters: length L_i and angle θ_i. Length L_i and angle θ_i can be obtained from the known feature point coordinates $A_i(x_i, y_i)$, $A_i'(x_i', y_i')$.

$$L_i = \sqrt{\left(x_i' - x_i\right)^2 + \left(y_i' + y_i\right)^2} \tag{1}$$

$$\theta_i = \tan^{-1} \frac{|x_i' - x_i|}{|y_i' + y_i|} \tag{2}$$

(4) Set $M = \{M_i : i = 1, 2, \ldots, n\}$ is composed of all M_i, and it is the initial calculation result of the structure's full-field displacement. The full-field deformation vector of the structure's surface can be extracted by solving the vector set M.

4.2 Full-Field Displacement Monitoring Results of Test Beam

According to the calculation method of the structural full-field displacement vector, we obtain the calculation result of the test beam and the chromatographic assignment is carried out according to the length of the displacement vector, as shown in Fig. 9.

The full-field displacement vector of the structure surface in Fig. 9 shows the holographic deformation characteristics of the structure, greatly expands the deformation monitoring data, further expands the structural deformation monitoring from single-point measurement to full-field displacement measurement.

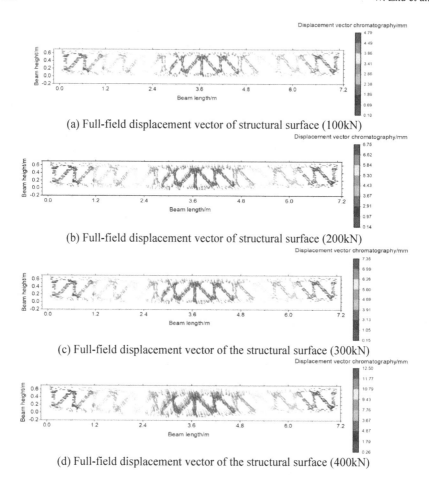

(a) Full-field displacement vector of structural surface (100kN)

(b) Full-field displacement vector of structural surface (200kN)

(c) Full-field displacement vector of the structural surface (300kN)

(d) Full-field displacement vector of the structural surface (400kN)

Fig. 9 Full-field displacement monitoring results of the test beam

4.3 Verification of Structural Displacement Vector

The full-field vector length of the structure surface near the dial indicator under each working condition is extracted and compared with the dial indicator. The verification results are shown in Table 2 (only 400kN is displayed).

The results show that the maximum absolute error of the lower edge vector is 0.24 mm and the maximum relative error is less than 5%.

Table 2 Comparison of Deformation Measurement Accuracy under 400kN

load class	Deflection Extraction Location	Measured value of dial indicatorR1/mm	vector magnitude R2/mm	R2-R1/ mm	Error ISI/R1/%
400kN	580	3.73	3.9	0.17	4.56%
	1080	5.88	6.16	0.28	4.76%
	1580	7.16	7.28	0.12	1.68%
	2080	8.19	8.22	0.03	0.37%
	2580	8.96	8.94	−0.02	0.22%
	3080	9.13	9.33	0.2	2.19%
	3580	9.47	9.51	0.04	0.42%
	4080	9.22	9.46	0.24	2.60%
	4580	8.94	9.18	0.24	2.68%
	5080	8.35	8.23	−0.12	1.44%
	5580	6.64	6.77	0.13	1.96%
	6080	4.65	4.74	0.09	1.94%
	6580	2.76	2.86	0.10	3.62%

5 Conclusion

Propose digital image processing technology to extract the natural texture feature points on the structural surface. Analyzed the feature points' displacement field calculation theory before and after deformation and proposed a structural full-field displacement monitoring method. Obtained structural full-field displacement monitoring results under the natural texture condition for the first time.

(1) The structural surface has rich natural texture features. The natural texture features of the structure surface can be extracted by the image scale-invariant feature transform (SIFT) algorithm to form the point source data of the structural full-field displacement monitoring.

(2) Arranging a fixed point can constrain the position of feature points on the structural surface before and after deformation. The full-field displacement vector of the structural surface can be obtained by calculating the relative position relationship between the feature points and the fixed point before and after deformation.

(3) The three-dimensional laser scanning verifies the accuracy of the structure's full-field displacement vector. The verification results show that the maximum absolute error of the full-field vector length is 0.24 mm, and the maximum relative error is less than 5%, indicating that the extracted displacement vector can accurately reflect the full-field displacement characteristics of the structure.

References

1. Abdel-Qader I, Abudayyeh O, Kelly ME (2003) Analysis of edge-detection techniques for crack identification in bridges. J Comput Civil Eng 17(4):255–263
2. German S, Brilakis I, Desroches R (2012) Rapid entropy-based detection and properties measurement of concrete spalling with machine vision for post-earthquake safety assessments. Adv Eng Inform 26(4):846–858
3. Zalama E, Gómez-García-Bermejo J, Medina R (2014) Road crack detection using visual features extracted by Gabor filters. Comput-Aided Civil Infrastruct Eng 29(5):342–358
4. Sinha SK, Fieguth PW, Polak MA (2003) Computer vision techniques for automatic structural assessment of underground pipes. Comput-Aided Civil Infrastruct Eng 18(2):95–112
5. Koch C, Brilakis IK (2011) Pothole detection in asphalt pavement images. Adv Eng Inform 25(3):507–515
6. Prasanna P, Dana KJ, Gucunski N (2016) Automated crack detection on concrete bridges. IEEE Trans Autom Sci Eng 13(2):591–599
7. Sarvestani AA, Eghtesad M, Fazlollahi F et al (2016) Dynamic modeling of an out-pipe inspection robot and experimental validation of the proposed model using image processing technique. Iran J Sci Technol Trans Mech Eng 40(1):77–85
8. Yeum CM, Dyke SJ (2015) Vision-based automated crack detection for bridge inspection. Comput Aided Civil Infrastruct Eng 30(10):759–770
9. Yang Y, Nagarajaiah S (2015) Dynamic imaging: real-time detection of local structural damage with blind separation of low-rank background and sparse innovation. J Struct Eng 142(2):04015144
10. Nagarajaiah S, Yang Y (2017) Modeling and harnessing sparse and low-rank data structure: a new paradigm for structural dynamics, identification, damage detection, and health monitoring. Struct Control Health Monit 24(1):e1851.1–e1851.22
11. Zhong Q, Wang T, Shiquan A (2018) Image seamless stitching and straightening based on the image block. IET Image Process 12(8):1361–1369
12. Cao JL, Xu L, Guo SS et al (2014) A new automatic seamless image stitching algorithm based on the gray value of edges. Appl Mech Mater 496–500:2241–2245

The Effect of Replacing Natural Aggregate with Geopolymer Artificial Aggregates on Air Voids of Hot Mix Asphalt

I Dewa Made Alit Karyawan, Januarti Jaya Ekaputri,
Iswandaru Widyatmoko, and Ervina Ahyudanari

Abstract High voids in hot mix asphalt (HMA) can increase sensitivity to moisture, resulting in premature failure of asphalt pavements. The use of artificial aggregates as a substitute for natural aggregates has been considered in this paper to optimize air voids that meet the requirements. The reason is that natural aggregates are non-renewable raw materials and their availability continues to decrease. This study aims to determine the effect of replacing some natural aggregates with artificial aggregates on air voids. This experiment assessed 2 sets of HMA standard gradations, namely the Bina Marga standard (Indonesian standard) and the Federal Aviation Administration (FAA) standard. For each set, 2 different raw materials are used, namely: 1) asphalt mixture using 100% natural aggregate and 2) asphalt mixture using a combination of 25% artificial aggregate and 75% natural aggregate. The analysis includes volumetric mixes, such as VIM (voids in mix), VMA (voids in mineral aggregates), and VFA (bitumen filled voids). The results of the analysis found that the replacement of natural aggregates with artificial aggregates affected the cavities formed in the asphalt mixture. HMA that uses natural aggregates, shows VMA, VFA and VIM that meet the requirements for pavement surface materials.

Keywords Artificial Aggregates · Void in Minerals Aggregate · Void Filled with Asphalt and Void in Mix

I. D. M. A. Karyawan (✉)
Department of Civil Engineering, Faculty of Engineering, Universitas Mataram, Jalan Majapahit 62, Mataram 83125, West Nusa Tenggara, Indonesia
e-mail: dewaalit@unram.ac.id

J. J. Ekaputri · E. Ahyudanari
Department of Civil Engineering, Faculty of Civil, Planning, and Geo-Engineering, Institut Teknologi Sepuluh Nopember, Kampus ITS Sukolilo, Surabaya 60111, East Java, Indonesia

I. Widyatmoko
Transportation and Infrastructure Materials Research, AECOM, 12 Regan Way, Nottingham NG9 6RZ, UK

© The Author(s) 2023   455
G. Feng (ed.), *Proceedings of the 9th International Conference on Civil Engineering*,
Lecture Notes in Civil Engineering 327,
https://doi.org/10.1007/978-981-99-2532-2_39

1 Introduction

Hot mix asphalt (HMA) material is easily applied in road constructions. However, inaccurate applications can lead to early damage to the road, before reaching the designed service life. Pavement age and the influence of heat, weather and traffic, cause changes in the durability of the pavement [1]. These changes can damage the function of bitumen as a binder and as a waterproof agent [2], resulting in decreased durability.

High air voids can also cause premature failure due to increased moisture sensitivity. Moisture damage can be defined as the loss of strength and durability in asphalt mixtures caused by the presence of water. This damage is mainly due to the aggregate stripping of HMA [3, 4]. Gradations and aggregate shapes affect the size of air voids in HMA. Pavement durability can be improved by using good material and the right gradation with the optimum air void content.

In this study, aggregates made with geopolymers using Suralaya fly ash materials. The artificial aggregates have physical properties broadly comparable to those of natural aggregate, but with the exception that they have higher water absorption value (3%) [5, 6, 7].

Study of the variation of natural aggregate with plastic with variations, 0%, 10%, 20%, 30%, 40%, and 50% in the mixture of asphalt concrete wearing course (AC-WC), showing the lowest VMA (void in mineral aggregate) and VIM (void in mix) found in mixtures that do not use plastic additives, as well as for the highest VFA (void filled with asphalt). Recommendations for reducing porosity can be done by increasing compaction energy in mixtures that use plastic as a natural aggregate substitute [8]. Some disadvantages of previous studies, require the study of the use of other materials as natural aggregate substitutes. Therefore the aim of this research is to find out the effect of replacing some natural aggregates with artificial aggregates on voids.

2 Methods

2.1 Design Experiment

In this experiment, 2 sets of HMA standard gradation were used, namely Bina Marga (Indonesian standard) [9] and gradations of the Federal Aviation Administration (FAA) [10]. The new HMA is designed for surface layers. Bina Marga gradations are intended for road pavements, while FAA gradations for airfield pavements. Samples were prepared in accordance with the Marshall mix design method. Five asphalt contents were used for each gradation and each composition. For each set, 2 different raw materials were adopted, namely: 1) asphalt mixture using 100% natural aggregate and 2) asphalt mixture using 25% artificial aggregate and 75% natural aggregate. So that the total number of samples is 3 specimens × 5 asphalt content × 2 variations in aggregate proportions × 2 gradation variations = 60 samples. Furthermore, the analysis is based on the volumetric measurement data. The volumetric values sought are VIM, VMA and VFA.

2.2 Material

Natural aggregates, artificial aggregates and asphalt for HMA are used. The aggregate used to make HMA is an aggregate made with Suralaya fly ash with 50° pan granulator slope and the ratio of alkali activator (Na_2SiO_3/NaOH) = 2.5 [5, 5]. Specific gravity material used in making Marshall HMA samples is shown in Table 1.

Marshall samples are made from a mixture of coarse aggregates, fine aggregates, filers and asphalt. The first one prepared is aggregate with gradations according to the design. The gradations used are shown in Fig. 1. Aggregate needs are shown in Table 2 and Table 3.

Based on Table 2 of the Bina Marga method, the percentage of coarse aggregate (CA) = 54%, fine aggregate (FA) = 39% and Filler = 7%. Estimated asphalt content is calculated using the equation:

$$Pb = 0.035 \ (\% \ CA) + 0.045 \ (\% \ FA) + 0.18 \ (\% \ FF) + k \qquad (1)$$

where:

Pb = initial asphalt content design is% of mixed weight
CA = coarse aggregate is% against aggregate held by sieve no. 8
FA = fine aggregate is% of aggregate escaped sieve no. 8 held by sieve no. 200
FF = filler is% against aggregate escapes sieve no. 200
k = constants ranging from 0.5 to 1.0

Using the formula (1), the asphalt content used for the Bina Marga gradation is:

$$Pb = 0.035 \ (54) + 0.045 \ (39) + 0.18 \ (7) + 1$$

Table 1 Specific gravity material properties for HMA

No.	Material Type	Unit	Bulk SG	Apparent SG	Effective SG
1	Natural aggregate				
	a. Coarse aggregate	gr/cm^3	2.54	2.70	2.62
	b. Fine aggregate	gr/cm^3	2.60	2.85	2.72
	c. Filler	gr/cm^3	2.75	2.75	2.75
2	Artificial aggregate				
	a. Coarse aggregate	gr/cm^3	1.85	2.09	1.97
3	Asphalt	gr/cm^3	-	-	1.05

Note: SG = specific gravity

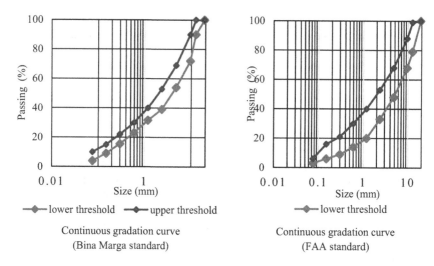

Fig. 1 Aggregate gradation for the Marshall for Bina Marga [11] and FAA Standard [10]

Pb = 1.89 + 1.76 + 1.26 + 1 = 5.91% round to 6.0%

Used five asphalt content, 5.0%, 5.5%, 6.0%, 6.5%, 7.0%. So that requires asphalt = 5.0% × 1200 g + 5.5% × 1200 g + 6.0% × 1200 g + 6.5% × 1200 g + 7.0% × 1200 g = 360 g. Each bitumen content is made in 3 samples, so it requires total asphalt = 3 × 360 g = 1080 g.

Based on Table 3 of the FAA method, the percentage of coarse aggregate (CA) = 57%, fine aggregate (FA) = 38.5% and Filler = 4.5%. Using formula (1), the asphalt content used for FAA gradation is:

Pb = 0.035 (57) + 0.045 (38.5) + 0.18 (4.5) + 1
Pb = 1.99 + 1.73 + 0.81 + 1 = 5.53% rounded off 5.5%.

Used five asphalt content, 5.0%, 5.5%, 6.0%, 6.5%, 7.0%. So that requires asphalt = 5.0% × 1200 g + 5.5% × 1200 g + 6.0% × 1200 g + 6.5% × 1200 g + 7.0% × 1200 g = 360 g. Each asphalt content is used in 3 samples, so it requires a total of asphalt = 3 × 360 g = 1080 g.

In the same way, based on Table 3 of the FAA method, the percentage of coarse aggregate (CA) = 57%, fine aggregate (FA) = 38.5% and Filler = 4.5%.

Using formula (1), the asphalt content used for FAA gradation is:

Pb = 0.035 (57) + 0.045 (38.5) + 0.18 (4.5) + 1
Pb = 1.99 + 1.73 + 0.81 + 1 = 5.53% rounded off 5.5%.

Five content of asphalt used is 5.5%, 6.0%, 6.5%, 7.0%, 7.5%. So that requires asphalt = 5.5% × 1200 g + 6.0% × 1200 g + 6.5% × 1200 g + 7.0% × 1200 g + 7.0% × 1200 g = 390 g. Each bitumen content is used in 3 samples, so it requires total asphalt = 3 × 390 g = 1170 g.

Table 2 Calculation of aggregate requirements for Marshall samples of Bina Marga method [9, 11]

Sieve number		Middle treshold (%)	Retained (%)	Composition (%)		Composition (gram)		Composition (%)		Composition (gram)		Composition (%)		Composition (gram)	
in	mm			AA	NA	AA	NA	AA	NA	AA	NA	AA	NA	AA	NA
				0%	100%	0%	100%	25%	75%	25%	75%	25%	75%	25%	75%
3/4	19.1	100.0	–	–	–	–	–	–	–	–	–	–	–	–	–
1/2	12.5	95.0	5.0	–	5.0	–	60	1.3	3.8	15.0	45	1.3	3.8	15.0	45
3/8	9.5	81.0	14.0	–	14.0	–	168	3.5	10.5	42.0	126	3.5	10.5	42.0	126
NO. 4	4.8	61.5	19.5	–	19.5	–	234	4.9	14.6	58.5	176	4.9	14.6	58.5	176
NO. 8	2.4	46.1	15.5	–	15.5	–	185	3.9	11.6	46.4	139	3.9	11.6	46.4	139
NO. 16	1.2	35.8	10.3		10.3		123		10.3		123		10.3		123
N0. 30	0.6	26.6	9.3		9.3		111		9.3		111		9.3		111
N0. 50	0.3	18.8	7.8		7.8		94		7.8		94		7.8		94
N0. 100	0.2	12.0	6.8		6.8		81		6.8		81		6.8		81
N0. 200	0.1	7.0	5.0		5.0		60		5.0		60		5.0		60
Pan	0.0	–	7.0		7.0		84		7.0		84		7.0		84
Total			100	–	100	–	1,200	13	87	162	1,038	13	87	162	1,038

Note: AA = Artificial aggregate; NA = Natural Aggregate

Table 3 Calculation of aggregate requirements for Marshall samples of FAA method [10]

Sieve number		Middle treshold (%)	Retained (%)	Composition (%)		Composition (gram)		Composition (%)		Composition (gram)	
in	mm			AA 0%	NA 100%	AA 0%	NA 100%	AA 25%	NA 75%	AA 25%	NA 75%
3/4	19.1	100.0	–	–	–	–	–	–	–	–	–
1/2	12.5	89.0	11.0	–	11.0	–	132	2.8	8.3	33.0	99
3/8	9.5	78.0	11.0	–	11.0	–	132	2.8	8.3	33.0	99
NO. 4	4.8	58.0	20.0	–	20.0	–	240	5.0	15.0	60.0	180
NO. 8	2.4	43.0	15.0	–	15.0	–	180	3.8	11.3	45.0	135
NO. 16	1.2	30.0	13.0		13.0		156		13.0		156
N0. 30	0.6	22.0	8.0		8.0		96		8.0		96
N0. 50	0.3	15.0	7.0		7.0		84		7.0		84
N0. 100	0.2	11.0	4.0		4.0		48		4.0		48
N0. 200	0.1	4.5	6.5		6.5		78		6.5		78
Pan	0.0	–	4.5		4.5		54		4.5		54
Jumlah	50.5		100	–	100	–	1,200	14	86	171	1,029

Note: AA = Artificial aggregate; NA = Natural Aggregate

2.3 Marshall Sample Preparation

Sampling is done with the following steps [11]:

1) Prepare asphalt and aggregate with the weight for each sample of 1200 g. The aggregate amount used refers to asphalt content. For example, for asphalt content of 6%, then used asphalt 6% × 1200 g = 72 g and aggregate 94% × 1200 g = 1128 g. Furthermore, the other asphalt content is done in the same way.
2) Perform mixing, starting with heating the aggregate at a certain temperature. Then add the asphalt that has been heated with the amount according to the design.
3) Conduct compaction by giving a load to the mixture in the mould of the marshal sample. Mash on both sides were given 75 times for heavy traffic designs.

2.4 Void Calculation

The voids discussed in this paper include 1) Void in minerals aggregate (VMA) is the volume of voids found between aggregated particles of a compacted pavement mixture. This void consists of air voids and effective asphalt content volume. The aggregate volume is calculated from bulk density; 2) Void in a mixture (VIM) is the total volume of air in between asphalt-coated aggregate particles in compacted pavement; 3) Void filled with asphalt (VFA) is part of the void that is between the VMA which is filled with effective asphalt. All of these voids are expressed in percent [11].

As pavement material for roads and airports, the required VMA value is at least 15%. VMA is calculated based on bulk density (Gsb) aggregate and is expressed as the volume percent of the compacted bulk mixture. VMA can also be calculated against the total mixture weight or the total aggregate weight (formula 2 and formula 3).

a) Against the total mixed weight

$$VMA = 100 - \frac{G_{mb} \times P_s}{G_{sb}} \qquad (2)$$

b) Against the total aggregate weight

$$VMA = 100 - \frac{G_{mb}}{G_{sb}} \times \frac{100}{100 + P_b} \times 100 \qquad (3)$$

where:

VMA = Void in minerals aggregate, percent volume bulk.
Gsb = Bulk specific gravity aggregate.
Gmb = Bulk specific gravity of mixed (AASHTO T-166) [12]

Pb = Asphalt content, percent total mixture.

Ps = Aggregate content, percent of total mixture

Void in a mixture (VIM) consists of air space between aggregate particles covered with asphalt. As a pavement material for the road, the required VIM value is at least 3.5% and a maximum of 5%. While for the airfield at least 3.5%. VIM is expressed in percent, can be determined by formula 4.

$$VIM = 100 \times \frac{G_{mm} - G_{mb}}{G_{mm}} \qquad (4)$$

where:

VIM = Void in a mixture, percent of the total mixture

Gmb = Bulk specific gravity of mixed (AASHTO T-166) [12]

Gmm = Specific gravity maximum of mix (ASTM 2041) [13]

Void filled with asphalt (VFA) is a percent of voids contained in VMA filled with asphalt, not including asphalt absorbed by aggregate. As a pavement material for the road, the VFA value for required pavement is at least 65%, while for the airfield at least 76% and a maximum of 82%. VFA is calculated by Formula 5.

$$VFA = \frac{100(VMA - VIM)}{VMA} \qquad (5)$$

where:

VFA = Void filled with asphalt, percent of VMA

VMA = Void in minerals aggregate, percent volume bulk.

VIM = Void in a mixture, percent of total mixture

3 Result

The sample void value data is obtained by calculating the test results data, using formulas 2–5. The complete calculation results are shown in Table 4 and Table 5.

Table 4 Dense graded Bina Marga [9]

Composition: 0% AA: 100% NA				Composition: 25% AA: 75% NA			
Asphalt content	VMA	VFA	VIM	Asphalt content	VMA	VFA	VIM
5.0	18.8	58.1	7.9	5.0	20.3	52.8	9.6
5.5	18.3	63.6	7.7	5.6	20.6	61.0	8.5
6.0	19.4	67.5	7.6	6.0	19.4	67.5	7.6
6.5	17.2	82.6	5.4	6.5	19.0	72.2	7.5
7.0	16.6	96.2	3.4	7.0	20.3	75.4	7.6

Note: AA = Artificial aggregate; NA = Natural aggregate; VIM = Void in mixture; VMA = Void in Mineral Aggregate; VFA = Void Filled with Asphalt

Table 5 Dense graded FAA [10]

Composition: 0% AA: 100% NA				Composition: 25% AA: 75% NA			
Asphalt content	VMA	VFA	VIM	Asphalt content	VMA	VFA	VIM
5.0	22.8	60.0	9.1	4.5	19.2	48.4	9.9
5.5	22.7	65.4	7.9	5.0	18.4	56.6	8.0
6.0	21.5	76.1	5.1	5.5	18.5	61.9	7.1
6.5	20.4	87.2	2.6	6.0	18.0	70.7	5.3
7.0	22.2	83.9	3.6	6.5	17.9	75.3	4.4

Note: AA = Artificial aggregate; NA = Natural aggregate; VIM = Void in mixture; VMA = Void in Mineral Aggregate; VFA = Void Filled with Asphalt

4 Analysis and Discussion

Figure 2 shows the value of Void (VMA, VFA and VIM) produced from Marshall samples with gradations of Bina Marga. Based on the specifications for VMA, both HMA with natural aggregates and that with 25% replacement of the natural coarse aggregates show compliance with the requirements. The VFA value for both HMA with natural aggregate and that with 25% replacement of the natural coarse aggregates do not meet the requirements for asphalt content <5.8%. This means that more than 5.8% of asphalt will be needed to provide the required asphalt film thickness to bind to the aggregate. VIM in HMA without natural aggregate replacement does not meet the requirements for asphalt content <6.5%, while that with 25% replacement of natural coarse aggregate does not show compliance for all asphalt content.

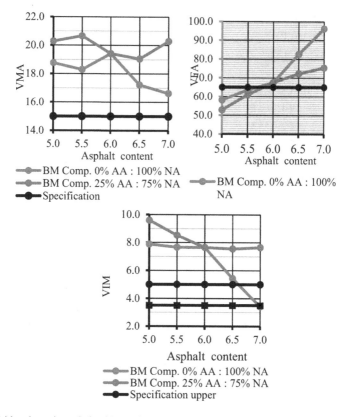

Fig. 2 Void on hot mix asphalt with continuous gradation (Bina Marga) for highway pavement [9]

At Fig. 3 shows the value of Void (VMA, VFA and VIM) produced from Marshall specimens with FAA gradations. Based on specification limits for VMA, both HMA with natural aggregate and that with a replacement of 25% natural coarse aggregate met the compliance requirements. VFA value for HMA with natural aggregate does not meet the specification requirements for asphalt content 5.25%–6.0%, while that with 25% replacement of natural coarse aggregate show no compliance for all asphalt contents. VIM in HMA with natural aggregate does not match the requirements for asphalt content <6.3%, while that with 25% replacement of natural coarse aggregate the compliance is met for all asphalt contents.

Refer to Fig. 2 and Fig. 3, Fig. 4 shows the interval of bitumen content based on voids that meet the requirements used as a binder. HMA which was made with 100% natural aggregate produces air void that meets the requirements for pavement material for both Bina Marga and FAA gradations. In the Bina Marga gradation, asphalt content of 6.7%–7% meet the requirements for VMA, VFA and VIM. As for the FAA, asphalt content at intervals of 6%–6.25% is eligible for VMA, VFA and VIM.

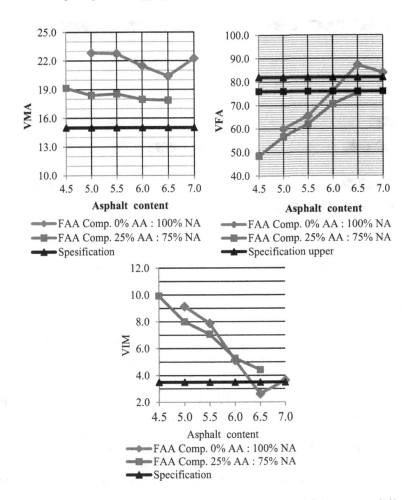

Fig. 3 Void on hot mix asphalt with continuous gradation (FAA) for airfield pavement [10]

The HMA made by replacing 25% of natural coarse aggregate with artificial aggregates, showed air voids that met the requirements for VIM to Bina Marga and VFA to FAA gradations. In the Bina Marga gradation, asphalt levels of 5.8%–7% will meet the requirements for VMA and VFA. As for the FAA, the asphalt level at the interval will be 4.5%–6.5% are eligible for VMA and VIM.

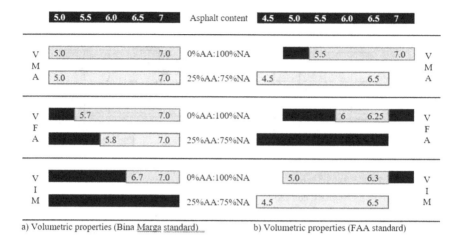

a) Volumetric properties (Bina Marga standard)   b) Volumetric properties (FAA standard)

Fig. 4 Summary of asphalt contents and volumetric properties requirements

5   Conclusions

Based on the discussion of the results of data analysis, related to the effect of replacing natural aggregates with artificial aggregates are as follows:

A) In both mixed aggregate gradations to Bina Marga and FAA standards, which use 100% natural aggregate, VMA, VFA and VIM are still within the requirements so that they can be used as pavement material.

B) Replacement of 25% natural aggregate with artificial aggregates on Bina Marga gradations and FAA gradations has an impact on air voids. Air voids that did not meet the requirements due to the replacement of natural aggregates with artificial aggregates, were VIM and VFA for Bina Marga and FAA standard gradations respectively.

Acknowledgements The authors express their gratitude and acknowledge those who contributed directly or indirectly to this project. Special thanks to the *Beasiswa Unggulan Dosen Indonesia-Dalam Negeri (BUDI-DN), through the Lembaga Pengelola Dana Pendidikan (LPDP)*, Ministry of Finance of the Republic of Indonesia, for their support through the Indonesian education scholarship program.

References

1. Arjun N (2019) Durability of bituminous pavements and factors affecting it. The constructor-civil engineering home for civil engineers. https://theconstructor.org/transportation/durability-bituminous-pavements/16209/

2. Lee D (1969) Special report: durability and durability tests for paving asphalt. The Engineeriri–Research Institute for the Iowa State Highway Commission, Iowa
3. Yilmaz A, Sargin Ş (2012) Water effect on deteriorations of asphalt pavements. J Sci Technol 2(1). https://www.researchgate.net/publication/267686527_Water_Effect_on_Deteriorations_of_Asphalt_Pavements
4. Kennedy TW, Roberts FL, Lee KW (1977) Evaluation of moisture effects on asphalt concrete mixtures. Transp Res Rec (3):134–143
5. Karyawan IDMA, Ekaputri JJ, Widyatmoko I, Ahyudanari E (2019) The effects of $Na_2SiO_3/NaOH$ ratios on the volumetric properties of fly ash geopolymer artificial aggregates
6. Yuliana H, Karyawan IDMA, Murtiadi S, Ekaputri JJ, Ahyudanari E (2019) The effect of slope granulator on the characteristic of artificial geopolymer aggregate used in pavement. J Eng Sci Technol 14(3):1466–1481
7. Karyawan IDMA, Ekaputri JJ, Widyatmoko I, Ahyudanari E (2020) The effect of various $Na_2SiO_3/NaOH$ ratios on the physical properties and microstructure of artificial aggregates. J Eng Sci Technol 15(2):1139–1154
8. Gunadi MAD, Thanaya INA, Negara INW (2013) Analisis Karakteristik Campuran Aspal Beton Lapis Aus (AC-WC) dengan Menggunakan Plastik Bekas (Analysis of asphalt concrete-wearing course characteristics by plastics as a partial replacement of aggregate). J Ilm Tek Sipil 17(2):191–201
9. Syarwan S, Hazmi F (2013) Kajian Gradasi Agregat Beton Aspal Lapis Aus (AC-WC) Terhadap Nilai Parameter Marshall Berdasarkan Spesifikasi Bina Marga Tahun 2010 (Study of aggregate gradation for wear course asphalt concrete against marshall parameter value based on Bina Marga Specif). Reintek J Ilmu Pengetah dan Teknol Terap 8(2):149–155
10. U.S. Department of Transportation Federal Aviation Administration, Advisory Circular. Airport Pavement Design and Evaluation AC No: 150/5320-6F (2016)
11. Departemen Pekerjaan Umum, Pedoman Perencanaan Campuran Beraspal dengan Pendekatan Kepadatan Mutlak (Guidelines for mixed paved planning with the absolute density approach). Lampiran No. 3 Keputusan Direktur Jenderal Bina Marga Departemen Pekerjaan Umum, no. 025 (1999)
12. American Association of State Highway and Transportation Officials (AASHTO) (2018) Bulk specific gravity (GMB) of compacted asphalt mixtures using saturated surface-dry specimens FOP for AASHTO T 166, pp 1–10
13. Standard test method for theoretical maximum specific gravity and density of bituminous paving mixtures. ASTM international, West Conshohocken, United States

Finite Element Analysis of Reinforced Hollow High-Strength Concrete Filled Square Steel Tubular Middle-long Column under Eccentric Load

Zhijian Yang, Jixing Li, Guochang Li, and Weizhe Cui

Abstract Traditional hollow concrete filled steel tube components are mainly used in the transformer power structure. In order to improve the mechanical properties and application scope of hollow concrete filled steel tube components, a new type of reinforcement hollow steel tube high-strength concrete combination structure is proposed. Given that the components in the construction system are frequently in an eccentric stress condition throughout the structural system, the finite element analysis program ABAQUS produces 20 eccentric compression middle-long column models. The overall force process of the composite members is investigated, and parametric analysis is done on the yield strength, eccentricity, steel content, and various lengths to slenderness ratios of the various steel components. According to the results of finite element analysis, the stress process of reinforced hollow high strength concrete filled square steel tube components is mainly divided into four stages: elastic section, elastic–plastic section, plastic strengthening section and descending stage. The configuration of steel bars in the composite components can significantly improve the ductility of the composite components. This new combination structure can be used as a wind-resistant column in industrial plants in the future, which has certain application prospects and economic benefits.

Keywords Hollow Concrete Filled Steel Tube · Eccentric load · Middle-Long Columns · Finite Element Analysis

1 Introduction

Steel bars may greatly increase the ductility of concrete filled steel tubular (CFST) in addition to increasing the final bearing capacity of composite components. The compressive strength of hollow steel tube concrete components is around 10% higher than that of conventional poured steel tube concrete components as a result of

Z. Yang (✉) · J. Li · G. Li · W. Cui
School of Civil Engineering, Shenyang Jianzhu University, Shenyang 110168, China
e-mail: faemail@163.com

© The Author(s) 2023
G. Feng (ed.), *Proceedings of the 9th International Conference on Civil Engineering*,
Lecture Notes in Civil Engineering 327,
https://doi.org/10.1007/978-981-99-2532-2_40

469

the steam health preservation procedure used in their fabrication [1]. Researchers both domestically and internationally have conducted a great deal of experimental work recently to enhance the mechanical characteristics of CFST members. In 2016, Hamidian et al. [2] conducted axial compression test study on 15 specimens fitted with spiral reinforcement. The results showed that the spiral reinforcement design outperformed conventional concrete-filled steel tube concrete columns. The test results are compared with ACI 318-11 and EC4-1994 to indicate a substantial improvement in the performance of concrete-filled steel tube columns after yielding. The test results and EC4 are well-aligned, and a conservative estimate of ACI has been made.

In 2018, Hasan et al. [3] tested the mechanical performance of reinforced concrete filled steel tube columns under axial loads and compared the performance of composite columns made of no more than two alternative configurations of steel bars welded into steel tubes and embedded in concrete. Because of the constraint effect between the stirrup and the steel tube, a new axial ultimate load model was proposed in order to accurately predict the member's ultimate bearing capacity. In 2020, Fujiang Xia et al. [4] showed through tests that the configuration of steel bars in the member. Instead of increasing the wall thickness of steel tubes, it is preferable to enhance the mechanical characteristics of concrete columns welded steel tube. In 2021, Chen Zongping [5] carried out eccentric compression tests on 18 specimens by adjusting the spiral steel bar diameter, longitudinal bar diameter and other parameters, and suggested the optimal steel blending design scheme by parameter analysis.

In 2022, Yuan [6] et al. conducted axial compression test research on concrete filled square steel tube columns with built-in spiral steel bar constraints, which proved that increasing the volume of HSS spiral bars improved the ultimate bearing capacity of components better than improving the ultimate bearing capacity of components by increasing the external steel tube of the same volume. In 2015, Lu et al. [7] carried out an experimental investigation on the RC column reinforced by SCC filled square steel tube under eccentric compression, and developed a design formula to compute the ultimate strength of the reinforced column under eccentric load. In 2020, Li [8] et al. conducted that the reinforcement of damaged RC square columns with a square steel tube sandwich can significantly improve the stiffness and load-bearing capacity of RC columns and significantly improve the ductility of the members by studying the reinforcement of damaged RC square columns with a square steel tube sandwich.. In 2018, Yang et al. [9] proposed a reinforced hollow steel tube high-strength concrete column to improve the mechanical properties of the member while reducing the self-weight of the member, while to a certain extent reducing the wet work at the construction site..

Using ABAQUS finite element software, the author simulates the functioning state of reinforced hollow high strength concrete filled square steel tube components under eccentric load, taking into account that in practical applications, hollow components are typically in an eccentric state under wind load, seismic action, or the entire structural system. The effects of parameters such as eccentricity, slenderness ratio, and steel content on the mechanical properties of components are studied, which provides a theoretical basis for practical engineering applications.

2 Model Design

In this article, a total of 20 eccentrically compressed mid-length members are designed. The members are often made of steel tube, sandwich concrete, and PHC column concrete. Spiral steel bars, prestressed tendons, and HRB400 steel bars are used in the construction of the tube columns.

Table 1 displays the different member parameters in detail. The major focus of this article is on the relationship between mechanical qualities of components and steel yield strength, eccentricity, steel content, and slenderness ratio. Figure 1 depicts the components' cross-sectional structure.

Table 1 Design parameters of specimens

Specimen Number	$t/$ mm	Stirrup/ mm	Prestressed tendon/mm	e/r	λ	f_y $fy/$ MPa	$fcu1$	$fcu2$	Deformed bar/mm
ELRHCFST-01	6	Φ4@45	6Φ7.1	0.2	17.32	355	60	80	6C16
ELRHCFST-02	6	Φ4@45	6Φ7.1	0.3	17.32	355	60	80	6C16
ELRHCFST-03	6	Φ4@45	6Φ7.1	0.4	17.32	355	60	80	6C16
ELRHCFST-04	6	Φ4@45	6Φ7.1	0.5	17.32	355	60	80	6C16
ELRHCFST-05	6	Φ4@45	6Φ7.1	0.6	17.32	355	60	80	6C16
ELRHCFST-06	6	Φ4@45	6Φ7.1	0.7	17.32	355	60	80	6C16
ELRHCFST-07	6	Φ4@45	6Φ7.1	0.8	17.32	355	60	80	6C16
ELRHCFST-08	6	Φ4@45	6Φ7.1	0.5	17.32	355	60	80	—
ELRHCFST-09	6	Φ4@45	6Φ7.1	0.5	17.32	235	60	80	6C16
ELRHCFST-10	6	Φ4@45	6Φ7.1	0.5	17.32	390	60	80	6C16
ELRHCFST-11	6	Φ4@45	6Φ7.1	0.5	17.32	420	60	80	6C16
ELRHCFST-12	6	Φ4@45	6Φ7.1	0.5	17.32	460	60	80	6C16
ELRHCFST-13	5	Φ4@45	6Φ7.1	0.5	17.32	355	60	80	6C16
ELRHCFST-14	7	Φ4@45	6Φ7.1	0.5	17.32	355	60	80	6C16
ELRHCFST-15	8	Φ4@45	6Φ7.1	0.5	17.32	355	60	80	6C16
ELRHCFST-16	9	Φ4@45	6Φ7.1	0.5	17.32	355	60	80	6C16
ELRHCFST-17	6	Φ4@45	6Φ7.1	0.5	20.78	355	60	80	6C16
ELRHCFST-18	6	Φ4@45	6Φ7.1	0.5	24.25	355	60	80	6C16
ELRHCFST-19	6	Φ4@45	6Φ7.1	0.5	31.18	355	60	80	6C16
ELRHCFST-20	6	Φ4@45	6Φ7.1	0.5	41.57	355	60	80	6C16

a. The yield strength of the spiral stirrup is 650 MPa, and the diameter of the distribution circle is 230 mm;

b. The yield strength of prestressed tendon is 1420 MPa;

c. The common steel bar model is HRB400, and the yield strength is 400 MPa

Fig. 1 Schematic diagram
of the component section

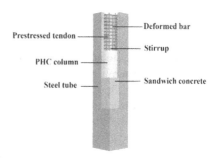

3 Numerical Model Building

3.1 Selection of Material Constitutive Relation

The sandwich concrete and PHC column concrete make up the middle-long column concrete portion of the reinforced hollow high strength concrete filled square steel tube. The concrete plastic damage model of ABAQUS is used for the concrete [10]. The stress state is analogous to a three-dimensional load since it is jointly restrained by the steel tube. The concrete uniaxial stress–strain model modified by Liu Wei is adopted for the stress state and constitutive relationship [11]. The prestress of prestressed tendon is applied by cooling technique [12], the steel tube employs the low carbon steel five-fold line model, and the steel bar uses the two-fold line model [13].

3.2 Establishment of Finite Element Model

The concrete used for the tube column is reinforced with spiral, prestressed tendon, and HRB400 steel bars. Tie restrictions are used in PHC column concrete and sandwich concrete. Sandwich concrete and steel tube are designed to use hard contact in the vertical plane, and the Coulomb friction model is used in the tangential direction. Then, as illustrated in Fig. 2, bind the tube string, sandwich concrete, and steel tube end surfaces to the end plate. The loading method is displacement loading, with the rotation of the top and bottom of the column set to zero in the X, Z direction. Next, set the displacement in the X, Y direction of the column's top and the X, Y, Z direction of the column's base to zero. For steel tubes, PHC columns, and sandwich concrete, C3D8R solid units are utilised. For steel bars, truss units are used.

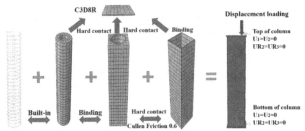

Fig. 2 Model building process

Fig. 3 Component
load–deflection curve of
middle section in literature
[14]

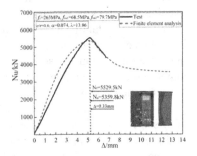

3.3 Validation of Finite Element Model

When the finite element analysis results are compared to the experimental data from the literature [14], it is discovered that the ultimate bearing capacity of the finite element simulation components is 3.15% lower than the ultimate bearing capacity of the test. Simultaneously, the deflection of the mid-height corresponding to the ultimate bearing capacity of the components is 0.33 mm different from the test. Figure 3 shows that the finite element analysis curve is in good agreement with the trend of the test curve, demonstrating the model's stability.

4 Finite Element Analysis Results

4.1 Analysis of the Whole Process of Force

Elastic phase (OA): Fig. 4 shows that when the load increases, the cross-sectional deflection of the elastic stage elements also increases. The components are in a full-section compression condition, and the curves are roughly linearly coupled at this point. Since the stress states of the steel tube, sandwich concrete, and PHC column

Fig. 4. ELRHCFST-4
member load-medium
section deflection curve

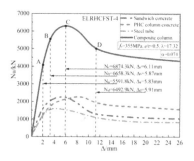

concrete are all different at this point, the steel tube does not have a significant restraint impact on the concrete.

Elastoplastic stage (AB): The element enters the elastic–plastic stage as it proceeds from characteristic point A to characteristic point B; at this point, the steel tube and concrete continue to support the majority of the load. The steel tube on the compression side commences to gradually enter the yield stage when it achieves characteristic point B.

Plastic strengthening stage (BC): The component undergoes the plastic strengthening stage from characteristic point B to characteristic point C at this point, and the sandwich concrete and PHC column concrete share the bulk of the internal force. At this stage, as the load increases, so does the growth rate of the segment deformation in the component.

Descending stage (CD): The maximum bearing capacity of the member has been attained at characteristic point C. The stiffness of the part continues to decrease when the load is applied. The sandwich concrete is currently being crushed in a portion of the segment on the compression side, and the PHC column concrete is also gradually gaining its functional capacity at this point.

4.2 Effect of Configuration Reinforcement on the Mechanical Properties of Components

According to Fig. 5, the ultimate bearing capacity of components constructed with HRB400 is enhanced by 3.46% when compared to components without HRB400. In accordance with the maximum bearing capacity, the middle section diverts by 0.43 mm more. Figure 5 shows that the ductility of the HRB400-configured components has also been greatly enhanced. Concrete structures made of hollow steel tubes are more resilient to elastoplastic deformation than components without steel bars.

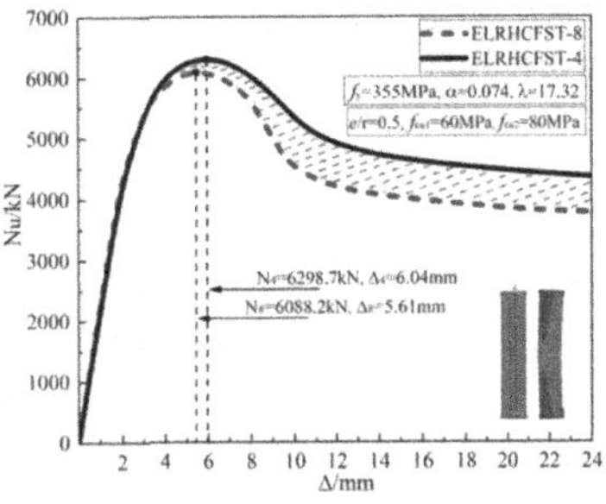

Fig. 5 The influence of HRB400 on the mechanical properties of components

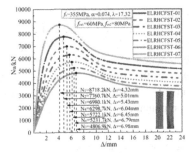

Fig. 6 Effect of eccentricity on load-mid-section deflection curve

5 Parameter Analysis

5.1 Effects of Eccentricity

The load-mid-heigth section deflection curves for components with various eccentricities are shown in Fig. 6. The final bearing capacity of the components reduces by 10.9%, 10.1%, 9.8%, 9.2%, 8.6%, and 8.1%, respectively, as the eccentricity goes from 0.2 to 0.8. The ultimate bearing capacity of the component constantly declines at a rate proportional to the increase in eccentricity. Figure 7 depicts the variation in elastic stiffness for the various eccentricity-related components. It is evident that both the component's elasticity and eccentricity both drop from 0.8 to 0.2. The stiffness rose by 11.6%, 19.6%, 20.2%, 23.3%, 31.6%, and 50.5%. The elastic stiffness of the component increases as eccentricity decreases, and the growth rate of the elastic modulus of the component likewise increases.

5.2 Effect of Steel Strength

Figure 8 illustrates the mid-heigth section deflection curve corresponding to the segments of the steel tube under different yield strengths. It can be seen that the increase in the yield strength of the steel has no obvious effect on the elastic stiffness

Fig. 7 Effect of eccentricity
on initial stiffness

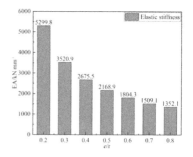

of the components. The ultimate bearing capacity of the components increased by
12.64%, 3.08%, 2.55%, and 3.24%, respectively, while the yield in steel strength
varied from 235 to 460 MPa. Figure 9 depicts the ultimate bearing capacity of the
component steel yield strength at various eccentricities. According to the investigation, the lower the eccentricity under the same conditions, the more noticeable
the improvement in steel yield strength on the ultimate bearing capacity of the
component.

Fig. 8 Effect of steel tube
yield strength on
load-medium section
deflection curve

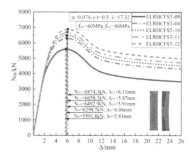

Fig. 9 Effect of steel tube
yield strength on bearing
capacity

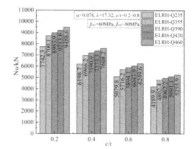

Fig. 10 Effect of steel content on load-to-middle section deflection curve

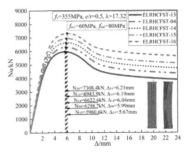

Fig. 11 Effect of steel content on bearing capacity

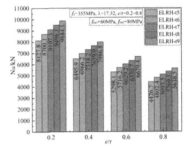

5.3 Effect of Steel Content

Figure 10 shows the load-mid-heigth section deflection curves for steel tube members with various wall thicknesses. With an increase in steel content, the member's elastic stiffness gradually rises, but elastic stiffness is also impacted by the addition of steel. Because the confinement effect of the steel tube on the concrete is improved by increasing the wall thickness of the steel tube, it is not particularly significant. The ultimate bearing capacity of the member rises by 5.67%, 5.14%, 5.25%, and 4.85%, respectively, while the steel composition varies from 0.061 to 0.112. The maximum bearing capacity of components with various steel contents at various eccentricities is shown in Fig. 11. It may be inferred that the greater the eccentricity of the components under the same parameters, the greater the effect of increasing the wall thickness of the steel tube on enhancing the components' ultimate bearing capacity.

5.4 Effect of Slenderness Ratio

The load-mid-heigth section deflection curves for members with different slenderness ratios are shown in Fig. 12. The effect of increasing the slenderness ratio on the elastic stiffness of the part becomes more obvious. The maximum bearing capacity of the member varies when the slenderness ratio increases from 17.32 to 41.57. The respective forces decreased by 1.55%, 1.64%, 2.94%, and 6.34%. The maximum

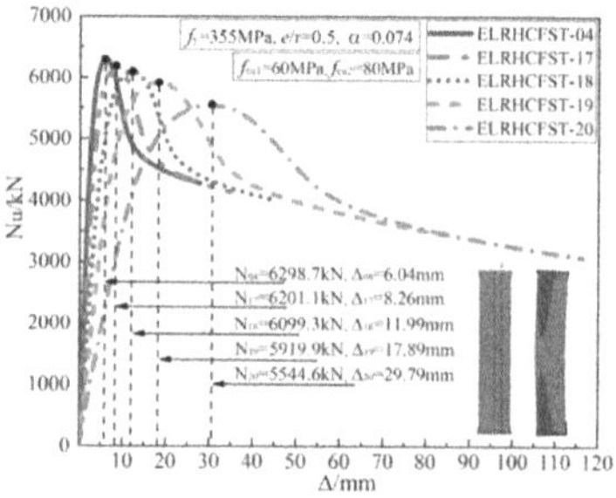

Fig. 12 Effect of slenderness ratio on load-medium section deflection curve

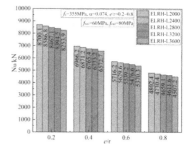

Fig. 13 Effect of slenderness ratio on bearing capacity

carrying capacity of components with various length-to-slenderness ratios at various eccentricities is shown in Fig. 13. The analysis inferred that the carrying capacity of components with various lengths and slender ratios decreases under the same conditions at various eccentricities.

6   Conclusion

(1) During the loading process, the different components of the reinforced hollow high strength concrete filled square steel tube mid-long column can work well together. The concrete of the PHC column can still play a significant role after the ultimate bearing capacity when the steel tube initially reaches the yield strength and exits the working state.

(2) When all other parameters kept constant, the change in eccentricity or slenderness ratio has the greatest impact on the elastic stiffness of the member. The change in steel content ratio has a minor effect on the elastic stiffness of the member, while the change in steel yield strength has almost no effect on the elastic stiffness of the member.

(3) When the eccentricity is small and all other parameters are kept constant, the effect of changing the steel yield strength or steel content on the final bearing capacity of the member is more obvious than the effect of altering the slenderness ratio.

Innovation

This innovative reinforced hollow concrete-filled steel tube composite member not only enhances the functionality of previous hollow concrete-filled steel tube components, but also, to a certain extent, expands the range of applications for hollow components, offering a certain benchmark for practical engineering.

Acknowledgements This research was supported by National Natural Science Foundation of China (52178148, 51808353), Excellent Youth Fund of Liaoning Province (2021-YQ-10), Fundamental scientific research project of Liaoning Provincial Department of Education (LJKZ0598).

References

1. Harbin Institute of Technology and China Academy of Building Science (2009) Technical specification of hollow concrete-filled steel tubular structures. China Architecture & Building Press, Beijing
2. Hamidian MR, Jumaat MZ, Alengaram UJ, Ramli Sulong NH, Shafigh P (2016) Pitch spacing effect on the axial compressive behaviour of spirally reinforced concrete-filled steel tube (SRCFT). Thin Wall Struct 100:213–223
3. Hasan HG, Ekmekyapar T, Shehab BA (2019) Mechanical performances of stiffened and reinforced concrete-filled steel tubes under axial compression. Mar Struct 65:417–432
4. Alifujiang X, Yierpanjiang A, Liu X (2020) An experimental study on axial compressive performance of reinforced concrete filled steel tubular column made of thin-walled steel tube and high-strength concrete. Prog Steel Build Struct 22:85–91
5. Chen ZP, Huang LZ, Tan QH (2021) Experimental study and analysis of concrete filledsquare steel tube columns with spiral reinforcementunder eccentric compression. Eng Mech 38:205–219
6. Yuan F (2022) Cao L and Li H 2022 Axial compressive behaviour of high-strength steel spiral-confined square concrete-filled steel tubular columns. J Constr Steel Res 192:107245
7. Lu YY, Liang HJ, Li S, Li N (2015) Numerical and experimental investigation on eccentric loading behavior of RC columns strengthened with SCC filled square steel tubes. Adv Struct Eng 18:295–309
8. Li S, Zhao Q, Lu Y-Y, Xiao L (2020) Study on the bias behavior of RC columns strengthened by square steel tube sandwich concrete. J Wuhan Univ Technol 42:51–56
9. Yang ZJ, Han JM, Lei YQ, Li GC, Liu SA (2018) New kind of hollow steel tube high strength concrete column with reinforcement: CN108590037A 2018-09-28
10. Han LH (2007) Concrete-Filled Steel Tube Structures: Theory and Practice. Science Press, Beijing
11. Liu W (2005) Research on mechanism of concrete-filled steel tubes subjected to local compression Ph.D dissertation of Fuzhou University
12. The Central People's Government of the People's Republic of China and Standardization Administration of the People's Republic of China 2010 Pretensioned spun concrete piles. Standards Press of China, Beijing

13. ACI 318 Committee (2011) Building code requirements for structural concrete and commentary. American Concrete Institute, Michigan
14. Yang ZJ, Peng SC, Li GC, Cong XL (2022) Finite element analysis of reinforced hollow high concrete filled square steel tubular stub columns under eccentric compression. J Shenyang Jianzhu Univ Nat Sci 38:655–663

Finite Element Analysis of Self-centering Reinforced Concrete Shear Walls

Nouraldaim F. A. Yagoub and **Xiuxin Wang**

Abstract Post-tensioned precast concrete walls are an attractive research trend in structural engineering, which replaces cast-in-place concrete walls in earthquake-prone buildings. Precast concrete walls use mild steel and high-strength post-tensioning steel for flexural resistance. Mild steel reinforcement yields in tension and compression, dissipating inelastic energy. Unbounded tendons are used inside the wall to give self-centering capability, which lowers residual displacements. At the wall-foundation, horizontal slots are installed equally. Meanwhile, the middle-wall concrete is still anchored to the base. A three-dimensional finite element (FE) model is developed in this study to assess the lateral load response of shear walls with horizontal bottom slots. The seismic performance of three distinct walls is evaluated using the Abaqus software FEA. The model is validated by comparing the experimental data accessible in the literature. Furthermore, we investigate the effects of bottom slit length, steel strand position, initial prestressing level, and concrete strength. All of these criteria are critical for constructing structures with the new concept. The results of this study show that the three-dimensional finite element model accurately predicts all of the above-mentioned properties, including the lateral force–displacement response and toe area damage of self-centering shear walls.

Keywords Self-centering Precast wall · Shear Walls · Nonlinear finite Element Analysis · Cyclic Loading · Concrete Structures

N. F. A. Yagoub (✉) · X. Wang
School of Civil Engineering, Southeast University, Nanjing 210096, China
e-mail: 233179921@seu.edu.cn

X. Wang
e-mail: wangdisks@163.com

N. F. A. Yagoub
Department of Civil Engineering, Faculty of Engineering Science, University of Nyala, Nyala, Sudan

X. Wang
Full Key Laboratory of Concrete and Prestressed Concrete Structures of Ministry of Education, Southeast University, Nanjing 210096, China

© The Author(s) 2023

G. Feng (ed.), *Proceedings of the 9th International Conference on Civil Engineering*, Lecture Notes in Civil Engineering 327, https://doi.org/10.1007/978-981-99-2532-2_41

481

1 Introduction

Shear walls are structural elements that provide both lateral force resistance and drift control to buildings during seismic events. Conventional concrete shear walls that are part of monolithic structures are highly likely to suffer extensive damage from flexural and shear cracking. Its toe crushing, rebar fracture, buckling, and residual lateral displacement respond to reversed cyclic loading during design intensity or higher seismic events. Reinforcing steel and base concrete collapse in concrete shear walls, dissipating energy. Self-centering (SC) earthquake-resistant systems are innovative. SC structural systems are designed to decompress at a certain level of lateral loading, contradicting conventional systems. The recently proposed self-centering (SC) structural systems are a feasible alternative to traditional structural systems as they can make structures usable and repairable after strong earthquakes. The SC systems have important advantages in terms of their overall earthquake performance. They can reduce damage to the main structural components to minimal levels and eliminate residual lateral deformations due to strong earthquakes. Recent research on the PRESSS (Precast Seismic Structural Systems) project has shown that precast concrete wall and frame structures that use high-strength post-tensioning (PT) steel and mild steel reinforcement have good seismic properties during a severe earthquake. In earthquake-prone regions, precast concrete walls have been created as an alternative to cast-in-place reinforced concrete (RC) shear walls for building lateral-force resisting systems [1–6]

These precast concrete walls incorporate unbonded post-tensioning (PT) tendons extending from the roof to the foundation level (see Fig. 1). The self-centering response of precast concrete walls is achieved by restoring forces provided by gravity loading and the PT force. Precast wall systems have been subject to numerous pseudo-static lateral load tests [1, 3, 7]and extensive analytical investigations [2, 3, 7–9] Restrope and Rahman [8] researched a hybrid shear wall while subjecting it to quasi-static reversed cyclic loading tests. Holden et al. [9] studied a hybrid shear wall with carbon fiber tendons and steel fiber concrete, and the reinforcements of the walls were reduced compared with conventional precast walls. Sritharan et al. [10] developed an end-column precast wall that was unbonded and post-tensioned to the base. PreWEC can be built at a low cost while limiting damage and providing self-centering capability.

Finite element (FE) analysis is widely adopted in earthquake engineering, particularly when dealing with systems with a high number of degrees of freedom (DOF), which represent the structural configuration of buildings. On the other hand, the simulation of the seismic response of RC structures through finite element analysis can be complex and challenging. This is mostly because of how the different mechanical responses of concrete and steel rebars cause them to work together in a complicated way. Although many tests on self-centering shear walls have been conducted, the authors have limited knowledge on how to numerically model these structures in order to represent the usual nonlinear response associated with self-centering shear

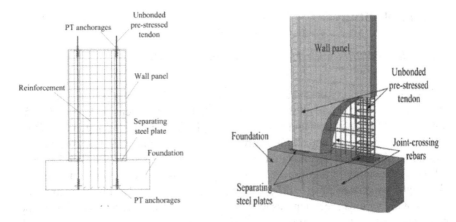

Fig. 1 Self-centering wall with horizontal bottom slots [12]

wall behavior. To anticipate the behavior of self-centering shear walls, most earlier studies used rotational spring, fiber, and multi-spring models [11].

While these simplified FE modeling techniques can estimate the global behaviors of self-centering shear walls, they are lacking in predicting the damage to the concrete of these walls under extreme loading conditions. To cope with these issues and bridge the gap in the modeling for self-centering, the current study proposes a novel model based on moment-rotation analysis using Abaqus to provide a reference and foundation for the design of self-centering concrete walls. This article uses ABAQUS FEA to model self-centering shear walls with horizontal bottom slots. Horizontal slots are made by inserting separating steel plates at the wall-foundation joint of prestressed shear walls and maintaining the reinforced concrete in the center of the wall width linked with the foundation. Additionally, this study examines the effects of slot length, steel strand positioning, initial pretension level, and concrete strength. Applying the new idea to engineering structures required consideration of all these factors. This work is organized as follows: Sect. 2 is intended to describe the method, including the procedures for modeling reinforced concrete post-tensioned precast shear walls. In Sect. 3, the analytical analysis is described, and the discussion and results analysis are described in Sect. 4. Finally, conclusions drawn from the study are summarized in Sect. 5.

2 Sample Study

2.1 Shear Walls with Horizontal Bottom Slots

This study used one typical shear wall and walls with horizontal bottom slots from [12]. The specimens were all the same size, with the wall dimension (2 * 1 * 0.125) mm. The reinforcement features of the examples for various arrangements (sw0 through sw1-3) in regard to elevation are shown in Fig. 2. The aspect ratio of 2.3 was chosen to obtain wall specimen reaction rather than shear slip ACI ITG-5.1 (ACI 2007). The quantity of reinforcement crossing the wall-foundation joint defined the slot length, which was equal to the dividing steel plate (Fig. 2). The initial prestressing values of the various specimens, as well as the slot lengths, are reported in Fig. 2. The reinforcement quantities across the joint in sw1-1, sw1-2, and sw1-3 were varied. Steel strands with total area of 140 mm^2 were employed in the prestressed element examinations. To improve the performance of concrete in compression, with closely spaced stirrups, the concrete at the specimens' toes was confined (spacing 50 mm). The strands' unbonded length was 3 m, which was equivalent to the specimen's overall height.

2.2 Experimental Setup

Figure 3 shows the experimental set-up. Out-of-plane braces prevented wall panel deformations. Table 1 reveals that vertical weights were added to the specimens by hydraulic jacks at the top. Figure 4 shows that all specimens were subjected to cyclic lateral displacement. The average concrete compressive strength of each specimen was 20.8 MPa. fy = 527 MPa, fu = 683 MPa for 6 mm reinforcement; fy = 448 MPa, fu = 576 MPa for 10 mm reinforcement. Figure 6 shows the yield stress of the steel strand is 1,740 MPa, while the ultimate tensile strength is 1,950 MPa.

3 Finite Element Modelling

ABAQUS software is used to generate a nonlinear three-dimensional (3D) finite element (FE) model of the self-centering reinforced concrete shear walls. The 3D FE model was developed to mimic the test scenario more precisely (Fig. 5). In all of the produced models, all concrete components, including the precast concrete wall and the concrete base, were constructed using an eight-node 3D brick element. ABAQUS software's concrete damage plasticity model, which is a common concrete structure analysis model, was used to define concrete material. This model includes concrete elements subjected to monotonic or cyclic loads.

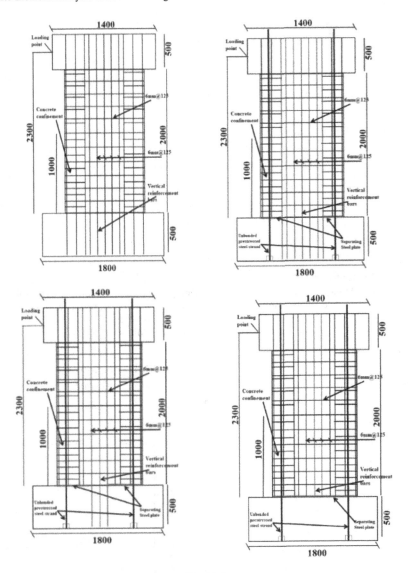

Fig. 2 The experimental reinforcement detailing [12]

In this investigation, the CDP input data was defined using a newly suggested a model of the damage made to softened concrete's plasticity by Feng et al. [13]. This model takes into account the impact of compress softens on the estimated stress–strain data as well as concrete material degradation. The specimen's ultimate concrete compressive strength is displayed in Fig. 7. Assume all concrete materials' ultimate strain is 0.003. The concrete damaged plasticity model in ABAQUS involves the definition of five criteria related to plasticity. These factors are the dilation angle ψ,

Fig. 3 Test setup [12]

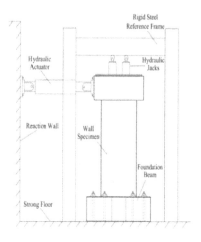

Table 1 Wall configurations tested [12].

Slot length (mm)	Distance from steel strands to wall centerline (mm)	Quantity of strands	Average initial prestress (MPa)	Extra vertical load by hydraulic jacks (kN)
Null	-	-	-	370
180	420	2	464	370
360	420	2	465	370
1000	420	2	475	370

Fig. 4 Cyclic loading used for test

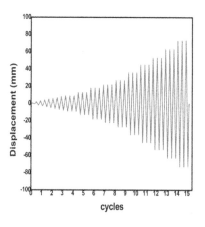

the flow potential eccentricity ϵ, the biaxial to uniaxial strength ratio b0/c0, the yield function Kc second stress invariant on the tensile to compressive meridian, and the viscosity parameter. These parameters were expected to be 30, 0.1, 1.16, 0.667, and 0.005. This assumption is based on program documentation suggestions (SIMULIA

Fig. 5 3 D view of the FE
model

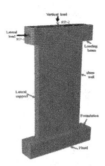

Fig. 6 Steel and strands
stress–strain

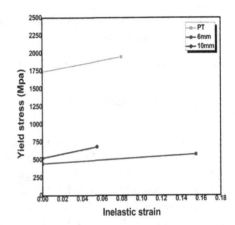

2008) and past research [16, 17]. The pre-stressed strands, energy dissipation bars, and vertical and horizontal reinforcements of the shear wall were simulated using the truss element T3D2

Unbonded tendons were embedded in a cap beam and foundation. Concrete had connected tendons, longitudinal and transverse reinforcing bars. Model the gap between a precast wall and foundation using surface-to-surface contact. 0.5 friction determined tangential contact behavior [15]. Hard contact was employed to avoid penetration upon contact between solid parts, allowing shear wall and foundation surfaces to easily separate and compress. To emulate fixing, the footing's bottom was completely constrained.

The produced models were analyzed in three steps to imitate the real sequence of load for the experiment operation. The initial stress was applied to the unbonded tendons as the first stage. The vertical load was then applied to the model using the same way as in the test. Finally, the predetermined lateral drifts were applied.

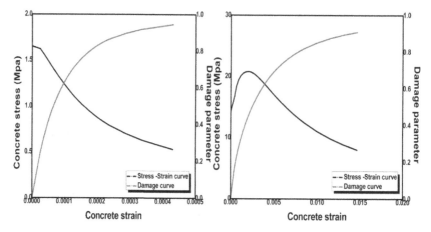

Fig. 7 Definitions of stress–strain for concrete instate of: **a** Compression; **b** Tension

3.1 Hysterical Behavior

Based upon numerical and experimental findings, the hysteretic responses of shear wall specimens Sw0 and Sw1-3 are shown in Figs. 8 and 9. After going through rebar failure, the specimens lost strength and failed. The black lines indicate the experimental test findings, and the red curves represent the numerical model results. Generally, the newly created F.E. models predict hysterical curves effectively. The results from the diagrams show that the concrete wall modeling in Abaqus is accurate enough.

Fig. 8 Hysteretic curves SW0

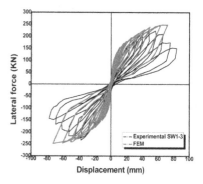

Fig. 9 SW1-3 Hysteretic
curves

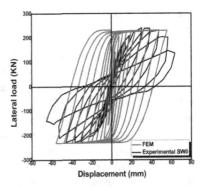

3.2 Damage Mode

The concrete wall FEM behaved similarly to test specimen. Wall toe stresses and strains were high; however, the restricted concrete area was strong enough to avoid crushing. The strain distribution and deformation mode simulation findings at 3.65% drift levels ($\Delta = 3.65\%$) are shown in Fig. 10, along with the plastic deformation of the wall panel. When the lateral force that was being applied to the (sw0 through sw1-3) specimens wase eliminated, for sw1-3 specimen, the gap closed in completely.

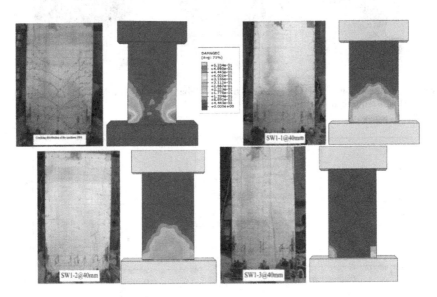

Fig. 10 Comparison between FEM and Test results Damage

3.3 Local Response

For the purpose of evaluating the FEM's accuracy, characteristics of the local response were examined.; further, the comparisons of envelope curves generated from peaks of the cycle are given in Fig. 12 to compare findings visually. The uplift shape at the wall toe of Units 1–3 of the FEM results and the Von-Mises stress contour of the steel bars reinforced by Sw0, were shown in Fig. 11. Local responding (a) Gap opening (b) SW0 Rebar model (c)Von-Mises stress contour of the steel bars reinforced sw-0.

Regarding all local response variables, the FEM findings generally exhibited a good match with the experimental results. The accuracy of the global lateral force–displacement response was most importantly confirmed by the precisely predicted local response characteristics. They gave the 3D finite element model used to model the self-centering concrete wall more confidence.

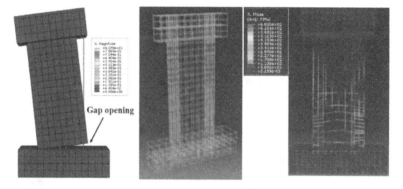

Fig. 11 Local responding (**a**) Gap opening (**b**) SW0 Rebar model (**c**) Von-Mises stress contour of the steel bars reinforced sw-0

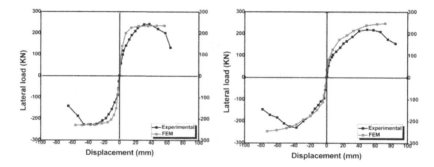

Fig. 12 FEM and test envelope curves

3.4 The Viscous Damping Ratio

Viscous damping ratios are used to measure a system's energy dissipation capacity. For a typical hysteretic loop, given in Fig. 13, Eq. 1 [16] was utilized to determine the wall's equivalent viscous damping (sw 0, sw 1-3). Figure 14 a comparison between both the equivalent viscous damping ratio of the hysterical analytical curve and the corresponding experimental data. As demonstrated in Fig. 14, the equivalent viscous damping ratio of computational models corresponds well to experiments, indicating that the model can predict the hysteretic response of such structural systems with reasonable accuracy. Moreover, the accurately predicted local response characteristics verified the accuracy found in the global lateral force–displacement response and increased confidence in the 3D FEM utilized to model the self-centering concrete wall Eq. 1:

$$\zeta_{ei} = \frac{1}{4\pi} \frac{E_i}{E_{si}} \tag{1}$$

Fig. 13 Typical hysteresis loop of SC-PSBCs

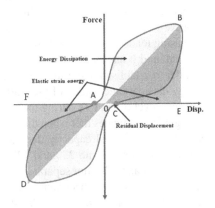

Fig. 14 Eq. viscous damping coefficient of the Sw0, Sw1-3 specimen

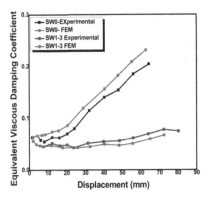

4 Parametric Studies

A parametric analysis of the effect of important structural variables on the earthquake characteristics of self-centering concrete walls with bottom slots was carried out using provided FEA models. The inverted cyclic model and its drift in Fig, 4 were used to validate and compare the influence of the parameters. Bottom slot size, location of steel strands, initial level of post - tensioned, and concrete compressive strength were studied. The effect of these parameters is shown below.

Figure 15 demonstrates the impact of concrete's compressive strength, Fc. The compressive strength of concrete should have a substantial effect on the crushing of concrete at the toes (bottom shear wall), which touches the foundation, and the rocking behavior of entire walls. Given that the proposed walls are for four-story low-rise buildings, the values of Fc were set between 30 and 60 MPa, and C75 concrete was used to confirm the influence of high-strength concrete on the hysteretic behavior of the walls, particularly their rocking behavior. It is demonstrated that increasing the concrete's strength slightly increases the shear walls' elasticity; however, the increase in concrete strength slightly affects the yield and ultimate loads. Figure 16 depicts the lateral load vs displacement for three different prestressing forces. It is observed that as the effective prestressing force f_{PTi} is raised, the lateral load values improve and have minor effects on the energy dissipation capacity. However, by increasing tendon forces, the values of lateral displacement related to yielding were reduced, and it has essentially no effect on behaviors under massive deformations. In order to establish the optimal balance between the self-centering capacity and the energy dissipation capacity of the wall, it is crucial to effectively control the initial prestress level of the PTs during the design stage.

Figure 17 indicates that when the tendons were positioned near to the specimen's centerline, the lateral capacity decreased. To investigate the influence of tendons position, specimens with 220 and 420 mm tendons, as well as specimens with two tendons 220 and 420 mm separated from the center line, were studied. The cross-sections of all the wall specimens were 2000, 1000, and 125 mm, with the first model (220 mm) interior, the second model (420 mm) exterior, and the third model (220 and

Fig. 15 Effect of unbounded tendon initial stress on the lateral force versus top lateral displacement of precast wall

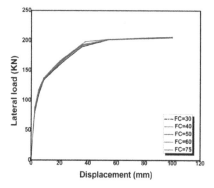

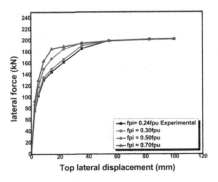

Fig. 16 Impact of concrete strength

420 mm) double. Figure 17a illustrates the cross-sectional model for all specimens; a prestressing force of 0.24 f_{Pu} i is induced in all cases, and the walls are examined under monotonic incremental loading. Figure '17b exhibits the lateral load vs drift relationships for the three spacemen's models of the shear wall. The base shear value is lower in the shear wall with interior tendons than in the shear wall with external tendons, but higher in double tendons. Both models have about the same effective stiffness (before system yield) and have minor influence on the energy dissipation capacity. Changing the tendon location, however, has the slightest impact on the self-centering property.

Figure 18 the aperture of the joint between the wall and foundation, which started approximately at the load stages, predominated the responses of (sw1-1 through sw1-3). When the joint-crossing rebars were put under tension and bonded with the wall panel's concrete, horizontal stresses were created. With an increase in the lateral load displacement, these stresses grew horizontally. However, as a result of that the nonlinear deformation was concentrated at the wall panel-to-base joint, the amount of stress was relatively low. Figure 10 depicts the pictures of sw0 through sw1-3 after the application of loads with respective drift ratios of 0.025, 0.45, 0.9, 1.35, 1.8, 2.65, 3.15, and 3.65%. It is discussed how the performance of the specimens is affected by various slit lengths. The deformation range decreased as the number

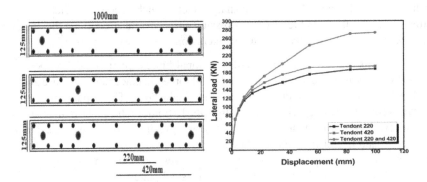

Fig. 17 Effect of s distance from centering of wall

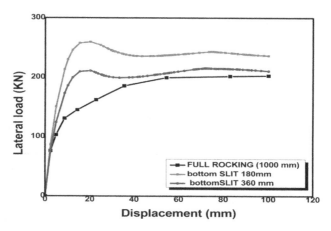

Fig. 18 Effect of bottom slits

of slits rose. Due to the positioning of the steel plates, sw1-1 through sw1-3 exhibit minimal deformations. The wall exhibited low strain because sw1-3 lacked rebars that crossed at joints. There was no considerable tension created because the breach opened. Therefore, the crushing of the concrete at the base of the walls is to fault for the decrease in strength. According to Fig.10, the damage shapes of FEM specimens with bottom slots matched those of the experiments. The specimen's maximum load can be enhanced by shortening the bottom slot.

5 Conclusions

This paper investigates the earthquake performance of self-centering concrete walls with bottom slots under seismic loads. A 3D detailed model was built to predict the lateral load behavior of a shear wall, which we tested against experimental data [12] Comprehensive parametric analyses were carried out after the numerical models were validated. An FEA model was used to analyze factors such as bottom slit length, steel strand location, the initial level of prestressing and concrete strength on the damage pattern, and lateral capacity. The main conclusions are drawn as below:

- The results indicate that the three-dimensional finite element model proposed in this study is accurate in predicting the lateral force–displacement response and the local response observed during the experimental testing. The impacts of concrete walls configurations, such as the length of the bottom slit, the place of the tendons, the level of initial post - tensioned, and the strength of the concrete, on the damage pattern and lateral capacity are shown.
- Wall-foundation gap interface causes high stresses and strains in a localized region of the wall toe, which were mitigated by the steel plate.

- The results in this paper show that the shear capacity increased slightly when strength of concrete was raised.
- Prestressed steel tendons added vertical restoring load for self-centering.
- Prestressing can affect shear wall response to seismic forces. According to this study, f-PTi affects wall behavior under minor and medium deformations but not large ones. Effective prestressing increases a structure's lateral deformation resistance. PTs must be designed with an appropriate starting prestress level. The wall's self-centering and energy dissipation capacities must be balanced.
- It is suggested that slits should be put in places where there is a lot of bending during an earthquake.
- As the length of the bottom slit got shorter, the number of cracks and the range of where they were found got smaller. The specimens' remaining drifts also got smaller; it confirms that the self-centering capacity performed as predicted.
- Making the bottom slot shorter, placing the steel strands closer to the specimen's center line, and raising the initial stress in the steel strands all will enhance the specimen's maximum load capacity.
- The corresponding viscous damping coefficient with the highest value is found in the traditional shear wall specimen.
- The outcomes also revealed that the specimen with bottom slits's ability to dissipate energy was not significantly affected by the placement of steel tendons or the initial prestress amount.

The analytical model accurately represented the total lateral load displacement, wall gap opening and closing behavior, PT tendon behavior, and stress on reinforced steel bars. Engineers can use the FEM model that was used in the research as a useful tool in their daily work. More information was required in order to fully understand the behavior of the concrete structures.

References

1. Priestley S, Nigel MJ, Sritharan S, Conley J, Pampanin R (1999) Preliminary results and conclusions from the PRESSS five-story precast concrete test building. PCI J 44(6):42–67
2. Smith BJ, Asce SM, Kurama YC, Asce M, Mcginnis MJ (2013) Behavior of precast concrete shear walls for seismic regions: comparison of hybrid and emulative specimens J Struct Eng139(11):1917–1927
3. Perez FJ Felipe J, Sause R, Pessiki S (2007) Analytical and experimental lateral load behavior of unbonded posttensioned precast concrete walls. J Struct Eng133(11):1531–1540
4. Sritharan S., Henry R, Ingham J (2011) Self-centering precast concrete walls for buildings in regions with low to high seismicity. Auckland
5. Kurama Y, Pessiki S, Sause R, Lu LW (1999) Seismic behavior and design of unbonded post-tensioned precast concrete walls. PCI J 44(3):72–89
6. Palermo A, Pampanin, S, Carr AJ (2005) Efficiency of simplified alternative modelling approaches to predict the seismic response of precast concrete hybrid systems (2005)
7. Erkmen AE, Bulent S (2009) Self-centering behavior of unbonded, post-tensioned precast concrete shear walls. J Earthq Eng13(7):1047–1064

8. Restrepo JL, Rahman A (2007) Seismic performance of self-centering structural walls incorporating energy dissipators. J Struct Eng133(November):1560–1570

9. Holden JB, Tony R, Jose M (2003) Seismic performance of precast reinforced and prestressed concrete walls. J Struct Eng129(3):286–296

10. Sritharan TKS, Aaleti S, Henry RS, Liu KY (2015) Precast concrete wall with end columns (PreWEC) for earthquake resistant design. Earthq Eng Struct Dyn 44(12):2075–2092

11. Palermo A, Pampanin S, Carr A (2005) Efficiency of simplified alternative modelling approaches to predict the seismic response of precast concrete hybrid systems. In: fibSymposium 2005: "Keep Concrete Attractive" vol 2, pp 1083–1088

12. Dang X, Lu X, Qian J, Zhou Y, Jiang H (2017) Experimental study of self-centering shear walls with horizontal bottom slits. J Struct Eng (United States)143(3), 1–14 (2017)

13. Feng DC, Ren XD, Li J (2018) Cyclic behaviour modelling of reinforced concrete shear walls based on the softened damage-plasticity model. Eng Struct166(7):363–375

14. Hashim N, Doaa A, Alyaa K (2020) Investigation of plastic hinge length of reinforced concrete wall. Pervasive Health Pervasive Comput Technol Healthc2:518–532

15. Chao K, Li H, Hong B (2017) Numerical study on the seismic performance of precast segmental concrete columns under cyclic loading. Eng Struct1(148):373–386

16. Elmenshawi A, Brown T (2010) Hysteretic energy and damping capacity of flexural elements constructed with different concrete strengths. Eng Struct 32(1):297–305

Effect of Cross-Section Shape on RC Specimen's Behavior Under Asymmetrical Impact Loading

Khalil Al-Bukhaiti, Liu Yanhui, Zhao Shichun, Han Daguang, and Hussein Abas

Abstract By applying asymmetrical lateral impact forces on RC specimens, the specimens' cross-sectional shape is analyzed. The effectiveness of the RC specimens' resistance to impact was examined using a drop hammer. Performing research on the factors that led to the failure of various RC specimen shapes and the dynamic responses they exhibited. In the experiment, eight circular and square specimens were used. Includes the method of failure, the impact force, and the deflection time history. The findings point to shear fractures between the point of impact and the adjacent support. The right side of the impact point has suffered significant damage, and the shear tests on all specimens failed. The peak impact force that square specimens can bear may be greater than circular ones. Protecting the concrete core and reducing maximum deflection are benefits of using a square specimen. When the ratio of stirrups is raised, there is only a little variation in the square specimen's damage range. This may cause a slight reduction in damage, but it is not significant. The plateau force of a circular specimen can be increased, but only a little. An increase in the stirrup ratio may increase energy use.

Keywords Asymmetrical impact · Absorbed energy · Shear fractures

K. Al-Bukhaiti · L. Yanhui (✉) · Z. Shichun · H. Abas
School of Civil Engineering, Southwest Jiaotong University, Chengdu 310061, Sichuan, China
e-mail: yhliu@swjtu.edu.cn

K. Al-Bukhaiti
e-mail: eng.khalil670@hotmail.com

H. Daguang
Faculty of Civil Engineering, Southeast University, Nanjing 210096, China

© The Author(s) 2023
G. Feng (ed.), *Proceedings of the 9th International Conference on Civil Engineering*,
Lecture Notes in Civil Engineering 327,
https://doi.org/10.1007/978-981-99-2532-2_42

497

1 Introduction

Reinforced concrete structures occupy most of the existing building structures in the world. Most reinforced concrete buildings have been in service for more than 30 years. At the same time, many reinforced concrete structures built before the 1950s are still in use in many European and American countries, including many transportation infrastructures [1]. Many reinforced concrete structures, especially those for civil infrastructure, have been in service beyond their usefulness. Reinforced concrete structures built for many years generally suffer from corrosion of steel reinforcement, cracking, and spalling of concrete [2]. These structures often carry loads higher than their maximum design-bearing capacity. The requirements for the seismic performance of structures also increase the need for reinforcement and repair buildings that have been in service for a long time [3–6]. Based on the above reasons, many scholars are devoted to exploring an economical technology capable of strengthening and repairing structures, which has made great progress in strengthening and repairing reinforced concrete structures in recent years. Since the beginning of the new century, infrastructure construction in China has grown rapidly, and reinforced concrete (RC) structure is one of the main structural forms. Due to their good mechanical properties and economy, RC components are widely used in bridges, urban underground spaces, and building structures. They are often used as load-bearing components of structures. During service and use, RC members may be impacted by automobiles, derailed trains, and airplanes. Suppose the impact resistance of RC members is weak or lacks impact resistance. In that case, the damage may have a catastrophic impact on the overall structure, resulting in huge economic losses and heavy casualties. In recent years, with the advancement of urban construction, more and more accidents have occurred due to the impact of building structures. For example, the JR train derailment accident in Japan [7] caused damage to buildings along the track, and many people died; the German high-speed rail derailment accident [8] caused the concrete bridge above the track to collapse. Accidents of ships hitting bridges have also occurred in Jiujiang Bridge piers, China, as shown in Fig. 1 according to the (http://www.chinadaily.com.cn/china/2007-06/15/content_8 95374.htm) Chinese net news website [9]. In impact accidents, RC members will show failure modes such as bending, shearing, and punching [10], as shown in Fig. 2.

Similar to static loading conditions, reinforced concrete (RC) members under impact loading will exhibit two failure modes: bending failure and shear failure [12]. After the plastic hinge is formed at the position where the member reaches the ultimate moment bearing capacity, the member undergoes bending failure. This failure mode has good flexibility and can absorb more impact energy during impact [13]. Contrary to the bending failure mode, the shear failure mode is brittle, and the component resistance cannot reach the ultimate bending moment capacity. This failure mode has poor flexibility, leading to premature failure of the component, which will seriously reduce the impact resistance of the component. However, the behavior of RC members under impact loading differs from that under static loading

Fig. 1 Accidents of ship on RC piers of Jiujiang Bridge in China [9]

Fig. 2 Partial diagram of failure of RC member under impact [10, 11]

conditions. The impact load will cause the material's strain rate effect, the component's inertial force, and the local effect caused by the impact load. The static test cannot reflect the impact resistance of the RC component under the impact load. Experimental studies have shown that: RC beams fail in bending under static conditions, but shear failure occurs under impact loads [14]. RC beams under impact loads may exhibit more localized damage (such as punching shear damage) [15, 16]. Therefore, there have been some experimental studies at home and abroad to explore the behavior characteristics of RC components under impact loads. Khalil et al. (2021) [17] conducted experimental research on the bearing capacity of ordinary concrete and CFRP-strengthening concrete columns after the impact. The RC column and the other four reinforced columns were placed under the same impact load in the test.

Then the static axial loading test was carried out on the impact-damaged column to obtain the remaining bearing capacity of the impact-damaged column to evaluate the effectiveness of different strengthening measures. The test results show that adding CFRP layers of energy-dissipating materials can improve the post-impact bearing capacity. The strength, deformation, elasticity, and confining effect of members can be affected by changes in cross-sections, reinforcements, and width-to-thickness ratios [18–21]. Recently, there have been a lot of studies that have a deep discussion of the effect of asymmetrical lateral impact load on RC specimens with and without FRP-strengthening materials [22–26]. Consequently, research on the failure mode and force parameters of cross-section shape RC specimens is limited. Furthermore, the study investigated the dynamic behavior of RC specimens using asymmetrical-span drop hammer impact testing. The main focus of the study is on the typical failure mode caused by an asymmetrical span drop hammer, the time history curve of deflection and impact force, and the effect of different cross-section shapes on RC specimens.

2 Materials and Methods

In the impact test, eight reinforced concrete specimens will have square and circular cross-sections. The test was conducted using a DHR-9401 drop hammer impact tester from the Taiyuan University of Technology. Using Chinese Pinyin initials, "F" represents square members, "Y" represents circular members, and "H" represents concrete specimens. Each test specimen is poured with the same material to achieve interoperability. Asymmetrical lateral impact forces are applied to specimens with different impact heights, longitudinal reinforcement ratios, and stirrup ratios to determine impact resistance. The specimen is supported at both ends, has a design length of 1500 mm, and supports lengths of 200 mm in the test. This collision is designed to mimic a collision between a high-speed railway locomotive and an RC column member. The impact point is located at 2/9 of the clear span, 200 mm from the right and 700 mm from the left. [27]. The cross-sectional dimension of the square specimen is 120 * 120 mm, while the diameter of the circular specimen is 114 mm. Figure 3 shows that square specimens contain 4 HRB235 steel bars, whereas circular specimens have 6 HRB235 steel bars evenly distributed along the ring direction. The stirrup has a diameter of 4 mm. The concrete cover thickness for square specimens is 20 mm, while for circular specimens is 12 mm.

The specimens' dimensions, reinforcing details, and parameter factors are included in Table 1. Frictional energy loss is included in the impact energy depicted in the Table, which was calculated using energy conservation principles. Under the asymmetrical impact scenario, this study investigates the RC members' failure mode, impact force, and deflection. Since the impact, members' responses have been related to strain rate, confinement effects, and other train-specific factors. As a consequence, two impact scenarios and several reinforcement configurations were considered.

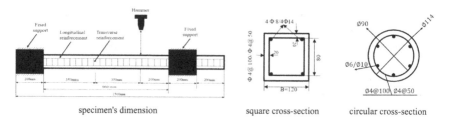

specimen's dimension   square cross-section   circular cross-section

Fig. 3 The specimen's schematic illustration

Table 1 Test specimen's design information

No.	Long. Reinf ratio	Volume stirrup ratio	Impact height (m)h	Impact velocity (m/s) $v = \sqrt{2gh}$	Impact energy (J) $E_{impact} = mgh$	Factors
FH1	2.01% (4 φ 8)	0.62% (φ 4@50)	1.0	4.42	2646	Height
FH2	2.01% (4 φ 8)	0.62% (φ 4@50)	2.0	6.26	5292	
FH3	6.16% (4 φ 14)	0.64% (φ 4@50)	2.0	6.26	5292	Long reinf ratio
FH4	2.01% (4 φ 8)	0.31% (φ 4@100)	2.0	6.26	5292	Stirrup ratio
YH1	1.67%(6⌀6)	1.26% (⌀4@50)	1.0	4.42	2646	Height
YH2	1.67%(6⌀6)	1.26% (⌀4@50)	2.0	6.26	5292	
YH3	4.61%(6⌀10)	1.26% (⌀4@50)	2.0	6.26	5292	Long reinf ratio
YH4	1.67%(6⌀6)	0.63% (⌀4@100)	2.0	6.26	5292	Stirrup ratio

2.1 Test Devise, Loading, and Measurement Scheme

The authors completed this impact test on the DHR9401 drop hammer impact-testing machine, as shown in Fig. 4. The DHR9401 drop hammer slideway height is 13.47 m, and the effective drop distance is 12.60 m. The test frame is a portal frame composed of two vertical lattice steel columns and rigid beams. The body falls smoothly, and the impact velocity of the drop hammer is very repeatable. The lower part of the support is fixed on the rigid platform, and a linear bearing is installed in the support. This bearing can restrain the displacement of the members in all directions except the axial direction to realize the boundary condition of fixed support. The impact hammer is made of (45# forged steel) and consists of three main elements, and the first element is a stiffened flat head constructed of large-strength chromium 15 (64HRC), which is a rectangular parallelepiped face, and the size is 80 mm × 30 mm × 80 mm. The striker's second element is the weight portion; this element contains the striker's main mass and provides a weight of (270) kg. The third element is that of load cells, which are used to calculate the impact force record. The impact hammer's total weight is 65 kg and could be lifted to the requisite height to generate varied velocities of impact (up to 15.7 m/s) and energy (see Fig. 4). A load cells sensor is placed between the impact head and the drop hammer to measure the impact force. The test specimen support is made of (Q235) steel and divided into upper and lower parts. Figure 4 shows a bolt connection between the upper and lower parts. The bolts provide pre-tightening force for the two parts of the support, effectively improving the friction between the support and the test specimen. It can restrain the specimens

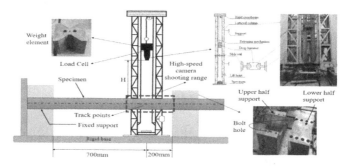

Fig. 4 Test equipment photos

from moving in the axial direction to a certain extent. A force sensor connects an impact head and drops a hammer for measuring the impact time-time history curve.

Data is collected at a million per second. The TDS420 dynamic data storage oscilloscope records and saves the strain amplifier signal. Furthermore, with a shooting frequency of 2500 frames/s, the high-speed camera in front of the specimen's impact site records the impact deformation and destruction process. A high-speed camera captures a deformation shot of the test specimen each 0.4 ms. Three-track white spots were marked on the specimen at 150 mm intervals to capture deformations (see Fig. 4). A collection of Microsoft Excel spreadsheets, Origin 19 pro, CAD design applications, and picture editing software handles the specimens' white spots data. Both qualitative and quantitative data were evaluated.

2.2 Material Properties

The concrete properties were measured using the "Standard for Test Methods of Mechanical Properties of Concrete" (GB50081-2010) on concrete cubes to determine the average compressive concrete strength. The quasi-static tensile tests on steel reinforcement bars were performed according to the "Metal Material Tensile Test Method" (GB/T228-2010) [28]. Table 1 summarizes the materials' properties and figures of testing procedures of materials (Table 2).

2.3 Details of Test and Analysis

The specimens are positioned in the impact test machine to demonstrate the structural specimen's dynamic response properties under the asymmetrical lateral impact (see Fig. 4). The specimen's compression zone concrete was shattered. Large bending

Table 2 Details on the materials of the test specimen

Materials		Parameters	Magnitude	Figure
Concrete		Comp. strengthf'_c	55MPa	Cube uniaxial comp. test
Reinforcement steel	$\emptyset 4mm$	Elastic modulus E_s	180GPa	Steel bar tensile test
		Ultimate strengthf_u	684MPa	
		Yield strengthf_y	520 MPa	
		Elongation rateδ	0.213	
	$\emptyset 6mm$	Elastic modulusE_s	200GPa	
		Ultimate strengthf_u	509 MPa	
		Yield strengthf_y	380 MPa	
		Elongation rateδ	0.227	
	$\emptyset 10mm$	Elastic modulusE_s	188GPa	
		Ultimate strengthf_u	412 MPa	
		Yield strengthf_y	325 MPa	
		Elongation rateδ	0.224	

fractures approach the compression zone. Once the ultimate deflection of the specimen exceeds 1.1% of the clear span, it is indicated that the specimen has failed related to bending.

3 Failure Modes

Figure 5 shows the final failure mode of specimens after impact load. It seems that all RC specimens have suffered a shear failure. Damaged concrete looks to have lost its bearing ability at impact. Arrows indicated the left support has flexural fractures, while the dotted line has a brittle shear failure. Observe how the drop height affects the RC specimens' failure patterns. The 1 m hammer specimen had a shear failure at the impact point and a vertical fracture at the left support. The impact point's bottom vertical crack did not propagate, but the support's left and right vertical cracks developed. Figure 5-a shows brittle shear failure with significant concrete cracking and buckling of steel reinforcing bars in the diagonal crack zone. A larger longitudinal and transverse reinforcement ratio improve the impact resistance of specimens.

On the other hand, Fig. 5-b shows the total damage of circular cross-section specimens is rather high. Compared to square shape specimens, the steel bar is stretched whenever longitudinal reinforcement fails between the impact point and the failure surface. In the final failure mode, square specimens seem to have more bearing ability than circular specimens. In summary, the cross-sectional shape affects the distribution of contact damage, which controls the pattern of specimens' failure and results in the difference between failure surfaces [29, 30].

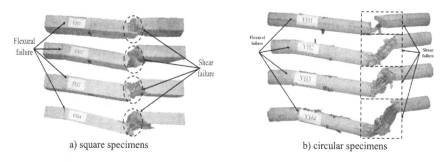

a) square specimens   b) circular specimens

Fig. 5 Failure modes for specimens after the end of the impact test

4 Test Results and Discussion

4.1 Time History of Impact Load Analysis

Figure 6 illustrates an impact force-time history curve that may be separated into different stages; a triangle pulse is produced by contacting the impact body with the specimen at zero time. In the stable stage, the impact force varies within a particular range and lasts for a significant duration. The impact force begins to attenuate after a steady stage and reaches zero. The peak impact force for square shape specimens is focused between 294 and 486kN in less than a second (4 ms), while circular shape specimens are concentrated between 209 and 305kN in less than a second (2 ms). FH1 and YH1 have the least impact force once compared to other specimens due to their lower impact height. The deformation of the specimens transfers and dissipates a large amount of kinetic energy. The peak impact force for specimen FH2, notable compared to other specimens, increases as impact velocity increases with stable fluctuations on the plateau value. This observation is consistent with prior impact test patterns in previous studies [31, 32]. Simultaneously, increasing impact height for YH2 might result in a lower plateau value and shorter impact duration. In the plateau stage, increasing the longitudinal reinforcement had a limited effect on the plateau impact force for FH3. Still, it enhanced the peak impact force and improved the plateau value for YH3. With extended impact duration in consideration, increasing the stirrup ratio causes a violently fluctuating plateau value for FH4 with a minimal influence on the peak impact value, which differs from prior studies [33, 34]. By increasing the peak impact force and decreasing the plateau value for YH4, the stirrup ratio has a greater influence on strengthening the specimen's stiffness than the reinforcement ratio.

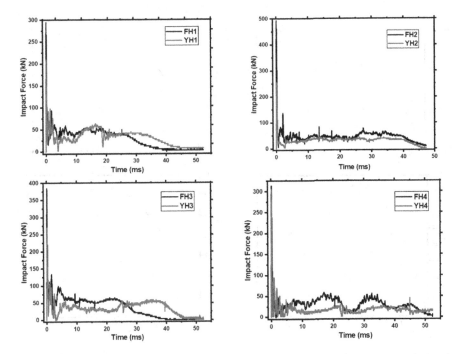

Fig. 6 Impact force time history curves

4.2 Time History of Deflection Analysis

Deflection time history curves were calculated by tracking the specimen's middle three spots with the high-speed camera (see Sect. 2.1, Fig. 4). The deflection data collected by the impact body starts once it contacts the specimen and ends when it stops falling. The deflection time history curves for specimens are shown in Fig. 7. There are two distinct stages in the vertical deflection curve of square shape specimens. The deflection increases from zero to the maximum value during the loading stage, then drops from the maximum to the residual value during the unloading stage. While there is just one stage (loading stage) in which the deflection goes from zero to the maximum value for circular shape specimens, data will be lost in the later stage owing to the concrete fragments collapsing. Table 3 summarize the impact, deflection, and duration findings for square and circular shape specimens.

For the FH2 specimens, both the peak and residual deflection increase with the drop hammer height. With the same reinforcing configuration, impact height affects deflection rates for YH2. As shown on the FH3 specimen, the reinforcement ratio improves the members' lateral stiffness, which was observable during the loading stage. Deflection change for YH3 is decreased and slowed by increasing the reinforcement ratio. The maximum deflection is slowed when the stirrup ratio is increased. The YH4 deflection value is increased by damaged concrete between the impact

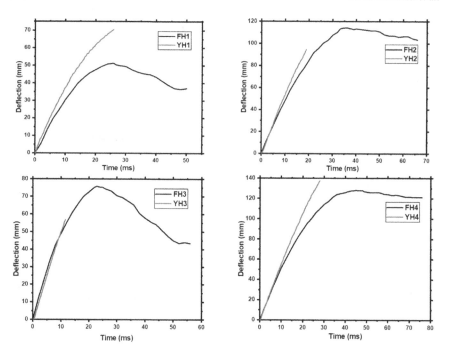

Fig. 7 Deflection time history curves

Table 3 Test specimen results collected from experimental outcomes

No	$F_{peak}(kN)$	$F_{plateau}(kN)$	$t_d(ms)$	$\Delta(mm)$
FH1	294.5	30.23	50	51.18
FH2	493	83.54	53	117.4
FH3	376	12.32	52	76.3
FH4	315	22.34	51	128.2
YH1	209.30	39.55	30	70.39
YH2	255.30	34.18	37	94.30
YH3	269.77	40.16	35	56.77
YH4	305.43	16.20	42	137.55

point and the right-end support. Compared to other specimens, the negative bending fracture of the YH4 distant bearing is practically penetrated, which differs from prior studies [35, 36]. Compared to previous specimens, the FH4 and YH4 impact specimens deflect significantly.

4.3 *Effect of Cross-Section Shape*

Figure 8 shows the impacts of various shapes cross-sections on peak impact force, plateau, duration, and deflection for the specimen parameters listed in Table 1 (see Sect. 2). The square specimen FH2 with an impact height of 2 m has the largest peak impact force because its squaring shape protects the concrete core body, as shown in Fig. 8-a. Since it transmits and dissipates much kinetic energy, the circular specimen YH1 with an impact height of 1 m has the least peak impact force. Simultaneously, Fig. 8-b shows the effect of plateau force on the specimens' shape cross-section. Although increasing the longitudinal reinforcement ratio had a limited influence on the plateau force, square specimen FH3 with a high longitudinal reinforcement ratio had the least plateau force. YH3's peak impact force and plateau value were improved as a result. At the same time, the plateau values for square specimen FH2 improve as the impact velocity increases. Due to the concrete collapsing of circular specimens, which have a smaller peak impact force and bearing capacity than square specimens. As seen in Fig. 8-c, the impact duration, square-shaped cross-section specimens consume longer than circular specimens during the impact period [37, 38]. As Fig. 8-d illustrates for specimen FH2, increasing impact velocity creates maximum deflection. The YH4 and FH4 distant bearings' negative bending fractures have been almost penetrated. As a result, increasing the stirrups ratio does not affect the deflection value for square and circular specimens [39].

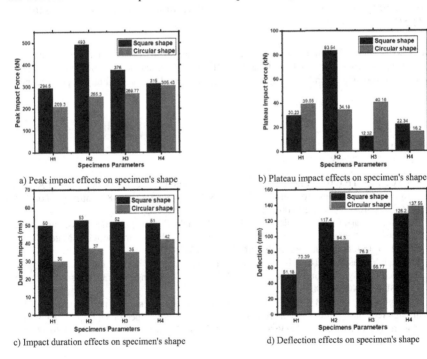

a) Peak impact effects on specimen's shape

b) Plateau impact effects on specimen's shape

c) Impact duration effects on specimen's shape

d) Deflection effects on specimen's shape

Fig. 8 Effect of specimen's shape cross-section

5 Conclusion

The experimental work on RC square and circular shape cross-section specimens under asymmetrical lateral impact loads was summarized in this study. The impact height, reinforcement, and stirrups ratio were investigated and explored. These factors have been compared in terms of (impact force and deflection record). Both square and circular-shaped RC specimens subjected to asymmetrical impact loading conditions tend to fail brittle, with diagonal cracks extending from the impact point down the right support. The drop hammer height affects crack width early in the impact stages for square shape specimens. The stirrups reinforcement ratio was one of the most important factors influencing square RC specimens' performance and failure mode. The results of this experiment demonstrated that decreasing the stirrups reinforcement spacing from 100 to 50 mm improves the stiffness of the circular shape specimen. The increased passive pressure from the stirrups then decreases the concrete's overall and local failure. Thereby minimizing the maximum deflection of the square shape specimens by 8.5%. In addition, longitudinal reinforcement greatly protects the concrete core and mitigates the specimen's maximum deflection of both shapes. The experimental outcomes presented in this work showed an increase in longitudinal reinforcement for square shape specimens from (2.01%) to (6.16%), indicating an increase in crack numbers. Simultaneously, these cracks' width is smaller in square shape specimens than in circular shapes containing a high longitudinal reinforcement ratio, resulting in a decrease in the maximum deflection of the square RC specimens by 41.5%.

Acknowledgements The authors would like to express many thanks to the authority of the National Natural Science Foundation of China (Grant no.52178168 and Grant no. 51378427) for financing this research work and several ongoing research projects related to structural impact performance.

References

1. Tunc G, Tunc TE (2022) Engineering the public-use reinforced concrete buildings of Ankara during the Early Republic of Turkey, 1923–1938. Endeavour 46(3):100832. https://doi.org/10.1016/J.ENDEAVOUR.2022.100832
2. Bertolini L, Carsana M, Gastaldi M, Lollini F, Redaelli E (2011) Corrosion assessment and restoration strategies of reinforced concrete buildings of the cultural heritage. Mater Corros 62(2):146–154. https://doi.org/10.1002/MACO.201005773
3. Manfredi V, Masi A (2018) Seismic strengthening and energy efficiency: towards an integrated approach for the rehabilitation of existing RC buildings. Buildings 8(3):36. https://doi.org/10.3390/BUILDINGS8030036
4. Harrington CC, Liel AB (2021) Indicators of improvements in seismic performance possible through retrofit of reinforced concrete frame buildings. Earthq Spectra 37(1):262–283. https://doi.org/10.1177/8755293020936707/ASSET/IMAGES/LARGE/10.1177_8755293020936707-FIG4.JPEG

5. Joseph R, Mwafy A, Alam MS (2022) Seismic performance upgrade of substandard RC buildings with different structural systems using advanced retrofit techniques. J Build Eng 59:105155. https://doi.org/10.1016/J.JOBE.2022.105155
6. Villar-Salinas S, Guzmán A, Carrillo J (2021) Performance evaluation of structures with reinforced concrete columns retrofitted with steel jacketing. J Build Eng 33:101510. https://doi.org/10.1016/J.JOBE.2020.101510
7. Niwa Y (2009) A proposal for a new accident analysis method and its application to a catastrophic railway accident in Japan. Cogn Technol Work 11(3):187–204. https://doi.org/10.1007/S10111-008-0112-5/FIGURES/13
8. Solomon B (2003) Electric Locomotives
9. China Daily: Bridge Collapses in Collision with Boat (2007). http://www.china.org.cn/english/China/214086.htm. Accessed 7 Feb 2023
10. Cao R, El-Tawil S, Agrawal AK, Xu X, Wong W (2019) Behavior and design of bridge piers subjected to heavy truck collision. J Bridg Eng 24(7):04019057. https://doi.org/10.1061/(ASCE)BE.1943-5592.0001414
11. Buth CE (2009) Analysis of large truck collisions with bridge piers: phase 1. Report of guidelines for designing bridge piers and abutments for vehicle collisions. https://static.tti.tamu.edu/tti.tamu.edu/documents/9-4973-1.pdf. Accessed 13 Feb 2023
12. Ho CM (2003) Inelastic design of reinforced concrete beams and limited ductilehigh-strength concrete columns. The University of Hong Kong (Pokfulam, Hong Kong). https://doi.org/10.5353/TH_B2750030
13. Thilakarathna HMI (2010) Vulnerability assessment of reinforced concrete columns subjected to vehicular impacts. Queensland University of Technology
14. Saatci S, Vecchio FJ (2009) Effects of shear mechanisms on impact behavior of reinforced concrete beams. Struct J 106(1):78–86. https://doi.org/10.14359/56286
15. Roehm C, Sasmal S, Novák B, Karusala R (2015) Numerical simulation for seismic performance evaluation of fibre reinforced concrete beam–column sub-assemblages. Eng Struct 91:182–196. https://doi.org/10.1016/J.ENGSTRUCT.2015.02.015
16. Fan W, Liu Y, Liu B, Guo W (2016) Dynamic ship-impact load on bridge structures emphasizing shock spectrum approximation. J Bridg Eng 21(10):04016057. https://doi.org/10.1061/(ASCE)BE.1943-5592.0000929
17. Al-Bukhaiti K, Yanhui L, Shichun Z, Abas H, Aoran D (2021) Dynamic equilibrium of CFRP-RC square elements under unequal lateral impact. Materials 14(13):3591. https://doi.org/10.3390/ma14133591
18. Liu Y, Al-Bukhaiti K, Abas H, Shichun Z (2020) Effect of CFRP shear strengthening on the flexural performance of the RC specimen under unequal impact loading. Adv Mater Sci Eng 2020:1–18. https://doi.org/10.1155/2020/5403835
19. AL-Bukhaiti K, Yanhui L, Shichun Z, Abas H (2022) CFRP strengthened reinforce concrete square elements under unequal lateral impact load, pp 1377–1387. https://doi.org/10.1007/978-3-030-91877-4_157
20. Abas H, Yanhui L, Al-Bukhaiti K, Shichun Z, Aoran D (2021) Experimental and numerical study of RC square members under unequal lateral impact load. Struct Eng Int 1–18. https://doi.org/10.1080/10168664.2021.2004976
21. Liu Y, Dong A, Zhao S, Zeng Y, Wang Z (2021) The effect of CFRP-shear strengthening on existing circular RC columns under impact loads. Constr Build Mater 302:124185. https://doi.org/10.1016/J.CONBUILDMAT.2021.124185
22. AL-Bukhaiti K et al (2022) Experimental study on existing RC circular members under unequal lateral impact train collision. Int J Concr Struct Mater 16(1):1–21. https://doi.org/10.1186/s40069-022-00529-5
23. Al-Bukhaiti K, Yanhui L, Shichun Z, Abas H (2022) Dynamic simulation of CFRP-shear strengthening on existing square RC members under unequal lateral impact loading. Struct Concr. https://doi.org/10.1002/suco.202100814
24. Yanhui L et al (2022) Comparative study on dynamic response of square section RC-members and CFRP-reinforced members under unequal transverse impact loading, vol 201. https://doi.org/10.1007/978-981-16-6932-3_12

25. Yanhui L et al (2022) Experimental and numerical study on unequal lateral impact behavior of Circular RC and CFRPRC components. In: IABSE Symposium Prague, 2022: Challenges for Existing and Oncoming Structures - Report
26. Yanhui L et al (2022) Failure mechanism analysis of circular CFRP components under unequal impact load. Structurae 22:1668–1676. https://doi.org/10.2749/NANJING.2022.1668
27. Kang XJ, Liu YH, Zhao L, Yu ZX, Zhao SC, Tang H (2019) Dynamic response analysis method for the peak value stage of concrete-filled steel tube beams under lateral impact. Adv Steel Constr 15(4):329–337. https://doi.org/10.18057/IJASC.2019.15.4.4
28. GB/T 228.1-2010: Translated English of Chinese Standard
29. Zhang C, Lin H, Qiu C, Jiang T, Zhang J (2020) The effect of cross-section shape on deformation, damage and failure of rock-like materials under uniaxial compression from both a macro and micro viewpoint. Int J Damage Mech 29(7):1076–1099. https://doi.org/10.1177/105678 9520904119
30. AL-Bukhaiti K et al (2023) Failure mechanism and static bearing capacity on circular RC members under asymmetrical lateral impact train collision. Structures 48:1817–1832. https://doi.org/10.1016/J.ISTRUC.2023.01.075
31. Saatci S, Vecchio FJ (2009) Effects of shear mechanisms on impact behavior of reinforced concrete beams. ACI Struct J. https://www.researchgate.net/publication/279897771_Effects_of_Shear_Mechanisms_on_Impact_Behavior_of_Reinforced_Concrete_Beams. Accessed 09 Feb 2021
32. Fan W, Yuan WC (2014) Numerical simulation and analytical modeling of pile-supported structures subjected to ship collisions including soil-structure interaction. Ocean Eng 91:11–27. https://doi.org/10.1016/j.oceaneng.2014.08.011
33. Kadhim MMA, Jawdhari AR, Altaee MJ, Adheem AH (2020) Finite element modelling and parametric analysis of FRP strengthened RC beams under impact load. J Build Eng 32:101526. https://doi.org/10.1016/J.JOBE.2020.101526
34. Demartinod XX, Wud XG, Xiaod XY (2017) Response of shear-deficient reinforced circular RC columns under lateral impact loading. Int J Impact Eng 109:196–213. https://doi.org/10.1016/j.ijimpeng.2017.06.011
35. Fujikake K, Li B, Soeun S (2009) Impact response of reinforced concrete beam and its analytical evaluation. J Struct Eng 135(8):938–950. https://doi.org/10.1061/(ASCE)ST.1943-541X.000 0039
36. Plauk G (1982) Concrete structures under impact and impulsive loading
37. Raval R, Dave U (2013) Behavior of GFRP wrapped RC columns of different shapes. Procedia Eng 51:240–249. https://doi.org/10.1016/J.PROENG.2013.01.033
38. Narule GN, Bambole AN (2018) Axial behavior of CFRP wrapped RC columns of different shapes with constant slenderness ratio. Struct Eng Mech 65(6):679–687. https://doi.org/10.12989/SEM.2018.65.6.679
39. Wei Y, Xu Y, Wang G, Cheng X, Li G (2022) Influence of the cross-sectional shape and corner radius on the compressive behaviour of concrete columns confined by FRP and stirrups. Polymers 14(2):341

Nonlinear Dynamic Behaviour of Hollow Piles Based on Axial Harmonic Loading

Surya Prakash Sharma, Shiva Shankar Choudhary, and Avijit Burman

Abstract The main objective of the current work is to examine the dynamic axial response of three pile group under machine-based harmonic loads. Field tests are carried out under axial harmonic excitations on a group pile with a pile length of 300 cm and an diameter of 11.4 cm in order to accomplish this objective. For various eccentric moments, the frequency versus amplitude responses of the group pile are measured. The field test results of the soil-pile system show non-linear behaviour as their resonant frequencies decrease and their resonant amplitudes disproportionally increase with eccentric forces. The inverse methodology proposed by Novak (1971) is used for the theoretical study. With this methodology, the changes in stiffness, damping, and effective mass of the piles under various eccentric moments are quantified by analyzing the frequency-amplitude response curves of field results. The theoretically back-calculated soil-pile system response curves are compared with the field results, and it is observed that the analytically predicted responses closely match the field responses. It is also found that the values of estimated damping of the pile group increased, while effective mass and average stiffness values decreased with an increase in eccentric moments.

Keywords Axial harmonic excitations · Soil-pile system · Field tests · Frequency-amplitude response

1 Introduction

In recent years, pile dynamics has drawn more attention in geotechnical engineering problems due to its successful application in machinery foundations and dynamic loading like wind or seismic conditions on heavy or large structures. However, numerous researchers are still investigating and researching the impacts of dynamic

S. P. Sharma · S. S. Choudhary (✉) · A. Burman
Department of Civil Engineering, National Institute of Technology Patna, Patna 800005, India
e-mail: shiva@nitp.ac.in

© The Author(s) 2023
G. Feng (ed.), *Proceedings of the 9th International Conference on Civil Engineering*,
Lecture Notes in Civil Engineering 327,
https://doi.org/10.1007/978-981-99-2532-2_43

behaviour and characteristics on the nonlinear response of piles with machine foundations. In the case of axial loading, limited research has been carried out for pile group to evaluate the dynamic properties of machine-based foundation systems. A few researchers conducted dynamic vibration tests on piles to examine the nonlinear behaviour of soil-pile systems in terms of frequency-amplitude responses in order to better understand the complex dynamic properties of machine foundations. Some other researchers [1–5], who have worked on pile dynamics, have conducted experiments and done analytical studies on piles to determine the stiffness and damping using a continuum approach method. From the literature, it can be inferred that, in comparison to the experimental findings under machine vibration, the theoretical solutions also indicate a reasonable assessment of nonlinear loading responses. Liu and his co-authors [6] investigated 3D Voigt model stiffness and damping coefficients to evaluate the dynamic pile-soil-pile resistance and, in the case of torsional vibration, the dynamic resistance in order to comprehend the dynamic response of soil-pile interaction in longitudinal vibration. The soil-pile system was studied under dynamic settings by Khalil [7] and from the obtained datasets it was found that the loading frequency significantly influences the dynamic stiffness and damping with generated amplitudes. A fundamental component of machine-based foundation systems, the prediction of dynamic frequency-amplitude response of group piles and its variation due to soil-pile stiffness and damping under dynamic loading needs further exploration because of limited research work. The existing theories, which have hardly ever been examined, require experimental (field or laboratory) verification. Therefore, in this study, machine-induced axial harmonic loading is investigated together with the dynamic properties of a three-pile group, using a nonlinear theoretical approach based on Novak [8] inverse method. The combined set of soil-pile configurations in this method are subjected to harmonic forces, which are generated by a mechanical oscillator and whose amplitude values increase with the square of the frequency. To determine the dynamic parameters of the soil-pile arrangement, the theoretical nonlinear response calculated from Novak's inverse method is compared with field test results. The values of restoring force, damping, and the effective mass of the three-pile group are also calculated from the dynamic response curve.

2 Site Location and Soil Properties

Forced vibration field tests are carried out on the campus of IIT Delhi in India. Various geotechnical tests are carried out on soil samples that are extracted from the borehole to determine their in-situ soil properties. Up to a depth of 6.5 m, two different soil layers are found [0 to 3.5 m (Sand-39%, Silt-43%, and Clay-18%), and 3.5 to 6.5 m (Gravel-3%, Sand-36%, Silt-42%, and Clay-19%)], with the first top layer having an average dry density of 15.33 kN/m^3 and a shear modulus of 1.3×10^4 kN/m^2. In the case of the second layer, the average dry density and shear modulus are found to be 16.72 kN/m^3 and 2.3×10^4 kN/m^2 respectively. From the soil test results, it is found that the soil particles are low in plasticity and represent a clayey silt type of soil.

Fig. 1 Schematic diagram
of three-pile group

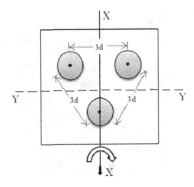

3 Field Testing Setup and Testing Procedure

3.1 Dynamic Axial Loading Tests

Axial harmonic tests are conducted on hollow steel three-pile groups (length of pipe = 300 cm, inner diameter = 11.1 cm, and thickness = 0.3 cm). The schematic diagram of the pile group with rotational forcing direction is shown in Fig. 1. The piles are embedded into the ground by SPT hammer in undersize boreholes which are constructed by a 100 mm diameter hand augur. The bottom end is closed with a circular plate to get the end bearing of the pile.

The axial loading is generated on the pile group with the help of a rotating mechanical oscillator. As per requirement, the axial magnitude of the vibration increases or decreases with the use of an outer valve of eccentricity (θ). This is the part of two counters rotating-mass inside the oscillator which is presented in Fig. 2. The produced eccentric moment ($m.e$) can be represented as follows:

$$m.e = (W/g).e = [2.59 \sin(\theta/2)]/g \ \text{NSec}^2 \tag{1}$$

where, W and m are the weight and mass of eccentric rotating parts inside the oscillator. e = eccentric distance of the rotating masses, and g = acceleration due to gravity.

The soil-pile responses of the pile group is obtained for four different eccentric moments. For an axially loaded field setup, first an oscillator is resting on a pile-cap

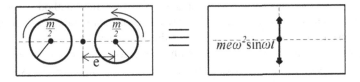

Fig. 2 Rotating masses inside the oscillator with forcing direction

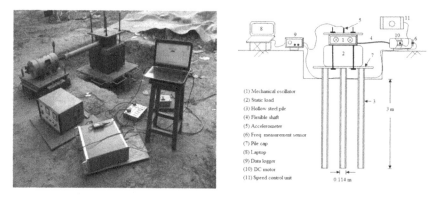

Fig. 3 Complete field test setup and schematic diagram of axial harmonic test

loading system in a horizontal manner so that the harmonic forces are generated in the axial direction. The axial harmonic loading tests are performed for a static load (W_s) of 14 kN (W_s = combined weight of the pile cap, steel plates, and mechanical oscillator). The steel plates are used to bring the dynamic responses of the combined system within the frequency range of 0 Hz to 50 Hz. The nonlinear responses of the combined soil-pile are recorded at varying frequency ranges. The data sets (time vs frequency and time vs acceleration) are transmitted and recorded on the laptop (with software package) through the data acquisition system. To measure the acceleration, an accelerometer is connected to the top and center of the soil-pile setup. To measure the frequency, the wire of the frequency measuring sensor is joined with a DC motor. During dynamic loading tests, the frequency-amplitude responses are recorded. The axial loading field test setup and schematic diagram are presented in Fig. 3.

4 Analytical Study

Nonlinear axial loading can be predicted by many approximate or asymptotic approaches when soil-pile dynamic characteristics of the system are given in terms of frequency, amplitude, loading forces etc. An inverse problem proposed by Novak [8] is one of the analytical approaches based on experimental data sets to measure the nonlinear dynamic responses of piles. In this approach, the machine-based inverse problem method of Novak with a field data set is used for the investigation of frequency-amplitude response curves of pile groups. In this method, steady state oscillation is vibrated by a dynamic force whose axial amplitude of the combined soil-pile system increases with the square of the frequency. In this method, the relationship between backbone curves (Ω) with respect to dynamic nonlinear response curves in terms of undamped natural frequencies is very important to evaluate the nonlinear properties of the soil-pile system. The proposed relationship between both parameters may be expressed as:

$$\Omega = \sqrt{\omega_1 \omega_2} \tag{2}$$

in which ω_1 and ω_2 are the frequencies at the points of interaction through field frequency-amplitude curve. In the case of elastic nonlinearity, the restoring force is linear under harmonic excitation under certain amplitude. If the function of steady amplitudes is assumed to be $F(A)$ during steady excitation, then the square of natural frequencies can be represented as follows:

$$\Omega^2(A) = \frac{F(A)}{Am_{eff}} \tag{3}$$

where, $\Omega(A)$ represents the natural undamped frequencies, m_{eff} represents the effective mass of the soil-pile loading system. Two points (ω_1 and ω_2) on the response curve corresponding to amplitude A_T (resonant amplitude) and the corresponding point on the Ω - curve with ω_T (resonant frequency) are used in the damping calculation formula (Eq. 4). If $A = A_T$ / SQRT 2 is chosen analogous to the often-used approach in soil-pile system, then the damping is:

$$D = \frac{1}{2} \frac{\omega_2 - \omega_1}{\omega_O} \left[2 \left(\frac{\Omega}{\omega_T} \right)^2 - 1 \right]^{-1/2} \tag{4}$$

In this case, mass is well known so that the behaviour of the pseudo-nonlinear restoring forces can be determined with the help of Ω-curve.

$$F(A) = Am_{eff}\Omega^2 \tag{5}$$

5   Comparison of Test and Analytical Results

The measured dynamic field response of the pile group is used to back calculate the frequency-amplitude response curves using nonlinear vibration theory. Further, the back-calculated response curves and the field response curves are compared. By intersecting the field response curves with a trace of lines, as illustrated in Fig. 4, the backbone curve is plotted for each of the four response curves, which are obtained from various eccentric moments. The resonant frequencies are found to decrease as eccentric moments are raised, and in the case of amplitudes, the resonant values are discovered to be non-proportional for both back-calculated theoretical and field results. This could be a representation of the nonlinear characteristics of the soil-pile system. The backbone curve pattern shows that the eccentric moments have an impact on the pile group stiffness characteristic. The nonlinear properties of the pile group are listed in Table 1. The tabular findings show that when the vibrating forces increase, the evaluated damping values increase while effective mass and average stiffness values decrease. These trends in the soil stiffness that are decreasing around the pile indicate a reduction in soil resistance. The decreasing pattern of additional effective mass indicates a partial break in the bond between the soil and the pile (or an increase in the soil-pile separation length) with the increase of eccentric moments or excitation forces. It is also observed that the field resonant frequencies and amplitudes are higher as compared to theoretical resonant values. These results may indicate the higher soil-pile stiffness in this theoretical case. However, in the case of field tests, the complete bond between soil and pile is not fully developed or the soil does not return to its natural condition after piles are left to rest for almost two months.

Using the previously described theory, the $F(A)$ and axial displacements are calculated from the measured frequency-amplitude curves of the pile group. For the different eccentric moments, the obtained data sets are used to plot $F(A)$ vs axial displacements (Fig. 5). It is discovered that as eccentric moments increase, the pile group overall stiffness decreases. For various eccentric moments, the values of the soil-pile system effective mass, damping, and stiffness are provided in Table 1.

Theoretical findings show that the nonlinear frequency-amplitude response curves match the outcomes of the field tests conducted under axial loading. Therefore, using

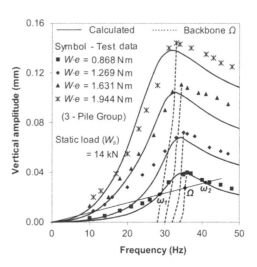

Fig. 4 Comparison between field and analytical response curves of three-pile group

Table 1 Measured dynamic parameter based on theoretical methodology

Eccentric moment (Nm)	Effective mass		Damping	Stiffness (kN/mm)
	Mass, m_{eff} (kg)	Mass coeff., ξ		
0.868	11,348	7.12	0.170	397.71
1.269	9412	5.72	0.175	295.57
1.631	8381	4.98	0.180	210.44
1.944	6579	3.70	0.185	160.23

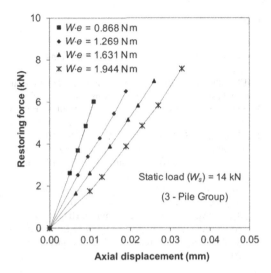

Fig. 5 Measured restoring force versus displacement of pile group

this inverse method and a system with a nonlinear restoring force and linear damping, a comprehensive and straight-forward theoretical investigation is made possible.

6 Conclusions

Under machine-based axial loading testing, the dynamic performance of a three-pile group is examined to understand the behaviour of the soil-pile system. According to the field test and analytical findings, the dynamic properties of the soil-pile system show a nonlinear pattern as their resonance frequencies fall and their amplitudes disproportionally rise with eccentric moments. With the use of the inverse back calculation method established by Novak, the field response curves are used to determine the variation of stiffness, damping, and effective mass at various eccentric moments. According to the theoretical calculation, the pile group stiffness decreases as the eccentric moments are increased. The decreasing behaviour in soil stiffness may occur due to a decrease in soil resistance around the pile. The experimental findings and the back-calculated soil-pile response curves demonstrate the close agreement between theoretical and field test data. Therefore, the comprehensive and straight-forward analytical application can be used through this inverse method as a realistic and versatile solution for dynamic analysis of soil-pile systems.

References

1. Elkasabgy M, El Naggar MH (2013) Dynamic response of vertically loaded helical and driven steel piles. Can Geotech J 50:521–535
2. Elkasabgy M, El Naggar MH (2018) Lateral vibration of helical and driven steel piles installed in cohesive soils. J Geotech Geoenviron Eng ASCE 144(9):1–8
3. Sinha SK, Biswas S, Manna B (2015) Nonlinear characteristics of floating piles under rotating machine induced vertical vibration. Geotech Geol Eng 33:1031–1046
4. Choudhary SS, Biswas S, Manna B (2020) Effect of pile arrangements on the dynamic coupled response of pile groups. Geotech Geol Eng 39(3):1963–1978
5. Kumar A, Choudhary SS, Burman A (2022) Machine induced vertical responses of single and pile group - experimental and theoretical study. Int J Geotech Earthq Eng 13(1):1–17
6. Liu X, Wang K, El Naggar MH (2020) Dynamic pile-side soil resistance during longitudinal vibration. Soil Dyn Earthq Eng 134:1–10
7. Khalil MM, Hassan AM, Elmamlouk HH (2020) Dynamic behavior of pile foundations under vertical and lateral vibrations: review of existing codes and manuals. HBRC J. 16(1):39–58
8. Novak M (1971) Data reduction from nonlinear response curves. J Eng Mech ASCE 97(4):1187–1204

Project Budget Comparison of a Conventional Building and a Seismically Isolated Building

Oguzcan Catlioglu⬤, Senem Bilir Mahcicek⬤, and Arcan Yanik⬤

Abstract There have been numerous studies in the past three decades about seismic isolation in the field of structural and earthquake engineering. However the studies that take into account economical aspects of implementing seismic isolation to the structures are very rare. In this study, firstly the costs of both conventional and seismically isolated buildings are calculated. Then, the seismically isolated and conventional buildings are compared in terms of their seismic performance and the project budget. Finally, the results of the seismic performance and the cost benefit are presented and discussed. It is obtained from this paper that, although the initial construction cost of the seismically isolated structure is obviously more than the conventional building, eventually in a possible earthquake scenario seismic isolation may provide financial advantage over the conventional building case.

Keywords Seismic Isolation · Project budget · Earthquake · Cost Analysis

1 Introduction

One of the most important challenge in civil engineering is to control the structural vibrations. The major aim of the earthquake codes and the specifications is protecting the structures from the hazards caused by earthquakes. There may be a loss of the serviceability of the building due to extreme accelerations and displacements that the structure may experience. These accelerations and displacements may exceed the limitations that are specified by the codes. Some types of structures are more crucial than the others. Important structures such as hospitals, power plants, municipality buildings, and fire stations have to be operational both during and after the earthquake movement. The usage of seismic isolation has been more common in these kinds of structures. Furthermore, in Turkey, seismic isolation is mandatory for

O. Catlioglu · S. B. Mahcicek · A. Yanik (✉)
Istanbul Technical University, Istanbul 34469, Turkey
e-mail: yanikar@itu.edu.tr

© The Author(s) 2023
G. Feng (ed.), *Proceedings of the 9th International Conference on Civil Engineering*,
Lecture Notes in Civil Engineering 327,
https://doi.org/10.1007/978-981-99-2532-2_44

521

the state hospitals according to Turkish Building Earthquake Code 2018 (TBDY-2018). Although the usage of seismic isolation is also advantageous for residential buildings, in Turkey seismic isolation implementation to those kinds of buildings recently started. Past earthquake experiences showed that many reinforced concrete structures experience damages partially or entirely during an earthquake.

Generally speaking, the seismic isolation concept is relatively new compared to the conventional design method. Because of this fact, we focused on the recent studies, especially published after 2015. More specifically, the research was conducted in terms of seismic and cost performance of seismically isolated structures and the conventional structures. Thanks to the literature review, the results show that the seismic isolation is used for different types of structures such as nuclear power plants, bridges, hospitals and buildings. The core objective of performance-based design of buildings is to transfer earthquake demands on the structural elements [1]. To achieve this, the predominant period of the structure should be long. For the high-rise building, the implementation of the seismic isolation is not beneficial and feasible because of the fact that the natural predominant vibration period of the high-rise buildings is already long. Seismic isolation is a commonly applied technique to reduce structural and nonstructural damage during severe earthquake excitation [2]. External energy dissipation devices and base isolators extend the natural period of the buildings and add additional damping for the building.

In the scope of this research, we investigated the research that contain the analysis of a 5 story building in terms of not only the earthquake performance but also the cost analysis. There are a few studies that cover the cost effectiveness of the utilization of the seismic isolation among the reviewed literature. Plenty of studies only deal with the initial and construction cost of the studies. The study of [3] shows that the cost benefit of base isolation is barely discoverable because of the high initial cost. Moreover, an indicator of structural seismic performance based on life-cycle cost for a 7-story hospital building was proposed in [4]. In the light of the literature review, despite the papers related to the seismic isolation, there is no concrete knowledge on direct comparison of seismically isolated and fixed base residential building in terms of earthquake and cost performance. On the other side, [5–7] were the valuable references for this review. While the two of them, [5, 7], focused on the comparison of base isolated three and six-story reinforced concrete buildings, the other one, [6] analyzed the earthquake performance and retrofitting cost of base isolated 4, 6, and 8 story reinforced concrete buildings. As a result of the literature review, it can easily be realized that the knowledge regarding to the comparison of fixed-based and isolated buildings in terms of both earthquake performance and cost benefit is still limited. According to this outcome, this research focuses on the lack of knowledge stated above. This study will compare two different types of buildings by considering the earthquake performance and cost.

This study concentrates on two different parameters that are earthquake performance and the project budget of a building. Initially, a 5-story reinforced concrete building is modeled according to TBDY 2018 which is the current earthquake code valid in Turkey. The seismic behaviour of the structures is analyzed to determine the earthquake parameters including lateral displacements, and inter-story drifts for the different cases with and without seismic isolation. SAP 2000 software is used for testing the models. On the other hand, this research focuses on the financial part of the structure as a secondary perspective. It is recommended that cost estimation should be performed before the start of a construction. Since every construction project is unique, the estimations differ from one project to another. In the very beginning of a project, the design method has to be determined. The design method is directly related to the project budget. The difference on the selection of the building material as steel, reinforced concrete or timber has a huge impact on the material cost of the project. Moreover, the usage of relatively new methods such as implementation of dampers and seismic isolators may significantly change the project budget. In this study, the project budget is calculated for cases that are; a conventional building and a seismically isolated building. The approach for cost estimation is chosen as activity-based cost analysis.

2 Numerical Example

In this section, information of the numerical example is presented. The example building performance is evaluated in terms of earthquake behavior and project cost. Moreover, structural system does not include shear walls. The model can be considered as a frame system. Frame system is basically defined as a structural system that contains only columns and beams. Effective stiffness factors of reinforced concrete structural system elements are chosen for columns and beams as 0.7 and 0.35 in a respective way. Figure 1 shows the 3D view of the model. It should be noted here that this paper is derived from the master thesis of the first author [8]. More information about this subject can be obtained from [8].

Fig. 1. 3D View of 5 story
fixed-base building

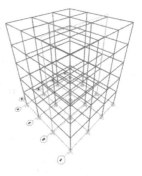

Fig. 2 Acceleration time
history of Duzce Earthquake

2.1 Structural Material and Dimensioning of the Building

The building material is chosen as reinforced concrete. The usage of concrete with a compressive strength less than 25 MPa (C25 concrete) is prohibited by the TBDY 2018 specification for all type of structures that are going to be constructed in Turkey. In this study the compressive strength of concrete is chosen as 30 MPa with an elasticity modulus of 32,000 N/mm^2. For the seismically isolated case, lead rubber isolators are chosen and modelled in SAP2000 as link/support elements. Rubber diameter of seismic isolator is 500 mm. Total overall height including external plates is 167 mm. Overall plate size is 550 × 550 mm. Total rubber thickness 77 mm. While the vertical stiffness is 885 kN/mm, the effective horizontal stiffness is 1.6 kN/mm. The building that is analyzed in this section is chosen as five-story residential reinforced concrete structure. Dimensions of the structure are chosen as 12 × 12 m in plan. There are 4 axes in both x and y-direction. The distance between axes is 4 m. Symmetric floor plan was preferred to eliminate eccentricity, the dimensional, and directional effects. While the dimensions of the columns are selected as 60 × 60 cm, beams are modelled as 30 × 60 cm. Lastly thickness of the slab is chosen as 15 cm.

2.2 Earthquake Used in the Dynamic Analysis

Duzce Earthquake (1999) that was occured on North Anatolian Fault (NAF) fault in Turkey with a moment magnitude of 7.2 is used in this study. Since NAF is a right lateral strike-slip fault, East–West component of this earthquake is chosen for the time history analysis. Acceleration time history of this earthquake is shown in Fig. 2.

2.3 Results of the Dynamic Analysis

The support condition is modelled as fixed support in Case 1, link support is preferred for modeling seismic isolation case which is Case 2. For Case 2, model is revised by alternating support condition. In earthquake engineering, the first phenomenon to take into account is the modal periods of the structure. Table 1 shows the first

Table 1 Modal periods and frequencies of the FB building

	Period (s)	Frequency (s⁻¹)
Mode 1	0.4535	2.2049
Mode 2	0.4535	2.2049
Mode 3	0.3972	2.5178

Table 2 Modal periods and frequencies of the SI building

	Period (s)	Frequency (s⁻¹)
Mode 1	1.9890	0.5028
Mode 2	1.9890	0.5028
Mode 3	1.8088	0.5528

3 modal periods and the frequencies of the 5-story fixed base building (FB), Table 2 presents the modal periods and frequencies of seismically isolated building (SI). While the fundamental period of the FB building is approximately 0.45 s, it is about 1.99 for the SI case.

One of the key parameters in analyzing a structure for earthquake engineering discipline is the displacement of the structure under an earthquake. To ensure the life safety of the residents, the story drift must be under control. Design engineers have to check the displacements to make sure if they are in the tolerable limits or not. The maximum story displacements and comparison of the inter story drifts are presented in Fig. 3 and Fig. 4 respectively. According to the results of the analysis, the displacement of the isolation level is 93 cm. Figure 3 shows the displacement comparison for both cases. The isolation level displacement is subtracted from each story while constructing this figure. As expected, the changes between each story displacements are not the same. The difference in the first story is about 8%. This ratio is increasing while elevating from the first to fifth story. The change between fifth story displacements is 41%. Seismic Isolation has a great impact on decreasing inter story drifts as it can be seen from Fig. 4.

Fig. 3 Comparison of maximum story displacement

Fig. 4 Comparison of inter story drifts

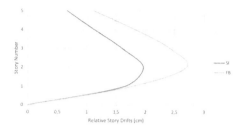

3 Cost Analysis

Without a doubt, cost is one of the most important phenomena all around the world in all sorts of business. Finance affects decisions significantly on different sectors. As well as the other industries, it is also important for the construction sector. It is not possible to construct a building without considering its financial part. As it was performed for the earthquake analysis part, cost comparison is made for two different cases which are the conventional and seismically isolated building. It is known that there are several methods for cost calculation. In this paper the activity base costing technique is chosen for comparison purposes.

3.1 Activity-Base Cost Calculation

Activity-Base Cost Calculation is one of the most important costing approaches that are used for the calculation of a construction project. In contrast to the problems related to the traditional cost methods, activity-based costing system has been developed as an alternative approach and has become widespread in time with the developing technology. The most important benefit of this system is that activity-based costing systems provide more realistic product cost information in comparison with the traditional cost methods.

3.2 Establishing the Work Breakdown Structure of the Case Project

Before the market research of unit prices, work breakdown structure is established for the 5-story residential building. Undoubtedly, applying seismic isolation has financial consequences. However, it is very beneficial to enhance the earthquake performance of structures. On the other hand, it increases the cost of the construction due to its advanced technology. Static project cost for seismically isolated building is higher than the fixed base building because the design engineers that work on seismically isolated buildings are rare. In addition, the design of a structure with

Fig. 5 Work breakdown
structure (WBS)

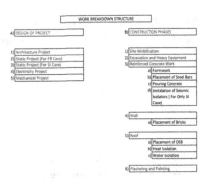

seismic isolation requires more efforts. Static project cost is calculated for both cases
(FB and SI) and presented in Fig. 5. The only difference between the FB case and
SI case is the installation of the seismic isolators to the related level of the building.
Seismic isolators must be implemented on each column above the foundation level.
Basically, SI structure needs 16 seismic isolators for installation. Furthermore, two
more isolators are necessary for the prototype testing before the construction starts.

3.3 Market Research and Pricing

After the work breakdown structure is completed, market research is carried out
for determining the unit prices. Unit prices are determined for the resources which
are materials and labor. Quantity calculations are made for structural construction
activities. Unit prices are multiplied by their related quantities to obtain the cost
of each activity. In contrast with the structural construction, fine construction cost
estimation is not carried out in a very detailed way. It is not completely possible
to calculate the cost of each fine construction activities. According to the investigation of a real building project, the ratio between the structural construction and
fine construction is calculated. Three different approaches that are mentioned just
before are applied to estimate the cost of the reference building. Considering the
date of May 24, 2021 1 $ was equal to 8.40 Turkish Lira ₺ and 0.82 EUR) the total
cost of the building is found as ₺1,139,788.00 ($135,750.5) according to the Turkish
Republic Ministry of Environment and Urbanization's Unit Price Chart, and it costs
₺1,365,297.50 ($162,608.9) by referring to the private sector prices. These prices
were calculated by considering the economic conditions in April 2020. To make the
cost up to date, the inflation rate of 17.14% that was announced by Turkish Central
Bank is used. The effect of inflation is included in the cost estimation. Updated
cost estimation is obtained as ₺1,597,612.63 ($190,278.1). It is predictable that the
number of the seismic isolators affected the project budget dramatically. Due to the
seismic design procedure, prototypes are manufactured and tested before the installation. As the structural system has 16 columns on each floor,18 pieces of seismic

Table 3 Total and m^2 costs (reference date: May 24, 2021)

Building Type	Cost Type	Ministry ($)	Private Sector ($)	Updated ($)
FB	Total	135,750.5	162,608.9	190,278.1
FB	m^2	188.5	225.8	264.3
SI	Total	191,685.1	218,633.6	246,408.5
SI	m^2	266.2	303.7	342.2

isolators are needed. Price of an isolator is 2500 EUR ($3050.8). It corresponds to the ₺25,625.00 by taking the EUR/₺ rate as 10.25 on 24 May 2021. Another cost item is the design of the static project cost. It can be known that the design of SI structure is more expensive than the design of FB structure. The difference between the design cost prices for each meter square is ₺5.50. These differences are added to the FB project cost. The comparison of the total and m^2 costs is shown at Table 3 in USD currency. It should be noted here that the isolators are not prone to corrosion which happens in some other types of materials due to the water and humidity. However, only the steel plates and connecting bolts which are attachment elements to the superstructure and the substructure should be checked periodically. In addition, since it does not affect the ultraviolet (UV) rays under the buildings, the rubber does not have any disadvantage depending on its UV resistance. For these reasons, the maintenance cost of lead core rubber isolators is dramatically lower than the other construction costs. Therefore, the maintenance cost of isolators is being excluded in this paper. It can be advised that the visual inspection shall be carried out after major earthquakes or within 10-year periods.

According to the ministry price list, seismic isolator cost is approximately 28% of the total construction cost. If the private sector prices are considered, the share of the isolation is being 25%. Lastly, seismic isolation expense is 22% of the total cost if the updated prices are being taken into account. Although the total construction area is relatively less and the number of the seismic isolator is 18, the ratio of SI Cost/Total Cost of 22% seems to be feasible. Conclusions obtained from this study are given below.

4 Conclusion

In this paper, it is obtained that seismic isolation implementation reduced the top story displacement (the most critical floor in terms of earthquake effects is the top floor) of the building about 69%. Financial results shows that seismic isolation increases the total project cost by 29.12% according to up-to-date prices. For the example building with seismic isolators, although the cost of the project is increased by 29.12%, the top story relative displacement is reduced by 69% under the seismic excitation. The cost increase is relatively high, however it should be noted that the benefit of seismic isolation implementation is very significant. The decrease in the inter story drift of each story shows that the structural elements are not experiencing any serious damages. Therefore, the failure of the structure due to the earthquake excitations is prevented. In addition, the functionality of the structure is sustained, and the residents of the building may not even feel the earthquake shaking. Furthermore, the reduction of the structural vibrations may save the equipment, electronical devices or valuable things that are in the building. These are considered as secondary advantages of seismic isolation usage. Another advantage is that the seismic isolation not only may prevent the collapse of the structural system but also may save the house furniture and valuable goods.

References

1. Amjadian M, Agrawal AK (2019) Seismic response control of multi-story base-isolated buildings using a smart electromagnetic friction damper with smooth hysteretic behavior. Mech Syst Signal Process 130:409–432
2. Yang TY, Zhang H (2019) Seismic safety assessment of base-isolated buildings using lead-rubber bearings. Earthq Spectra 35(3):1087–1108
3. Dong Y, Frangopol DM (2016) Performance-based seismic assessment of conventional and base-isolated steel buildings including environmental impact and resilience. Earthq Eng Struct Dyn 45:739–756
4. Dang Y, Han J, Li Y (2015) Analysis of the seismic performance of isolated buildings according to life-cycle cost. Comput Intell Neurosci 2015, Article ID 495042, 7 p
5. Mitropoulou C, Lagaros ND (2016) Life-cycle cost model and design optimization of base-isolated building structures. Front Built Environ 2, Article 27
6. Cardone D, Gesualdi G, Perrone G (2019) Cost-benefit analysis of alternative retrofit strategies for RC frame buildings. J Earthquake Eng 23(2):208–241
7. Han R, Li Y, Lindt J (2017) Probabilistic assessment and cost-benefit analysis of nonductile reinforced concrete buildings retrofitted with base isolation: considering mainshock-aftershock hazards. ASCE-ASME J Risk Uncertainty Eng Syst Part A Civil Eng 3(4):1–15
8. Catlioglu O (2021) Earthquake performance and project budget comparison of a conventional building and a seismically isolated building, MSc thesis, Istanbul Technical University, Istanbul

Research on Construction Method of Soft Soil Foundation Pit with L-Shaped Small Steps

Yanhui Cao, Jinhua Ye, and Gang Hao

Abstract A layered, segmented, and block building technique was created during the construction of the cut-and-cover foundation pit. Traditional foundation pit excavation often necessitates both support erection and mechanical excavation at the same time, however owing to the intersection of support and excavation mechanical operation, which may decrease building efficiency, and the construction requirements are higher due to the low mechanical characteristics of the soil in the silty soil layer. Therefore, the construction method of an L-shaped short step with small steps soft soil foundation pit is proposed, based on the mechanical characteristics of the silty soil layer and the construction operation characteristics of the foundation pit. After excavation, the support is erected using a numerical calculation of the stability of the foundation pit. According to engineering practice, the L-shaped short step with small steps soft soil foundation pit building technique can significantly facilitate basement pit deformation, improve construction efficiency, and be popularized.

Keywords Silty soil layer · Foundation pit · L-shaped short steps with small steps

1 Introduction

The excavation process of the foundation pit will inevitably cause the deformation of the surrounding strata. There has been a lot of practical work done to control the deformation of the foundation excavation. The excavation of large-area deep foundation pits in soft soil is currently possible using segmental and layered excavation methods [1]. The construction process for excavating a silt-deep foundation pit with prestressed pipe piles was thoroughly explained by Jiang Hui [2]. Currently, there are two types of foundation pit excavation: unsupported excavation and supported

Y. Cao (✉) · J. Ye · G. Hao
Beijing Municipal Road and Bridge Co., Ltd., Beijing 100045, China
e-mail: 15650751272@163.com

Y. Cao
Beijing University of Technology, Beijing 100124, China

© The Author(s) 2023
G. Feng (ed.), *Proceedings of the 9th International Conference on Civil Engineering*,
Lecture Notes in Civil Engineering 327,
https://doi.org/10.1007/978-981-99-2532-2_45

excavation [3]. According to Wang Wei [4], mechanical excavation is designed for large-scale foundation pit excavation work, and it is preferable to excavate in layers or open excavations with a depth of 5 m or more. A grab excavator is used to excavate the bottom layer of soil from the cantilever bridge, and a small bulldozer is used to pile the earth in the foundation pit. Wu Minhong [5] demonstrated that the space–time impact should be considered while determining the foundation pit excavation design. The size of the foundation pit area, the design of the enclosing structure, the excavation depth, and the engineering environmental conditions should all be evaluated. Based on the Shanghai Metro foundation pit project, Sun Jiuchun [6] analyzed the small-scale pot excavation method for soft soil deep foundation pits and developed a finite element analysis model utilizing the soft-soil creep model taking time effect into account. Wang Jiangtao et al. [7] analyzed soil deformation in the x and y directions during foundation pit excavation. They were able to establish the multi-dimensional deformation law of the foundation pit soil that was retrieved during excavation. Numerical simulation of various foundation pit excavation depths was done by Zhang Hui et al. [8]. Additionally, researchers compared the on-site monitoring data with the changes in the pile body's surface settlement, supporting axial force, and lateral displacement.

Cai Dongming et al. [9] presented a "double-layer gantry reinforcement system" targeted anti-uplift plan for the excavation of the subway tunnel's foundation pit, as well as the associated sectional, layered, and troughed progressive foundation pit excavation scheme. Liu Yi and colleagues [10] used a two-dimensional finite element numerical simulation method and the soil elastoplastic model (ST model) to analyze the support process of the foundation pit. They compared the calculation results with the practical engineering monitoring data and the design values calculated by the beam-spring model. Using the finite element software Plaxis 3D, Yang Zhonghong et al. [11] simulated the process of building deep foundation pits on silty soft soil in coastal areas. The impact of supporting structures on the deformation of foundation pits was then investigated. According to Lin Zhibin et al. [12], soft soil creep would significantly affect when foundation pits are excavated and deformed. Du Jiazhi [13] fully takes into consideration that many variables, such as the physical and mechanical properties of the soil, the geometrical requirements of the foundation pits, and the excavation methods, affect the numerical model. Xiong Yuanlin [14] studied the supporting effect of the foundation pit engineering envelope in soft soil area, and optimized the detailed parameters of the supporting structure. At present, the research on the excavation of cut-and-cover foundation pits is mostly an introduction to the formal segmentation and blocking, and rare reports on how to excavate and support each piece of earth in the detailed foundation pit, how to solve the cross problem between mechanical excavation and support erection, and the timing of support excavation. Moreover, the existing results show that the numerical analysis method can better reflect the excavation process of foundation pit.

In this paper, based on a tunnel project in Nanjing, the finite element analysis model is established by using the soft soil creep model. Taking into account the time effect, the influence of support erection timing, and refining the specific construction technology for earthwork excavation and support erection have been discussed.

2 Engineering Background

Nanjing Hengjiang Avenue SG-2 is located in Pukou District, Nanjing. It belongs to the floodplain landform unit of the Yangtze River. Inadequate geological conditions coexist with distributions. The main construction consists of a pipe gallery and an open tunnel.

The excavated partial foundation pit, for example, has a total length of 240 m, a width of approximately 60 m, and an overall excavation depth of approximately 10 m. The foundation pit enclosure construction uses the SMW engineering approach, piles + internal support system, and is reinforced in the strips and edges. The SMW engineering technique piles are φ850@600SMW three-axis mixing piles with a steel insert of H7003001324 mm. Two internal supports are positioned vertically: the first is an 800-by-800-mm concrete support placed on top of the foundation pit, and the latter is a Φ609 double steel pipe support located at a depth of approximately 5 m.

3 L-Shaped Short Step with Small Steps. Construction Technology

3.1 Process Overview

The standard stepped excavation method takes a long time to excavate muddy foundation pits. The machine's back and forth movement increases the number of interruptions to the enclosure structure and soil, which has a significant impact on the creep of the muddy foundation pit topsoil. At the same time, the support system in the foundation pit has a significant effect on the operational space in the pit as well as the mud transportation parameters. To address the aforementioned issues, the building method of L-shaped short and small stairs is proposed. Figure 2. depicts the technical principle. The soil is modified with precipitation and soil hardening subsequent to the excavation of the earthwork in the pit. A 3 m-wide platform is prepared for excavator operation at a depth of approximately 5 m (in the center of the foundation pit) at a slope of about 1:1. Whereas, B-B sections employ a slope with a minor slope for grading excavation, and the slope is typically not greater than 1:3. Section A-A is being excavated by bulldozer type at the same time as section B-B, and steel supports will be constructed when the bottom slope foot of the pit is excavated to the location of the steel support projection line. The bottom covering construction is completed once the steel support has been built. Engineering practice has confirmed that this method is adequate for the construction of muddy long-span and long-distance foundation pits with excavation depths ranging from 7 to 12 m (Fig. 1).

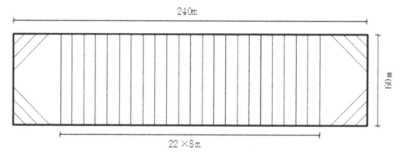

Fig. 1 Plane diagram of foundation pit

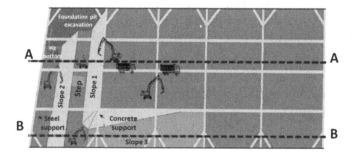

Fig. 2 Plan of foundation pit excavation

3.2 A-A Cross-Section Excavation Method

Figure 3. depicts the earthwork excavation process for the A-A section, as well as the construction stages:

(1) In Fig. 3(a), the A-A profile of the foundation pit adopts a 1:1 slope excavation, and the PC320 excavator excavates vertically downward at the excavation surface as the PC320 excavator enters the foundation pit. The PC320 excavator will excavate the dirt to a depth of 3 m, and the PC400 excavator will then pour the soil into the mud truck..

(2) Fig. 3(b) illustrates how to construct a 3 m wide step at 5 m after the foundation pit has been excavated out to a depth of 5 m.

(3) The foundation pit is excavated to its bottom using a PC320 excavator, as shown in Fig. 3(c). The excavator at the bottom of the pit transports the soil to the excavator on the 5 m step, where it is relocated to the excavator in the upper part of the foundation pit. The upper excavator is then immersed into the muck truck, and the excavation is functional.

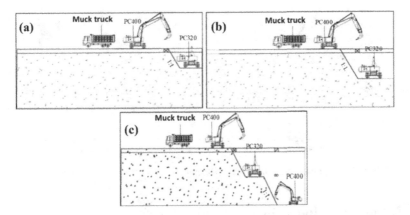

Fig. 3 Schematic diagram of earthwork excavation at section A-A: (a) Preliminary excavation, PC320 excavator into the foundation pit, (b) when excavating to 5 m, (c) continue to slope excavation below 5 m

Fig. 4 Excavation sequence of section B-B

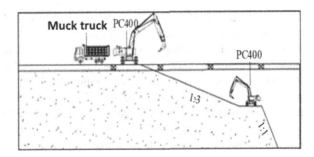

3.3 B-B Cross-Section Excavation Method

Figure 4. depicts the schematic diagram of the excavation of the B-B section, which is analogous to the excavation method of the A-A section but is carried out with a slope of approximately 1:3 on the side of the excavation direction.

3.4 Excavation Angle

Finally, the excavation is completed to the pit's corner, and to achieve the capping angle, the B-B profile is excavated simultaneously along the foundation pit's long-distance direction and span direction. Figure 5. illustrates the accurate construction steps. The mud truck is excavated at the pit's edge.

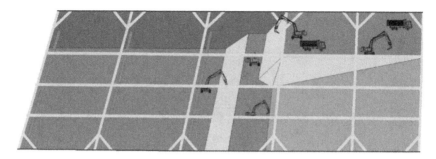

Fig. 5 Schematic diagram of foundation pit retracting

4 Numerical Model Building and Analysis

4.1 Model Building

In this study, a three-dimensional model is created using the finite element MIDAS GTS software. The internal support utilizes the beam element for simulation, while the soil simulation method adopts the HS (modified Mohr–Coulomb parametric) model. The equivalent stiffness approach is used in the SMW building method pile [15]. The pile length is 23 m, the excavation depth of the foundation pit is 10 m, and it is an underground diaphragm wall. Create a model with the dimensions 340 × 160 × 50 m (length x width x height). Displacement and angle impose a complete constraint on the model boundaries. The concrete support is made of C30 concrete, which has a Poisson's ratio of 0.2 and an elastic modulus of 30GPa. Figure 6. depicts the model of a foundation pit. Table 1. displays the physical characteristics of the soil layer.

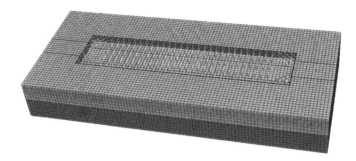

Fig. 6 Model of foundation pit

Table 1 Physical and mechanical parameters of soil layers

Layer No	Soil	Soil thickness /m	γ/kN/m³	c/kPa	φ/°	μ	E_s/MPa
1	Muddy silty clay	4.5	17.5	10.3	11.8	0.42	3.46
2	Powdered sand and soil	3.0	18.9	2.5	30.5	0.39	11.61
3	Muddy silty clay	4.0	17.5	10.3	11.8	0.42	3.46
4	Powdered sand and soil	3.0	18.8	2.2	32.6	0.30	11.76
5	Muddy silty clay	3.0	17.4	5.1	11.8	0.42	3.71
6	Fine sand	8.0	18.7	2.1	32.4	0.28	12.44

4.2 Numerical Analysis Results

Due to the large excavation range of the foundation pit after the final closing angle, the deformation of the support structure is the largest. Figures 7, 8 and 9. shows the closing angle state model and calculation and analysis results of the excavation process of the foundation pit. It is evident that the side of the supporting structure with the larger excavation surface (B-B section) has a significant deformation and supporting axial loads. The axial force and maximum displacement of the supporting structure, in both, are 766.5 kN and 34.69 mm, respectively. The early warning value of the horizontal displacement of the foundation pit is determined to be 55 mm in accordance with the DGJ32/J189-2015 "Technical Regulations for Building Foundation Pit Engineering Monitoring in Nanjing Area". The specific circumstance and analogous engineering experience suggest that the early warning value of the concrete support axial force is 800 kN. When the foundation pit is excavated using the construction method of L-shaped short and small steps, the overall foundation pit is stable.

Fig. 7 Numerical model of L-shaped short step and small pace construction after foot retraction state

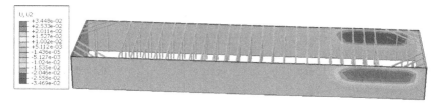

Fig. 8 Deformation of envelope structure

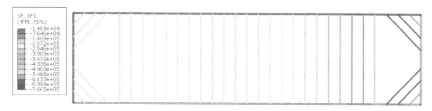

Fig. 9 Axial force of concrete bracing piece

5 Influence of Steel Support Erection Timing on Supporting Structure

During excavation, the foundation reinforcement, SMW engineering method piles, and concrete supports are all performed throughout the construction phase. The steel support is excavated during the L-shaped short steps and small step distance technology construction process from the slope foot of the pit bottom to the extrapolation of the steel support on the base.

The deformation and stability of the foundation pit are significantly impacted by the time of the construction of the line and the steel support. The efficiency of the construction process may be increased by optimizing the time of the erection of steel supports, which will also increase the foundation pit's overall safety. The steel support is simulated and calculated based on the L-shaped short step small step building method immediately following the steel support's erection at intervals of 2, 4, 6, 8, 10, and 12 h. When the erection time of the steel support is not taken into account, as shown by the calculation results in Table 2. and Fig. 10. The displacement of the foundation pit and the axial force of the concrete support are both less than the value based on the erection time. It reveals that the erection time of the steel support has a significant impact on the displacement and deformation of the foundation pit.

With increasing erection time, the axial force of the concrete support for the foundation pit and the displacement of the enclosed structure gradually rise, and the increase rate decreases gradually. The creep deformation is mostly concentrated in the initial stages after the excavation is completed. Therefore, early construction of the steel support can effectively reduce creep. with the effect of deformation on foundation pit deformation. When the steel support was constructed within 12 h from

Table 2 Calculation results

S. No	Support erection timing	Maximum horizontal displacement (mm)	Maximum concrete support axial force (kN)
1	Erected slope-bottom excavation	34.69	766.5
2	Erected slope-bottom excavation 2 h	35.84	784.1
3	Erected slope-bottom excavation 4 h	36.13	789.3
4	Erected slope-bottom excavation 6 h	36.29	793.5
5	Erected slope-bottom excavation 8 h	36.40	796.4
6	Erected slope-bottom excavation 10 h	36.49	799.1
7	Erected slope-bottom excavation 12 h	36.56	801.6

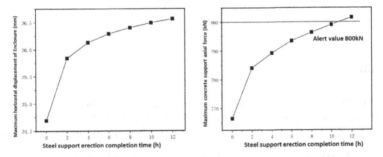

Fig. 10 Influence of brace erection time on stress and deformation of supporting structure: (a) Maximum horizontal displacement of enclosure structure-erection time relationship, (b) Concrete support axial force maximum - erection time relationship

the excavation of the foundation pit to the bottom of the slope. The axial force of the concrete support exceeded 800kN, but it was still within a safe range when it was erected within 10 h. Therefore, the L-shaped short steps and small steps construction method is used. The steel support erection completion time in this project should be controlled within 10 h.

6 Conclusions

Due to the limitations of the traditional excavation method, interior support construction cannot be finished while foundation pit excavation is being done. The construction efficiency is poor, which is detrimental to silty soil creep management. The construction method of an L-shaped short step and small step foundation pit was designed to resolve the problem of delayed construction efficiency during the excavation of long-span and long-distance foundation pits in silty soil layers. The finite element method was used to study the deformation of the supporting structure during excavation, and the impact of steel support erection time was investigated. According to practical application and calculated values, the L-shaped short step and tiny step construction method may increase excavation construction and excavation efficiency under foundation pit safety parameters. The displacement and supporting axial force of the foundation pit can meet the specification requirements. Furthermore, it is suggested that the steel support structure in the foundation pit be constructed earlier taking into account all of the creep characteristics of the silty soil foundation pit and the connection of the building process. To avoid the excavation surface being unsupported for an extended period of time, resulting in severe deformation of the foundation pit.

References

1. Xin Q (2011) Research and application of foundation pit support, dewatering and earthwork excavation technology in Coastal area. Constr Sci Technol 15:78–81
2. Jiang H (2009) Construction method of deep mud foundation pit excavation with prestressed pipe pile. Sci Technol Innov Herald (21):41
3. Deng Z (2004) Research progress of excavation and support methods for foundation pit J Wuyi Univ (Nat Sci Ed) (04):37–42
4. Wang W (2012) Discussion on construction method and matters needing attention of foundation pit excavation. Sci Technol Innov Appl (21):203
5. Wu M (2011) Discussion on key points of excavation and support of foundation pit in soft soil area. Value Eng 30(06):50
6. Sun J, Bai T (2021) Research and application of small scale basin excavation method for iron deep foundation pit in soft soil. J Undergr Space Eng 17(04):1244–1252
7. Wang J, Chen F, Zhu L, et al (2022) Research on soil deformation monitoring and application of deep and large foundation pit based on multi-dimensional deformation measurement. J Municipal Technol 40(03):195–198
8. Zhang H, Zhou G, He S, et al (2021) Numerical simulation of metro station foundation pit excavation based on FLAC3D J Municipal Technol 39(04):115–118+127
9. Cai D, Bi Q (2021) Reasonableness analysis of anti-uplift design of subway tunnel under foundation pit excavation. J Municipal Technol 39(03):72–77
10. Liu Y, Wang X, Li H et al (2021) Numerical analysis of surface deformation characteristics after foundation pit support. J Municipal Technol 39(02):134–137
11. Yang Z, Tian Q (2021). Construction method of deep foundation pit in coastal soft soil under safety protection. Undergr Water 43(05):232–234
12. Lin Z, Zhang B, Yang D (2020) Simulation of foundation pit excavation in soft soil considering creep and seepage. Modern Tunnelling Technol 57(01):91–98

13. Du J (2015) Research on time-dependent characteristics and evaluation method of disturbance displacement in excavation of large foundation pit in soft soil. Shanghai Jiao Tong University, Shanghai
14. Xiong Y (2021) Research on deformation law of foundation pit excavation and optimization of supporting structure parameters in soft soil area. Xi 'an: Xi 'an University of Science and Technology
15. Zhang J (2017) Deformation simulation and support optimization of deep foundation pit excavation for comprehensive pipe gallery in Coastal area. Fuzhou University, Fuzhou

Study on Structural Performance and Design Method of Rectangular Steel Plate Water Tank

Gan Tang, Zelong Jia, Peng Li, Ziheng Ye, and Junchun Dou

Abstract According to the engineering requirements of the integrated sewage treatment equipment, the structural design of the steel plate water tank used in the integrated sewage treatment equipment is carried out. Based on the investigation of actual projects and theoretical analysis, proposed the failure mode of steel plate water tank structure and the corresponding design guidelines. The finite element modeling calculation of the steel plate water tank project example is carried out to obtain the calculation results of the stress–strain level and displacement of the water tank structure under the stress state, so as to analyze the structural performance of the steel plate water tank structure and judge whether the water tank structure reaches the failure state accordingly. According to the design guidelines to adjust the structural arrangement and member size to ensure the safety of the water tank structure in the process of use, providing design methods for the design of steel plate water tank structure.

Keywords Steel Plate Tank · Finite Element Analysis · Structural Design

1 Introduction

With the increase of water consumption in cities and towns, the harmless and resourceful treatment of domestic sewage can effectively reduce pollution and improve the utilization of water resources. Urban sewage treatment plant and the corresponding pipeline network construction investment is huge, and sewage treatment system has gradually changed from large-scale centralized to small-scale decentralized, in which integrated sewage treatment equipment has been widely applied and developed. The main structure of the integrated sewage treatment equipment is the water tank, which is divided into several sewage treatment areas inside the tank and arranged with the corresponding sewage treatment equipment. The steel

G. Tang (✉) · Z. Jia · P. Li · Z. Ye · J. Dou
Department of Civil Engineering and Airport Engineering, Nanjing University of Aeronautics and Astronautics, Nanjing, China
e-mail: tanggan@sina.com

© The Author(s) 2023 543
G. Feng (ed.), *Proceedings of the 9th International Conference on Civil Engineering*,
Lecture Notes in Civil Engineering 327,
https://doi.org/10.1007/978-981-99-2532-2_46

plate tank has the advantages of high strength, light weight, easy construction, easy maintenance, etc. It is suitable for temporary integrated sewage treatment equipment [1].

At present, there is no special specification for the design and manufacture of steel plate water tank, the manufacturer is mostly designed and manufactured according to experience, although some practical engineering problems can be solved, but the bearing capacity of the steel plate water tank, the mechanical response of each part of the tank under load, the possible failure mode of the tank and the overall and local structural analysis and design of the tank are not thoroughly studied. In this paper, the mechanical properties of the steel plate water tank under the state of stress are analyzed by finite element calculation. The structural design criterion of steel plate water tank is proposed, and based on this criterion, the structural design of the steel plate water tank used in integrated wastewater treatment equipment is carried out to verify the safety of the tank structure [2].

2   Project Profile and Structure Selection

2.1   Project Profile

The water tank structure design for the above-ground integrated sewage treatment equipment with a capacity of 4.0 m^3/h, the rectangular steel tank is 3500 mm long, 3500 mm wide and 5300 mm high, and the design working water level is 5000 mm. The structural safety level is Grade II, the design service life is 25 years, the seismic protection category is the standard protection category, the equipment is suitable for the I and II sites with seismic protection intensity of 8° and all kinds of sites in the area of 7 degrees and below 7° [3].

2.2   Structure Selection

The main body of the water tank structure is a box-shaped thin-walled structure composed of tank plates and reinforcing elements welded together [4]. The horizontal and vertical reinforcement ribs set in the outer wall of the tank, the tank plate is divided into smaller cells to increase the stiffness and reduce the deformation, in addition, the reinforcement ribs placed in the outer wall is also convenient for the inner wall of the tank for the anti-corrosion and maintenance. According to the distribution characteristics of hydrostatic pressure, the horizontal stiffening ribs take the distribution method of upper sparse and lower dense. The vertical stiffening ribs take equal spacing distribution, the spacing in the horizontal direction is 500 mm. Set up two rows of tension rods inside the tank in the vertical distribution corresponding to the horizontal stiffening ribs, in order to reduce the span of the horizontal stiffening ribs, so that the cross-sectional force of the horizontal stiffening ribs is more uniform [5]. Three internal partitions are arranged inside the water tank to separate different functional partitions.

3   Failure Modes and Design Criteria

Failure judgment basis and design criteria are the prerequisites for structural design. Based on the existing engineering cases of rectangular steel plate water tank and theoretical derivation, several structural failure modes of rectangular steel plate water tank and the corresponding structural design criteria are proposed [2].

3.1   Overall Slippage or Overturning of the Water Tank Structure

As a single structure placed on the ground, the water tank is subjected to wind load, relying only on the bottom of the box and the ground sassafras force to resist the horizontal force generated by the wind load. In the empty tank state, the pressure of the structure on the ground is reduced, resulting in a reduction in the maximum static friction force, which may lead to the phenomenon of overall structural slippage. In addition, when the height to width ratio of the tank structure is large, the wind area of the structure is relatively larger, and the overall structure may overturn the phenomenon.

When the tank structure overall slippage or overturning, may produce damage to the equipment and external pipe line, in order to prevent the structure from such failure, in the design of the structure, with reference to the stability design of the independent foundation, the tank structure as a whole to resist horizontal sliding and overturning test. If the test is passed, it means that the water tank structure can be directly rested on the ground and will not slip or overturn as a whole. Conversely, the support should be designed to connect the bottom of the tank structure with the ground to resist the forces that cause the overall slip and overturning of the structure in the horizontal direction.

3.2   Tank Plates Tear Damage

In the case of ensuring the welding quality and the strength of the stiffening ribs, the water tank plates are divided into grids by the stiffening ribs, and the plates are subjected to greater hydrostatic pressure near the bottom, and are more likely to be torn by excessive deformation. When the plate is in the elastic range, the plate will not tear and will not bulge outward significantly, which can ensure the normal operation of the water tank [6].

In the design of the water tank plate, should ensure that the water tank plate parts are basically within the elastic range. Calculate the deflection value of the plate under normal use and the equivalent force strength of the plate under the load carrying capacity limit state through finite element calculation. If the deflection value of the plate is less than b/100 and the equivalent force strength on the plate is less than the yield strength of the material, the plate meets the design requirements. Conversely, the thickness of the plate should be adjusted and the calculation should be performed again until the design guidelines are met.

3.3 Bending Damage to the Stiffening Ribs or Fracture of the Internal Tension Rods

The water tank is a typical thin-walled box-type structure, and its ability to bear internal pressure is very weak. The stiffening ribs and internal tension rods act as the structure skeleton to bear the force from the tank plates, while restraining the plates from excessive lateral deformation. The vertical stiffening ribs are designed as flexural members, the horizontal stiffening ribs are designed as tensile-bending members, and the internal ties are designed as axial tension members.

When the stiffening ribs are damaged by bending or internal tension rod fracture, the adjacent plate will assume a greater range of water pressure, thus producing a greater local equivalent force strength and more likely to occur plate tearing damage. In the design of the stiffening ribs and internal tension rods, the elastic design criterion is adopted, and the calculation is carried out by finite element software, and the steel structure calibration of the stiffening ribs and internal tension rods is applied to Chinese standards to obtain the stress ratio of the rods under different working conditions. If the stress ratio of the rods is less than 1.0, the rods meet the design requirements [7]. Conversely, the structural dimensions of the rods should be adjusted and the calculations should be performed again until they meet the requirements.

4 Anti-slip and Anti-Overturning Test

4.1 Anti-slip Test

The structural anti-slip test is performed by the following equation.

$$K_1 = \frac{\mu N}{F} \tag{1}$$

Table 1 Results of horizontal sliding resistance test

Work conditions	N(kN)	F(kN)	M(KN.m)	e_0(m)	K_1	K_2
1D + 1Wx	61.1	40.8	117.9	1.93	0.9	0.9
1D + 1Wy	61.1	40.8	108.0	1.77	0.9	1.0
1D + 1Ex-d	61.1	5.5	29.4	0.48	6.7	3.6
1D + 1Ey-d	61.1	5.5	19.5	0.32	6.7	5.5
1D + 1SS + 1Ex-s	717.1	57.1	161.0	0.22	7.5	7.8
1D + 1SS + 1Ey-s	717.1	57.2	151.2	0.21	7.5	8.3
1D + 1SS + 1Wx	717.1	40.8	117.9	0.16	10.6	10.6
1D + 1SS + 1Wy	717.1	40.9	108.1	0.15	10.5	11.6

In the formula: K_1 is the horizontal sliding safety factor, requiring $K_1 \geq 1.2$; N is the sum of the vertical forces acting on the structure; F is the sum of the horizontal forces acting on the structure; μ is the friction coefficient of the structure base plate and foundation, taken as 0.6.

According to Table 1, it can be concluded that the water tank structure can meet the anti-sliding requirements under the full water working condition. However, under the wind load, the tank structure in the empty tank state cannot meet the requirements of anti-sliding.

4.2 Anti-overturning Test

The structural overturning test is performed by the following equation.

$$K_2 = \frac{y}{e_0} \qquad (2)$$

In the formula: K_2 is the overturning stability coefficient, requiring $K_2 \geq 1.5$; y is the distance from the center of gravity of the structure to the side of the maximum pressure, y is taken as 1.75 m in this project; e_0 is the eccentricity distance of the combined external forces, $e_0 = M/N$.

According to Table 1, it can be concluded that the water tank structure can meet the requirements of overturning resistance under the working condition of full water. Under the wind load, the tank structure in the empty tank state cannot meet the overturning resistance requirements.

Fig. 1 Support design
drawing

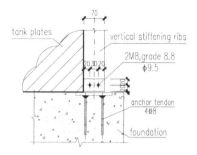

4.3   Support Design

Comprehensive consideration, in order to ensure the stability of the structure in the state of empty tank, the bottom of the vertical rib is connected with the foundation. The nodal reaction force is extracted for the design of the support articulation node, and the design pullout force of the support node is 5 kN and the shear force is 13 kN. The design of the support is shown in Fig. 1.

5   Finite Element Calculation Analysis of Steel Plate Water Tank Structure

5.1   Modeling of Water Tank Structure

Using SAP2000 finite element calculation software for structural design, considering the force characteristics and geometric structural features of the tank plates and tank reinforcement elements, the thin shell unit is applied to establish the tank plates, and the frame unit is applied to establish the stiffening ribs and tension rods. The stiffening ribs, ties and plates are made of Q355 steel [8]. The transverse stiffening ribs are box-type $100 \times 80 \times 4$ section, the vertical stiffening ribs are rectangular 70×6 section, and the tension rods are rectangular 50×4 section. The thickness of the tank plate is 5 mm, taking into account the rusting effect of sewage on the steel plate, the corrosion margin must still be reserved in the case of anti-corrosion coating. In the structural finite element analysis, the thickness of the tank steel plate is 4 mm.

For simplicity, and taking into account the actual form of connection between the components, the following assumptions are made in the finite element model: The bottom of the vertical stiffening ribs are hinged to the foundation to simulate the restraint of the support on the tank. Each node of the bottom plate is equipped with a support that only constrains downward displacement to simulate the support of the ground to the tank. The stiffening ribs are solidly connected to the box plate. Since the steel plate is thin, the joint action of the stiffening ribs and the plate is not considered

Fig. 2 Water tank finite
element model

in the design. The horizontal and vertical stiffening ribs are solidly connected to each other and together form the skeleton of the water tank structure. Taking into account the calculation accuracy and calculation speed, the plate is divided into 100 mm × 100 mm calculation units, and the junction between the plate and the bar is processed as a local subdivision unit. The finite element model is shown in Fig. 2.

5.2 External Force Datum

In the finite element calculation, D denotes the constant load; SS denotes the hydro-static pressure under full water working condition; Wx and Wy denote the wind load in X-direction and Y-direction respectively; Ex-d and Ey-d denote the seismic action in X-direction and Y-direction in the empty box state respectively. Ex-s and Ey-s denote the seismic action in X-direction and Y-direction under full water working condition respectively. The structure is symmetrical, the effect of wind load and earthquake in X and Y negative directions on the structure is the same as the effect in positive direction. According to *Load code for the design of building structures* and *Structural design code for special structures of water supply and waste water engineering*, load combinations are carried out to consider the effects of different working conditions on the structure [9, 10].

- Self-weight of the structure: Automatically accounted for by the software.
- Water load: The tank structure only considers the role of internal lateral hydro-static pressure. The standard value of lateral water pressure at the bottom of the equipment is 53.55 kPa. The hydrostatic pressure is triangularly distributed along the height of the wall plate.
- Wind load: Calculate the wind load borne by the water tank structure in accordance with *Load code for the design of building structures*. The basic wind pressure is taken as 0.75 kN/m^2 [9].
- Seismic action: When considering the seismic effect of the tank structure, the effect of internal water storage cannot be ignored. This design calculates the

seismic action according to Article 6.2 of *Code for sesismic design of outdoor water supply, sewerage, gas and heating engmeering* [11].

6 Water Tank Structure Analysis and Design

6.1 Structural Analysis and Design of Tank Plates

The tank plate in the working condition mainly bears the hydrostatic pressure inside the tank, considering the restraint effect of the stiffening ribs on the plate, the force form of the plate is bi-directional support of the bending member. The displacement cloud diagram and stress cloud diagram of the tank plate are shown in Fig. 3. and Fig. 4, respectively.

The calculation results show that the maximum span deflection of the tank plate is 3.65 mm in the limit state of normal use, which is less than b/100 = 5 mm, and the plate meets the stiffness requirement. In the limit state of bearing capacity, the maximum tensile stress of the box plate is 357 MPa and the maximum compressive stress is 150 Mpa. There are small areas in the plate where the stress exceeds the material strength design value of 305 MPa, mainly at the connection with the stiffening ribs there is stress concentration, but does not exceed the material strength limit value of 470 MPa, the plate meets the strength requirements [8].

Fig. 3 Displacement cloud diagram of tank plate

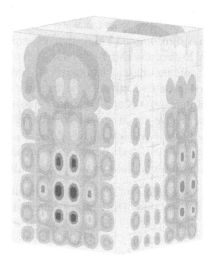

Fig. 4 Stress cloud diagram
of tank plate

6.2   Structural Analysis and Design of Tank Strengthening Members

The tank reinforcement member acts as the structural skeleton and assumes the role of the load transferred by the slab in the slave range. The displacement cloud diagram of the tank strengthening member and the stress ratio of the member are shown in Fig. 5. and Fig. 6, respectively.

The calculation results show that the maximum spanwise deflection of horizontal stiffening ribs is 0.31 mm, which is less than b/500 = 1 mm in the limit state of normal use. The maximum spanwise deflection of vertical stiffening ribs is 0.53 mm, which is less than b/500 = 1 mm. The stiffening ribs meet the stiffness requirements. The

Fig. 5 Displacement cloud
diagram of the member

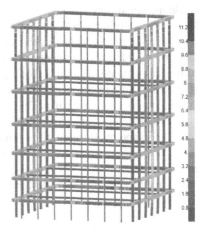

Fig. 6 stress ratio of the member

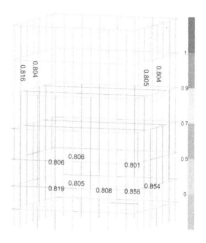

maximum stress ratio of the tank reinforcement member is 0.855 in the load carrying capacity limit state. The tank reinforcement member meets the strength requirement [8].

In summary, the stiffness and strength of the water tank panels and tank strengthening members meet the design requirements, and the design scheme is feasible.

7 Conclusion

In this paper, the structural design of the steel plate water tank used in integrated sewage treatment equipment is carried out, the force performance of the tank structure is analyzed, the failure mode of the steel plate water tank and the corresponding design guidelines are proposed, and design ideas are provided for the structural design of the steel plate water tank. SAP2000 software is applied to the finite element modeling calculation of the steel plate water tank structure, and the force performance analysis and structural failure judgment of the tank as a whole, the tank plate and the skeleton of the tank structure are carried out respectively. After calculation and analysis, the structural design of the steel plate water tank is carried out according to the design criteria proposed in this paper. Due to the light weight of the steel plate water tank structure, it is necessary to design the support at the bottom of the tank to resist the overall slip and overturning of the structure. By analyzing the structural mechanical properties such as stress level and displacement, suitable member sizes are selected to ensure the stiffness, strength and stability of the structure and meet the design requirements. The structural design method of the water tank can provide reference for other similar projects.

References

1. Zhang W (2013) Structural optimization design of mobile sewage treatment plant Tianjin University of Science and Technology
2. Chen XN (2005) Mechanical response characteristics and structural design of fabricated stainless steel water tank Zhejiang University
3. GB 50011–2010 (2010) Code for seismic design of buildings (Beijing: China architecture and building press)
4. Zeng LM (2001) Structural analysis and calculation of FRP monolithic veneer structure rectangular water tank. Fiber Compos Mater (01):18–9
5. Li BL, Shen QM, Sun GL, Shao QN, Huang JJ Wang R (2021) Structural design and calculation analysis of aluminum alloy civil water storage tank. Alum Process (05):31–34
6. Cao Y, Chen YK (2009) Stress distribution and reinforcement of side plate of box-shaped storage tank. Petrochem Equipment 38(01):55–57
7. Tang G, Wang FW (2015) Architectural steel structure design (Beijing: National defense industry press)
8. GB 50017–2017 (2017) Standard for design of steel structures (Beijing: China construction industry press)
9. GB 50009–2012 (2012) Load code for the design of building structures (Beijing: China architecture and building press)
10. GB 50069–2002 (2002) Structural design code for special structures of water supply and waste water engineering (Beijing: China architecture and building press)
11. GB 50032–2003 (2003) Code for sesismic design of outdoor water supply, sewerage, gas and heating engmeering (Beijing: China architecture and building press)

Study on Seepage Mechanism and Stability of Unsaturated Slope Based on Trefftz Method

Yan Su, Lingjun Yang, Chuan Lin, Guolin Guo, Yanfeng Tang, Yangmin Lin, Xiudong Xie, and Lidan Hong

Abstract This paper proposes a space–time Trefftz method (STM) to study the seepage mechanism and stability of unsaturated slopes. The groundwater flow under transient conditions is important in engineering practice for solving practical problems such as assessing the stability of unsaturated soil slopes. Based on the transient groundwater equation, we derived the Trefftz basis functions by splitting the variables. The solutions are approximated using Trefftz basis functions in the space–time domain. The Stabl software is subsequently employed to analyze the stability of the slope under the rainfall recharge condition with the combined reservoir water level fall. The results demonstrate that the steeper the hydraulic slope drop under combined reservoir water level fall and rainfall infiltration, the more unstable the slope becomes.

Keywords Space–Time Trefftz Method · Unsaturated Slope · Seepage Mechanism

1 Introduction

Slope instability occurs at the surface of the ground often affecting human activities and the building environment with catastrophic consequences. According to the results of geological hazard surveys in 290 counties and cities nationwide, slope instability accounts for 51% of geological hazards, while rainfall-induced landslides account for about 90% of the total [1]. Numerous scholars have studied unsaturated

Y. Su · L. Yang · C. Lin (✉) · X. Xie
College of Civil Engineering, Fuzhou University, Fuzhou 350108, Fujian, China
e-mail: linchuan@fzu.edu.cn

G. Guo
Fujian Institute of Water Resources and Hydropower Research, Fuzhou 350001, Fujian, China

Y. Tang · Y. Lin
Fujian Water Resources and Hydropower Survey, Design and Research Institute Co., Ltd., Fuzhou 350013, Fujian, China

L. Hong
School of Engineering, Taiwan Ocean University, Keelung 20224, China

© The Author(s) 2023
G. Feng (ed.), *Proceedings of the 9th International Conference on Civil Engineering*,
Lecture Notes in Civil Engineering 327,
https://doi.org/10.1007/978-981-99-2532-2_47

seepage [2] and slope stability analysis [3]. For reservoir bank slopes, the degree of influence of coupled rainfall effect on slope stability varies when the reservoir water level changes, and the unstable seepage of groundwater is the main factor affecting slope stability. Therefore, it is necessary to analyze and study the influence of precipitation-induced groundwater level change on the stability of soil slopes. At present, the experimental method and the numerical analysis method are mainly used for groundwater seepage in unsaturated soils [4]. Most of the traditional numerical methods for solving groundwater seepage problems are discretized by finite difference method [5], finite element method [6] to create meshes. Nevertheless, for complex study areas, such methods encounter the problem of difficult grid creation, and the problem boundary needs to be simplified to be analyzable [7]. In recent years, many scholars have studied meshless methods that are exempted from the establishment of a mesh [8], the collocation Trefftz method does not require the creation of a grid and integration of the boundary [9]. The basis function satisfies the control equation, and the points are only distributed on the boundary. The coefficient matrix satisfying the boundary conditions is obtained through the boundary points to solve the problem, which reduces the computational difficulty and ensures computational accuracy, has strong advantages for complex shape computational domain problems [10].

For solving the transient problem, this study instead of using the traditional time marching method, solves the problem based on the space–time coordinate system through the concept of space–time matching method, which effectively simplifies the calculation process and avoids error accumulation. In this paper, the groundwater seepage mechanism of unsaturated soil slopes under the action of rainfall is analyzed by using STM. The slope analysis model is established based on the limit equilibrium theory to calculate the safety coefficient at different water levels considering the influence of groundwater seepage, to investigate the influence of reservoir level changes combined with rainfall on the stability of unsaturated soil slopes.

2 Seepage Governing Equation and Unsaturated Slope Stability

2.1 Governing Equation

The governing equation for two-dimensional transient homogeneous isotropic seepage considering the complementary drainage effect of seepage is the Dupuit-Boussines equation as follows

$$\partial^2 h^2 / \partial^2 x + \partial^2 h^2 / \partial^2 y + w = (1 / D)(\partial h^2 / \partial t) \tag{1}$$

where $D = k/S_s$ (m²/s) denotes hydraulic diffusion coefficient, h denotes total head (m), k is hydraulic conductivity, S_s is the volumetric specific storage, m/s), $w =$

$2I/k$ expressed twice the ratio of subsurface subsidy rate and permeability coefficient for the dimensionless number, I indicates the aquifer unit time, unit area of vertical recharge and discharge, recharge is positive discharge is negative.

For simulating the transient flow of water in unsaturated soil slopes, the initial and boundary conditions are given as follows

$$u(x, y, 0) = h_0^2(x, y, t), (x, y, t = 0) \in \Omega_t \tag{2}$$

$$u(x, y, t) = h_D^2(x, y, t), (x, y, t) \in \Gamma_D \tag{3}$$

$$\partial u(x, y, t)/\partial n = (\partial u(x, y, t)/\partial x)n_x + (\partial u(x, y, t)/\partial y)n_y \\ = h_N^2(x, y, t), (x, y, t) \in \Gamma_N \tag{4}$$

where h_0 is the initial total head, h_D is Dirichlet boundary data, h_N is Neumann boundary data, Ω_t donates the space–time domain, Γ_D denotes the boundary that satisfies the Dirichlet boundary condition, Γ_N denotes the boundary that satisfies the Neumann boundary condition, $\vec{n} = (n_x, n_y)$ denotes the outward normal component.

2.2 Unsaturated Slope Stability

Janbu method was proposed by Janbu in 1973 to solve the safety coefficient by iteration. In this paper, we use the STABL program in (Janbu Circular) to calculate the slope stability, and use the two-dimensional limit equilibrium analysis to calculate the safety coefficient.

3 Numerical Algorithms

3.1 The Space–Time Trefftz Method

STM is to take time as a dimension, and the original N-dimensional space becomes (N + 1)-dimensional spatio-temporal domain. The transient seepage governing equation considering the complementary drainage effect is expressed in the polar coordinate system as follows

$$\frac{\partial^2 u(r, \theta, t)}{\partial r^2} + \frac{1}{r}\frac{\partial u(r, \theta, t)}{\partial r} + \frac{1}{r^2}\frac{\partial^2 u(r, \theta, t)}{\partial \theta^2} + w = \frac{1}{D}\frac{\partial u(r, \theta, t)}{\partial t}, (r, \theta, t) \in \Omega_t \tag{5}$$

where r is radius, t is time, θ is polar angle, The T-basis of the governing equation is obtained by using the variable separation method. To improve the order of STM and to eliminate the influence of the matrix pathology at the same time, the characteristic length R_0 is introduced, which effectively avoids the phenomenon of non-convergence of the basis function with the expansion of the level [11]. The T-basis of the simply domain of slope seepage under the Dirichlet boundary condition after adding the characteristic length as follow

$$u(r, \theta, t) = \sum_{p=1}^{m} \sum_{q=1}^{s} \bar{\alpha}_{pq} T_{qp}(r, \theta, t) = 1 - \frac{wr^2}{4} \sum_{p=1}^{m} \exp(-Dp^2 t) A_p J_0(\frac{pr}{R_0}) + \sum_{q=1}^{s} B_{2q}(\frac{r}{R_0})^{-q} \cos(q\theta)$$

$$+ \sum_{q=1}^{s} B_{3q}(\frac{r}{R_0})^{-q} \sin(q\theta) + \sum_{p=1}^{m} \sum_{q=1}^{s} \exp(-Dp^2 t) J_q(\frac{pr}{R_0})(A_{6pq} \cos(q\theta) + A_{7pq} \sin(q\theta)) \quad (6)$$

where J_0 and J_q denote the order 0 and q first class Bessel functions, respectively, $p = 1, 2, \cdots m$ and $q = 1, 2, \cdots, s$ denote the order, m and s denote the order of T-basis function, $\mathbf{T}_{pq}(r, \theta, t)$ denotes the basis function of Dirichlet boundary condition, $\bar{\alpha}_{pq} = \alpha$ denotes undetermined coefficient and $A_{1p}, B_{2q}, B_{3q}, A_{6pq}, A_{7pq}$ denote the coefficients to be determined. The basis function $\mathbf{T}'_{pq}(r, \theta, t)$ of Neumann boundary conditions as follow

$$T_{pq}(r,\theta,t) = \begin{cases} -\frac{r}{2}w(\cos(\theta)n_x + \sin(\theta)n_y), -\exp(-Dp^2 t)p\left(J_1(\frac{pr}{R_0})\cos(\theta)n_x + J_1(\frac{pr}{R_0})\sin(\theta)n_y\right), \\ q\left[\left((\frac{r}{R_0})^{q-1}\cos(q\theta)\cos(\theta) + (\frac{r}{R_0})^{q-1}\sin(q\theta)\sin(\theta)\right)n_x + \\ \left((\frac{r}{R_0})^{q-1}\cos(q\theta)\sin(\theta) - (\frac{r}{R_0})^{q-1}\sin(q\theta)\cos(\theta)\right)n_y\right], q\left[\left((\frac{r}{R_0})^{q-1}\sin(q\theta)\cos(\theta) - (\frac{r}{R_0})^{q-1}\cos(q\theta)\sin(\theta)\right)n_x + \\ \left((\frac{r}{R_0})^{q-1}\sin(q\theta)\sin(\theta) + (\frac{r}{R_0})^{q-1}\cos(q\theta)\cos(\theta)\right)n_y\right], \\ \exp(-Dp^2 t)\frac{q}{2}\left[\left((J_{q-1}(\frac{pr}{R_0}) - J_{q+1}(\frac{pr}{R_0}))\cos(q\theta)\cos(\theta) + qJ_q(\frac{pr}{R_0})R_0\sin(q\theta)\frac{\sin(\theta)}{r}\right)n_x + \\ \left(J_{q-1}(\frac{pr}{R_0}) - J_{q+1}(\frac{pr}{R_0}))\cos(q\theta)\sin(\theta) - qJ_q(\frac{pr}{R_0})R_0\sin(q\theta)\frac{\cos(\theta)}{r}\right)n_y\right], \\ \exp(-Dp^2 t)\frac{q}{2}\left[\left((J_{q-1}(\frac{pr}{R_0}) - J_{q+1}(\frac{pr}{R_0}))\sin(q\theta)\cos(\theta) - qJ_q(\frac{pr}{R_0})R_0\cos(q\theta)\frac{\sin(\theta)}{r}\right)n_x + \\ \left(J_{q-1}(\frac{pr}{R_0}) - J_{q+1}(\frac{pr}{R_0}))\sin(q\theta)\sin(\theta) + qJ_q(\frac{pr}{R_0})R_0\cos(q\theta)\frac{\cos(\theta)}{r}\right)n_y\right] \end{cases}$$

$$(7)$$

where J_1 and J_{q-1} denote the order 1 and q-1 first class Bessel functions, respectively.

3.2 Slope Seepage and Stability Solving Process

In this paper, instead of solving the transient problem using the traditional time marching scheme, the problem is solved in the spatio-temporal coordinate system, combined with the concept of spatio-temporal matching points, the initial conditions are discretized as boundary conditions. 1) The governing equation is Eq. (5). 2) The boundary condition is Eqs. (2–4). 3) Initial and boundary condition discretization. The initial value of the study area is discretized into n_v points, and the boundary points are discretized into n points in the spatio-temporal coordinate system where the Dirichlet boundary points are n_f and the Neumann boundary points are n_g, $n = n_f + n_g$. As shown in Fig. 1, the solid circle point is the initial condition point and

Fig. 1 Spatial and temporal
distribution of the study area

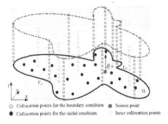

○ Collocation points for the boundary condition ● Source point
● Collocation points for the initial condition   Inner collocation points

the hollow circle point is the boundary condition point. Depending on the boundary points, the boundary points are brought into the numerical solution of the governing equations of the Dirichlet and Neumann boundary conditions, respectively.

The numerical solution of the initial conditions $V(r_i, \theta_i, t)$ into the basis function is expressed as follows

$$V(r_i, \theta_i, t = 0) = \sum_{p=1}^{m} \sum_{q=1}^{s} \overline{\alpha}_{pq} T_{qp}(r_i, \theta_i, t = 0) \tag{8}$$

Numerical solutions satisfying Dirichlet boundary conditions $F(r_i, \theta_i, t_i)$ and Neumann boundary conditions $G(r_i, \theta_i, t_i)$ are expressed as follows

$$F(r_i, \theta_i, t_i) = \sum_{p=1}^{m} \sum_{q=1}^{s} \overline{\alpha}_{pq} T_{qp}(r_i, \theta_i, t_i), \quad G(r_i, \theta_i, t_i) = \sum_{p=1}^{m} \sum_{q=1}^{s} \overline{\alpha}_{pq} T'_{qp}(r_i, \theta_i, t_i),$$
$$\tag{9}$$

4) Solving undetermined coefficient. The system of linear algebraic equations consisting of initial and boundary conditions is expressed in the form of $A\alpha = B$. A matrix is a scale $(n_v + n)(bb = 1 + 2s + 3 + 2ms)$ matrix formed by the Trefftz basis, the α matrix is a scale $bb = 1 + 2s + 3 + 2ms$ matrix of undetermined coefficient, and **B** matrix is a scale $n_v + n$ matrix of boundary values formed by using the boundary conditions. **A** matrix is obtained by substituting the initial boundary conditions and boundary positions into the Trefftz basis, and **B** matrix is obtained by the boundary values, the matrix of undetermined coefficient is obtained by left division of the matrix by $\alpha = A^{-1} B$. 5) Discrete for the study area. The internal distribution of the study area as shown in Fig. 1, the computational domain is discretized, the internal points satisfy the governing equations, and the coordinates of the computational points and the coefficients to be found are substituted into Eq. (6) to find the head values $h(r_i, \theta_i, t_i) = \sqrt{u(r_i, \theta_i, t_i)}$ of the points in the computational study domain.

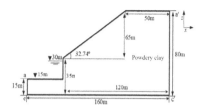

Fig. 2 Schematic diagram
of the study area

4   Numerical Results and Discussion

In this case, transient groundwater seepage with recharge is considered, the governing equation is as in Eq. (5), and the initial and boundary conditions are described below.

4.1   Calculation Model and Soil Parameters

The study area is as shown in Fig. 2. Figure 2 shows the slope of powdery clay with part of the foot of the slope cut off. The soil has a specific water storage coefficient of $S_s = 6 \times 10^{-5} \mathrm{m}^{-1}$, saturation permeability coefficient $k_s = 4 \times 10^{-3} \mathrm{mh}^{-1}$, saturation weight $r_s = 20$ kN/m^3, dry weight $\gamma = 18$ kN/m^3, cohesion c $= 23$ kPa, and internal friction angle $\varphi = 65^0$. The bottom of the slope of the soil body is impermeable layer, the left side ac is the reservoir water level boundary, and the right side $a'c'$ is the watershed.

4.2   Impact of Rainfall Infiltration on Slope Stability

The left ac initial reservoir level is 30 m with continuous precipitation over a 24 h period. In order to increase the reservoir capacity, its reservoir level decreases from 30 m mean to 15 m within 24 h. The right side is a watershed with no flow boundary condition at the bottom.

Figure 3 shows that in the case of no rainfall infiltration, the reservoir water level decreases from 30 to 15 m in 24 h. The groundwater at the waterfront first decreases rapidly, and due to its hysteresis, the water level at the waterfront decreases faster and slower as the water level decreases into the slope. With the change of time, the slope drop gradually increases. As shown in Fig. 4, the vertical free aquifer recharge under rainfall infiltration is $q = 1.2 \times 10^{-4} \mathrm{mh}^{-1}$. When the reservoir water level drops uniformly from 30 to 15 m in 24 h, the water level at the waterfront decreases rapidly with the reservoir water level, and the slope is near the slope with a gentler hydraulic slope drop. When the stabilization subsidy amount is large rain slope within the outflow water, the water level within the slope rises, but the outflow amount within the slope is close to the subsidy amount, the groundwater level rises

Fig. 3 Variation of free
liquid level with time in the
case of no replenishment

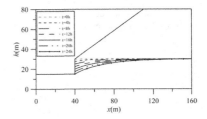

Fig. 4 Free liquid level
versus time for precipitation
recharge situation

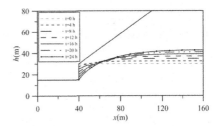

slowly and tends to stabilize. The stability of the slope is analyzed by substituting the dynamic groundwater level with and without rainfall infiltration into STABL, which shows that the soil remains stable after 16 h without rainfall, and then the stability of the slope increases due to the effect of decreasing soil weight is greater than the effect of increasing slope drop on the slope. With rainfall, the hydraulic drop in the slope increases faster than without rainfall, and the stability decreases and tends to be unstable.

4.3 Impact of Reservoir Water Level on Slope Stability

In the case of rainfall infiltration the vertical free aquifer recharge is $q = 1.2 \times 10^{-4} \mathrm{mh}^{-1}$. The bottom is the no-flow boundary condition, and the right is the watershed; if the reservoir level is lowered to 15 m before the onset of rainfall and the internal seepage reaches a steady state, the reservoir level is kept constant at the onset of rainfall. As Fig. 5 shows that the infiltration of stable rainfall, the hydraulic slope drop in the slope increases. As Fig. 6 shows that the initial reservoir level on the left side at the onset of rainfall decreases uniformly from 25 to 15 m in 24 h, and the water level in the slope changes with time.

As shown in Fig. 7, the stability of its slope gradually decreases under constant rainfall conditions when the reservoir water level is lowered to 15 m in advance and is kept constant during the rainfall process. When the reservoir water level decreases uniformly from 25 to 15 m in 24 h, the safety coefficient of its side slope decreases steadily under the rainfall infiltration condition, and the rate of its decrease is approximately the same as the rate of decrease of the safety coefficient under the constant reservoir water level. Figure 8 shows the change of safety coefficient with

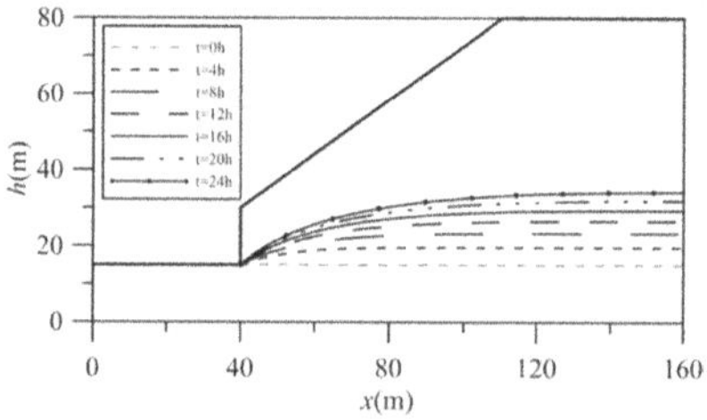

Fig. 5 Variation of free
liquid level with time in the
case of stable reservoir water
level

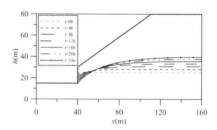

Fig. 6 Free liquid level
change pattern in uniform
decline of reservoir water
level

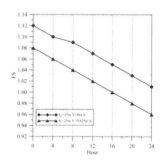

time in Fig. 4 and Fig. 6, where the reservoir level decreases from 30 and 25 m
to 15 m respectively in 24 h under the same rainfall infiltration conditions. The
rate of decrease of safety coefficient with time is approximately the same for both
conditions. However, the safety factor for the initial water level of 30 m is lower than
that for the initial water level of 25 m.

Fig. 7 The effect of
reservoir water level with
time on slope stability

Fig. 8 Influence of initial
reservoir water level on slope
stability

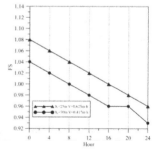

5 Conclusion

(1) The Space–Time Trefftz Method does not need to divide the grid, but only needs to lay points on the boundary. The Dupuit-Boussines equation is solved by STM to analyze the transient free liquid surface problem without iteration, saving a lot of time, and compared with traditional numerical methods, STM avoids the error accumulation problem caused by the time marching method.

(2) The stability of the bank slope is significantly influenced by the combined effect of rainfall and reservoir water level change. To ensure the slope stability, the reservoir water level should be declined according to the predicted rainfall intensity and duration when the rainfall comes, to avoid the slope instability caused by the high reservoir water level due to rainfall or the rapid decline of reservoir water level during the rainfall.

Acknowledgements This research was supported by the National Natural Science Foundation of China (Grant No. 52109118), the Young Scientist Program of Fujian Province Natural Science Foundation (Grant No. 2020J05108). The authors would like to thanks Fuzhou University, Fujian Institute of Water Resources and Hydropower Research and Fujian Water Resources and Hydropower Survey, Design and Research Institute for their technical support.

References

1. Li Y, Meng H, Dong Y, Hu SE (2004) Types and characteristics of geological disasters in China: based on survey results of geological disasters in counties and cities. Chin J Geol Hazard Control 15(2):29–34 (In Chinese)
2. Dang MR, Chai JR, Xu ZG, Qin Y, Cao J, Liu FY (2020) Soil water characteristic curve test and saturated-unsaturated seepage analysis in Jiangcungou municipal solid waste landfill, China. EngGeol 264:105374
3. Su AJ, Feng MQ, Dong S, Zou ZX, Wang JG (2022) Improved statically solvable slice method for slope stability analysis. J Earth Sci 33(5):1190–1203 (2022)
4. Ku CY, Liu CY, Su Y, Xiao JE (2018) Modeling of transient flow in unsaturated geomaterials for rainfall-induced landslides using a novel spacetime collocation method. Geofluids 2018:1–16
5. Omar PJ, Gaur S, Dwivedi SB, Dikshit PKS (2019) Groundwater modelling using an analytic element method and finite difference method: an insight into Lower Ganga river basin. J Earth SystSci 128(7):195
6. Suk H, Chen JS, Park E, Kihm YH (2020) Practical application of the galerkin finite element method with a mass conservation scheme under dirichlet boundary conditions to solve groundwater problems. Sustainability 12(14):1–18
7. Li G, Ge J, Jie Y (2002) Free surface seepage analysis based on the element-free method. Mech Res. Commun. 30(1) 9–19
8. Ku CY, Hong LD, Liu CY, Xiao JE, Huang WP (2021) Modeling transient flows in heterogeneous layered porous media using the space–time Trefftz method. Appl. Sci. 11(8):3421
9. Su Y, Huang LQ, Yang LJ, Benisi GH, Lin C (2022) Trefftz method applied to the problem of pumping well test counter-calculation. Pol. J. Environ. Stud. 31(3):2837--2849
10. Motamedi AR, Boroomand B, Noormohammadi N (2022) A Trefftz based meshfree local method for bending analysis of arbitrarily shaped laminated composite and isotropic plates. Eng. Anal. Bound. Elem. 143:237–262

11. Liu CS (2007) An effectively modified direct Trefftz method for 2D potential problems considering the domain's characteristic length. Eng. Anal. Bound. Elem. 31(12):983–993

Energy Spatial Distribution of Structure Noise Radiated from U-beam Slabs

Li Zhou, Tianqi Zhang, and Yanyun Luo

Abstract At present, the application of U-beams in urban rail transit is gradually increasing, but researches about the bridge structural noise are still insufficient. This article took U-type girder as the research object and studied the structural noise distribution on the basis of energies in different frequency band. Firstly, the dynamic response of the bridge under vehicle load was acquired with the combination of finite element simulation method and field test. Secondly, the structure-acoustic coupling finite element-infinite element model of the U-beam was established, and noise fields of the whole U-beam, the floor and the wing slabs were calculated respectively. By calculating the energies contribution of the sound field point in different frequency bands, the spatial distribution of the radiation noise of the floor and the wing slab in the above frequency bands were obtained. The results show that in the frequency range of 20–200 Hz and the main frequency band (40– 63 Hz), the floor plays a major role in most areas of the whole noise field, and the main influence domain of wing slabs is located in the sector area around wing slabs. In the frequency range of 100–160 Hz, the floor mainly contributes to the acoustic energy of the sector region above and below the U-beam.

Keywords Urban Rail Transit U-Beam · Structure-Borne Noise · Energy Contribution of Slabs

1 Introduction

During the operation of urban rail transit elevated lines, the bridge structure noise generated by vehicle moving mainly concentrated in the low frequency bands [1, 2]. Characteristics of the noise are easy diffraction, long propagation and slow attenuation. Accordingly, the life and work of surrounding residents are affected [3]. Box girder is the most widely used in urban rail transit, and related bridge structure noise

L. Zhou (✉) · T. Zhang · Y. Luo
Institute of Rail Transit, Tongji University, Shanghai, China
e-mail: Li.Z.TJU@hotmail.com; zhouli201007@163.com

© The Author(s) 2023
G. Feng (ed.), *Proceedings of the 9th International Conference on Civil Engineering*,
Lecture Notes in Civil Engineering 327,
https://doi.org/10.1007/978-981-99-2532-2_48

565

has received much researches. For the U-beam, which is very different from the box girder, the research on the U-beam radiation structure noise is insufficient as its application is gradually increasing.

Most existing researches on the structure noise radiated from U-beam mainly focuse on the spectral characteristics and propagation laws, and have made some achievements. Xiaodong Song, et al. [4–6] took the concrete U-beam as the research object, calculated the bridge noise when the train passed by using the 2.5-D acoustic infinite element model, and studied its spectral characteristics and spatial distribution law. Linya Liu et al. [7–9] took the 30 m simply supported U type bridge as the research object, calculated and analyzed the structural noise radiated by the U-beam with the combination of finite element method and acoustic transfer vector method. Jianglong Han [10] used the combining method of vibration power flow, 2-D acoustic simulation and the principle of vibration power equivalence to obtain the radiated sound pressure of the rails and U type bridge. Combined with the field measured noise data of a U-beam, the noise distribution characteristics were analyzed. Minjie Gu and Liu [11] adopted the power flow method to calculate the vibration responses of rails and bridges under vehicle loads. Through the acoustic finite element infinite element method, the generation and propagation of bridge structure noise were analyzed, and the differences in vibration and radiated noise between U-shaped beams and box girder were compared. Jianglong Han et al. [12] used the modal superposition method to calculate the dynamic responses of the vehicle-rail-bridge coupling system. Combined with the modal acoustic transfer vector method, the structural noise of the bridge is calculated. Wu and Liu [13] compared the noise radiation of U beam and box girder, and found that the radiated sound power in the low frequency band of box girder is smaller than that of U beam when the roughness excitation is same.

Generally, there are two main methods to reduce bridge structure noise. One is to reduce the track stiffness, which can reduce bridge structure noise level by lowering bridge vibration, but it possibly brings negative effects such as wheel rail vibration and noise increase [14–16]. The other is to optimize the geometric section of the bridge. At present, the research on the contribution of each slab of U-beam to the sound field of bridge structure noise and the corresponding spatial distribution is not enough. This work is very necessary for contrapuntally reducing the noise level in sensitive areas. Thus, this paper took the U-beam of rail transit as the research object, used the finite element simulation method to establish the vehicle-track-beam dynamic coupling model and structure noise model. Combined with the U-beam vibration test, the frequency spectrum distribution characteristics of vibration and radiated noise of U-beam under vehicle load were analyzed, and the sound field distribution of U-beam radiated structural noise in the 20–200 Hz frequency band is obtained. By changing the boundary conditions of the U-beam radiation noise model, the sound field distribution of the radiation noise of the floor slab and wing slab were acquired. Accordingly, acoustic energy spatial distributions in different frequency bands of different slabs were analyzed.

Fig. 1 Finite element model
of vehicle-track-bridge

2 Dynamic Response Characteristics of U-beam Under Vehicle Load

2.1 Vehicle-Track-U-Beam Finite Element Model

As shown in Fig. 1, the vehicle-track-U-beam dynamic coupling model is established in ABAQUS finite element software. The vehicle body and bogies are simplified as rigid bodies, and the motions of nodding, floating, shaking, and rolling are considered. Wheelsets, rails, track slabs and the bridge are all simulated by C3D8R element. The fastener and bridge supporters are represented by multi-body connection element "cartesian". The vertical stiffness of the fastener is 5×10^7 N/m, and the damping is 5000 N/(m/s). The vertical stiffness of the bridge supporter is 1.26×10^9 N/m. The contact relationship between wheel and rail is defined with surface-surface contact method, and the corresponding normal contact law satisfies Hertz's nonlinear theory [17]. The American track irregularity spectrum is selected as the system excitation, and spatial samples of irregularities (as shown in Fig. 2) obtained from the transformation of the irregularity spectrum by the trigonometric series method [18] are imported to the rail surface. Figure 3 and Fig. 4 show the vibration frequency response results of the U-beam floor slab and wing slab under vehicle load respectively. The main frequency band of floor slab vibration is 40–70 Hz. Besides, there are local peaks in the frequency band of 100–140 Hz. The frequency distribution characteristics of wing slab vibration are basically consistent with that of the floor slab, but the vibration amplitude in the main frequency band of 40–70 Hz is smaller than that of the floor slab.

2.2 Field Test

The U-beam of ordinary monolithic track bed in the actual operation line of urban rail transit is selected as the test section. As shown in Fig. 5, vibration observation points are arranged at the midpoint of the floor slab in the middle of the bridge span, and a vibration measuring point is mounted on the rail bottom as the signal trigger. It is worth mentioning that, considering the complex noise environment (including not only the bridge structure noise, but also the wheel-rail radiated noise, collector noise,

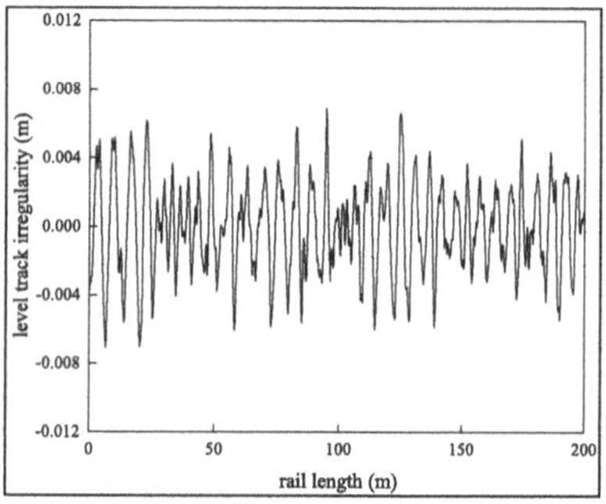

Fig. 2 Level track
irregularity of track

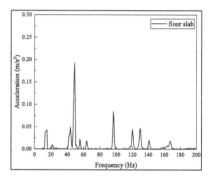

Fig. 3 Vibration
acceleration of floor

etc.,) around the U-beam during the train passing, this paper did not arrange noise measuring points in the test. Figure 6 shows the vibration octave curve of the U-beam floor slab collected from the test. Obviously, within 200 Hz, the main frequency band of the U-beam floor slab vibration is 40–80 Hz, and the sub main frequency band distributed in the 100–160 Hz frequency band. At the same time, the vibration octave curve of U-beam floor slab obtained by simulation is also drawn in Fig. 6. It is found that the distribution of the main frequency band of the simulation results is slightly different from that of the measured results, and the secondary main frequency band is basically consistent. It can be considered that the finite element simulation model can represent the frequency spectrum characteristics of U-beam vibration under the train load.

3 Spatial Distribution of Radiation Noise of U-beam

The relationship between the vibration and the radiated structural noise of the bridge meets weak coupling law. In another word, slabs vibration has a direct impact on the radiated structural noise, while the effect of the structural noise on the slab vibration can be ignored. Therefore, the finite-infinite element model of the U-beam structural noise shown as Fig. 7 is built in ABAQUS finite element software. The

Fig. 4 Vibration
acceleration of wing slab

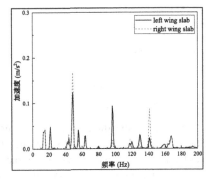

Fig. 5 Layout of measuring
points for vibration test (unit:
mm)

time-domain vibration responses of bridge panels obtained in Section I are applied
as the excitation source of the acoustic model to the acoustic-structure coupling
interface. In the acoustic model, AC3D8R element is used to simulate air medium,
and the corresponding density and bulk modulus of elasticity are 1.2 kg/m³ and 1.39
× 10⁵ Pa respectively. In order to ensure the calculation accuracy, the maximum
element size shall not exceed 1/6 of the wavelength corresponding to the upper
limit of the analysis frequency [19]. In this article, the size of elements is around
0.1 m. Furthermore, to avoid the influence of reflected waves generated by the outer
boundary, a layer of infinite elements is applied to the outer boundary, and the normal
direction of the elements is from the inside to the outside. As shown in Fig. 8, the
midspan section of U-beam is 20 × 14.8 m range is used as the sound field observation
area, and a total of 81 acoustic fields points in 9 rows and 9 columns are selected as
the observation points.

Fig. 6 Comparison between
measurement and simulation
of floor vibration

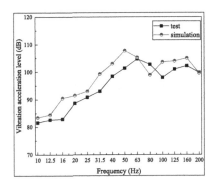

Fig. 7 Finite
element-infinite element
model of bridge
structure-borne noise

Figure 9 shows the near-field structure noise spectrum results (20–250 Hz) of the floor slab and two wing slabs obtained by simulation. It can be found that the near-field noise spectrum distribution characteristics of all panels are basically consistent. The main frequency band concentrates in the range of 40–63 Hz, and the secondary frequency band is from 100 to 160 Hz. The frequency distribution characteristics are basically consistent with the calculated slab vibration spectrum. The difference is that the near-field structure noise in the main frequency band of the floor slab is greater than that of the wing plates. Figure 10 shows the sound pressure level nephogram (20–250 Hz) of the mid span section of the U-beam. This nephogram is not the acoustic field distribution at a certain time during vehicle operation, but the sound field distribution result obtained by linear interpolation method after calculating the total sound pressure level of each sound field observation point during the whole process. The cloud chart mainly displays the overall distribution of the noise of the U-beam structure during the whole process of the train passing. Obviously, the upper and lower regions of the floor slab (corresponding plane coordinate position is −2.5– 2.5 m in the horizontal axis and 9–10.8 m in the vertical axis) are noise "hot spots". The noise sound pressure level within the range of 40–80° outside the wing slabs on both sides is relatively small. Within the space region below the U-beam floor slab, the sound pressure level decreases gradually with the increasing distance from the beam.

Fig. 8 Output point of
bridge radiated noise field
(unit: m)

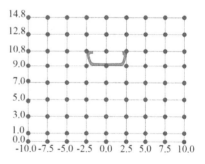

Fig. 9 Near-field noise
spectrum of slabs

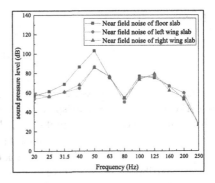

4 Energy Spatial Distribution of Noise Radiated From Each Panel

On the basis of the built U-beam structure-noise model, structure noise respectively radiated from the floor slab and wing slabs can obtained by changing the boundary conditions loaded by vibration excitation. Specifically, by selecting the floor slab and wing slabs as the vibration response mapping area separately, the U-beam floor slab radiation noise model and the wing slab radiation noise model can be built, and the sound pressure level of the radiation noise field caused by the floor slab and wing slabs are respectively calculated.

The corresponding cloud diagrams of the total sound pressure level (20–250 Hz) of during the whole process of vehicle passing are drawn as Fig. 11 and Fig. 12. The sound field distribution of the floor slab is close to the distribution characteristics of the whole U-beam, and there are also noise "hot spots" in the upper and lower region of the floor slab. Due to the shielding effects of two wing slabs, the sound pressure level of the noise in the sector area near two wing slabs is low. The sound field distribution of the wing slab is obviously different. The "hot spots" of the noise are mainly distributed in the fan-shaped area formed by taking the wing slabs on both sides as the center, and center angle is about 160°. With the increase of the distance from the wing slab, the sound pressure level of the field points gradually decreases. At the same time, in the sector area with the vertical line of the floor slab as the axis, the sound pressure level of the field point is smaller and gradually decreases with the increase of the vertical distance from the wing slab.

From the perspective of sound energy, the contribution of panel radiated noise is defined as the ratio of the sound energy of panel radiated sound field points to that radiated from U-beam in unit time. The sound energy per unit area passing through the direction perpendicular to the propagation direction in unit time is sound intensity I, and I is proportional to the square of sound pressure P, therefore, the contribution of panel radiated noise η can be calculated according to formula (1):

$$\eta = \frac{I_{slab}}{I_{U\text{-}beam}} = \frac{P_{slab}^2}{P_{U\text{-}beam}^2} \tag{1}$$

Besides,

$$L_p = 10\log_{10}(\frac{P}{P_0})^2 \tag{2}$$

where, L_p is sound pressure level, and P_0 is the reference value

The energy contribution spatial distribution of the floor slab and the wing slab can be obtained through the formula (1) and (2) on the basis of the obtained field point sound pressure level values of noise radiated from the U-beam, floor slab and the wing slab. It is worth noting that the adopted calculation method of panel radiated noise contribution does not take the phase factor of noise radiated by different panels into account, so the total contribution of panel radiated noise maybe exceed 100%. However, the calculation method of this contribution can reflect the contribution of each panel to the total structure noise field from the trend.

Energy contribution nephograms of the floor slab and wing slab are shown as Fig. 13 and Fig. 14 respectively. It can be seen that in the frequency range of 20–ral noise energy in the sector areas with 40–80° extension of both wing slabs are mainly contributed by the wing slab vibration, while the structural noise radiated by the floor slab has a weak influence in this area. Figure 15 and Fig. 16 respectively shows the contribution of the radiated noise from the floor slab and wing slabs on both sides in the main frequency band (40–63 Hz). The distribution tendency of the cloud chart is basically the same as that of panels in the 20–200 Hz frequency band. The main contribution areas of the floor slab are the upper sector area of the U-beam and the sound field below the U-beam bottom plate, and the wing slabs contribute most of the energy of the sound field in the sector area near two wing slabs.

Figure 17 and Fig. 18 respectively shows the radiated noise energy contributions of the floor slab and wing slabs in the sub dominant frequency band (100–160 Hz). It can be seen that in this frequency band, the main contribution area of the floor slab is the upper sector area of the U-beam and the sector area below, while the energy contributions in other areas of the sound field are significantly reduced. Spatial areas with more than 50% contribution of the radiated noise from the wing slabs increased significantly. Main contribution areas distributed in the sector area with the U-beam

Fig. 10 Nephogram of U-beam structure-borne noise field

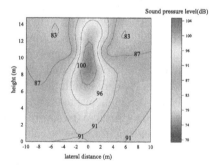

Fig. 11 Nephogram of noise radiated by floor

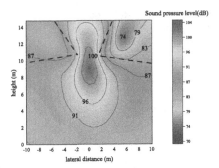

Fig. 12 Nephogram of noise radiated by wing slabs

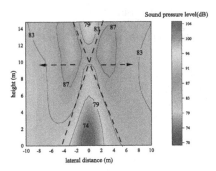

as the center, the center line parallel to the bottom plate, and the center angle of about 160°.

In general, in the 20–200 Hz frequency band and the main frequency band of U-beam structure noise, the radiated noise from the floor slab contributes most energies of the whole bridge structure noise field, especially the sector area above the floor slab and the whole field area below the U-beam. The radiated noise from the wing slabs is the main source of the sound field energy of the sector area diagonally above the wing plates of the U-beam. In the sub dominant frequency band (100–160 Hz) of the structural noise, the wing slab contributes most energies of the acoustic field of

Fig. 13 Spatial distribution nephogram of floor contribution (20–200 Hz)

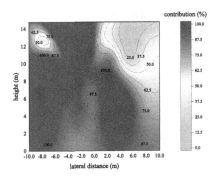

Fig. 14 Spatial distribution
nephogram of wing slabs
contribution (20–200 Hz)

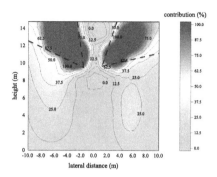

Fig. 15 Spatial distribution
nephogram of floor
contribution (40–63 Hz)

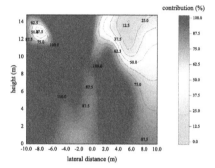

Fig. 16 Spatial distribution
nephogram of wing slabs
contribution (40–63 Hz)

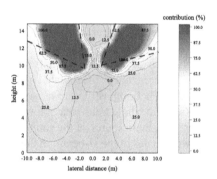

Fig. 17 Spatial distribution
nephogram of floor
contribution (100–160 Hz)

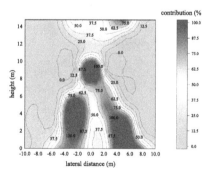

Fig. 18 Spatial distribution nephogram of wing slabs contribution (100–160 Hz)

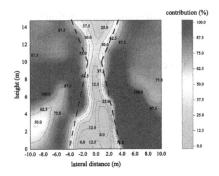

the bridge structure noise, and the main influenced area by the floor slab is located in the sector area directly above and below the U-beam.

5 Conclusion

In this paper, the urban rail transit U-beam is taken as the research object. The vehicle-track-bridge dynamic coupling model and acoustic-structure coupling model are established in the finite element software. Combined with the test data, the respective radiated noise fields of the floor slab and wing slabs are calculated and analyzed. By introducing the contribution calculation factor, the energy contribution distribution of the floor slab and wing slabs in different frequency bands is acquired. Obtained conclusions are as follows:

(1) In the frequency band of 20–200 Hz, the radiated noise from the U-beam floor contributes most energies of the whole bridge structure noise, and the main influence areas are located below the U-beam floor and directly above the U-beam floor. The main contributed region of the structural noise radiated by the wing slabs vibration is the sector area of 40–80° extension of the wing slabs. The structural noise radiated by the bottom plate has a weaker influence on the sound field of this area.

(2) In the main frequency band (40–63 Hz) of the U-beam structural noise, the energy contribution distribution trend of the floor slab and wing slabs are basically the same as that of the 20–200 Hz frequency band. In details, the floor slab mainly affects the sound pressure level of the sector area directly above the U-beam and that of the lower sector area, and two wing slabs affect the sound field size of the sector area diagonally above the U-beam.

(3) In the sub dominant frequency band (100–160 Hz) of the U-beam structural noise, the floor slab mainly contributes to the sound energy of the sector section directly above and below the U-beam, and the contribution area is greatly

reduced compared with that of main frequency band. The area where the contribution amount of the radiated noise of the wing slabs exceeds 50% is significantly increased. The region is mainly distributed in the sector area with the U-beam as the center, the center line parallel to the bottom plate, and the center angle about 160°.

References

1. Li XZ, Yang DW, Zheng J, et al (2018) Review on vibration and noise reduction of rail transit bridges. China J Highway Transp 31(07):55−75+136
2. He W, H Kewen, Zou C, et al (2021) Experimental noise and vibration characteristics of elevated urban rail transit considering the effect of track structures and noise barriers. Environ Sci Pollut Res 21(03):1–19
3. Liu DY, Wang LZ, Li X et al (2019) Psychoacoustic evaluation index of near-field environmental noise in an intercity railway. Railway Stand Des 63(03):169–174
4. Song XD, Wu DJ, Li Q (2015) A 2.5-dimensional infinite element based method for the prediction of structure-borne low-frequency noise from concrete transit bridges. J Vibrat Eng 28(06):929–936
5. Song XD, Li Q, Wu DJ (2018) Prediction of low-to-medium frequency structure-borne noise radiated from rail transit concrete bridges. J China Railway Soc 40(03):126–131
6. Song XD, Li Q (2019) Numerical study on vibration and noise reduction of rail transit concrete U-shaped bridges. J Southe Univ (Natural Science Edition). 49(03):460–466
7. Liu LY, Qin JL, Lei XY, et al (2018) Low frequency noise of a trough girder structure based on acoustic transfer vector method. J Vibrat Shock 37(19):132–138+152
8. Liu LY, Qin JL, Lei XY, et al (2018) A study on optimization of the structure-borne noise from a trough girder based on response surface methodology. J Vibrat Shock 37(20):56–60+80
9. Liu LY, Qin JL, Liu QM et al (2018) Prediction and optimization of structure-borne low-frequency noise from a rail transit trough girder. J China Railway Soc 40(08):107–115
10. Han JL, Li Q, Gu MJ (2022) Influence investigation of the rail-cum-road viaduct on noise spatial distribution from rail transit traffic. J Vibrat Eng 35:188–195
11. Gu MJ, Li WQ, Li Q (2019) Influence of section types on noise from elevated rail transit lines. J Southw Jiaotong Univ 54(04):715–723
12. Han JL, Wu DJ, Li Q (2013) Calculation and analysis of structure-borne noise from urban rail transit trough girders. Eng Mech 30(02):190–195+202
13. Wu TX, Liu JH (2012) Sound emission comparisons between the box-section and U-section concrete viaducts for elevated railway. Noise Control Eng J 60(4):450–457
14. Sheng X, Zhao CY, Wang P, et al (2018) Effects of fastener stiffness of monolithic bed track on vertical rail sound power characteristics. J Cent South Univ (Science and Technology) 53(05):928–936+1094
15. Zhang X, Su B, Li XZ (2015) Effects of fastener stiffness and damping on structure-borne noise of railway box-girders. J Vibrat Shock 34(15):150–155
16. Cui XL, Chen GX, Yang HG (2017) Influence of wheelset structure and fastener stiffness on rail corrugation. J Cent South Univ (Science and Technology) 52(01):112–117
17. Zhu ZH, Wang LD, Gong W et al (2017) Comparative analysis of several types of vertical wheel/rail relationship and construction of an improved iteration model for train-track-bridge system. J Cent South Univ (Science and Technology) 48(06):1585–1593
18. Chen CJ, Li HC (2006) Track irregularity simulation in frequency domain sampling. J China Railway Soc 03:38–42
19. Miao XH, Qian DJ, Yao XL et al (2009) Sound radiation of underwater structure based on coupled acoustic-structural analysis with ABAQUS. J Ship Mech 13(02):319–324

The Problem of the Local Stress/strain Modes in the Matrix of Fibrous Composites

A. G. Kolpakov and S. I. Rakin

Abstract On the basis of numerical solutions of the cellular problem of the theory of elasticity, an analysis was made of the deformation modes of the binder cross-reinforced with fibers of the composite at the microlevel. All possible types of local deformation of the binder of the fiber-reinforced composite, predicted in the simplified model in (Kalamkarov A L, Kolpakov A G: Analysis, Design and Optimization of Composite Structures. Wiley, Chichester (1997)), are found in numerical calculations. It is found that the local stress–strain state of the binder depends on the cross-sectional shape of the fibers. In particular, for fibers with a circular cross-section, one of the modes is strongly weakened. Numerical calculations do not show the presence of modes other than those described in (Kalamkarov A L, Kolpakov A G: Analysis, Design and Optimization of Composite Structures. Wiley, Chichester (1997)). If, in the case of square fibers, the assessment of the stress–strain state value is possible using simplified formulas from (Kalamkarov A L, Kolpakov A G: Analysis, Design and Optimization of Composite Structures. Wiley, Chichester (1997)), then for fibers of other sections, it is necessary to use a computer to calculate the stress–strain state.

Keywords Fiber-reinforced composite · matrix · strength · local stresses · averaged stresses · homogenization

A. G. Kolpakov (✉)
Siberian Transport University, D Koval'chuk St., 191, Novosibirsk, Russia
e-mail: algk@ngs.ru

S. I. Rakin
SysAn, A. Nevskogo St., Novosibirsk, Russia
e-mail: rakinsi@ngs.ru

G. Feng (ed.), *Proceedings of the 9th International Conference on Civil Engineering*,
Lecture Notes in Civil Engineering 327,
https://doi.org/10.1007/978-981-99-2532-2_49

1 Introduction

Problems of calculation and design of fibrous composite materials began to be intensively studied in the 60 s–70 s of the last century [2, 3]. At present, the relevance of the problems has only increased [4]. For example, in connection with the development of hydrogen energy technologies, the problem arose of improving, creating and designing high-pressure vessels made of composite materials for storing hydrogen fuel. The hydrogen gas storage pressure is 5000–10,000 psi (340–680 atmospheres) [4–6]. It seems, the composite gas reservoirs are the best solution to the problem which attract attention both the engineers and researchers. An increase in the role of fibrous composites (in terms of the volume of their use as well as in terms of the responsibility of structural elements made from composites, is also observed in other areas of mechanical engineering. The prospect of mass use of products made of fibrous composite materials raises the question of testing of existing and developing new (in particular, based on the use of computers) methods for their calculation.

Reviewing the strength criteria of composite materials used in the modern design systems, one finds that the most of them are based on the motion of the local deformation modes of composite materials. The use of the concept of the local deformation modes of composite arose at the initial stages of the study of composites and was especially widely used in obtaining the classical criteria for the strength of composite materials in the 1960s-70 s [7–27].

The notion of the deformation modes was based on the notes that the local stress–strain state in composite is strongly inhomogeneous. Naturally, the question arises: do exist any regularity in the inhomogeneous stress–strain state? Or the inhomogeneity is arbitrary and non-predictable? It was found that in many composites the stress–strain state has specific forms correlated with the local geometry and the local mechanical properties of the composite. These specific forms of the local stress–strain state were named the local deformation modes of composite. Note that the currently used theoretical criteria for the strength of fibrous composites, in fact, are based on taking into account certain modes of local stress–strain state [7–27]. In [1], it was pointed out that the primary question is about local stress–strain state modes in a composite (the existence and characteristics of such modes), and the question about destruction modes is secondary, the solution of which follows from the solution of the first question.

The calculation of the local (microscopic) stress–strain state in composite material is a non-trivial problem because this requires solving a problem of the elasticity theory (the explicit solution to the problem is known only for layered composites [28]). Therefore, the desire of researchers to indicate the "basic" types of stress–strain state that determine the strength of the composite was understandable. Note that the question of the existence of such "basic" types of stress–strain states was usually not even raised. It was also not discussed how many "basic" types of stress–strain state can exist (if they exist) in composites in general and in composites of a particular type in particular.

In this paper, composites "hard fibers in a soft binder" of periodic structure are considered. Such kind composites have found wide application in engineering. For composites of periodic structure, methods of the homogenization theory are applicable [29–31]. In this case, it is necessary to indicate the multicomponent effect of the homogenization in contrast media predicted in [32, 33], whose numerical study showed that a very general type of stress–strain state occurs in the matrix of these composites.

In [1] (see also [34, 35]), a scheme for computation of the microscopic stress–strain state in the composites "hard fibers in a soft binder" was proposed for square cross-section fibers. In [1], there were listed the "main" types of the local stress–strain states – the local stress–strain state modes exist, which may exist in the fiber-reinforced composite made of pre-pregs. In addition, a method for their approximate calculation of the "main" stress–strain state nodes was proposed. True, the analysis in [1] was only possible for the square-section fibers. For fibers of other cross-section geometry (for example, circular), the local stress–strain state can only be computed numerically. The modes described in [1] included previously known modes [7–27] as well as new modes.

After the analysis presented in [1], questions arose: are all the deformation modes predicted in [1] realized and are there any other modes. In particular, what is the answer to this question for fibers of a cross-section other than considered in [1]?

For the reasons indicated above, an exhaustive answer to these questions can only be given by numerical calculations. Other, for example, experimental, study of local stress–strain state modes in cross-reinforced composites seems to the author practically unrealizable. Experimental methods (see, e.g., [36–38]) do not provide exhaustive information either on the types of modes or on the characteristic values of the stress–strain state.

In numerical calculations, one encounters mathematical and programming problems. The main problem is related to the fact that the problem of constructing a periodic FEM mesh in regions of complex geometry, as far as the author knows, still remains practically unsolved [39–41].

The popular software systems used in the calculation of composites and the strength criteria used in them are given in Table 1 [7–27, 36]. The strength criteria that the developers of the complexes indicated are also given; their list may vary depending on the development of software systems, changes in the composition of modules, as well as the licensing policy of developer firms, etc. We note that the listed in Table 1 strength criteria are based on the use of the local stress–strain state modes. One can verify this by looking through the corresponding references list [7–27].

Table 1 The strength criteria used in popular software systems

	Ansys	Abaqus	MSC. Nastran	Nastran	Nastran in-CAD	Helius	Comsol	Hyper Works
max stress	+	+	+	+	+	+	+	+
max strain	+	+	+	+	+	+	+	+
Tsai-Wu	+	+	+	+	+	+	+	+
Tsai-Hill	+	+	+	+	+	+	+	+
Azzi-Tsai-Hill	−	+	−	−	−	−	+	+
Hoffman	+	+	+	+	+	−	+	+
Hashin	+	−	−	−	−	+	−	−
Puck	+	+	+	+	+	+	+	+
LaRC	+	−	−	+	+	+	−	−
special	Cuntze	Yankinson -Cowin	−	−		Christensen	Norris	Yamada -Sun

2 The Homogenization Method as Applied to an Orthogonally Fiber-Reinforced Composite

Let us describe the structure of the considered composite. The composite is obtained by stacking layers of unidirectional fibers - prepregs. In a layer, the fibers are parallel to each other. The layers are stacked parallel to each other (for definiteness, we assume the layers are perpendicular to the Ox_2-axis, see Fig. 1). The direction of fiber lying in the s-th layer is described by the direction vector $\mathbf{v}_s$. The fibers have diameter εR, half the distance between the fibers in the layers εh, half the distance between the layers of fibers $\varepsilon \delta$, Fig. 1. The gaps between the fibers are filled with a binder that forms the matrix. In the work under consideration, the "ideal" connection between the fibers and the matrix is assumed.

The characteristic size ε is small compared to the size of the structure (material sample). It is usually assumed that the sample size is of the order of 1, then $\varepsilon \ll 1$. In addition to the original (macroscopic) variables, "fast" (microscopic) variables $\mathbf{y} = \mathbf{x}/\varepsilon$ are introduced.

Since the fibers in the layers and the layers are periodic, entire structure is periodic. Denote εP the periodicity cell of composite. The periodicity cell may be chosen in

Fig. 1 Orthogonally stacked fibers (two adjacent layers of fibers shown) and PC (fibers and binder) in "fast" variables

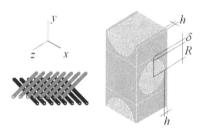

different ways. One of the possible choices is displayed in Fig. 1. The periodicity cell (PC) has the form of a rectangle $P = [0, h_1] \times [0, h_2] \times [0, h_3]$. Faces of P are perpendicular to the Ox_s-axis. The faces are rectangles $\Gamma_s = \{\mathbf{x} : x_s = 0\}$ and $\Gamma_s + h_s \mathbf{e}_s$. . Figures are given in "fast" (microscopic) variables. Let us consider the simplest composite obtained by stacking periodically alternating layers of orthogonally oriented fibers. We assume that the fibers in one layer are parallel to the Ox-axis, in the other - to the Oz-axis ($\mathbf{v}_1 = \mathbf{e}_1$ and $\mathbf{v}_1 = \mathbf{e}_3$; $\mathbf{e}_1, \mathbf{e}_2, \mathbf{e}_3$ - coordinate vectors), Fig. 1. We confine ourselves to considering the simplest packing, keeping in mind that this case allows to focus on the presentation of the mechanical aspects of the problem without cluttering them up with purely computational questions.

The solution to the elasticity theory problem of composite material in the framework of the homogenization theory [30, 31] is sought in the form

$$\mathbf{u} = \mathbf{u}_0(\mathbf{x}) + \varepsilon \mathbf{u}_1(\mathbf{x}/\varepsilon), \tag{1}$$

where $\mathbf{u}_0(\mathbf{x})$ is the homogenized (macroscopic, overall) solution, and $\varepsilon \mathbf{u}_1(\mathbf{x}/\varepsilon)$ is the corrector (the function $\mathbf{u}_1(\mathbf{y})$ is periodic with PC P) [1]. The corrector makes a small contribution to displacements (1), but it makes not a small contribution to the local stresses [1]. It is proved in [30, 31] that the corrector has the form $\varepsilon \frac{\partial u_{0k}}{\partial x_l}(\mathbf{x})\mathbf{N}^{kl}(\mathbf{x}/\varepsilon)$, where $\mathbf{N}^{kl}(\mathbf{y})$ is the solution of the so-called periodicity cell problem (PCP) (see PCP problem (3) below).

We introduce the following function:

$$a_{ijkl}(\mathbf{y}) = \begin{cases} a_{ijkl}^F \text{ in fibre} \\ a_{ijkl}^M \text{ in binder} \end{cases} \tag{2}$$

where a_{ijkl}^F - elastic constants of fibers, a_{ijkl}^M – elastic constant matrices. Formula (2) defines (in "fast" (microscopic) variables) the distribution of elastic constants over cell εP.

We consider a linear problem in the elasticity theory of composites. This is due, in particular, to the fact that the purpose of the article is to consider the deformation modes from [1] obtained for a linear problem. In this case, the PCP has the following form [1]: find the solution $\mathbf{N}^{kl}(\mathbf{y})$ to the following boundary value problem:

$$\begin{cases} (a_{ijkl}(\mathbf{y})N_{k,ly}^{mn} - a_{ijmn}(\mathbf{y}))_{,jy} = 0 \text{ in } P, \\ \mathbf{N}^{kl}(\mathbf{y}) \text{ periodic in } P \end{cases} \tag{3}$$

The solutions to the problem (3) differ from each other by displacements of a rigid body. The differences can be excluded in different ways. In the homogenization theory, it is assumed that the average value of $\mathbf{N}^{kl}(\mathbf{y})$ over PC P is zero [30, 31]. In numerical computations, it is convenient to fix a proper point of PC. It is important that any method of excluding displacements of a rigid body from problem (3) leads to the same results when calculating local strains and stresses in composite, since these quantities depend on the derivatives of the function.

In (3) and below, indices mn =11, 22, 33 correspond to macroscopic tension (along the corresponding axis). Indices mn =12, 13, 23 correspond to macroscopic shear (in the corresponding plane).

We introduce the function $\mathbf{Z}^{mn}(\mathbf{y}) = \varepsilon_{mn}(\mathbf{x})[\mathbf{N}^{mn}(\mathbf{y})(\mathbf{y}) + y_m\mathbf{e}_n]$ and write (3) in the form

$$
\begin{cases}
\left(a_{ijkl}(\mathbf{y})Z_{k,ly}^{mn}\right)_{,iy} = 0 \text{ in } P, \\
\mathbf{Z}^{mn}(\mathbf{y}) - \varepsilon_{mn}(\mathbf{x})y_m\mathbf{e}_n \text{ periodic on } P
\end{cases}
\tag{4}
$$

Denote $[f(\mathbf{y})]_s = f(\mathbf{y}+h_s\mathbf{e}_s) - f(\mathbf{y})$, where $\mathbf{y} \in \Gamma_s$ ($\mathbf{y} \in \Gamma_s$ and $\mathbf{y} \in \Gamma_s+h_s\mathbf{e}_s$ are the opposite faces of PC P). In this notation, we can write the periodicity condition in (3) in the form $[f(\mathbf{y})]_s = 0.$. Then (4) can be rewritten in the following form

$$
\begin{cases}
\left(a_{ijkl}(\mathbf{y})Z_{k,ly}^{mn}\right)_{,iy} = 0 \text{ in } P, \\
[\mathbf{Z}^{mn}(\mathbf{y})]_s = \varepsilon_{mn}(\mathbf{x})[y_m\mathbf{e}_n]_s
\end{cases}
\tag{5}
$$

(no sum in m, n in (5)).

The solutions to the PCs (problems (3), (4) or (5)) provide us with the detailed information about the local (microscopic) stress–strain state in composite. For example,

$$
\sigma_{ik}^{mn}(\mathbf{y}) = a_{ijkl}(\mathbf{y})Z_{k,l}^{mn}(\mathbf{y})
\tag{6}
$$

are the local stress in the composite subjected to macroscopic deformations $\varepsilon_{mn}(\mathbf{x})$ [30, 31].

At the time of the writing of the works [1, 7–27], the computers available (in public domain for engineers) made it possible to solve a PC problem with a simple geometry. For example, for a system of unidirectional fibers (in this case, complex variables method can be used [42, 43], which also assume sufficient massive computations [42, 43]). But the mentioned computers had not sufficient to solve the problems corresponding to the cross-reinforced or particles-filled composites. At present, the power of computers has reached a level that allows solving a general type of problem. It should be noted that the progress in the field is restricted not only by the power of the computer, but also the theoretical programming problems, for example, the problem of the generation of periodic meshes. As an illustration of the current state of the art, we mention the recent monograph [42], which contains many programs for solving problems in the theory of composites. However, in relation to the PC of the form (5), [42] is limited only to the formulation of this problem. The solutions in [42] are given only for unidirectional fibers.

3 Numerical Analysis of the Local Stress–Strain State in the Periodicity Cell

For a cross-reinforced composite, even for the simplest reinforcement scheme under consideration, the local stress–strain state turns out to have a general form when any loads are applied to the composite (even loads of the simplest type). Let us consider the main types of the overall (the homogenized) deformations and present the calculations of the corresponding local deformations. Due to the symmetry of the problem, out of six cases (three axial and three shear deformations), only four types of averaged deformations listed below can be considered.

Let us present the results of the numerical solution of the PC for the case under consideration. The solution was carried out for a composite with the following characteristics. Fibers: Young's modulus $E_f = 1.7 \cdot 10^{11}$ Pa, Poisson's ratio $v_f = 0.3$. Binder: Young's modulus $E_b = 2 \cdot 10^9$ Pa, Poisson's ratio $v_b = 0.36$. These characteristics correspond to steel and epoxy resin.

In this paper, the theoretical problem of construction of FEM mesh is not considered, bearing in mind that the construction of a finite element mesh is carried out by predefined procedures. At the same time, the author had to develop an APDL program for setting periodic boundary conditions (an imperfection of ANSYS in this part is well known [42]). The convergence of the numerical solution of the finite element method for the elasticity theory problem is a well-known fact [21], the question in this case is the size of the finite elements. The selection of the finite elements size was carried out by dividing their characteristic size in half until the difference in the numerical solutions became less than 2–3%.

For the visual representation of the distribution of local stresses/strains, it is convenient to use the scalar characteristic of the stress tensor. The authors use the von Mises stress. Although the von Mises stress is often used in the formulation of the strength criterion [44, 45], here it is chosen as a visual scalar characteristic of the stress tensor. In combination with the image of the deformed periodicity cell, it makes it possible to clearly see the main type of local deformation.

If one wants to carry out strength calculations, one should keep in mind the discussion in the literature [36, 45–47] on the von Mises criterion, the Tresca criterion and the modification of the Mises and Drucker-Prager criteria for rubbers (including epoxy ones). In engineering calculations, apparently, the best is to use a criterion developed for a particular rubber [36] (if any) or an experimental criterion. Such criteria can be programmed and used together with other (in particular, with our proposed) methods.

3.1 Square Cross-Section Fibers

The model of a composite reinforced with square fibers was used in [1] in the analysis of typical types (modes) of local stresses in composites reinforced with fibers.

Therefore, we will begin our consideration also with this type of fiber. For square cross-section fibers, stress concentrators appear near 90 degrees angles. It is inconvenient when analyzing the stress–strain state in the entire cell. To avoid these stress concentration, square fibers with fillet corners were considered.

In [1], the layers of fibers were placed one above the other and connected with a fragment of a matrix. The size of the matrix fragment in [1] coincides with the cross section of the fibers. As a result, it was possible to carry out numerical calculations for this case in the explicit form.

In the calculations, the cross section of the fibers $H \times H$, filet radius R. Other dimensions are the following: half the distance between the fibers h, half the distance between the fiber layers δ. Two types of periodicity cell were considered: «narrow» and «wide». For the «narrow» periodicity cell $H = 0.96$, $h = 0.07$, $\delta = 0.02$, $R = 0.25$. For the «wide» $H = 0.92$, $h = 0.88$, $\delta = 0.04$, $R = 0.15$. Note: indices 1, 2 and 3 of the tensor components correspond to the variables x, y, and z respectively.

Tension along Ox-axis. Macroscopic Deformations ε_{11}

«Narrow» PC. The results of numerical calculations are presented in Fig. 2. Sections of fiber $H \times H$, $H = 0.96$, fillet radius R $= 0.25$. Other dimensions - half the distance between the fibers $h = 0.07$, half the distance between fiber layers $\delta = 0.02$. In calculations, the macroscopic strain $\varepsilon_{11} = 0.2/1.1 = 0.18$. Dimensions are given in dimensionless variables.

«Wide» PC. The results of numerical calculations are presented in Fig. 2. Sections of fiber $H \times H$, $H = 0.92$, fillet radius $R = 0.15$. Other dimensions - half the distance between the fibers $h = 0.88$, half the distance between fiber layers $\delta = 0.02$. In calculations, the macroscopic strain $\varepsilon_{11} = 0.4/2.68 = 0.14$. The maximum stresses in the binder are observed between the fibers of one layer and are associated with a

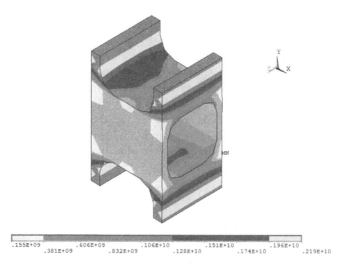

.155E+09   .606E+09   .106E+10   .151E+10   .196E+10
      .381E+09   .832E+09   .128E+10   .174E+10   .219E+10

Fig. 2 Stretching along the *Ox*-axis.

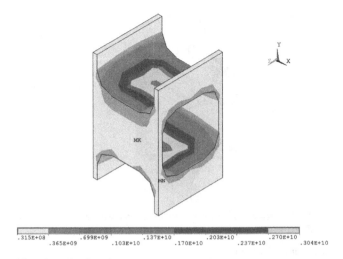

.315E+08   .699E+09   .137E+10   .203E+10   .270E+10
      .365E+09   .103E+10   .170E+10   .237E+10   .304E+10

Fig. 3 Stretching along the Oy-axis

strong stretching of the matrix between the fibers, which are much more rigid than the matrix. These types of microscopic deformation of the binder are described in [1]

Tension along Oy-axis. Macroscopic Deformations ε_{22}

«*Narrow*» *PC*. In calculations, the macroscopic strain $\varepsilon_{22} = 0.2/2 = 0.1$. Sections of fiber size 0.96×0.96, fillet radius $R = 0.25$. Other sizes: $h = 0.07$, $\delta = 0.02$. The results of numerical calculations are presented in Fig. 3. The maximum stresses in the binder are observed between the fibers of one layer and are associated with a strong stretching of the matrix between the rigid fibers.

«*Wide*» *PC*. In calculations, the macroscopic strain $\varepsilon_{22} = 0.4/2 = 0.2$. Fiber sections $H \times H = 0.92 \times 0.92$, fillet radius 0.15. Other sizes: $h = 0.88$, $\delta = 0.04$. The maximum stresses in the binder are observed between the fibers of one layer and are associated with a strong stretching of the matrix between the rigid fibers. The maximum stress according to Mises is $0.103E + 10$ Pa.

Shift in Oxz-plane Macroscopic Deformations ε_{13}

«*Narrow*» *PC*. In calculations, the macroscopice deformation $\varepsilon_{13} = 0.05$. Fiber sections size 0.96×0.96, the fillet radius $R = 0.25$. Other sizes: $h = 0.07$, $\delta = 0.02$. The results of numerical calculations are presented in Fig. 4. The maximum stresses in the binder are observed between the fibers of one layer and are associated with the shear of the binder between the fibers of one layer. This case is described in [1].

«*Wide*» *PC*. In calculations, the macroscopic strain $\varepsilon_{13} = 0.4$. Fiber Sections 0.92×0.92, fillet radius $R = 0.15$. Other sizes: $h = 0.88$, $\delta = 0.04$. The maximum stresses in the binder are observed between the fibers of one layer and are associated with the twisting of the binder between the fibers of different

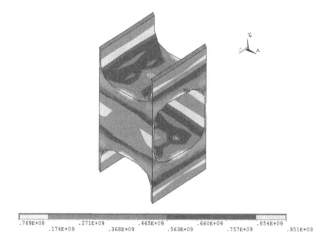

.769E+08   .271E+09   .465E+09   .660E+09   .854E+09
       .174E+09   .368E+09   .563E+09   .757E+09   .951E+09

Fig. 4 Interfiber shear a layer of fibers

layers. The results of numerical calculations are presented in Fig. 5. This type of microscopic deformation of the binder is described in [1].

Shift in Oxy-plane.
«Narrow» PC. In calculations, the macroscopic strain $\varepsilon_{12} = 0.2$. Fiber Sections 0.96×0.96, fillet radius $R = 0.25$. Other sizes: $h = 0.07, \delta = 0.02$. The maximum stresses in the binder are observed between the fibers of one layer and are associated with the twisting of the binder between the fibers of different layers.

«Wide» PC. In calculations, the macroscopic strain $\varepsilon_{13} = 0.8$. Fiber sections 0.92×0.92, fillet radius $R = 0.15$. Other sizes: $h = 0.88, \delta = 0.04$. The maximum stresses in the binder are observed between the fibers of one layer and are associated with the twisting of the binder between the fibers of different layers.

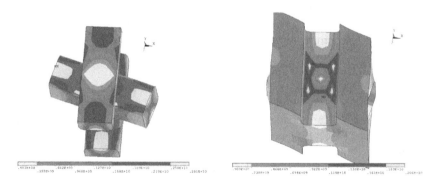

Fig. 5 Mutual rotation of the fibers and torsion of the binder between the layers of fibers («wide» PC)

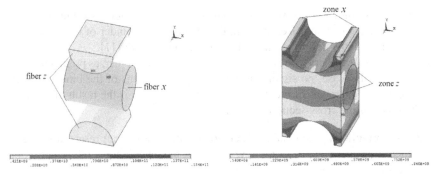

Fig. 6 Local stresses in the fibers (left) and in the matrix (right) at macroscopic deformation ε_{11} (stretching along the Ox-axis)

3.2 Circular Cross-section Fibers

Fibers of various cross-sections are mentioned in the literature. Above, the fibers are rectangular in cross section. The most common are, however, fibers of a circular cross section. Let's see what types of local stresses are typical for a fiber with a circular cross section.

In the calculations, the distance between the fibers is taken $h = 0.1$, distance between fiber layers $\delta = 0.1$. Cell dimensions $h_1 = 1.1$, $h_2 = 2$, $h_3 = 1.1$. Fiber radius 0.45. Since this problem depends only on the relative dimensions, all dimensions can be referred to the radius of the fibers.

Tension along Ox- axis. Macroscopic Deformations ε_{11}. Figure 6 shows the von Mises stresses σ_M. In calculations, the macroscopic strain $\varepsilon_{11} = 0.1/1.1 = 0.091$. Due to the linearity of the problem, ε_{11} can be chosen arbitrarily. The choice of ε_{11} = 0.091 (9.1%) here and below is due only to convenience for the programmer.

In the binder, there is a stress–strain state of general type. The maximum von Mises stresse takes place in the fiber x and in the matrix in the zone x, see Fig. 6 The maximum von Mises stress takes place in the zone where the distance between the fibers is minimal. The fibers are circular in cross section. As a result, the stress quickly decreases with distance from the zone x. Thus, there is stress localization effect similar to one discussed theoretically in [48, 49] and demonstrated in numerical calculations [50–52] for rigid bodies in a soft binder.

The problem of stretching along the Oz-axis in this case is solved by rotating the PC around the Oy-axis.

Tension along Oy-axis. Macroscopic Deformations ε_{22}. In calculations, the macroscopic strain $\varepsilon_{22} = 0.1/2 = 0.05$. According to the results of numerical computations, it can lead to the increasing of the von Mises stress in the fibers and the binder of one layer.

Shift in Oxz-plane Macroscopic Deformations ε_{13}. The results of numerical calculations are presented in Fig. 7. In calculations, the macroscopic strain $\varepsilon_{13} = 0.1/1.1$

= 0.091. According to the results of the numerical computations, see Fig. 7, it leads to the increasing of von Mises stress in the fibers and the binder of one layer. This type of microscopic deformation of the binder is described in [1].

In this case, the fibers in the layers rotate relative to each other, as predicted in [1]. However, in this case, no significant torsional deformations of the binder element connecting the fibers of different layers are observed. This is the result of a change in the shape of the cross-section of the fibers.

Shift in Oxy-plane. The results of numerical calculations are presented in Fig. 8. Symmetric problem - shift in the Oyz-plane. In the calculations, the macroscopic strain $\varepsilon_{12} = 0.1$

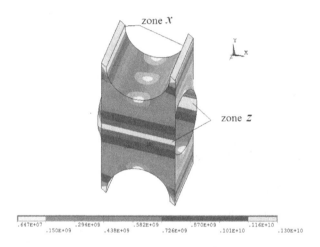

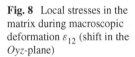

Fig. 7 Local stresses in the matrix at macroscopic deformation ε_{13} (shear in the Oxz-plane)

Fig. 8 Local stresses in the matrix during macroscopic deformation ε_{12} (shift in the Oyz-plane)

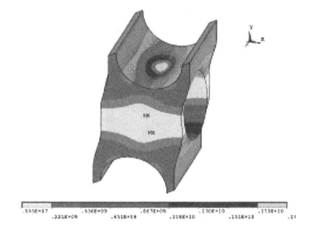

4 Conclusions

The performed solutions of the periodicity cell problems of the homogenization theory showed that all potentially possible types (modes) of local deformation of the binder fibrous composite predicted in [1] are found in numerical calculations.

The local stress–strain state in the binder essentially depends on the cross-sectional shape of the fibers. So for a composite reinforced with circular fibers, some types of local deformation of the binder are small. If, in the case of square cross-section fibers, the stress–strain state estimation is possible using explicit formulas from [1], then for fibers of other cross-sections, it becomes necessary to use a computer to calculate the stress–strain state.

If a general macroscopic deformation ε_{ij} is applied to the composite, then the local deformations ε_{kl}^{loc} in the binder are given by the linear combination

$$\varepsilon_{kl}^{loc} = \varepsilon_{kl}^{loc\,ij}\varepsilon_{ij} \tag{7}$$

Here $\varepsilon_{kl}^{loc\,ij}$ is the local deformations in the binder, corresponding to the application macroscopic deformation ε_{ij}. Examples of the deformation calculation $\varepsilon_{kl}^{loc\,ij}$ are given above.

When the macroscopic strain ε_{ij} applied to the composite is changed, the microscopic strains change $\varepsilon_{kl}^{loc\,ij}$ quite difficultly, for example, they can go from one to another. When using computer technology, for example, the ANSYS software package [53], the calculation of quantities (7), after all $\varepsilon_{kl}^{loc\,ij}$ are calculated, does not present a problem.

In this article, the author aimed to substantiate the maximum set of local stress–strain state modes in cross-reinforced composites of the type described above, predicted in [1].

The issue of composite fracture modes has been widely discussed and continues to be actively discussed, see, e.g., [4, 5, 7–27, 36]. In [1], it is indicated that the primary issue is the local stress–strain state modes in the composite (the existence and characteristics of such modes), and the issue of destruction modes is secondary, the solution of which follows from the solution of the first question. In [1], an example of obtaining a multimode strength criterion for a model composite is given.

The largest number of modes is observed in a composite reinforced with square cross-section fibers. In this case, the number of modes is six. For circular cross-section fibers, one mode weakens, up to the possibility of neglecting this mode. Accordingly, the number of modes becomes equal to five.

The alternative method of construction of the homogenized (macroscopic, overall) strength criteria is the numerical construction of the homogenized strength criteria [54–56].

References

1. Kalamkarov AL, Kolpakov AG (1997) Analysis, design and optimization of composite structures. Wiley, Chi Chester (1997)
2. Kelly E (ed) (1994) Concise encyclopedia of composite material. Pergamon, Oxford
3. Broutman LJ, Krock RH (1984) Composite materials, vol 1–8. Academic Press, New York
4. Pinho ST, Dávila CG, Camanho PP, Iannucci L, Robinson P (2005) Failure models and criteria for FRP under in-plane or three-dimensional stress states including shear non-linearity NASA/TM-2005-213530
5. https://www.energy.gov/eere/fuelcells/hydrogen-storage. Storage-Basics, U.S. DoE
6. Kolpakov AG, Rakin SI (2020) Homogenized strength criterion for composite reinforced with orthogonal systems of fibers. Mech Mater 148:103489
7. Hashin Z, Rotem AA (1973) Fatigue failure criterion for fiber reinforced materials. J Com Mater 7(4):448–464
8. Narayanaswami R (1977) Evaluation of the tensor polynomial and Hoffman strength theories for composite materials. J. Comp. Mater 11(4):366–377
9. Parry TV (1981) Wroski A S : Kinking and tensile, compressive and interlaminar shear failure in carbon-fiber-reinforced plastic beams tested in flexure. J Mater Sci 16(4):439–450
10. Soden D, Leadbetter D, Griggs R, Eckold GC (1978) The strength of a filament wound composites under biaxial loading. Composites 9(4):247–250
11. Tsai SW, Wu EM (1971) A general theory of strength for anisotropic materials. J Comp Mater 5(1):58–67
12. Norris CB (1950) Strength of orthotropic materials subjected to combined stress: Report No. air 18126. Department of Agriculture, Forest Products Laboratory, Madison
13. Hill R (1950) The mathematical theory of plasticity. Oxford University Press, London
14. Azzi VD, Tsai SW (1965) Anisotropic strength of composites. Exp Mech 5(9):283–288
15. Yamada SE, Sun CT (1978) Analysis of laminate strength and its distribution. J Comp Mater 12(3):275–284
16. Hoffman O (1967) The brittle strength of orthotropic materials. J Comp Mater N1:200–206
17. Cowin SC (1986) Fabric dependence of an anisotropic strength criterion. Mech Mater 5(3):251–260
18. MSC Laminate Modeler Version 2008 r2 (2008) User's Guide. MSC Software Corp.
19. Mascia NT, Simoni RA (2013) Analysis of failure criteria applied to wood. Eng Failure Anal N35:703–712
20. Puck A, Schneider W (1969) On failure mechanisms and failure criteria of filament-wound glass fiber/resin composites. Plastic Polymer Tech 33–43
21. Puck A (1969) Festigkeitsberechnung an Glasfaser/Kunststoff-Laminaten bei zusammengesetzter Bean spruchung. Kunststoffe 59(11):780–787
22. Knops M (2008) Analysis of failure in fiber polymer laminates. Springer, Heidelberg
23. Puck A, Schurmann H (1998) Failure analysis of FRP laminates by means of physically based phenom enological models. Comp Sci Tech 58:1045–1067
24. Puck A, Kopp J, Knops M (2002) Failure analysis of FRP laminates by means of physically based phenomenological models. Comp Sci Tech 62:1633–1662
25. Hashin Z, Rotem AA (1973) Fatigue failure criterion for fiber reinforced materials. J. Comp. Mater. 7(4):448–464
26. Hashin Z (1980) Failure criteria for unidirectional fiber composites. J Appl Mech N47:329–334
27. Christensen RM (1997) Stress based yield/failure criteria for fiber composites. Int J Solids Struct N34:529–543
28. Annin BD, Kolpakov AG (1987) Designing laminated composites with specified deformation-strength characteristics. Mech Comp Mater 23(1):50–57
29. Bakhvalov NS, Panasenko G (1989) Homogenization: averaging processes in periodic media. Kluwer, Dordrecht
30. Sanchez-Palencia E (1980) Non-homogeneous media and vibration theory. Springer, Berlin

31. Panasenko G (2005) Multi-scale modeling for structures and composites. Springer, Berlin
32. Panasenko GP (1991) Multicomponent homogenization for processes in essentially nonhomogeneous structures. Sb. Math. 69(1):143–153
33. Panasenko GP (1990) A numerical-asymptotic multicomponent averaging method for equations with contrast coefficients. Comput Math Math Phys 30(1):178–186
34. Annin BD, Kalamkarov AL, Kolpakov AG: Analysis of local stresses in high modulus fiber composites. In: Localized damage computer-aided assessment and control, pp 131–144. V.2. Computer Mechanics Publ., Southampton
35. Annin BD, Kalamkarov AL, Kolpakov AG, Parton VZ (1998) Calculation and design of composite materials and structural elements. Nauka, Novosibirsk
36. Grinevich DV, Yakovlev NO, Slavin AV (2019) The criteria of the failure of polymer matrix composites (review) Trudi VIAM 7(79) (in Russian)
37. Okoli OI, Smith GF (1998) Failure modes of fibre reinforced composites: the effects of strain rate and fibre content. J Mater Sci 33:5415–5422
38. Sun L, Jia Y, Ma F, Sun S, Han ChC (2009) Mechanical behavior and failure mode of unidirectional fiber composites at low strain rate level. J Comp Mater 43(22):2623–2637
39. Yvonnet J (2019) Computational homogenization of heterogeneous materials with finite elements. Springer, Cham (2019). https://doi.org/10.1007/978-3-030-18383-7
40. Barbero EJ (2013) Finite element analysis of composite materials using ABAQUS. CRC Press, Boca Raton
41. Rakin SI (2022) Strength analysis of fiber composites with ANSYS. AIP Conf Proc 2647:060040
42. Gluzman S, Mityushev V, Nawalaniec W (2018) Computational analysis of structured media. Academic Press, Amsterdam
43. Mituyshev V, Drygas P (2019) Effective properties of fibrous composites and cluster convergence. Multiscale Model Simul 17(2):696–715
44. Lubin G (ed) (1982) Handbook of composites. Van Noxtrand, New York
45. Rudawska A (2021) Mechanical properties of selected epoxy adhesive and adhesive joints of steel sheets. Appl Mech 2:108–126
46. Hu Y, Xia, Z, Ellyin F (2006) The failure behavior of an epoxy resin subject to multiaxial loading. In: ASCE 2006 Pipeline conference, pp 1–8, 1 July
47. De Groot R, Peters MC, De Haan YM, Dop GJ, Plasschaert AJ (1987) Failure stress criteria for composite resin. J Dent Res 66(12):1748–1752
48. Keller JB, Flaherty JE (1963) Elastic behavior of composite media. Comm Pure Appl Math 26:565–580
49. Kang H, Yu S (2020) A proof of the Flaherty-Keller formula on the effective property of densely packed elastic composites. Calc Var 59:22
50. Kolpakov AA, Kolpakov AG (2010) Capacity and transport in contrast composite structures: asymptotic analysis and applications CRC Press, Boca Raton
51. Kolpakov AA (2007) Numerical verification of the existence of the energy-concentration effect in a high-contrast heavy-charged composite. JEPTER 80(4):166–172
52. Rakin SI (2014) Numerical verification of the existence of the elastic energy localization effect for closely spaced rigid disks. J Engng Phys Thermophys 87:246–252
53. Thompson MK, Thompson JM (2017) ANSYS mechanical APDL for finite element analysis. Butterworth-Heinemann, Oxford
54. Kolpakov AG, Rakin SI (2023) The homogenized strength criterion for reinforced plate with orthogonal systems of fibers. Comput Struct 275:106922
55. Kolpakov AG, Rakin SI (2021) Local stresses in the reinforced plate with orthogonal sytems of fibers. Compos Struct 265:113772
56. Kolpakov AA, Kolpakov AG (2005) Solution of the laminated plate design problem: new problems and algorithms. Comput Struct 83(12–13):964–975

Numerical Evaluation of Punching Shear Capacity Between Bonded and Unbonded Post-tensioned Slab Using Inverted-U Shape Reinforcement

Milad Khatib⬤ and Zaher Abou Saleh⬤

Abstract The Both bonded and unbonded prestressing tendons may be used to reinforce post-tensioned (PT) concrete members. The binding state of tendons may have an impact on how well different types of PT concrete elements function in flexural and shear loads. Inverted-U shape reinforcement was used experimentally to enhance the behavior of different structural elements. The obtained results confirmed the proposed model's accuracy for both PT slabs and beams (bonded, and unbonded). This study's major goal is to compare numerically PT concrete slabs provided with inverted-U shape reinforcement with the two different tendon systems and evaluate their performances due to applying punching shear load. To do this, the PT slab that has already been tested is reexamined numerically. By using, a nonlinear finite element, the results were carried out utilizing visualization tools. For a better study of the behavior of bonded and unbonded PT slabs, the obtained numerical results, and the previous experimental one are compared. Good correlations are shown.

Keywords Bonded Tendons · Unbonded Tendons · Post-tension Slab · Inverted-U Shape Reinforcement · Nonlinear Finite Element

1 Introduction

Today, many structures require prestressed concrete in order to use the minimum concrete slab thickness and, reduce of both cracks and deflection. Pretensioned, unbonded, and bonded PT prestressed concrete are the three types of prestressed concrete construction. The study of structural mechanisms in PT concrete members

M. Khatib (✉)
Research Supervisor Civil Engineering, ISSEA-Cnam, Beirut, Lebanon
e-mail: milad.khatib@isae.edu.lb

Z. A. Saleh
Co-Founder & CEO LTC, Jiyeh, Lebanon

G. Feng (ed.), *Proceedings of the 9th International Conference on Civil Engineering*,
Lecture Notes in Civil Engineering 327,
https://doi.org/10.1007/978-981-99-2532-2_50

595

still requires to be considerably improved in several aspects, like shear-flexure inter-actions in PT slab. Moreover, investigations comparing the differences between the slabs, bonded and unbonded, were rarely discussed, and published. The behaviors of PT concrete beams and slabs one-way with various tendon-bonding parameters have only been compared in a few studies.

1.1 Review

Freyssinet had develop the first application of post-tensioning for the construction of a Normandy marine port in France. This technique was established in the United States in the 1950s [1].

The guidelines for assessing the vibration serviceability of thin post-tensioned concrete floors have shown to be inaccurate. Some researchers validated a set of empirically rules for performing out such analysis. It has been shown that the elastic modulus of essential columns contributes significantly to the total elastic modulus of the floor, and that they should not be treated as pin supports that allow unrestricted rotation. The usage of bar finite elements to replicate the columns worked well [2].

A potentially weak PT beam was kept after the entire dismantling of a prestressed concrete bridge in southern France for non-destructive testing (NDT), then reviewed all the outcomes of four alternative approaches [3]. By focusing on ambiguous zones to get accurate local measurements or by enhancing measurement reliability, the adoption of a different approach should give a more appropriate response while removing certain ambiguities. The issues that these methods have not fully addressed are then identified. The collection of demands that these strategies have thus far failed to satisfy comes as the final phase.

Later, five internal connections were tested experimentally. Each one was consisted of two PT beams (top-and-seat-angle wide flange) connected to a column. These connections were seismically loaded on steel moment resistant frames. The number of PT strands were considered as the studied factors, and the initial post-tensioning force. The connections were fabricated, in such a way, that the energy was dissipated in the angles; while the other structural parts remain elastic. The obtained results showed that PT connectors have significant cyclic strength and ductility. The preliminary elastic stiffness is equivalent to that of a welded connection, and the connection has almost no structural damage after extreme inelastic drifted cycles [4].

A numerical model was suggested to estimate the response of unbonded PT beams. The effects of monotonic and repetitive loads are both examined. The model's unique-ness may be seen in the estimation of prestressed concrete serviceability following cracking, which is based on a non-linear macro finite element model. It was distin-guished by its homogenous average inertia. A second novelty was the estimation of total elongation for the unbonded tendons at all stress phases: cracking, service-ability, and ultimate. The results demonstrate that the model predicts the bending stiffness of the beam during loading cycles and the stable remaining deflection after unloading related to crack development. Both of bearing capacity and deflection at failure estimations are also quite precise [5].

A clear evaluation was conducted between members with the two distinct tendon systems and examine their structural behavior in detail. Actually identified PT beams, one-way slabs, and slab-column connections were reevaluated, and extensive nonlinear finite element calculations were undertaken utilizing new modelling approaches. A set of test results demonstrates that the established model was trustworthy for both bonded and unbonded PT parts. The findings of the earlier experimental and present numerical investigations are utilized for better examining the behavior of bonded and unbonded PT components [6].

Punching shear reinforcement is a practical method for enhancing the strength and deformation resistance of flat slabs. Such reinforcement have been produced in many forms in the past. With "Inverted-U Shaped Reinforcement," an unbonded PT slab, was exposed numerically to punching loads, and examined [7]. The outcomes were compared with those from an earlier experimental study. There was a good relation between all the obtained results (numerically and experimentally).

An experimental PT beams provided with bonded and unbonded tendons were studied [8]. These tendons were referred as "mixed tendons". A numerical evaluation for the obtained results were achieved using finite element model. The proposed numerical technique reported was proven to accurately estimate ultimate flexural strength, mode of failure, and tendon stress within 5%. The final displacement was predicted quite well (around 15%) for the mixed tendons.

PT concrete beams prevent shear via a variety of mechanisms. Special shear reinforcement can be utilized to improve the shear capacity of such beams. A conducted experiment studied the effect of inverted-U shape reinforcement on the shear strength of bonded PT concrete beams. The results were compared to another type of bonded PT concrete beams, which provided with conventional type of reinforcement against shear (closed stirrups). The results indicate a 13% improvement in shearing strength. There are important correspondences between projected and actual experimental outcomes [9].

Four unbonded PT beams were tested. During the tests, several data were collected. The obtained results were compared to the currently used methods in Russia. Tests and calculation results demonstrate that Russian analytical techniques are underestimated for the unbonded PT beams design [10].

In order to enhance the concrete shear strength of PT beams utilizing various shear reinforcing methods, several tests and computational analysis were carried out. An investigation for the benefit of inverted-U shaped reinforcement in bonded PT beams, experimentally, was conducted. Recently, numerical investigation was carried to explore the impact of these reinforcement on the shear behavior of two types bonded PT beams [11]. The numerical outcomes showed that the ACI 318–14's restriction on the nominal shear reinforcement for bonded pre-stressed concrete beams was overly cautious [12].

2 Objectives of the Investigation

The aim of this study is to apply numerically a punching load on two types of bonded PT slabs, the first one provided with inverted-U shaped reinforcement, while the other provided with stud reinforcement, and then analyze their behaviors in terms of load deflection relationship as well as average shear strength around the critical shear zone. A previous numerical unbonded PT slab subjected to punching load was achieved.

Unbonded tendons provide unique structural benefits rather than bonded tendons, the post-flexural catenary capacity. It is nearly hard to fail an unbonded tendon under tension according to applied force. This component provides apparent benefits under extreme loading or in reducing punching shear failure. Tests have revealed that slabs with unbonded tendons contain catenary capacities three to four times the requirement at factored load [13].

Since the bonded circumstances seem to be the worst situation for the unbonded ones, then a finite element program "ABAQUS" is applied to provide numerical conclusions for this situation, and compare it with the unbonded one [14]. Those findings were examined with those previously obtained through an experimental evaluation of the punching shear strength utilizing unbonded PT concrete slabs with "Inverted-U Shaped Reinforcement" and the ACI specifications for punching shear strength.

3 Numerical Unbonded Investigation

A numerical research, using finite element analysis FEA, was suggested [7]. Set of inverted-U shaped reinforcement (Fig. 1) were inserted in the punching shear zone of unbonded PT concrete slab (PT-A). The slab structure is vertically symmetric about an axis at centerline, and had dimensions $1.8 \times 1.8 \times 0.01$ m. (Fig. 2).

These dimensions were set to yield at a nominal punched shear stress (approximately 25 tons) by using ACI formula for stud shape ($0.66 \sqrt{f_c'}$ at a distance d/2 from the column face). An upward statically force was exerted across a central surface (10×10 cm). Six tendons were placed in each direction. The PT forces (146.5 kN) were supplied in a separate phase. The flat concrete slab was prevented from sliding by four fixed supports, which situated 90 cm apart from center and extended 45 cm on

Fig. 1 Hairpin shaped reinforcements

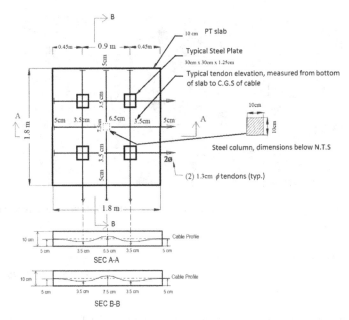

Fig. 2 Unbonded PT Tendon layout in PT slab [7]

every edge. Longitudinal and transverse reinforcing steel bars were placed under the four steel supports (Fig. 3).

The purpose of this numerical research, using ABAQUS, was to analyze the behavior of unbonded PT flat slab provided with inverted-U shaped reinforcement in the center area, and subjected to punching shear force as indicated in Fig. 4. The

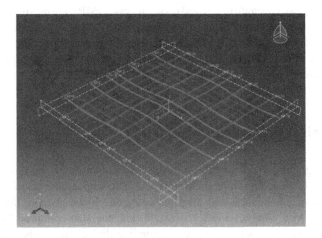

Fig. 3 PT Tendons and Reinforcement Layout

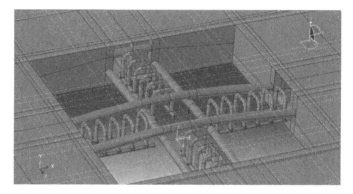

Fig. 4 Unbonded PT Slabs Reinforcement Provide with Inverted-U Shaped Reinforcement (PTU-A)

Fig. 5 Stud Reinforcement

investigation was also undertaken on another unbonded PT concrete slab (PT-B) provided with stud reinforcement (Fig. 5) to evaluate, and compare the results.

ABAQUS is a useful program for FEA, initially introduced in 1978. The study of a structural system is accomplished through three processes. Firstly, specify the FE model and external conditions to be applied to it (Pre-processing). Then, this FE model will be simulated (Analysis solver or Simulation). Finally, analyze the generated data (Post-processing). By utilizing the visualization module, numerous outcomes can be examined by showing the results, such as contour lines, deformed shapes, and X–Y plots.

The compressive strength of utilized concrete was 30 MPa; the prestressing strands were 12.5 mm diameter complying to ASTM standard A421, with a required ultimate strength of 1860 MPa. Bars of 10 mm were utilized for modelling of both 6 hairpins reinforcement (PT-A) and 6 studs (PT-B) for each direction. The shear reinforcement of both specimens have such a yielding strength of 420 MPa.

4 Numerical Unbonded Investigation

A smoothly grading crack forms when the load increases on the unbonded slab with inverted-U shaped reinforcement (PTU-A). The sequential cracks development signifies that major, flexural and diagonal ones, developed at more than yielding of steel bars. At a compression force of 290 kN, the slab collapsed because the cracks expanded towards top, with corresponding deflection equals to 19.30 mm (Fig. 6). The slab collapsed in flexural mode prior to punching shear.

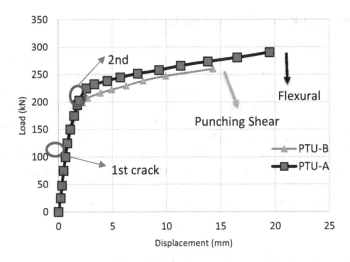

Fig. 6 The Load–Deflection Curves for PTU-A & PTU-B

Regarding the second unbonded slab type with stud reinforcement (PTU-B), it was found that the fractures appeared in a sudden manner. With increased load, further shear fractures formed. As the observed load reached to 260 kN, the top of the slab cracked with maximum deflection corresponds to 14.32 mm (Fig. 6). This slab collapsed in shear mode.

All the numerical results for the two types of unbonded slabs (PTU-A) and (PTU-B) are reported in Table 1.

It was clear that the load capacity for the (PTU-A) was increased about 10.34% than the (PTU-B). In addition, (PTU-A) resisted the punching shear stresses more than (PTU-B) by 10.50%. However, the (PTU-A) failed in flexural mode, while the (PTU-B) failed in shear mode.

The obtained results for PTU-A were improved than PTU-B by 10.34% concerning failure load, and 10.50% regarding nominal shear stresses.

Table 1 Numerical results for "Unbonded" PT Slabs

Specimen designation	Reinforcement type	Failure loads (kN)	Deflection (mm)	Nominal shear stress (MPa)	Failure type
PTU-A	Inverted-U	290	19.30	$4.76 = 0.87\sqrt{f'c}$	Flexural
PTU-B	Stud	260	14.32	$4.26 = 0.78\sqrt{f'c}$	Punching Shear
% difference		10.34%	25.80%	10.50%	

5 Numerical Bonded Investigation

The greater part of PT tendons being used in unbonded PT slabs. In the United States only, and by the mid of 2006, more than 50% of the residential structures, use the PT concrete. Moreover, it was expected that over 375,000 residences will built using PT method annually. Furthermore, slabs with bonded tendons are widely utilized in other regions of the world [15]. By the end of 2012, it was estimated that more than 2.5 billion ft^2 with unbonded PT slabs in use in the United States [16].

The FEA for bonded PT concrete slab provided with inverted-U shaped reinforcement (PTB-A) is proposed and developed, which will include material parameters, element types, and analysis processes. The same development will be followed for the bonded PT concrete slab provided with stud reinforcement (PTB-B).

5.1 Concrete and Steel Material Modeling

All the utilized materials characteristics in the model are similar to that in the prior numerical model. A mix of 3D solid and truss components are used to model the simulated structures.

Concrete is modelled by utilizing the installed "damaged plasticity model" in ABAQUS [14]. The theory of "Concrete Damaged Plasticity" (CDP) was distinguished in 1989 [17], and later confirmed in 1998 [18]. By using it, the three-dimensional stress–strain controlling equations are derived from the uniaxial stress–strain relationship of concrete. The uniaxial stress–strain compression relationship is computed based on an available empirical approach [19]. Recently, many investigations have shown that the damage plasticity model provides an accurate technique for simulating concrete behaviour in both of the tension and the compression [20–22]. The tensile relationship is considered straight until cracking, and, for post-cracking behavior, is modelled using tension stiffening. The tension stiffening effect is considered minor enough that the stress following cracking gradually falls to a minimum value at about nearly twice the cracking strain. This concept was demonstrated to be appropriate for PT slab [23]. Just for simplicity, the nonlinear stress–strain curves are divided into piecewise linear parts as illustrated in (Fig. 7).

For mild steel, the behavior is perfectly elasto-plastic. However, the tendons had a nonlinear behavior one. For the PT tendons, the nonlinear stress–strain relationship is computed based on empirical approach [24] for seven wire strands (Grade 270).

5.2 Elements

Two types of ABAQUS elements are utilized in the constructed models. The rebar reinforcement are modelled by applying a sequence of linear truss elements (T3D2:

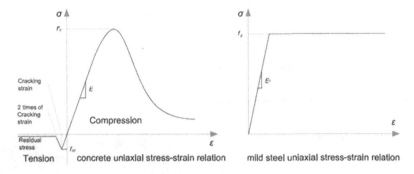

Fig. 7 Uniaxial Stress–Strain Models for FEM

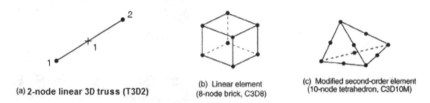

(a) 2-node linear 3D truss (T3D2)

(b) Linear element
(8-node brick, C3D8)

(c) Modified second-order element
(10-node tetrahedron, C3D10M)

Fig. 8 Elements Abaqus

three-dimensional, and two nodes, Fig. 8a). This element can be utilized to model the tendons. However, it does not have external surface, so we cannot make a surface cohesion with the concrete.

Then, the solid element (C3D8R: three-dimensional, eight nodes, and linear with reduced integration, Fig. 8b) is the solution for this problem. In addition, this element is used to model the concrete.

The two reinforcement types against shear (inverted-U and stud) are modelled using the element (C3D10M: three-dimensional, and ten nodes modified, Fig. 8c).

Modeling Interaction Between Reinforcement and Concrete. All used rebars, inverted-U and stud reinforcement are considered as fully bonded. The embedding approach is used in the modelling to fulfill the perfect bond requirement. The bonded aspect of prestressing tendons, and large tendons sliding are modelled using the contact formulation [23].

The bonding conditions mechanism may be reproduced by utilizing a contact formulation between tendons and concrete (surface-to-surface contact, Fig. 9).

The tangential behavior of the surface contact is designed to be frictionless for the PT period, enabling the tendon to glide easily. To replicate the grouting process, the modeling approach for the tangential contact concept is changed from "frictionless" to "rough". Mesh slab is shown in Fig. 10, while fixed support conditions and applied loads are shown in Fig. 11.

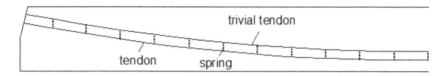

Fig. 9 Modeling Details for Unbonded PT Tendons [23]

Fig. 10 Meshed PT Slab

Fig.11 Boundary
Conditions PT Loaded Slab

6 Numerical Bonded Results

All the obtained results for modeled bonded slabs provided with both of inverted-U shaped reinforcement (PTB-A), and stud reinforcement will be discussed as follows:

6.1 Bonded PT Slabs with Inverted-U Shaped Reinforcement: PTB-A

As the load increases, a uniform scaled fracture appears on the load–deflection curve (Fig. 12) and the distorted colored outline contour plots (Figs. 13(a), and 13(b)).

However, at mid-span, a significant deflection occurred. The crack developments indicate that considerable, flexural and diagonal fractures occur further than the yielding value of reinforcement rebars. The slab collapsed in flexural mode with an applied load equal to 270 kN with corresponding deflection of 8.43 mm.

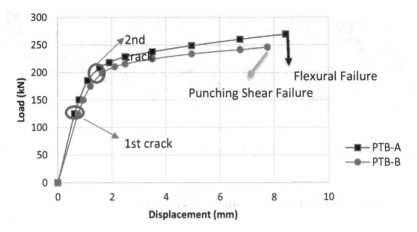

Fig. 12 The Load–Deflection Curves for PTB-A & PTB-B

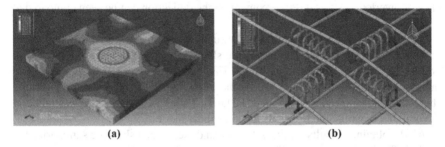

Fig. 13 a Directional Deformation for Bonded PTB-A. b Directional Deformation for Bonded PTB-

6.2 Bonded PT Slabs with Stud Shaped Reinforcement: PTB-B

Based on the obtained numerical results (Fig. 12) and the distorted colored contour plots (Fig. 14(a) and 14(b)), it appears that the fractures occurred suddenly. The shear fractures occurred as the strain increased. The slab model collapsed in shear mode with a deflection of 7.76 mm at 246 kN, the highest reached load.

In bonded cases, PTB-A resists punching load greater than PTB-B by 10.72%, while sustains more nominal shear stresses by 32.50%.

(a)   (b)

Fig. 14 a Directional Deformation for Bonded PTB-B. b Directional Deformation for Bonded PTB-B

7 Comparison Between Numerical Results

The displayed load deformation curves in Fig. 13 are based on the obtained numerical findings from for both of the bonded models, PTB-A and PTB-B. The positive influence of inverted-U shaped reinforcement on the deflection can be seen compared to stud reinforcement.

Table 2 summarize the results of nominal shear stresses at the boundary of punching shear on the top surface of the slab for bonded PT slabs provided with both of "inverted-U shaped" and "stud". The load capacity of (PTB-A) was clearly increased by 8.88% over (PTB-B). Furthermore, the (PTU-A) resist the punching shear stresses by 10.34% more than (PTU-B). The (PTU-A) nevertheless failed in flexural mode, whilst the (PTU-B) failed in shear mode.

All the obtained results for the unbonded and the bonded slabs are summarized in Table 3. The load capacity for PTU-A was decreased by 6.90% comparing to PTB-A, and the nominal shear stress was reduced by 8.20% (Fig. 15). While, for the slab PTU-B, there is a reduction of both the capacity load about 5.38% from PTB-B, and 30.75% for the nominal shear stresses (Fig. 16).

Table 2 Numerical results for "Bonded" PT Slabs

Specimen Designation	Reinforcement Type	Failure Loads (kN)	Deflection (mm)	Nominal Shear Stress (MPa)	Failure Type
PTB-A	Inverted-U	270	8.43	$4.37 = 0.79 \sqrt{f'c}$	Flexural
PTB-B	Stud	246	7.76	$2.95 = 0.54 \sqrt{f'c}$	Punching Shear
% difference		8.88%	7.94%	32.50%	

Table 3 Numerical Comparison between "Unbonded" and "Bonded" PT Slabs with ACI provisions

Specimen Designation	Reinforcement Type	Numerical Results			ACI Equation
		Failure Loads (KN)	Deflection (mm)	Nominal Shear Stress (MPa)	Nominal Shear Stress (MPa)
PTU-A	Inverted-U	290	19.30	$4.76 = 0.87 \sqrt{f'c}$	Not Applicable
PTB-A		270	8.43	$4.37 = 0.79 \sqrt{f'c}$	
% difference			6.90%	56.32%	8.2%	
PTU-B	Stud	260	14.32	$4.26 = 0.78 \sqrt{f'c}$	$0.66 \sqrt{f'c} = 3.614$
PTB-B		246	7.76	$2.95 = 0.54 \sqrt{f'c}$	
% difference			5.38%	45.81%	30.75%	

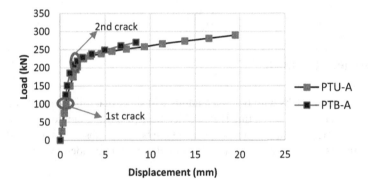

Fig. 15 The Load–Deflection Curves for PTU-A & PTB-A

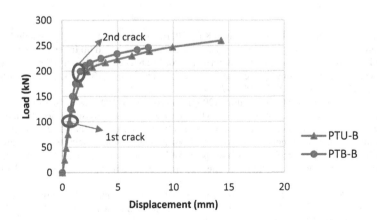

Fig. 16 The Load–Deflection Curves for Bonded PTB-B & PTU-B

8 Discussions and Conclusions

Despite the absence of studies on the real stress behavior of bonded slabs, and their construction methods outside the US, it appears reasonable that ACI 318 should impose a same quantity of reinforcement for bonded PT slabs, as is needed in unbonded PT slabs.

The slab provided with inverted-U shaped reinforcement resist punching load more than the slab provided with stud reinforcement, in both cases of unbonded (10.34%) and bonded (8.88%).

Although, the slab provided with inverted-U shaped reinforcement has nominal shear stress higher than the slab provided with stud reinforcement, in both cases of unbonded (10.50%) and bonded (32.50%). These improvements in the studied values were due to confinement of the concrete.

Finally, it is predicted that the increased quantity of bonded reinforcement necessary to meet the parameters specified herein would be minor, posing no major economic penalty for the use of bonded tendons in PT slabs while providing a large performance gain.

9 Further Research

Checking concrete slabs with bonded tendons will show that this proposal is conservative; however, until such testing is available, these proposals should be preceded to confirm the same performance level, and safety in bonded slabs as has been indicated by experiments and performance in slabs provided with unbonded tendons.

References

1. Xercavins P, Demarthe D, Shushkewich K (2010) Eugene Freyssinet – His Incredible Journey to Invent and Revolutionize Prestressed Concrete Construction. 3rd fib International Congress
2. Pavic A, Reynolds P, Waldron P, Bennett K (2001) Dynamic modelling of post-tensioned concrete floors using finite element analysis. Finite Elem Anal Des 37(4):305–323. https://doi.org/10.1016/S0168-874X(00)00045-7
3. Dérobert X, Aubagnac C, Abraham O (2002) Comparison of NDT techniques on a PT beam before its autopsy. NDT E Int 35(8):541–548. https://doi.org/10.1016/S0963-8695(02)00027-0
4. Garlock MM, Ricles JM, Sause R (2003) Experimental studies on full-scale post-tensioned seismic-resistant steel moment connections. Book Stessa, 1st edn ISBN:9780203738290
5. Vu NA, Castel A, François R (2010) Response of PT concrete beams with unbonded tendons including serviceability and ultimate state. Eng Struct 32(2):556–569. https://doi.org/10.1016/j.engstruct.2009.11.001
6. Kang TH-K, Huang Y, Shin M, Lee JD, Cho AS (2015) Experimental and numerical assessment of bonded and unbonded post-tensioned concrete members. ACI Struct J 112(6):735–748. https://doi.org/10.14359/51688194

7. Khatib M, Abou Saleh Z, Baalbaki O, Temsah Z (2018) Numerical punching shear analysis of unbonded post-tensioned slabs with inverted-u shaped. KSCE J Civil Eng 22:4490–4499. https://doi.org/10.1007/s12205-018-1505-5
8. Brenkus N, Tatar J, Hamilton H, Consolazio G (2019) Simplified finite element modeling of PT concrete members with mixed bonded and unbonded tendons. Eng Struct 179:387–397
9. Khatib M, Abou Saleh Z (2020) Enhancement of shear strength of bonded post-tensioned beams using inverted-U shaped reinforcements.Case Stud Construct Mater 13. https://doi.org/10.1016/j.cscm.2020.e00370
10. Gavrilenko A, Barkaya T (2021) Experimental and theoretical study of post-tensioned unbonded beams. IOP Conf Ser Mater Sci Eng 1030(1):012077. https://doi.org/10.1002/suco.202000774
11. Khatib M, Abou Saleh Z, Baalbaki O, Hamdan Z (2022) Numerical shear of post-tensioned beams with inverted-U shaped reinforcements. Mag Civil Eng 110(2). https://doi.org/10.34910/MCE.110.6
12. ACI Committee 318 (2011) Building code requirements for structural concrete (ACI 318–11) and Commentary. American Concrete Institute, Farmington Hills, 503 pp
13. Freyermuth CL (1989) Structural integrity of buildings constructed with unbonded tendons. Concr Int 11(3):56–63
14. ABAQUS, 6.13 (2013) ABAQUS/CAE user's guide, Providence
15. Rogers J (2006) PT concrete in the residential industry. Concrete Constr Mag
16. Bondy KB (2012) Two-way PT slabs with bonded tendons. PTI J 8(2):43–48
17. Lubliner J, Oliver J, Oller S, Oñate E (1989) A plastic-damage model for concrete. Int J Solids Struct 25:229–326
18. Lee J, Fenves GL (1998) Plastic-damage model for cyclic loading of concrete structures. J Eng Mech 124:892–900
19. Carreira DJ, Chu KH (1985) Stress-strain relationship for plain concrete in compression. ACI J Proc 83(6):797–804
20. Hamoda A, Basha A, Fayed S et al (2019) Experimental and numerical assessment of reinforced concrete beams with disturbed depth. Int J Concrete Struct Mater 13(55). https://doi.org/10.1186/s40069-019-0369-5.
21. Seok S, Haikal G, Ramirez JA, Lowes LN, Lim J (2020) Finite element simulation of bond-zone behavior of pullout test of reinforcement embedded in concrete using concrete damage-plasticity model 2 (CDPM2). Eng Struct 221(15). https://doi.org/10.1016/j.engstruct.2020.110984
22. Liu J, Shi C, Lei M, Wang Z, Cao C, Lin Y (2022) A study on damage mechanism modelling of shield tunnel under unloading based on damage–plasticity model of concrete. Eng Fail Anal 123. https://doi.org/10.1016/j.engfailanal.2021.105261
23. Kang TH-K, Huang Y (2012) Nonlinear finite element analyses of unbonded post-tensioned slab-column connections. PTI J 8(1):4–19
24. Devalapura RK, Tadros MK (1992) Critical assessment of ACI 318 Eq. (18–3) for prestressing steel stress at ultimate flexure. ACI Struct J 89(5):538–546

Use of Electrical Resistivity Tomography (ERT) for Detecting Underground Voids on Electrical Pylon Installation Sites: Case Studies from Labé Prefecture, Republic of Guinea

Ahmed Amara Konaté, Oumar Barou Kaba,
Mohamed Samuel Moriah Conté, Muhammad Zaheer, Baye Mbaye Thiam,
Fassidy Oularé, and Moussa Diallo

Abstract This study was carried out within the framework of the Gambia River Basin Development Organization Energy Project (OMVG), which brings together Gambia, Guinea, Guinea-Bissau and Senegal. The construction of a high voltage interconnection network fed by a hydro-electricity power station to supply power to OMVG member states must be done on stable ground. During the construction of the installation pits for the pylons in Labé Prefecture, cavities in the geological formations were discovered on the initial planned route. Regarding to geotechnical context of subsurface cavities risk it is mandatory to know their exact location in study areas to reduce the potential degree of risk. A field study was conducted to investigate the lithology, as well as in detecting the presence of apparent faults. Geophysical data was collected using Two Dimensional ERT with different electrode spacing to delineate associated subsurface cavities on each site and each cavity respectively. The target depth was 20 m. The result shows that the observed cavities were because of the dissolution of certain minerals constituting the weathering crust due to the infiltration and the seasonal variation in the level of the watercourses. The cavities were almost entirely empty such as in the cases of pylons 213 and 210 indicated that the weathering crust is due to the groundwater infiltration and the seasonal variations in the river flow level. The finding of this study showed that the ERT method have a good applicability in the detection of underground cavities in the study area.

A. A. Konaté (✉) · O. B. Kaba · M. S. M. Conté · F. Oularé · M. Diallo
Laboratoire de Recherche Appliquée en Géoscience et Environnement, Institut Supérieur Des Mines Et Géologie de Boké, BP 84, Boké, Baralandé, Republic of Guinea
e-mail: konate77@yahoo.fr

M. Zaheer
School of Civil Engineering and Mechanics, Lanzhou University, Lanzhou 730000, Gansu, China

B. M. Thiam
GEOTEC Afrique, BP 96, Conakry, Republic of Guinea

© The Author(s) 2023
G. Feng (ed.), *Proceedings of the 9th International Conference on Civil Engineering*,
Lecture Notes in Civil Engineering 327,
https://doi.org/10.1007/978-981-99-2532-2_51

Keywords Labé Prefecture · Underground Cavity · Electrical Resistivity
Tomography

1 Introduction

Our basement is crossed by a considerable number of underground natural cavities
which are linked to the human activities. Engineering structures such as electrical
pylon installation require subsurface investigation to map and assess the presence of
cavities, which reduces geotechnical hazards risks such as subsidence or complete
collapse [1]. The risk of collapse requires firstly the identification and localization
of the origin of cavity by the geophysical tool fits. From a technical point of view,
an underground cavity is characterized by: its geological context, i.e. the physical
characteristics of the rock; its origin anthropic or natural cavity; and its dimensions
(the geometry of the cavity itself).

Applied geophysics is one of the reconnaissance tools used to investigate a cavity.
It involves a number of different techniques that help to identify the anomalies in
the physical and chemical properties of the subsurface including the propagation
of electromagnetic, gravity, acoustic, and electrical, or magnetic signals [2]. These
methods based on the different principles, which depend on the context and faced
problem, their advantages and their limitations. One of these techniques is electrical
resistivity tomography (ERT), which involves for the determination of the subsurface distribution of electrical resistivity by making measurements on the ground
surface. Electrical resistivity is a bulk physical property of materials that describes
how difficult it is to pass an electrical current through the various materials.

ERT as applied to geotechnical engineering problems, have been extensively used
as a method for the detection of cavities ([2–15, 15]). ERT offers a rapid and cost-
effective imaging of the shallow subsurface with acceptable resolution [6]. Despite
the applicability success, ERT investigation often leads to the non-unique models of
the subsurface. This is because the inversion of electrical resistivity data is a nonlinear problem, while the solutions were obtained using a linearized forward model
[12]. Consequently, it is necessary to increase a priori information on the environment
and the cavities sought.

This study is carried out within the framework of the Gambia River Basin
Development Organization Energy Project (OMVG), which brings together Gambia,
Guinea, Guinea-Bissau and Senegal. Referring to [16], the purpose of the OMVG
Energy Project is to contribute to the socio-economic development of the OMVG
member countries through increased access of the population to electricity. It aims
to enable energy exchanges and increase the quality of electricity supply in OMVG
member countries by giving renewable, clean and cost-competitive energy. The
project is financed by numerous institutions including: the African Development
Bank (AFDB), the World Bank, and the ECOWAS.

The OMVG involves the development of the Sambangalou hydro-electricity dam
with an installed capacity of 128 MW and the construction of an interconnection

network for the evacuation of energy, comprising 1,677 km of 225 kV lines, 15 high voltage /medium voltage transformer stations, and 2 load dispatch centers [16]. The electrical interconnection or electricity transmission is ensured by high voltage lines supported by pylons. The construction of a high voltage interconnection network fed by a hydro-electricity power station to supply power to OMVG Member States must be done on stable ground. During the construction of the installation pits for the pylons in Labé Prefecture, Republic of Guinea, cavities in the geological formations where identified on the initial planned route. The length of the route is about 12 km. Thirty-seven (37) sites planned for the locations of electricity pylons were investigated. The spacing between the pylons varies from 300 to 500 m. Early discovery of cavities in the areas where these were viewed can considerably decrease the risk to infrastructure, personnel, project time lines and the budgets.

This study aims to find the presence of cavities likely to cause surface damage on the location of electricity pylons along a high voltage line route in Labé Prefecture in Republic of Guinea. The main objective of this paper is to study the extent of these cavities and where these features were located in the study area. A geological field observations supplemented by a geophysical campaign was carried out. Selected sites are shown in this paper as an example of the capabilities of the methods employed. Results are promising, and the ERT method combine with the geological field observations proved to be a helpful approach to define and detect cavities.

2 Materiel and Methods

2.1 Presentation of Study Area

Labé is in the northwest of the Republic of Guinea, in Middle Guinea, also called the Fouta Djallon highlands, more precisely in the Labé region, 308 km from the capital Conakry (see Fig. 1).

It is administratively bounded by the regions of Pita, Mali, Gaoual, Tougué, Lélouma and Koubia. Its relief rich in mountainous massifs and its abundant rainfall are dominant factors in Middle Guinea where the study region is located, constitutes the source of several rivers in the sub-region. The study area is located in the central part of the Fouta Djallon, which corresponds to the watershed line of the Niger basin, going towards the east and the basins of the rivers flowing into the Atlantic Ocean to the West [17]. Given its particular relief and its significant hydrography, the region has been the subject of several studies ([17, 19, 18, 19]). The study area is located on the geological sheet of Labé, which consists of two structural-tectonic zones that differ from each other by the conditions of their accumulative sedimentation until before the Paleozoic [17].

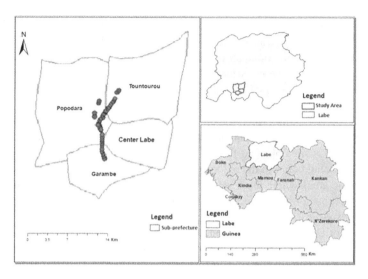

Fig. 1 Presentation of study area

2.2 Geological Field Observation

A field study was consisted in studying the lithology, the geomorphology as well as in detecting the presence of apparent faults on the developed route. The geological field observations were done by researchers from Higher Institute of Mining and Geology of Boke, Guinea.

2.3 Electrical Resistivity Tomography Method

ERT is based on an ABEM acquisition system which contains the measurement protocols. The electrical device comprises forty (40) copper or nickel electrodes, spaced 5 m apart; two (02) cables (yellow) to connect the electrodes to the acquisition system; Dipole–Dipole array configuration was chosen to detect the cavities. This Dipole–Dipole is recommended for vertical structures in an area that is not too noisy, with a fairly sensitive resistivity meter and good contact with the ground ([2, 8, 12, 15, 20–22]). 48 ERT profiles of 100 m, and a test board on the cavities were carried out respectively on each site and on each cavity. The target depth was 20 m. The correct field/electrode coupling was monitored by wetting the electrode. The results of the electrical method are pseudo-section representing the apparent resistivities of the subsoil as a function of depth (20 m) in a vertical plane plumb with the line of surface electrodes. The inversion of the pseudo-section makes it possible to obtain a 2D image of the resistivities of the basement. The apparent resistivity pseudo-section inversion software used were Res2Dinv and Erigraph. The

Fig. 2 Geological map of
the study area

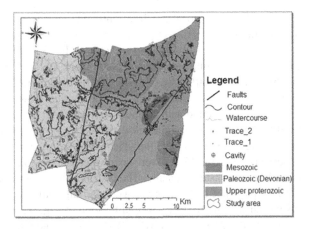

Legend
/ Faults
∿ Contour
⌣ Watercourse
, Trace_2
. Trace_1
⊕ Cavity
▓ Mesozoic
▒ Paleozoic (Devonian)
▓ Upper proterozoic
C3 Study area

0 2.5 5   10 Km

geophysical data acquisition and processing were made by the geophysical survey
staff from the university of Thiès, Senegal. The presence of a cavity was obtained
by an electrical profile. A significant increase in resistivity indicated that the cavity
is filled with air (vacuum) and a significant drop in electrical resistivities values
reflected that the cavity is filled with water. Thus, this study approach is used for the
detection of potentially unstable cavity or zone. Any geotechnical horizons, which
were showing higher values of electrical resistivities such as greater than 30,000
Ω-m indicates that the cavities were filled with air. Very low resistivities values such
as less than 200 Ω-m reflects that the cavities were filled with water (Fig. 2).

3 Results and Discussions

The study area is bounded by two major normal faults to the East and West trending
northeast which control the emplacement of Mesozoic basic and ultramafic intrusions (dunite, dolerite, basalt, gabbro, and pyroxene) as can be seen in Fig. 2. The
emplacement of these major faults was not generate a fault network on the surface
as well these were not bring deformations to the surrounding formations apart from
the contact metamorphism. In the study area, there are no cracks or joints of tectonic
origin observed at the limit of the excavations and their environment. However, cracks
were observed in places because of the chemical and/or mechanical weathering. It
was found that intrusions of dolerites in the beds of watercourses, in particular one
of the tributaries of the Kakrima river near the village Leila in the district of Nadhel.

The presence of these dolerites in this stream would explain the relationship
between the tectonics and the valleys/streams. Also, it was visible on the geological
map of the region in places of rivers controlled by these two major faults as shown in
Fig. 2. The Paleozoic (Ordovician, Silurian) tabular formations in the study area are
essentially of low dips between 10° and 15°. These formations experienced brittle

tectonics due to the opening of the continent to the ocean in the Mesozoic. No ductile deformations were observed.

Mottled zone was observed in at P210 and P213 as shown in Fig. 3. It corresponds to the transition zone between the parent rock and the lateritic gravel levels. It was subdivided into mottled soaked zone and mottled indurated zone. Field observations at P210 and P213 showed the presence of cavities. However, no genetic or functional relationship was found with the faults network (Fig. 2). This was probably the result of the dissolution of certain minerals that make up the weathering crust due to infiltration and underground waters and the seasonal variation in the level of rivers. The cavities observed in P210 and P213 were near watercourses. Figure 4 shows the ERT profiles of P210 pylon installation site. The ERT profiles show a very heterogeneous terrain with the presence of cavities and were partially saturated soils. The large cavity was visible on the site footprint as can be seen in Fig. 3a, which was appeared clearly on profile P2_210. Around pylon 210, It was conducted an additional ERT profile (named P_210') centered 50 m north of P210 to verify the lateral extension of the cavity which was visible on the surface in the study area. It was noted that the cavities make border between the pylon to the North and South. Figure 5 shows that the ERT profiles on the P213 pylon installation site. It was noted that on profile P1_213, higher resistivity as compared to a cavity visible in Fig. 3a. This cavity has a diameter of less than 5 m and a thickness of about 2.5 m. The air-filled cavity will give high resistivity values than those to water filled or water-saturated sediments ones [23, 24]. On P3_ 213 in Fig. 5, it was also found that another structure of low resistivities (between 200 and 400 Ω.m) which could also correspond to a cavity filled with muddy soil. However, it is far enough from the right-of-way of the pylon, on the South-West side, not to require any treatment.

The discoveries of underground cavities by using traditional methods such as local surveys are somehow difficult because of the occurrence of heterogeneities and random dispersal of the cavities. Geophysical investigations may integrate all information on a larger scale and therefore it will be highly acceptable criterion for the subsurface investigations.

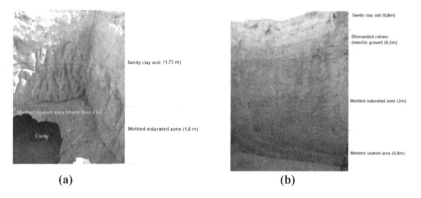

(a)   (b)

Fig. 3 *a* P210 lithology. *b* P213 lithology.

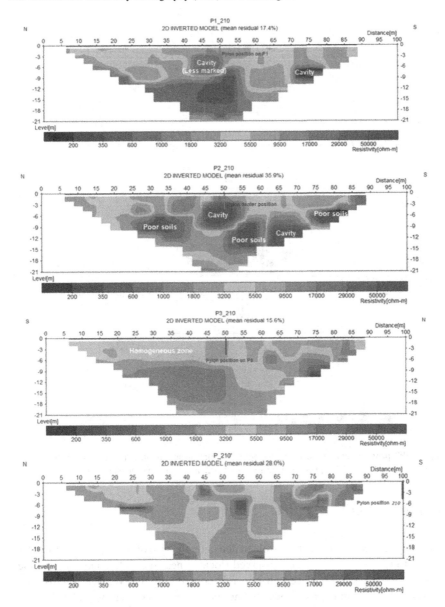

Fig. 4 ET profiles of the P210 pylon installation site. In order to locate and characterize the cavity zones, four ERT profiles (referred as P1, P2, P3 and P) were measured on Pylon 210

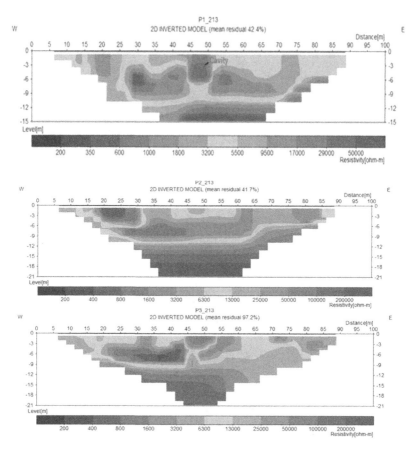

Fig. 5 ERT profiles obtained on the P213 pylon installation site. In order to locate and characterize the cavity zones, three ERT profiles (referred as P1, P2, and P3) were measured on Pylon 213

4 Conclusion

The study carried out made it possible to detect on the grip of some pylons structures similar to cavities. The cavities observed (located near the watercourses) were the result of the dissolution of certain minerals constituting the weathering crust due to infiltration and underground waters and the seasonal variation in the level of the watercourses (floods and low water levels). In the study area, there are no cracks or joints of tectonic origin observed at excavations sites and their environment. However, it was observed cracks at some places because of the chemical and/or mechanical weathering.

The cavities are entirely empty, as in the case of pylons 213 and 210. Mottled zone was observed at Pylon 210 and Pylon 213, which corresponds to the transition zone between the parent rock and the lateritic gravel levels. Field observations at P210 and P213 showed the presence of cavities but no genetic or functional relationship with the faults. They are probably the result of the dissolution of certain minerals that make up the weathering crust because of the groundwater infiltration and the seasonal variation in the level of rivers. The air-filled cavity will give high resistivity values than those to water filled or water-saturated sediments ones.

Acknowledgements The authors thank GEOTEC Afrique, Conakry, Republic of Guinea, for their permission to use the geological and geophysical data of the case study

References

1. De Bruyn IA, Bell FG (2001) The occurrence of sinkholes and subsidence depressions in the far West Rand and Gauteng Province. S Afr Eng Implications: Environ Eng Geosci 7:281–295. https://doi.org/10.2113/gseegeosci
2. Martınez-Lopez J, Rey J, Duen¯ as J, Hidalgo C, Benavente J (2013) Electrical tomography applied to the detection of subsurface cavities. J Cave Karst Stud 75:28–37 https://doi.org/10.4311/2011ES0242
3. Militzer H, Rösler R, Lösch W (1976) Theoretical and experimental investigations for cavity research with geoelectrical resistivity methods. Geophys Prospect 27:640–652
4. Cardarelli E, Di Filippo G, Tuccinardi E (2006) Electrical resistivity tomography to detect buried cavities in Rome: a case study. Near Surf Geophys 4:387–392
5. Khan AS, Khan SD, Kakar DM (2013) Land subsidence and declining water resources in Quetta Valley, Pakistan. Environ Earth Sci 70:2719–2727
6. Mohamed M, Fouzan A (2013) Application of 2-D geoelectrical resistivity tomography for subsurface cavity detection in the eastern part of Saudi Arabia. Geosci Front 4:469–476
7. Lynda DL, Bogdan P, Peter S, Bernard B, Isabelle C (2008) Geophysical detection of underground cavities. In: Symposium Post-Mining 2008, Feb 2008, Nancy, France. pp. NC. Ineris-00973287
8. Partha PM (2018) Detection of Cavity Using Electrical Resistivity Tomography (ERT) at Patherdih Jharia Coal Field, Dhanbad, India. Univ J Geosci 6(4):114–117. https://doi.org/10.13189/ujg.2018.060402
9. Putiska R, Nikolaj M, Dostal I, Kusnirak D (2012) Determination of cavities using electrical resistivity tomography. Contrib Geophys Geodesy 42:201–211
10. Terijo A, Chavez RE, Urbieta J, Flores-Martinez L (2015) Cavity detection in the Southwestern Hilly Portion of Mexico City by Resistivity Imaging. J Environ Eng Geophys 130–139 (2015)
11. Barbosa MK, Braga MA, Gama MFP, Paula RG, Brandi LV, Dias LSDO (2020) Electrical resistivity contrast in the geotechnical assessment of iron caves, N4EN mine, Carajas, Brazil. Geophys 85:B1–B7
12. Doyoro YG, Chang PY, Puntu JM (2021) Uncertainty of the 2D resistivity survey on the subsurface cavities. Appl Sci 11:31–43. https://doi.org/10.3390/app11073143
13. Bharti AK, Singh KKK, Ghosh CN (2022) Detection of subsurface cavity due to old mine workings using electrical resistivity tomography: a case study. J Earth Syst Sci 131:39. https://doi.org/10.1007/s12040-021-01781-1
14. Kenfack, JAA, Gwet H, Wirngo HM, Tchawa P, Tagne T, Rufis F (2022) Detection of Underground Water Cavities in Urban Areas by Electrical Resistivity Tomography https://ssrn.com/abstract=4081815
15. Osama H, Al-Jumaily AM, Abed KKA (2022) Using 2D resistivity imaging technique to detect and delineate shallow unknown cavities in Al-Haqlaniyah area. West Iraq Iraqi J Sci 63(3):1091–1102
16. Abbas SS, Ahmed MM, Abed AM (2022) Determination of civil engineering problems using resistivity methods in Ramadi. West Iraq. Iraqi Geol J 55(1F):160–165

17. African Development Bank Group (AFDB) 2015 (2015): OMVG Energy projet - Gambia River basin development organization. https://www.afdb.org/fileadmin/uploads/afdb/Docume nts/Boards-Documents/Multinational_-_AR-Gambia_River_Basin_Development_Organizat ion.pdf

18. Coulibaly V (1979) Essai d'élaboration d'une notice explicative de la feuille Labé (p. 106) [Mémoire de fin d'études supérieures]. Institut Polytechnique Gamal Abdel Nasser de Conakry - Faculté de Géologie - Mines de Boké

19. Cole A (1984) Etudes des eaux souterraines de la ville de Tougué en vue de l'alimentation de cette localité en eau potable [Mémoire de fin d'études supérieures]. Institut Polytechnique Gamal Abdel Nasser de Conakry - Faculté de Géologie - Mines de Boké

20. Camara L (1089) Possibilités d'application des méthodes géophysiques pour la recherché des eaux souterraines à Labé [Mémoire de fin d'études supérieures]. Institut Polytechnique Gamal Abdel Nasser de Conakry - Faculté de Géologie - Mines de Boké

21. Mamedov VI, Bouféév YV, Nikitine YA (2010) Géologie de la République de Guinée (Université d'Etat de Moscou, vol. 1)

22. Zhou W, Beck BF, Adams AL (2002) EAective electrode array in mapping karst hazards in electrical resistivity tomography. Environ Geol 42:922–928

23. José ABR, Marco APF, Yuri AC, Maximiliano BH (2019) Electrical resistivity tomography for the detection of subsurface cavities and related hazards caused by underground coal mining in Coahuila. Geofísica Internacional 58(4):279–293

24. Sutter EM, Barounis N (2021) 2021Underground void detection by applying the electrical resistivity tomography (ERT) method for a limestone quarry in Northland, NZ. Conference, Proceedings. https://fl-nzgs-media.s3.amazonaws.com/uploads/2022/06/Sutter_ NZGS2020_Submission_Ref-0081_Create-1.pdf

25. Ogungbe AS, Olowofela JA, Da-Silva OJ, Alabi AA, Onori EO (2010) Subsurface characterization using electrical resistivity (Dipole-Dipole) method at Lagos State University (LASU) Foundation School Badagry. Adv Appl Sci Res 1:174–181

Analysis of the Effect of the Deep Excavation of a High-Rise Building on the Deformation of the Ground Surface and Diaphragm Wall

Hong Nam Nguyen

Abstract Study on the influence of the deep excavation by top-down construction method on the deformation of the adjacent ground surface and diaphragm wall has been analyzed in detail for a deep foundation pit of a high-rise building at 27 Lang Ha, Hanoi. The analysis was implemented by the finite element method. The influencing factors such as wall stiffness, wall depth, ground surface load, and initial groundwater level were carefully investigated by applying a parametric study, taking two cases of short-term and long-term working conditions. The simulated results show that the settlement of the ground surface adjacent to the deep excavation depends significantly on the surface load and the initial groundwater level. Under the long-term condition, the calculated ground surface settlement has a larger value than that under the short-term condition. The simulated horizontal displacement and the bending moment of the diaphragm wall depends significantly on the initial groundwater level and the large surface load.

Keywords Deep Excavation · Finite Element · Settlement · Diaphragm Wall · Bending Moment

1 Introduction

Working analysis of a deep foundation pit constructed by top-down method was performed in this study. The selected project for analysis is the 27 Lang Ha building. The foundation pit of the building includes 2 basements, with a depth of 8.9 m. The plan size of the foundation pit is 30 × 40 m. The building consists of 27 floating floors. The construction method in the form of top-down. A typical design cross-section is shown in Fig. 1 [1].

H. N. Nguyen (✉)
Division of Geotechnical Engineering, Faculty of Civil Engineering, Thuyloi University, 175 Tay Son Street, Dong Da District, Hanoi, Vietnam
e-mail: hongnam@tlu.edu.vn

© The Author(s) 2023
G. Feng (ed.), *Proceedings of the 9th International Conference on Civil Engineering*,
Lecture Notes in Civil Engineering 327,
https://doi.org/10.1007/978-981-99-2532-2_52

Fig. 1 Typical design
section

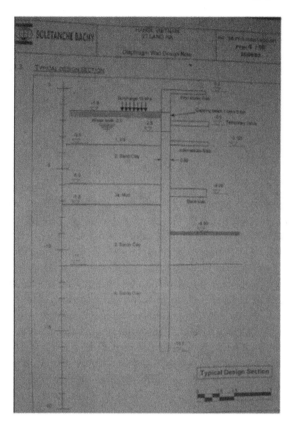

In order to consider the working of the above-mentioned foundation pit, a parametric study has been implemented in detail considering the influence of factors
such as diaphragm wall depth, diaphragm wall stiffness, surface load, and initial
groundwater level. The results of numerical simulation analysis, combined with
field monitoring data, are the basis for evaluating effective design and construction
solutions.

2 Modeling

The finite element method was used to study the problem. The software Plaxis 2D
version 8.6 [2] was employed to simulate the problem according to the plane strain
problem model.

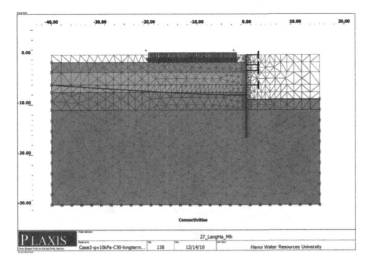

Fig. 2 The simulated section

2.1 Geometry

The analyzed cross section is shown in Fig. 2. A finite element mesh consists of 15 node plane strain elements. The total number of elements is greater than 1000 elements. The number of elements varies depending on the simulation case. Assuming the problem is symmetric, a half of the problem was considered.

2.2 Soil Properties and Initial Groundwater Level

The ground foundation consists of 5 soil layers, from top to bottom, including the filling 1, sandy clay 2, mud 3a, sandy clay 3 and 4. The soil was modelled by Mohr–Coulomb model. The model parameter values are shown in Tables 1 and 2 for short-term and long-term working conditions, respectively. Note that due to incomplete geological data, some parameter values were assumed empirically.

The initial groundwater level is located 2 m below the natural ground (at an elevation of −2.0 m). During the construction of the foundation pit, to ensure that the foundation pit is dry, it is necessary to carry out a suction pump to lower the groundwater level to the bottom of the excavation at each stage.

Table 1 Parameters of Mohr–Coulomb model (short term condition)

Soil and Interfaces Info

Mohr–Coulomb

ID	Name	Type	γ_{unsat} [kN/m³]	γ_{sat} [kN/m³]	k_x [m/day]	k_y [m/day]	ν [-]	E_{ref} [kN/m²]	c_{ref} [kN/m²]	φ [°]	ψ [°]
1	3a. Mud	UnDrained	16.0	16.0	0.0086	0.0086	0.30	3000.0	15.0	0.0	0.0
2	2. sandy clay	UnDrained	17.5	17.5	0.0864	0.0860	0.25	12000.0	30.0	0.0	0.0
3	3. sandy clay	UnDrained	19.0	19.0	0.0864	0.0864	0.30	18002.0	50.0	0.0	0.0
4	4. sandy clay	UnDrained	19.0	19.0	0.0864	0.0860	0.30	30000.0	85.0	0.0	0.0
5	1. fill	Drained	19.0	19.0	0.0086	0.0086	0.25	15000.0	0.5	25.0	0.0

Table 2 Parameters of Mohr–Coulomb model (long term condition)

Soil and Interfaces Info

Mohr–Coulomb

ID	Name	Type	γ_{unsat} [kN/m³]	γ_{sat} [kN/m³]	k_x [m/day]	k_y [m/day]	v [-]	E_{ref} [kN/m²]	c_{ref} [kN/m²]	φ [°]	ψ [°]
1	3a. Mud	Drained	16.0	16.0	0.0086	0.0086	0.30	3000.0	0.5	22.0	0.0
2	2. sandy clay	Drained	17.5	17.5	0.0864	0.0860	0.25	12000.0	0.5	24.0	0.0
3	3. sandy clay	Drained	19.0	19.0	0.0864	0.0864	0.30	18002.0	0.5	25.0	0.0
4	4. sandy clay	Drained	19.0	19.0	0.0864	0.0860	0.30	30000.0	0.5	25.0	0.0
5	1. fill	Drained	19.0	19.0	0.0086	0.0086	0.25	15000.0	0.5	25.0	0.0

2.3 Diaphragm Wall and Shoring System

The diaphragm wall is made of cast-in-place concrete, using bentonite solution to support the wall. The concrete has a strength of C30. The wall thickness is 0.6 m. The wall is modelled according to a linear elastic model with parameters given as follows: $E = 2.57 \times 10^7$ kN/m², $EA = 1.540 \times 10^7$ kN/m, $EI = 4.626 \times 10^5$ kNm²/m, w $= 9$ kN/m/m, $\nu = 0.2$. Slabs and temporary struts can be simulated by single-ended anchor struts with stiffness $EA = 3{,}500 \times 10^7$ kN.

2.4 Surface Load

The surface load is assumed to be evenly distributed on the ground with strength q $= 10$ kPa.

2.5 Construction Stages

The construction sequence is assumed as follows:

Stage 1: Construction of the diaphragm wall to a depth of -16.5 m.

Stage 2: Construction of the top slab, then digging the foundation pit to elevation -1.5 m, applying surface load q $= 10$ kPa.

Stage 3: Digging the foundation pit to the elevation -3.5 m, constructing the middle slab and installing a temporary strut system, pumping to lower the groundwater level to the level of the bottom of the foundation pit.

Stage 4: Digging the foundation pit to elevation -6.0 m, constructing the bottom slab, pumping to lower the groundwater level to the level of the bottom of the foundation pit.

Stage 5: Digging the foundation pit to elevation -8.9 m, pumping to lower the groundwater level to the level of the bottom of the foundation pit.

2.6 Parameter Studies

The influence factors such as diaphragm wall stiffness (C30, C40), diaphragm wall depth (H = 13.5 m, 16.5 m, and 19.5 m), surface load (q = 10,20, 50 kPa), and initial groundwater level ($-2.0, -3.5, -6.0$ m) were studied in detail. The ground soils were considered with two cases of short-term and long-term conditions, corresponding to undrained and drained geotechnical experimental data, respectively (see Tables 1 and 2). The total number of analyzed cases was 30 cases.

3 Simulation Results and Discussion

3.1 Ground Settlement

The results of calculation of ground settlement at the final excavation stage (stage 5) under short-term working condition are shown in Figs. 3a, b, c with values q = 10, 20 and 50 kPa, respectively. We considered the stiffness of the diaphragm wall by employing concrete grades C30 and C40. The results show that the differences in settlement values simulated with two concrete strengths C30 and C40 are not large.

Similarly, under long-term working condition, the results of simulated ground settlement at the final excavation stage are shown in Fig. 4. The results show that the stiffness of the wall does not significantly affect the settlement of the ground surface. Compared with the simulated settlement values under the short-term condition (Fig. 3), those under the long-term condition have larger values.

Figure 5a and b shows the effect of the depth of the diaphragm wall on the surface settlement at the final excavation stage under short-term and long-term working conditions, respectively. The calculation results reveal that the wall depth does not significantly affect ground surface settlement, especially under short-term conditions.

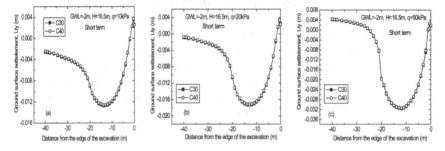

Fig. 3 Effect of wall stiffness on the surface settlement (short term)

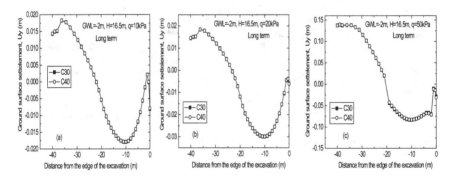

Fig. 4 Effect of wall stiffness on the surface settlement (long term)

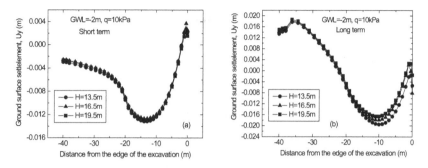

Fig. 5 Effect of depth of diaphragm wall on the surface settlement

The ground surface settlement under the long-term condition (Fig. 5b) is larger than that under the short-term condition (Fig. 5a).

The effect of surcharge load on the surface settlement (final stage) is shown in Figs. 6a and b, respectively, under short-term and long-term working conditions. The calculation results show that the surface load has a significant influence on the ground settlement, especially under short-term conditions (Fig. 6a). The larger the load, the greater the settlement of the ground surface. Under the long-term condition (Fig. 6b), the ground settlement has a larger value than that under the short-term one (Fig. 6a).

Figure 7 shows that the initial groundwater level (GWL) has a significant influence on the ground surface settlement, especially under long-term conditions. The shallower the initial groundwater level, the greater the ground settlement value due to the large distance between the initial groundwater level and the foundation pit bottom elevation. In the long-term condition (Fig. 7b), the ground surface settlement has a larger value than that in the short-term condition (Fig. 7a).

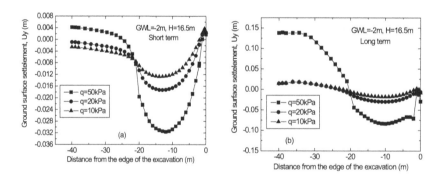

Fig. 6 Effect of surcharge load on the surface settlement

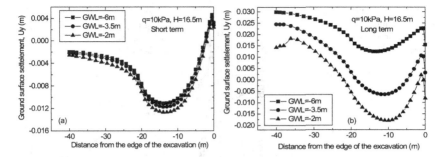

Fig. 7 Effect of initial groundwater level on surface settlement

3.2 Diaphragm Wall's Horizontal Displacement

Figure 8 shows that in the comparison of diaphragm walls with different concrete strengths C30 and C40, the concrete strength stiffness has a small influence on the horizontal displacement of the diaphragm wall.

Figure 9 shows that the surface load has a significant effect on the horizontal displacement of the diaphragm wall. However, when the load value is small (q = 10, 20 kPa), there is not too much difference in the horizontal displacement caused by the surface load.

The initial groundwater level has a significant influence on the wall displacement value (Fig. 10) under long-term working conditions. The shallower the initial groundwater level, the larger the horizontal displacement value, especially the wall segment below the elevation of the excavation bottom. It is noted that the horizontal displacement of the diaphragm wall could be reduced if the small strain soil stiffness and its stress state dependency could be taken into consideration.

Fig. 8 Effect of wall stiffness on the horizontal displacement of the diaphragm wall

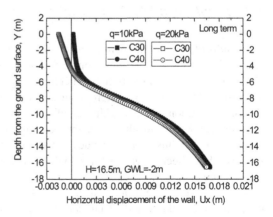

Fig. 9 Effect of surface load
on the horizontal
displacement of the
diaphragm wall

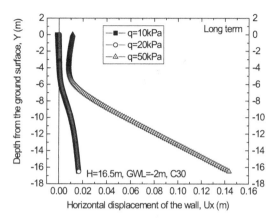

Fig. 10 Effect of initial
groundwater level on the
horizontal displacement of
the diaphragm wall

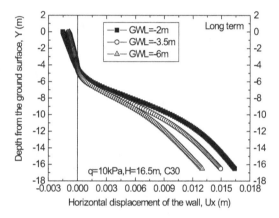

3.3 Diaphragm Wall's Internal Forces

The results of bending moment in the diaphragm wall at the final construction stage
(excavating to the bottom elevation of the foundation pit) are shown in Figs. 11,
12, 13 for the walls with different concrete strengths of C30 and C40. Comparisons
are also made considering different surface loads q = 10, 20 and 50 kPa, for both
short-term and long-term working conditions. In general, in all comparison cases,
the more rigid the wall, the larger the bending moment in the wall. However, the
difference in bending moment values is not large for walls with the aforementioned
stiffness. The effect of wall stiffness will be further investigated with employing a
wider range of strength values.

As can be seen in Fig. 14a, the surface load has a negligible influence on the desired
moment value in the wall (C30) under short-term conditions. However, under long-
term conditions, a large surface load (q = 50 kPa) has a significant effect on the

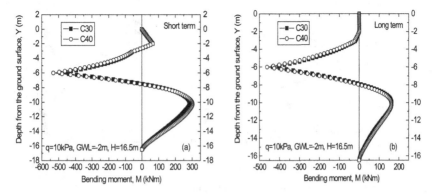

Fig. 11 Effect of the concrete strength of diaphragm wall on the bending moment (q = 10 kPa)

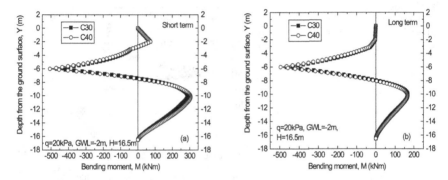

Fig. 12 Effect of the concrete strength of diaphragm wall on the bending moment (q = 20 kPa)

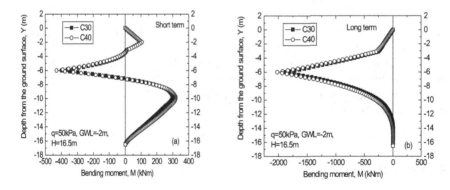

Fig. 13 Effect of the concrete strength of diaphragm wall on the bending moment (q = 50 kPa)

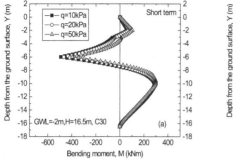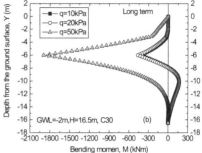

Fig. 14 Effect of surcharge load on the wall bending moment

Fig. 15 Effect of initial
groundwater level on the
wall bending moment (long
term)

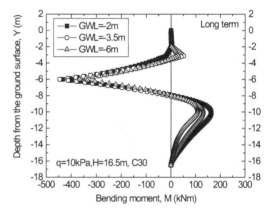

bending moment in the wall with the same stiffness (Fig. 14b). The above results are
also suitable for walls with greater stiffness (C40).

The influence of the initial groundwater level on the bending moment in the wall
is shown in Fig. 15. In general, the shallower the groundwater level, the greater the
value of the bending moment in the wall, especially for the part of the wall below
the elevation of the bottom of the pit.

4   Conclusions

Study on the influence of the deep excavation of the 27 Lang Ha high-rise building
by top-down construction method on the deformation of the adjacent ground surface
and the diaphragm wall has been analyzed in detail.

The analysis was implemented by the finite element method, employing the plane
strain problem. A parametric study has been carefully analyzed, by considering
influencing factors such as wall stiffness, wall depth, ground surface loading, and

initial groundwater level, with considering two cases of short-term and long-term ground working. The total number of analyzed cases was 30 cases. Following are the analysis results.

The settlement of the ground surface adjacent to the deep excavation depends significantly on the surface load and the initial groundwater level. In particular, the larger the surface load and/or the shallower the initial groundwater level, the larger the settlement of the ground surface. Under the long-term condition, the surface settlement has a larger value than that under the short-term condition. Within the scope of the present study, the depth and the concrete strength of the diaphragm wall did not largely affect the settlement of the adjacent ground surface.

The horizontal displacement and the bending moment values in the diaphragm wall depends significantly on the initial groundwater level and the large surface load. In general, the shallower the initial groundwater level, the greater the horizontal displacement and the greater bending moment values in the wall, especially the part of the wall located below the bottom elevation of the foundation pit.

References

1. Soletanche B (2002) Diaphragm wall design of 27 Lang Ha building, Hanoi
2. Plaxis BV (2006) Plaxis 8 Professional Version, The full manual, Plaxis, Delft, The Netherlands

Grid-on-Grid Transformation for Integrating Spatial Reference System of Multi-source Data

Maan Habib

Abstract The accessible, up-to-date, reliable, and usable data are considered sustainability tools for developing spatial data infrastructure. Geospatial data come from multi-sources and are georeferenced using an appropriate mapping reference system. Artificial satellite positioning data are now defined on a global geocentric frame, whereas traditional geodetic networks were built on a national datum. Hence, three-dimensional (3D) coordinate transformations are required for data harmonization using control points that can be caused some discrepancies between the physical reality and represented positions. In practice, grid-on-grid conversion is a mathematical model matching GNSS observations and official spatial data through two common datasets to minimize the datum-to-datum transformation errors. This research conducts a comparative analytical study of the conformal polynomial algorithms for map-matching with global coordinates utilizing least-squares estimation. The findings indicated that the proposed approach provides superior performance and employs any area with high accuracy.

Keywords Conformal polynomial · GNSS · Least-squares adjustment · Grid-on-grid transformation

1 Introduction

A high-precision coordinate system is crucial for establishing a spatial data infrastructure and a paperless land administration system to promote long-term sustainable socio-economic growth [1, 2]. The satellite-based geodetic techniques have opened a new era for determining the high positional quality of points on the topographical surface in a geocentric three-dimensional system using precise observations of artificial celestial bodies. In the past, most national maps were produced by traditional surveying methods and relied on a non-geocentric local datum selected to deliver

M. Habib (✉)
Faculty of Engineering, Cyprus Science University, Girne, Cyprus
e-mail: maan.habib@gmail.com

© The Author(s) 2023
G. Feng (ed.), *Proceedings of the 9th International Conference on Civil Engineering*,
Lecture Notes in Civil Engineering 327,
https://doi.org/10.1007/978-981-99-2532-2_53

635

the best possible match of the earth's figure within a defined geographical area [3]. The global navigation satellite system (GNSS) is utilized to build more accurate geodetic control points and other applications within the mapping industry [4, 5]. Accordingly, developing a mathematical algorithm connecting two spatial reference frames is a critical challenge to guarantee the consistency of the coordinates. In reality, several researchers have extensively addressed the three-dimensional coordinate transformation in the literature to change between GNSS-derived coordinates and national terrestrial ones using the common points between two datasets [6–12]. In contrast, the discrepancies in triangulation stations are inevitable because a local datum is oriented astronomically at an initial point, the ellipsoidal surface is not earth-centered, and its rotational axis is not coincident with the earth's axis [13]. In addition, the horizontal control scale change generates a stretch in the related lines of the network [14].

Grid-on-grid transformation is applied to define a mathematical relationship between two sets of georeferenced spatial data obtained from separate sources with different grid reference systems [15]. It generally converts two-dimensional coordinates described by axes rotations, origin transitions, and a scale change [16]. Several methods have been investigated in the literature to address this issue by determining the parameters of the prediction model through control points in both systems [17–19]. The number of training sample data and their geographical distribution pattern significantly impact the accuracy of estimating algorithm parameters [20, 21]. As a result, redundant observations and the least-squares technique give the optimum solution for calculating polynomial coefficients and evaluating their accuracy and root mean square error (RMSE) of data points [22]. This research investigates different two-dimensional conformal transformation models to determine which algorithm performs best in this context and develops a spatial tool in the Microsoft Visual Studio environment to solve the addressed issue by minimizing the sum of errors' squares. In addition, quantitative and qualitative methods are utilized to evaluate the findings by employing statistical analysis and descriptive illustrations.

2 Materials and Methods

It is necessary to cope with an increasingly broad spectrum of positioning information acquired from multi-sources such as terrestrial surveying and GNSS receiver observations. The location of a feature in space is related to a wide range of national and global datums. Hence, a comprehensive understanding of the various coordinate systems and their proper transformations is essential to guarantee an accurate evaluation of reference frame variations to deliver high-quality products in geospatial data processing activities. A transformation model is commonly applied to change GNSS-based coordinates to those in a national datum through two datasets of reference stations defined in both systems. Its parameters characterize the relationship between various spatial reference systems, significantly correlated with the diverse and local nature of deformations inherent in geodetic frameworks.

Fig. 1 Spatial distribution
of 1st and 2nd order Swiss
triangulation points

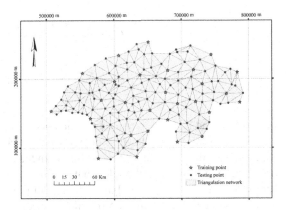

2.1 Case Study Description

The Swiss coordinate system (CH1903) is a geodetic reference frame for the national network in Switzerland for various mapping purposes. It is a cylindrical projection based on the Bessel ellipsoid and the Mercator projection. The geodetic network in Switzerland consists of several permanent control points on the earth's surface that have been accurately observed using traditional and advanced geodetic techniques. Figure 1 depicts the spatial distribution of first- and second-order Swiss triangulation points across Switzerland. The density and distribution of the points can significantly impact the data's accuracy, as well as the suitability of the data for different types of applications. The Federal Office of Topography swisstopo is responsible for maintaining and managing the geodetic network and the CH1903 coordinate system in Switzerland.

2.2 Grid-on-Grid Transformation

A coordinate transformation is a process of changing the point coordinates from one reference system to another [23], which can be broken into combining axes rotations, origin shifts, and scaling factors. The transformation parameters are initially calculated using the least-squares surface fitting approach. Hence, the other point coordinates are converted utilizing the interpolation algorithm. The two-dimensional conversion model is the most prevalent in surveying and mapping applications. On the other hand, geospatial data is acquired from various sources and georeferenced using a proper grid system. With the advent of artificial satellite positioning data, this issue has become critical for integrating GNSS observations and official spatial data to reduce the discrepancies between the physical reality and represented positions. The general formula of grid-on-grid transformation using the polynomial functions is presented in Eq. 1 [24].

$$E' = \sum_{i=0}^{u} \sum_{j=0}^{u-i} a_{ij} E^i N^j$$

$$N' = \sum_{i=0}^{u} \sum_{j=0}^{u-i} b_{ij} E^i N^j \tag{1}$$

In which a_{ij} and b_{ij} are polynomial coefficients, and u its degree.

The six terms up to the first degree of the polynomials surface describe the well-known affine transformation in Eq. 1. Indeed, a conformal transformation of coordinates preserves local angles that can be satisfied by applying the Cauchy - Riemann conditions on the polynomials model [25]. The number of the conformal polynomial coefficients (m) is proportional to the polynomial degree. The fourth-degree conformal algorithms are given in Eq. 4, where first-order functions define the similarity (Helmert) transformation and the quantities beyond that, referred to as conformal deformation.

$$\frac{\partial E'}{\partial E} \equiv \frac{\partial N'}{\partial N}$$
$$\frac{\partial E'}{\partial N} \equiv -\frac{\partial N'}{\partial E} \tag{2}$$

$$m = 2 \times u + 2 \tag{3}$$

$$E' = a_o + a_1 E - b_1 N$$
$$+ a_2 (E^2 - N^2) - 2b_2 E N$$
$$+ a_3 (E^3 - 3N^2 E) + b_3 (N^3 - 3E^2 N) \tag{4}$$

$$N' = b_o + a_1 N + b_1 E$$
$$+ 2a_2 E N + b_2 (E^2 - N^2)$$
$$+ b_3 (E^3 - 3N^2 E) - a_3 (N^3 - 3E^2 N) \tag{5}$$

A linear scale factor and meridian convergence at any point are typically illustrated in the theory of map projections utilizing relations 6 and 7 [26].

$$K = \sqrt{\left(\frac{\partial E'}{\partial E}\right)^2 + \left(\frac{\partial N'}{\partial E}\right)^2} \tag{6}$$

$$\tan \theta = -\frac{\partial N'}{\partial E} \bigg/ \frac{\partial E'}{\partial E} \tag{7}$$

where

$$\frac{\partial E'}{\partial E} = a_1 + 2a_2 E - 2b_2 N + 3a_3 \left(E^2 - N^2\right) - 6b_3 E N \tag{8}$$

$$\frac{\partial N'}{\partial E} = b_1 + 2a_2 N + 2b_2 E + 3b_3 \left(E^2 - N^2\right) + 6a_3 E N \tag{9}$$

3 Results and Discussion

Errors in coordinate conversion can impact the mapping between target locations and pointing positions. Local distortion describes how the transformation from one coordinate system to another affects the spatial relationships between nearby points. It is an essential concept in geographic information systems (GIS) and spatial data analysis, as it helps to understand how changes in the coordinate system influence spatial data accuracy. Figure 2 describes the displacement vectors of distortion for control stations, with their original location serving as a starting point and the transformed position as an ending point. The vector's length illustrates the distance from the reference point to the point, whereas the vector's direction defines the direction of the displacement. As shown in Fig. 2, some points moved in the southeast direction and others in the northwest, indicating that the movement of the earth's crust (also known as tectonic movement) can cause displacement of points on the earth's surface.

Figure 3 shows the displacement in the easting coordinates for three various polynomial transformation models of different degrees: first, second, and third. These models are utilized to describe the relationship between the displacement in the east–west direction and some other variables. As the polynomial degree increases, the displacement value tends to decrease for a given coordinate. This trend can be seen by comparing the curves representing the different models in the graph. This observation is also apparent when examining the displacement in the northing coordinates, as shown in Fig. 4, where the third-order polynomial model yields the lowest

Fig. 2 Displacement vectors of triangulation point between the original and transformed coordinates

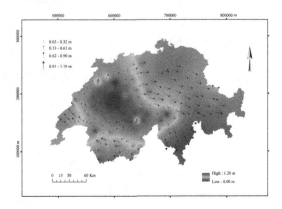

displacement values and the best fit to the data. Moreover, the results' statistical distribution is investigated to understand further the influence of the polynomial degree on the displacement value. This is illustrated in the box plots depicted in Fig. 5, which show each model's range, median, and quartiles of displacement values. These plots demonstrate that the first-degree polynomial always yields the highest displacement values, indicating that it has the lowest performance among the three models. On the other hand, the third-order polynomial consistently provides the best capabilities for the northing and easting coordinates, with the lowest displacement values and the most.

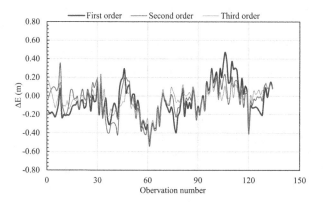

Fig. 3 Differences in easting coordinates after applying addressed models

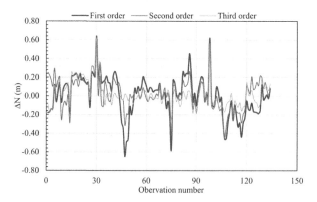

Fig. 4 Differences in northing coordinates after applying addressed models

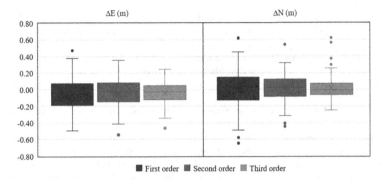

Fig. 5 Displacement in easting and northing coordinates after applying addressed models

4 Conclusion

This research investigates the effectiveness of conformal polynomial algorithms for making maps consistent with global coordinates using least-squares estimation to minimize the difference between the observed data and the model's predicted values. The study aims to determine the accuracy and reliability of these algorithms for map-matching with global coordinates and compare their performances. Based on the statements above, the following conclusions are drawn:

(1) The grid-on-grid transformation method using conformal polynomial algorithms was effective in minimizing errors in converting geospatial data from multiple sources.

(2) The study showed that increasing the polynomial degree decreased displacement values and improved performance.

(3) The third-order polynomial model was the most accurate and reliable for easting and northing coordinates.

(4) This approach is useful for harmonizing data from different reference systems and can be applied in any area with high accuracy.

References

1. Habib M (2022) Fit-for-purpose conformal mapping for sustainable land administration in war-ravaged Syria. Heliyon, e09384
2. Fazilova D (2022) Uzbekistan's coordinate system transformation from CS42 to WGS84 using distortion grid model. Geod Geodyn 13(1):24–30
3. Villar-Cano M, Marqués-Mateu Á, Jiménez-Martínez M J (2019) Triangulation network of 1929–1944 of the first 1: 500 urban map of València. Surv Rev52
4. Hill AC, Limp F, Casana J, Laugier EJ, Williamson M (2019) A new era in spatial data recording: low-cost GNSS. Adv Archaeol Pract 7(2):169–177
5. Abdalla A, Mustafa M (2021) Horizontal displacement of control points using GNSS differential positioning and network adjustment. Meas 174:108965

6. Yang Y (1999) Robust estimation of geodetic datum transformation. J Geod 73(5):268–274
7. You RJ, Hwang HW (2006) Coordinate transformation between two geodetic datums of Taiwan by least-squares collocation. J Surv Eng 132(2):64–70
8. Schaffrin B, Felus YA (2008) On the multivariate total least-squares approach to empirical coordinate transformations three algorithms. J Geod 82(6):373–383
9. Mahboub V (2012) On weighted total least-squares for geodetic transformations. J Geod 86(5):359–367
10. Even-Tzur G (2018) Coordinate transformation with variable number of parameters. Surv Rev 52
11. Abbey DA, Featherstone WE (2020) Comparative review of molodensky-badekas and burša-wolf methods for coordinate transformation. J Surv Eng 146(3):04020010
12. Kalu I, Ndehedehe CE, Okwuashi O, Eyoh AE (2022) Estimating the seven transformational parameters between two geodetic datums using the steepest descent algorithm of machine learning. Appl Comput Geosci 100086
13. Ogi S, Murakami M (2000) Construction of the new japan datum using space geodetic technologies. in towards an integrated global geodetic observing system (IGGOS), pp. 103–105. Springer, Berlin
14. Torge W, Müller J(2012) Geodesy. Geod. de Gruyter
15. Janssen V (2009) Understanding coordinate reference systems, datums and transformations. Int J Geoinform 5(4):41–53
16. Konakoglu B, Cakır L, Gökalp E (2016) 2D coordinate transformation using artificial neural networks. Int Arch Photogram Remote Sen Spatial Inf Sci 42:183
17. Vincenty T (1987) Conformal transformations between dissimilar plane coordinate systems. Surv Map 47(4):271–274
18. Mercan HÜSEYİN, Akyilmaz O, Aydin C (2018) Solution of the weighted symmetric similarity transformations based on quaternions. J Geod 92(10):1113–1130
19. Dąbrowski PS, Specht C, Specht M, Burdziakowski P, Makar A, Lewicka O (2021) Integration of multi-source geospatial data from GNSS receivers, terrestrial laser scanners, and unmanned aerial vehicles. Can J Remote Sens 47(4):621–634
20. Harwin S, Lucieer A (2012) Assessing the accuracy of georeferenced point clouds produced via multi-view stereopsis from unmanned aerial vehicle (UAV) imagery. Remote Sens 4(6):1573–1599
21. Habib M (2021) Evaluation of DEM interpolation techniques for characterizing terrain roughness. CATENA 198:105072
22. Amiri-Simkooei A, Jazaeri S (2012) Weighted total least squares formulated by standard least squares theory. J Geod Sci 2(2):113–124
23. Maling DH (2013) Coordinate systems and map projections. Elsevier
24. Habib M, A'kif MS (2019) Evaluation of alternative conformal mapping for geospatial data in Jordan. J Appl Geod 13(4):335–344
25. Thomas PD (1852) Conformal projections in geodesy and cartography, vol. 4. US Government Printing Office
26. Habib M (2008) Proposal for developing the Syrian stereographic projection. Surv Rev 40(307):92–101

Printed in the United States
by Baker & Taylor Publisher Services